MW01674999

Biodegradation, Pollutants and Bioremediation Principles

Editors

Ederio Dino Bidoia

Department of Biochemistry and Microbiology
Sao Paulo State University (UNESP)
Rio Claro-SP, Brazil

Renato Nallin Montagnolli

Department of Natural Sciences, Mathematics and Education
Agricultural Sciences Centre – Federal University of Sao Carlos (UFSCar)
Araras-SP, Brazil

CRC Press
Taylor & Francis Group
Boca Raton London New York

CRC Press is an imprint of the
Taylor & Francis Group, an **informa** business
A SCIENCE PUBLISHERS BOOK

Cover credit: The picture was taken by Renato Nallin Montagnolli. Reproduced on the cover by his permission.

First edition published 2021
by CRC Press
6000 Broken Sound Parkway NW, Suite 300, Boca Raton, FL 33487-2742

and by CRC Press
2 Park Square, Milton Park, Abingdon, Oxon, OX14 4RN

© 2021 Taylor & Francis Group, LLC

CRC Press is an imprint of Taylor & Francis Group, LLC

Reasonable efforts have been made to publish reliable data and information, but the author and publisher cannot assume responsibility for the validity of all materials or the consequences of their use. The authors and publishers have attempted to trace the copyright holders of all material reproduced in this publication and apologize to copyright holders if permission to publish in this form has not been obtained. If any copyright material has not been acknowledged please write and let us know so we may rectify in any future reprint.

Except as permitted under U.S. Copyright Law, no part of this book may be reprinted, reproduced, transmitted, or utilized in any form by any electronic, mechanical, or other means, now known or hereafter invented, including photocopying, microfilming, and recording, or in any information storage or retrieval system, without written permission from the publishers.

For permission to photocopy or use material electronically from this work, access www.copyright.com or contact the Copyright Clearance Center, Inc. (CCC), 222 Rosewood Drive, Danvers, MA 01923, 978-750-8400. For works that are not available on CCC please contact mpkbookspermissions@tandf.co.uk

Trademark notice: Product or corporate names may be trademarks or registered trademarks and are used only for identification and explanation without intent to infringe.

Library of Congress Cataloging-in-Publication Data

Names: Bidoia, Ederio Dino, editor. | Montagnolli, Renato Nallin, editor.
Title: Biodegradation, pollutants and bioremediation principles / editors, Ederio Dino Bidoia, Department of Biochemistry and Microbiology, Sao Paulo State University (UNESP), Rio Claro-SP, Brazil, Renato Nallin Montagnolli, Department of Natural Sciences, Mathematics and Education, Agricultural Sciences Centre--Federal University of Sao Carlos (UFSCar), Araras-SP, Brazil.
Description: First edition. | Boca Raton : CRC Press, Taylor & Francis Group, 2021. | Includes bibliographical references and index.
Identifiers: LCCN 2020041781 | ISBN 9780367259389 (hardcover)
Subjects: LCSH: Bioremediation.
Classification: LCC TD192.5 .B545 2021 | DDC 628.5--dc23
LC record available at https://lccn.loc.gov/2020041781

ISBN: 978-0-367-25938-9 (hbk)

Typeset in Times New Roman
by Radiant Productions

Preface

Environmental pollution is a major concern as its long-term impacts are still not fully known. The microbiological concepts and the fundamentals of bioremediation are at the core of every biotechnology-based environmental cleaning-up protocol. This book is designed to present a broad compendium in bioremediation techniques and biodegradation research. Chapters provide both legacy and up-to-date reviews, and approaches to mitigate pollution effects and address current issues. The concept for this book is to provide academic and industry researchers with an introduction to the current state of biodegradation and environmental assessment studies, and to expand current knowledge on bioremediation protocols. While the book is primarily focused toward the environmental sciences researcher, the range of techniques demonstrated in the book also provides an introduction of bioremediation methods to researchers outside of this field.

The selected collaborating authors are renowned for their microbiology expertise and provide an in-depth reference for students and specialists. A wide variety of pollutants (e.g., heavy metals, petroleum products, organo-halogenated, pesticides, and emerging pollutants) in various scenarios (e.g., coastal areas, sewage and urban pollution, extreme environments) provide a diverse, yet concise source of information on the most recent efforts in bioremediation. This book presents the challenges and encourages future researchers to look for a more sustainable and cleaner environment.

The authors thank all those who have contributed significantly in understanding the different aspects of environmental sciences and submitted their chapters. We hope that our book will prove of equally high value to advanced undergraduate and graduate students, research scholars, and professionals.

Araras, Brazil, 2020

Renato Nallin Montagnolli
Ederio Dino Bidoia

Contents

1

Alkylphenols and Alkylphenol Ethoxylates

Their Impact on Living Organisms, Biodegradation, and Environmental Pollution

Tomasz Grześkowiak,[1] *Andrzej Szymański,*[1]
Agnieszka Zgoła-Grześkowiak,[1,*] *Beata Czarczyńska-Goślińska*[2]
and *Robert Frankowski*[1]

Introduction

Alkylphenols (APs) and short-chain alkylphenol ethoxylates (APEs) containing 1, 2, and 3 ethoxylene units per molecule are known for their toxic and endocrine disrupting properties. However, despite bans and restrictions introduced by different countries (including the European Union, Canada, and the USA), APs and their ethoxylates are still widely used on a global scale because of their ease of production, low cost, and numerous applications.

In the present chapter, the influence of these compounds on living organisms is presented, including their toxic properties towards different organisms and disruption of the endocrine system, resulting in numerous harmful effects, such as lowered reproductive performance, or even cancer. Different factors affecting

[1] Poznan University of Technology, Institute of Chemistry and Technical Electrochemistry, Berdychowo 4, 60-965 Poznan, Poland.

[2] Poznan University of Medical Sciences, Department of Pharmaceutical Technology, Grunwaldzka 6, 60-780 Poznan, Poland.

Emails: civ@tlen.pl; andrzej.szymanski@put.poznan.pl; bgoslinska@ump.edu.pl; robert.z.frankowski@doctorate.put.poznan.pl

* Corresponding author: civ@o2.pl

endocrine disrupting properties are discussed, including length of both ethoxy and alkyl chains, branching of the alkyl chain, and its substitution position in the aromatic ring. Biodegradation of APs and APEs in both aerobic and anaerobic conditions is also presented, and studies on their bacterial and fungal degradation are discussed. Experiments on removal of these contaminants are described and their environmental fate is shown. Contamination of the environment is discussed separately for water and sediments.

Production and usage of alkylphenols and their ethoxylates

Alkylphenols, mainly nonylphenol (NP) and octylphenol (OP), are widely manufactured chemicals, which are produced during alkylation of phenol with nonene (obtained from trimerization of propene) or octene (obtained from dimerization of butene) in the presence of a catalyst. APs can be transformed to alkylphenol ethoxlyates, which are commonly used surfactants (Reed 1978, Groshart et al. 2001). NP is used mostly for production of nonylphenol ethoxylates (NPEs), and in plastic industry (e.g., for production of resins and tri(4-nonylphenyl) phosphite)—a heat stabilizer), and also in the production of oximes (used for manufacturing metals) (Groshart et al. 2001, SUBSPORT 2013). On the other hand, OP is rarely used in the synthesis of octylphenol ethoxylates (OPEs), and it is estimated that 98% of OP is used in the production of phenolic resins (OSPAR 2006). NPEs are used in many applications, including industrial cleaning, textile and leather processing, paper production, metal processing, extraction of crude petroleum and natural gas, paints and lacquer manufacturing, as well as pest control products (Groshart et al. 2001, SUBSPORT 2013). OPEs are mainly used in emulsion polymerization, but also in the production of octylphenol ether sulphates, textile and leather processing, pesticide formulations, and paints (OSPAR 2006). As a result of their widespread usage and problematic biodegradation, APs and APEs are directed to the environment, and are ubiquitous in both surface waters and sediments (Fig. 1.1).

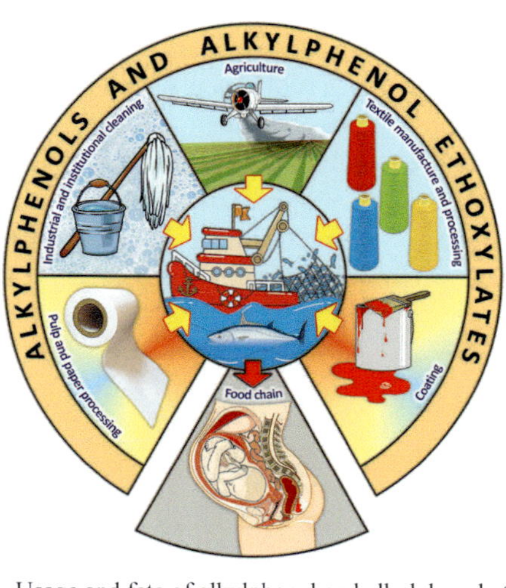

Figure 1.1. Usage and fate of alkylphenol and alkylphenol ethoxylates.

Global production of APEs at the edge of the new millennium was estimated at half a million tons, and it still rises, mainly due to their increased production in Brazil, China, India, and Russia (Renner 1997, Bergé et al. 2012), because usage of NP and NPEs in Europe and North America is a subject of very restrictive law regulations (Directive 2003, Canadian Act 2001, US EPA 2005). These regulations were introduced after a number of studies showed incomplete biodegradation of APEs (Giger et al. 1984, Ahel and Giger 1985, Potter et al. 1999), endocrine disrupting properties of their biodegradation products (including short-chain APEs, as well as APs) (Jobling and Sumpter 1993, White et al. 1994, Soto et al. 1995), and high accumulation of short-chain NPEs and NP in the environment (Ding and Tzing 1998, Shang et al. 1999), while OPEs and OP were present at lower concentrations (Snyder et al. 1999, Loyo-Rosales et al. 2003) due to much lower usage of OPE surfactants (Groshart et al. 2001, OSPAR 2006).

Impact of alkylphenols and their ethoxylates on living organisms

The toxicity of alkylphenols and their ethoxylates was already extensively studied in the 90s. Numerous results were collected for fish, crustaceans, and other organisms. Available data shows that for 4-nonylphenol, lethal concentration killing 50% of tested subjects (LC50) falls between 100 and 1000 μg/L (Fig. 1.2) (Servos 1999, Canadian Water Act 2002), and similar data was also gathered for 4-octylphenol (Brook et al. 2005). Alkylphenol ethoxylates were also found to be toxic (Servos 1999), but their toxicity considerably depends on the number of ethoxy groups in the molecule (Canadian Water Act 2002, Whitehouse 2002). Thus, the Canadian Environmental Quality Guideline on nonylphenol and its ethoxylates gives toxic equivalency factors (TEFs) for alkylphenols and their ethoxylates. For both NP and OP TEF = 1, for their ethoxylates with 1 to 8 ethoxy groups TEF = 0.5, for ethoxylates with 9 or more ethoxy groups TEF = 0.05, and for short-chained carboxylated ethoxylates TEF is also equal to 0.05 (Canadian Water Act 2002). Moreover, it was also found that toxicity of NPEs for different species (fish, invertebrates, algae) decreases exponentially with increasing ethoxy chain length (Whitehouse 2002). Therefore, the LC50 values of APEs often fall to the mg/L level (Warhurst 1995, Servos 1999).

Apart from their toxic properties, alkylphenols and their short-chain ethoxylates also belong to xenobiotics, which may be harmful to the development and reproduction of humans, wildlife, and aquatic organisms (Darbre 2015). These compounds have the ability to disrupt the functioning of the endocrine systems of humans and wildlife (Warhurst 1995). The first evidence of estrogenic potential of alkylphenols was reported by Dodds and Lawson in 1938 (Dodds and Lawson 1938). Later, these observations were confirmed in the research of Mueller and Kim (Mueller and Kim 1978), who proved the ability of various alkylphenols to displace the prebound estradiol from the estrogen receptor of uterine cytosols, and also to prevent estradiol from binding to the receptor. Moreover, the toxicity increased with the length of the hydrophobic chain. For hydrophilic 4-ethoxyphenol or 4-(α-OH-isopentyl)phenol, no activity of inhibiting estradiol binding to the receptor was found (Mueller and Kim 1978).

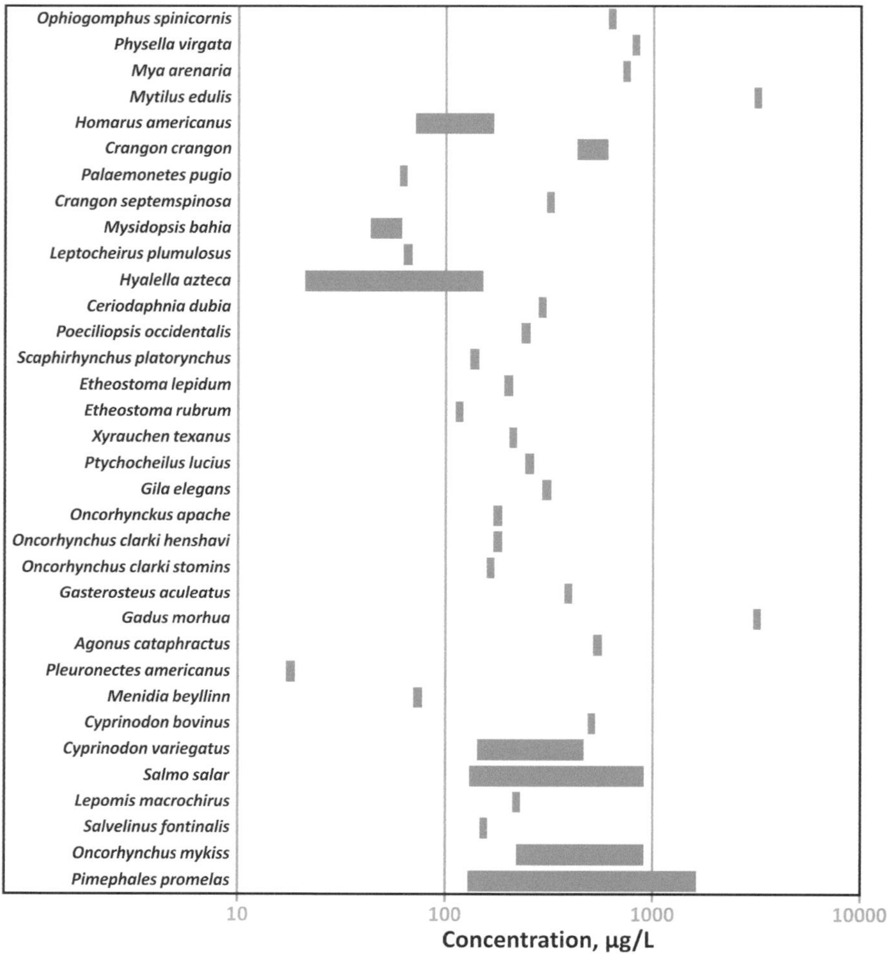

Figure 1.2. Toxicity of 4-nonylphenol in 96-hours tests, expressed as the LC50 values (Servos 1999).

The inadvertent exposure of marine and pelagic organisms to synthetic estrogens became apparent. Meier et al. (Meier et al. 2007) tested the effects of APs on the reproduction of Atlantic cod (*Gadus morhua*). Fish fed with a range of concentrations of 4-*tert*-butylphenol, 4-*n*-pentylphenol, 4-*n*-hexylphenol, and 4-*n*-heptylphenol were compared to unexposed fish and to fish fed with paste, including 17β-estradiol. The results showed the detrimental effect on both female and male fish. In female cods, impaired oocyte development, decreased estrogen levels, and delay in spawning time were observed. In male cods, impaired testicular development and reduced number of spermatozoa were revealed. Therefore, a negative influence on the reproductive health of cod populations was reported. Gimeno et al. (Gimeno et al. 1998) investigated the impact of 4-*tert*-pentylphenol on male adult common carp (*Cyprinus carpio*) during spermatogenesis. They observed the increased production of a typically female protein vitellogenin (Vtg) and the inhibition of the spermatogenesis. Similarly, NP had the potential to induce Vtg in fish. When sediments collected from Tokyo Bay near the sewage treatment plant were fractionated and administered to male mummichogs (*Fundulus heteroclitus*), Vtg was measured after two weeks. The highest level of Vtg was observed for the fraction including the highest concentration of NP (Kurihara et al. 2007). Kinnberg et al. (Kinnberg et al. 2000) also studied the estrogenic effects of NP in comparison to 17β-estradiol in male platyfish *Xiphophorus maculatus* based on Vtg synthesis and testis structure. They found a dose-dependent reduction in the gonadosomatic index and dose-dependent effects on the testis morphology. In other studies performed by Gray and Metcalfe (Gray and Metcalfe 1997) on Japanese medaka, *Oryzias latipes*, a feminization effect of the male fish was revealed. After a three-month exposure with aqueous solutions of NP at concentrations of 10, 50, and 100 mg/L, 50% of the male fish in the middle concentration treatment and 86% of the males in the highest concentration treatment developed an intersex condition manifested by both testicular and ovarian tissue in the gonad.

Not only fish but also the Pacific oyster *Crassostrea gigas* are affected by NP. The researchers found that a single two-day exposure of larvae at their key development stage to this compound at 1 and 100 μg/L resulted in severe long-term consequences of sexual functions. The sex ratio changed towards females, and an increased incidence of hermaphroditism was observed. Production of gametes of poor quality led to poor embryonic and larval development of the next generation (Nice et al. 2003).

The impact of OP on reproductive organs was studied both in male and female rats. It was found that OP acted biphasically on androgen biosynthesis. For the studies, cultured dispersed testicular cells from neonatal rats were used as a source of fetal Leydig cells. Cultured cells were exposed for 24 hours to varying concentrations of OP (1 to 2000 nM) together with 10 mIU/mL human chorionic gonadotropin. The lower concentrations of OP (1 and 10 nM) increased testosterone levels (approximately 10 to 70% above control), whereas higher OP concentrations (100 to 2000 nM) progressively decreased testosterone from peak levels to approximately 40 to 80% below control at the highest OP concentration (Murono et al. 1999). The effects of OP on uterine contractions in immature rats were investigated by An et al. (An et al. 2013). They used a collagen gel contraction assay to analyze the contractibility of rat uterus, and noticed that OP increased the oxytocin-related pathway, while the prostaglandin-related signaling decreased. OP significantly reduced the contractility, which indicated adverse effects of OP on uterine contraction of immature rats. Moreover, OP reduced adipogenic enzyme expression, which led to reduced fat deposition in adipocytes and body weight in pregnant rats. The altered fat metabolism may affect the nutrition balance during pregnancy, and therefore can result in pregnancy metabolic complications, such as ketosis, marasmus, and diabetes mellitus (Kim et al. 2015).

Other studies on OP and NP were performed to test their ability to generate reactive oxygen species harmful to the brain tissue of male Wistar rats. The rats were given either OP or NP (25 mg/(kg day)) for 45 days. The rats of the control group were given olive oil orally. During investigations, oxidative damage of the brains of the rats was observed due to decreased levels of reduced glutathione and higher concentrations of malondialdehyde in comparison to the control group (Aydogan et al. 2008).

Zumbado et al. (Zumbado et al. 2002) investigated the acute hepatotoxicity exerted by NP and OP on rat liver. They examined the livers from animals treated with APs, and found the percentage increase of mitotic activity and Ki-67-labeling index and the presence of abnormal mitosis and c-mitosis. Moreover, decrease of concentration of the microsomal binding sites for estrogens after 10 days of treatment with APs was revealed. These results indicate that APs have the ability to induce cell proliferation, spindle disturbances, and modulate the expression of membrane receptors.

The researchers examined the effects of APs on mice models. Sabbieti et al. (Sabbieti et al. 2011) studied the impact of NP on mouse primary calvarial osteoblasts. They found that NP decreases osteoblast viability through the control of programmed cell death. Additionally, the survival effects of estrogen on osteoblasts are eliminated in the presence of AP. Also, the studies on the effect of NP on the induction of DNA strand breaks in mouse somatic cells were performed. The experiments included injections of NP alone or in combination with ionizing radiation. Male and female mice were irradiated with X-rays, injected with NP, or treated with both procedures for 2 weeks for 5 days a week. Liver, spleen, femora, lungs, and kidneys were examined, and in all males organ damage was observed, whereas in females, only the kidneys were affected. Increased DNA damage of all organs was found for the higher doses of NP. The genetic material of the female mouse somatic cells appeared to be less sensitive than that of male mouse somatic cells to the induction of damage by NP, probably due to higher level of estrogens in females. The results indicate the probability of interaction of NP with irradiation or other chemicals, which can even lead to cancer (Dobrzyńska 2014).

Also, studies on the effects of NP, OP, and butylphenol in mice with eczema were performed, and demonstrated that APs contributed to the increase in the incidence of atopic dermatitis. The treatment of APs increased the levels of IgE and antigen-specific IgG1 in the serum, the expression in lesions of inflammatory cytokines, interleukin-4, and monocyte chemotactic protein-3. NP, OP, or butylphenol are supposed to enhance T-helper 2-type immune responses in NC/Nga mice, which exacerbate mite allergen-induced atopic dermatitis-like skin lesions (Sadakane et al. 2014). Lee et al. (Lee et al. 2016) examined the effect of OP given to pregnant mice on the basis of mRNA analysis of their placentas. They revealed that expression of calcium transporting genes (PMCA1, TRPV6), copper transporting genes (CTR1, ATP7A), and iron transporting genes (IREG1, HEPH) were affected by the treatment. TRPV6, PMCA1, CTR1, ATP7A, and HEPH mRNA expressions were decreased in OP groups compared to the vehicle control. OP was supposed to modulate the calcium, copper, and iron ion transporting channels during the pregnancy, which could disrupt ionic homeostasis. For that reason, exposure to OP should be avoided by pregnant women (Lee et al. 2016).

The adverse effects of NP on the reproductive performance of male Brown Tsaiya ducks (*Anas platyrhynchos*) were also studied. The ducks were orally treated with

1, 10, and 250 mg NP/kg-BW/d for 14 weeks. Then semen quality, fertilization rate, specific factors in blood plasma, and weights of organs were measured. The treatment was continued for 12 months on some ducks from each group to notice tissue changes. The results demonstrated that larger amounts of NP with long-term exposure can lead to chronic toxicity of testicular tissues, which impairs spermatogenesis. A short-term exposure with large doses of NP may affect testosterone secretions (Cheng et al. 2016).

The detrimental impact of OP on male testes of bank voles, seminal vesicles, and endocrine functions of these animals kept under short (30 days) or long (60 days) photoperiod was investigated. It was demonstrated that relative weights of the testes and seminal vesicles significantly decreased, as a result of OP administration. More evident changes were found in the group exposed 60 days to OP, which resulted from decreased testosterone synthesis and increased endogenous estrogen production (Hejmej et al. 2011).

Chen et al. (Chen et al. 2009) investigated the association of the urinary NP concentration with the development of secondary sexual characteristics in 786 Taiwanese pubertal students. In the study, 15% of the Taiwanese girls between 10.5 and 10.9 years old reported having undergone menarche. After considering the effect of age and body mass index, an inverse correlation with the increase of urinary NP levels was found. It indicated the possible impact of NP on pubertal development. Moreover, the pubertal girls seemed to be more affected by early maturation related to the exposure to endocrine disruptive chemicals than pubertal boys (Chen et al. 2009).

There are also studies on the association between APs and allergic diseases. The results of these investigations on cell lines suggest that the onset, progression, and severity of allergic diseases may be related to exposure to APs (Suen et al. 2012).

The hazardous effect of APs and their derivatives, not only on aquatic organisms and mammals, including humans, but also on plants, was studied. The toxicity of different concentrations of NP, nonylphenol-4-ethoxylate (NP4E), and nonylphenol-10-ethoxylate (NP10E) on wheat seedlings using hydroponic experiments was investigated. NP appeared to be the most toxic to wheat, followed by NP4E and NP10E. NP inhibited 50% root and shoot growth at the lowest concentration. The adverse effects of NP on wheat were found in case of germination, shoot length, root length, chlorophyll, lipid peroxidation, and enzymatic activities. Due to the results of quantitative real time PCR, NP was confirmed as the most toxic to wheat plants. As far as toxicity of NPEs was concerned, it decreased with the increase of the ethoxy chain length (Zhang et al. 2016).

Endocrine disrupting properties of ethoxylated alkylphenols were also studied. Jobling and Sumpter compared the estrogenic potential of 17β-estradiol with that of 4-*tert*-butylphenol, 4-*tert*-octylphenol, 4-nonylphenol, and ethoxylated nonylphenol (Jobling and Sumpter 1993). The authors found that the mean potency of alkylphenols and their ethoxylates was dependent on both alkyl and ethoxy chain length. The lower the length, the higher potency was observed. The potency of 4-*tert*-butylphenol was 6,000 times lower than that of 17β-estradiol, and those of 4-OP and 4-NP were 27,000 and 111,000 times lower, respectively. On the other hand, the potency of 4-nonylphenoldiethoxylate and 4-nonylphenoxycarbocylic acid were both only about 1.5 times lower than that of NP, and the potency of NPE with 9 ethoxy groups on average (i.e., the commercial product Tergitol NP9) was 45 times lower than that of NP (Jobling and Sumpter 1993). The authors also found that for ortho- and meta-substituted *tert*-butylphenol, and for NPE with 40 ethoxy groups, Vtg production from cultured hepatocytes was comparable to control values (Jobling and Sumpter 1993).

The influence of chain length and branching was also investigated by Routledge and Sumpter who (on the contrary to Jobling and Sumpter's study) found that endocrine activity of 4-*tert*-substituted alkylphenols increases up to 8 carbon atoms in alkyl chain, and then substantially lowers for 4-*tert*-nonylphenol (Routledge and Sumpter 1997). The influence of branching was also tested, and it was found that for short alkyl chains, the less branched AP isomers were more estrogenic than the more branched APs, while for long alkyl chains (with 6–8 carbon atoms) this pattern was inversed (Routledge and Sumpter 1997). Higher estrogenic properties of more branched long chain APs compared to their less branched isomers were later confirmed for NP isomers in a study, including its 12 isomers (Gabriel et al. 2008).

The comparison of estrogenic properties of alkylphenols and their ethoxylates was also studied on rainbow trout (*Oncorhynchus mykiss*) by inhibition of its testicular growth. The effect was found to be more pronounced for OP (which inhibited testicular growth by 50%) than for NP, as well as both 4-nonylphenoldiethoxylate and 4-nonylphenoxycarbocylic acid. The degree of testicular growth inhibition appeared to be influenced by the estrogenic potential of these compounds, as there was a highly negative correlation between the estrogenic potencies and the size of testes (Jobling et al. 1996).

In another study on the estrogenic properties of APs and APEs, White et al. found that these compounds were capable of stimulating Vtg gene expression in trout hepatocytes, gene transcription in transfected cells, as well as the growth of breast cancer cell lines (White et al. 1994). The authors proved that the action of alkylphenols is mediated by the estrogen receptor, as it was dependent on its presence and it was blocked by estrogen antagonists. The activity of OP was higher than that of NP, and it was also found that the estrogenic activity of OPEs was decreasing with increasing ethoxy chain length. Moreover, there was no influence of ethoxy group oxidation on the activity of APEs (White et al. 1994).

Biodegradation of alkylphenol ethoxylates and alkylphenols

Biodegradation processes of surfactants, especially anionic and non-ionic-leading in terms of production and consumption, have been extensively tested by researchers since the 60s. The papers in this field, published until the early 1980s, were collected and comprehensively described for all surfactant classes by Swisher in his monograph, in which a lot of attention was devoted to biodegradation studies on alkylphenol ethoxylates (APEs), too (Swisher 1987). An extended list of papers on the biodegradation of APEs published until the early 90s and their discussion can be found in the Talmage monograph, devoted to the impact on people and the environment of alcohol ethoxylates and APEs (Talmage 1994). Interest in biodegradation as a way to reduce the total stream of surfactants directed to the environment has focused on particularly troublesome surfactants, showing high toxicity and high persistence in the natural environment, which also includes APEs. The negative impact of a large stream of surfactants flowing into the surface waters through wastewater treatment plants resulted in regulations which were introduced as early as in the late 70s (the earliest in Germany), allowing widespread use only of such anionic and nonionic surfactants, the primary biodegradation of which is at least 80% (BGBl 1977). This helped to reduce the harmful effects of these xenobiotics on living organisms and the environment. This regulation did not include cationic and amphoteric surfactants due to their limited production and consumption. Until the 90s, the requirement of at least 80% primary biodegradation was the basic criterion for the approval of surfactants for use. APEs also met this criterion—the primary biodegradation, usually exceeding 90%, was confirmed for them by the results of many studies carried out since the introduction of the regulation in 1977, until the 1990s (Pitter and Fuka 1979, Tabak and Bunch 1981, Brüschweiler et al. 1983, Birch 1984, Kravetz et al. 1984, Patoczka and Pulliam 1990). However, on the other hand, toxicity tests for APEs clearly showed that the products of their primary biodegradation: APEs with low ethoxylation degree—APnE ($n < 4$), their carboxylated analogs—APnEC ($n < 3$), and free alkylphenols (APs), have higher toxicity than parent APEs (usually having an ethoxylation degree in the range of $n = 5–40$). Therefore, unlike other classes and groups of surfactants, the primary biodegradation products of which showed fewer toxic effects than the parent surfactants, for APEs meeting the criterion of at least 80% of the level of primary biodegradation was not an effective way to minimize their harmful effects on living organisms and the environment. This criterion began to lose significance for the assessment of the environmental impact of APEs in the first half of the 90s, when several important publications appeared in a short time, indicating that the products of primary biodegradation of APEs, i.e., APs, short-chain APEs (AP1E, AP2E), and short-chain acidic biodegradation products of APEs (AP1EC, AP2EC) belong to xenoestrogens, i.e., chemical compounds that interfere with the proper functioning of the endocrine system of living organisms (Jobling and Sumpter 1993, White et al. 1994, Soto et al. 1995). There was a breakthrough in the approach to APEs biodegradation testing and interpretation of results, then. Model experiments aimed at quickly determining the primary biodegradation of APEs, in which one of the chemical methods for determining ethoxylates was most often used as the analytical stage (CTAS—Cobalt Thiocyanate Active Substances (Tabak and Bunch 1981, Kravetz et al. 1984, Patoczka and Pulliam 1990) or BiAS—Bismuth Active Substances (Pitter and Fuka 1979, Brüschweiler et al. 1983, Birch 1984)) lost their importance. While increasingly, APEs biodegradation experiments were planned with a broadly developed analytical control based on the use of such techniques as HPLC (Mann and Boddy 2000), MALDI-MS (Sato et al. 2001) or GC-MS (Ball et al. 1989), and monitoring of the intermediate products of APEs biodegradation in environmental conditions, e.g., in sewage treatment plants (Ahel et al. 1994). These studies were a contribution to determining the mechanisms and paths of the biodegradation of APEs, which, supplemented and expanded on the basis of research results obtained in later years, are now widely accepted. Paying attention in the 90s to problems arising from the xenoestrogenic properties of the intermediate products of APEs biodegradation, contributed to the rapid development of biodegradation studies of this group of compounds, with a particular focus on the biodegradation of APs. First of all, the biodegradation of the nonylphenols (NPs) and octylphenols (OPs) most commonly found in the aquatic environment was dealt with. In this initial period, a number of biodegradation experiments were carried out for APs, both in model conditions (Ekelund et al. 1993, Corti et al. 1995) and in environmental samples (Marcomini et al. 1989). Often, only the loss of APs in time was analyzed (e.g., by HPLC (Marcomini et al. 1989)) or the degree of their total biodegradation was determined (e.g., by measuring the amount of $^{14}CO_2$ produced from the ring-labelled APs (Ekelund et al. 1993)), to confirm knowledge of the high resistance of APs to biodegradation. On the other hand, studies in which APs biodegradation intermediate products were determined (e.g., 4-acetylphenol-4-NP biodegradation product, analyzed by HPLC, NMR, FT-IR, and MS techniques (Corti et al. 1995)), were among the first to contribute to understanding the mechanisms and paths of APs biodegradation, currently in force.

Numerous studies performed for APEs in the last 20 years of the 20th century showed a relatively fast and deep primary biodegradation of these compounds (Tabak and Bunch 1981, Brüschweiler et al. 1983, Birch 1984, Kravetz et al. 1984, Ball et al. 1989, Patoczka and Pulliam 1990, Ahel et al. 1994, Mann and Boddy 2000, Sato et al. 2001). On the other hand, the above-mentioned and many other experiments conducted in those years confirm the tendency to accumulate in various elements of the environment of some intermediate products of APEs biodegradation (APs, short-chain APEs with one and two ethoxylene groups—AP1E and AP2E, as well as their carboxylated analogs—AP1EC and AP2EC), which clearly indicates their much greater resistance to biodegradation than the parent APEs (Stephanou and Giger 1982, Reinhard et al. 1982, Giger et al. 1984, Ahel et al. 1986, 1987, 1994, Ball et al. 1989, Sato et al. 2001). Combining the above information with the documented negative impact of APs, their short-chain ethoxylates and carboxylated analogs on living organisms (including the human organisms), and in particular, with their xenoestrogenic activity (Jobling and Sumpter 1993, White et al. 1994, Soto et al. 1995), it can be stated that the primary biodegradation of APEs leads to intermediate products that are more harmful to humans and the environment than the parent substances. A deeper analysis of the results of the above-mentioned publications in the field of the biodegradation of APEs gives evidence for the existence of both oxidative and non-oxidative biodegradation pathways, in which there are intermediate products indicating the existence of a biodegradation mechanism by shortening the ethoxylene chain, which leads to the formation of short chain APEs by non-oxidative route and their carboxylated analogs by oxidative route (Stephanou and Giger 1982, Reinhard et al. 1982, Giger et al. 1984, Ahel et al. 1986, 1987, 1994, Ball et al. 1989, Sato et al. 2001). This metabolic pathway is considered the most important and dominant for APEs, both in laboratory biodegradation experiments and during their biodegradation in sewage treatment plants (Ahel et al. 1987, Holt et al. 1992). Further biodegradation of these intermediates leads to the formation of APs, which are highly resistant to biodegradation. Interestingly, in 2003, the possibility of biodegradation of APEs in the process of enzymatic hydrolysis (according to the so-called central fission mechanism) was confirmed, in which the corresponding APs and poly(ethylene glycols) (PEGs) are formed directly in one stage (Frańska et al. 2003). The three mechanisms of APEs biodegradation listed above are graphically shown in Fig. 1.3. It is widely accepted that due to the formation of intermediate products with increased resistance to further biodegradation (APs, AP1E, AP2E, AP1EC, and AP2EC), the level of total biodegradation of parent APEs in biological wastewater treatment plants reaches a maximum of about 40% in relation to the APEs mass entering the treatment. This means that about 60% of the APEs mass introduced with raw sewage gets into the environment in the stream of treated sewage as partially degraded akylphenolic substances (Ahel et al. 1994).

The general outline of the APEs biodegradation pathways (Fig. 1.3) is now accepted without major reservations. The key role in the APEs biodegradation mechanism of shortening the ethoxylene chain by individual ethoxylene groups as a result of enzymatic hydrolysis (dominates in anaerobic conditions) or enzymatic hydrolysis combined with oxidation (dominates in aerobic conditions) is widely

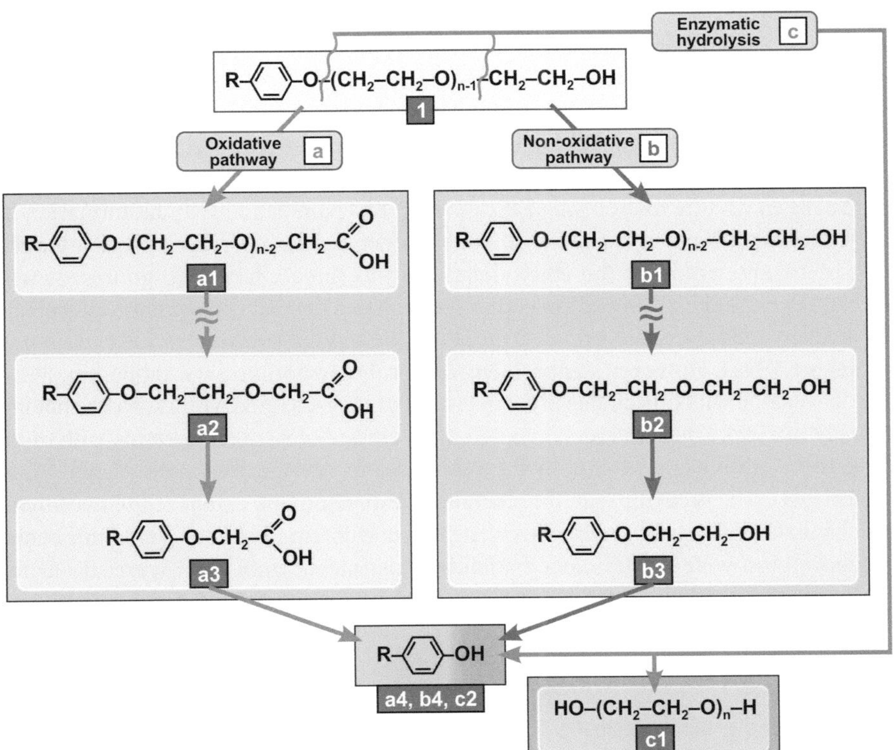

Figure 1.3. Main biodegradation routes of APEs. (1) parent APEs; (a1, a2, a3) APECs with a diminishing ethoxylation degree; (b1, b2, b3) partially biodegraded APEs with a diminishing ethoxylation degree; (c1) PEGs; (a4, b4, c4) APs as stable, intermediate products of the biodegradation of the ethoxylene chain according to three paths: (a) oxidative, (b) non-oxidative, (c) enzymatic hydrolysis according to the central fission mechanism.

recognized. An overview of the environmental fate of the most important intermediate products of APEs biodegradation by shortening the ethoxylene chain (APs, AP1E, AP2E, AP1EC, and AP2EC) and information on the level of their concentration in the various elements of the environment can be found in the review papers (Ying et al. 2002, Ying 2006, Hernandez-Raquet 2012, Olkowska et al. 2013, Grześkowiak et al. 2018). The widespread occurrence of these compounds is also confirmed by the latest publications in the field of their monitoring, both in the APEs model biodegradation tests (Wild and Reinhard 1999, Sciubba et al. 2014), in various elements of the environment (e.g., sewage sludge and sludge-amended soils (Venkatesan and Halden 2013), raw and treated wastewater in sewage treatment plants (Wu et al. 2017), river water sediments (Chokwe et al. 2016), river water (Rocha et al. 2016), surface water in heavily urbanized areas (Cladière et al. 2010)), as well as in various raw materials and products (e.g., recycled paper sludge (Hawrelak et al. 1999), plastic containers (Guart et al. 2011)), and in drinking water and food (e.g., bottled water and milk powder (Loyo-Rosales et al. 2004)).

For quite a long time, there was no consensus among researchers as to the real significance and extent of biodegradation processes leading to the formation of carboxylated intermediate products of APEs ethoxylene chain biodegradation according to the shortening mechanism. Until around the mid-90s it was thought, based mainly on the results of the monitoring of APEs biodegradation products in the wastewater treatment plant, conducted by the Giger research group (Giger et al. 1987) and on the model laboratory tests of Yoshimura (Yoshimura 1986) and the McCarty research team (Ball et al. 1989), that biodegradation of APEs through oxidative enzymatic hydrolysis of the ethoxylene chain, leading to the formation of alkylphenoxypolyethoxy carboxylic acids (APECs), can only take place after the hydrolytic shortening of the ethoxylene chain to three ethoxylene groups. It was thought that the shortening of long ethoxylene chains ($n > 3$) by enzymatic hydrolysis occurs very quickly, which prevents the slow oxidation of hydroxyl group to carboxyl (Talmage 1994). However, the problem was probably not too slow oxidation, but it was associated with difficulties in the detection of long-chained APECs. Nevertheless, already in 1994, Ahel et al. documented the presence of such compounds with more than three ethoxylene groups in the molecule, in sewage treatment plants (Ahel et al. 1994). The breakthrough for changing views about the extent of the hydrolytic mechanism of shortening the ethoxylene chain with simultaneous oxidation of the alcohol group were model biodegradation experiments from 2001, carried out for APEs in river water samples (Jonkers et al. 2001). Using HPLC-ESI-MS and HPLC-ESI-MS-MS techniques to determine NP9E biodegradation products, Dutch researchers demonstrated the presence in treated wastewater NPnEC containing up to $n = 14$ ethoxylene groups in the molecule. These results became the basis for recognizing the mechanism of oxidative enzymatic hydrolysis of the ethoxylene chain of APEs, which are as equally important as the mechanism of enzymatic hydrolysis. The latest publications, in which modern analytical techniques with increasingly better operating parameters are used for monitoring biodegradation products, confirm the widespread formation of APECs during biodegradation of APEs under aerobic conditions. It is worth emphasizing that the formation of acidic products, also for isomers containing up to about 20 ethoxylene groups in the molecule, has been documented in model biodegradation tests performed for nonylphenol monopropoxyethoxylates (NPPEs) (Zgoła-Grześkowiak et al. 2015) and nonylphenol propoxylates (Zgoła-Grześkowiak et al. 2014). There is also information about the possibility of the formation, during the biodegradation of APEs, of products containing carboxyl groups in the alkyl chain and both in the ethoxylene and alkyl chain (Di Corcia et al. 1998, Jonkers et al. 2001). Such processes began to be studied more widely only after 2000, when the interest in biodegradation of endocrine-active APs increased. It is worth noting that the products of simultaneous oxidation of APEs in the alkyl and ethoxylene chain have not yet been tested for their possible endocrine properties, as they have not yet been found to be as widespread as APEs biodegradation products.

One of the most controversial issues to be resolved is the answer to the question whether biodegradation of APEs is really possible according to the central fission mechanism. To date, the scientific community accepts with caution evidence indicating the presence of products derived from such a path of biodegradation (i.e., APs and PEGs (Frańska et al. 2003)), in the model tests of APEs biodegradation. However, more articles confirming this possibility have been published in recent years. Researchers from Poznan University of Technology conducted diesel biodegradation experiments in the presence of ethoxylated *tert*-octylphenol (Triton X-100), using a bacterial consortium isolated from petroleum-contaminated soil, capable of hydrocarbons biodegradation (Wyrwas et al. 2011). It turned out that under aerobic conditions, the consortium tended to use APEs as the preferred carbon source, converting them into APs and PEGs in the first stage according to the mechanism of central fission of the molecule. Easy access of microorganisms to alternative organic carbon in the form of APEs caused a reduction in the biodegradation level of diesel oil to about 60%, compared to about 80% in an experiment without the addition of APEs (Wyrwas et al. 2011). Meanwhile, Zgoła-Grześkowiak et al. studied primary biodegradation and the formation of various intermediates of biodegradation of nonylphenol monopropoxyethoxylates (NPPEs) (Zgoła-Grześkowiak et al. 2015). The authors conducted two biodegradation tests under aerobic conditions, using sewage sludge from the municipal sewage treatment plant as inoculum in one of them, and river water in the other one. In the test with activated sludge, complete primary biodegradation of NPPEs was achieved, whereas in the river water test, the degree of biodegradation was only about 60 percent. In both tests, intermediate products derived from both oxidative and non-oxidative biodegradation pathways of NPPEs were identified by HPLC-MS. The oxidative pathway showed the presence of NPPE derivatives carboxylated in an alkoxylene chain, which indicates the course of biodegradation under these conditions according to the oxidative shortening mechanism of the alkoxylene chain, i.e., in a manner considered to be the most typical for compounds of this structure. On the other hand, the analysis of products formed in the non-oxidative biodegradation path gave very interesting results, since monopropoxy poly(ethylene glycols) were identified among them, which indicates the course of biodegradation of NPPEs under these conditions by enzymatic hydrolysis, according to the mechanism of central fission of the molecule (Zgoła-Grześkowiak et al. 2015).

Over the past two decades, there has been a definite rise in attention to the biodegradation studies of free APs due to the demonstration of their endocrine activity in the mid-90s. However, data from environmental monitoring indicate continuously that, apart from APs, also APEs, including their endocrine active short chain derivatives (AP1E, AP2E) and their corresponding acidic oxidation products (AP1EC, AP2EC) are still present in the wastewater (Di Corcia et al. 1994, Ding and Tzing 1998, Wu et al. 2017), in sewage sludge and sludge-amended soils (Venkatesan and Halden 2013), river waters (Ding and Tzing 1998, Heemken et al. 2001, Rocha et al. 2016) and river bottom sediments, and coastal marine waters (Heemken et al. 2001). In addition, research also confirms the accumulation of these compounds in living organisms (Staples et al. 1998, Vidal-Liñán et al. 2015, Staniszewska et al. 2017, Lv et al. 2019). Therefore, it is important to constantly search for suitable microorganisms or methods that lead to effective and complete degradation of APEs and APs. Hence, biodegradation tests of parent APEs have been continued and intensively developed also in recent years. For several decades, until the end of the 20th century, the bacterial biodegradation processes of APEs were studied almost exclusively. Currently, apart from biodegradation studies using bacteria, research is increasingly appearing in regard to the use of algae and fungi for biodegradation of APEs. A new trend is also the search for more effective technological solutions for biodegradation processes (e.g., intensification of research on the introduction of a membrane bioreactor into the sewage treatment practice).

The authors of several publications point out that the use of a membrane bioreactor (MBR) significantly improves the efficiency of biodegradation and generally—removal of APEs from wastewater (Li et al. 2000, Lubello and Gori 2005, Terzic et al. 2005). Li et al. obtained in the research conducted in a conventional sewage treatment plant with activated sludge the elimination of APEs at the level of 87 percent. Treating wastewater from the same treatment plant in the membrane bioreactor system, they achieved APEs removal efficiency in the range of 91–97 percent. Other researchers have achieved in the laboratory model test using MBR, the elimination of NPEs and their biodegradation products, NP1EC and NP2EC, at the level of 98%, while in the classic bioreactor with activated sludge, only at 54% (González et al. 2007). However, pilot studies did not confirm such significant improvement (75% removal of alkylphenolic substances in a conventional reactor, and on average about 86% in experiments with two membrane reactors) (González et al. 2008). In summary, comparative studies of the removal of alkylphenolic compounds in conventional activated sludge technology and in MBR showed better removal of parent compounds and metabolites of low polarity (NP, NP1E, NP2E) in MBR. Acidic intermediate biodegradation products (NP1EC, NP2EC) were formed in both MBR and classic activated sludge, but their concentration was reduced only to a small extent—with significant concentrations, they were released with treated wastewater.

The vast majority of model biodegradation tests are conducted under standard conditions provided for them. Generally, there are few such works, also in the context of APEs biodegradation tests, in which significant parameters would be intentionally changed to verify the impact of specific physicochemical properties of

APEs on their biodegradation. One of the few such works performed in recent years concerned the verification of the impact of complexation of APEs with metal ions on the observed effects of their biodegradation. Frańska et al. (Frańska et al. 2009) studied the biodegradation of ethoxylated *tert*-octylphenol commercially available under the name Triton X-100. The biodegradation process was carried out under aerobic conditions, with significant modification of the conditions of the standard OECD 301E screening test by additionally introducing various inorganic salts in a significant concentration into the biodegradative mixture. An activated sludge taken from the municipal sewage treatment plant was used as the inoculum. Compared to standard conditions, various inorganic salts were added in significant concentration to the biodegradable mixture. In the case of the addition of a salt containing a metal cation toxic to microorganisms (e.g., lead), biodegradation did not occur, as expected. An interesting effect was observed, however, when metal salt harmless to microorganisms (e.g., potassium) was added in high concentration. Under these conditions, the *tert*-octylphenol ethoxylate homologue containing 6 ethoxylene units in the ethoxylene chain was found to be extremely resistant to oxidative shortening. The authors explained that the occurrence of this effect is a consequence of the creation by the Triton X-100 homolog containing 6 ethoxyelene groups, the most stable complex with a potassium cation, in the creation of which of an extreme oxygen atom of the ethoxylene chain also participates. Such a complex shows high resistance to attack by oxidative bacteria (Frańska et al. 2009). The results of this experiment indicate that conclusions based on model biodegradation studies should be drawn with extreme caution, because the model biodegradation system is naturally simplified—it usually contains a single source of energy for microorganisms, while minimizing any additional physicochemical factors (e.g., the susceptibility for complexation).

The discovery of endocrine activity of APs with long alkyl chains (mainly nonylphenols—NPs and octylphenols—OPs) intensified the research to improve the efficiency of their biodegradation in the 90s. This was due to the need to improve the level of their removal from wastewater, as they posed a serious threat to living organisms and the environment, being raw materials for high-volume production of APEs, and ultimately also intermediate products of their biodegradation, which are exceptionally resistant to attack by microorganisms—both in sewage treatment plants, as well as in the environment.

The unsatisfactory results of early tests on the biodegradation of NPs by bacteria were explained by the exceptional stability of a technical mixture consisting of 85% of isomers containing a quaternary α-carbon in an alkyl substituent (Wheeler et al. 1997), the presence of which was the impediment to their bacterial oxidation according to one of the typical mechanisms, ω-oxidation or β-oxidation (Van Ginkel 1996). The good biodegradation results of NPs in soil using mixed microbiological cultures (Topp and Starratt 2000) seemed to support the hypothesis that only the complex microflora are capable of biodegradation of branched side NP chains. To confirm this thesis, it was also argued that in the 90s, only single bacterial strains capable of biodegradation of linear alkyl substituents were known, e.g., *Candida aquaetextoris* (Vallini et al. 1997), but it soon became apparent that many individual strains also have the potential for biodegradation of NP isomers with branched alkyl substituents, e.g., isolated from the municipal sewage treatment plant *Sphingomonas* sp. strain TTNP3, *S. cloacae*, and *Sphingobium xenophagum* strain Bayram (Tanghe et al. 1999, Fujii et al. 2001, Gabriel et al. 2005) and Sphingomonas sp. YT, isolated from river sediments (de Vries et al. 2001).

Due to the large variety of bacteria capable of biodegradation of APs with long alkyl substituents (mainly NPs and OPs), it is difficult to identify any universal biodegradation mechanism. Due to the fact that the majority of NP-degrading bacteria belong to sphingomonads and closely related types, the pathways characteristic of these bacteria can be considered typical for bacterial biodegradation of NPs and OPs. The biodegradation of APs using the bacteria *Sphingomonas* sp. strain TTNP3 has been particularly well studied. The quite widely accepted path for biodegradation of APs containing branched alkyl substituents (with a quaternary α-carbon), using the bacterium *Sphingomonas* sp. strain TTNP3, is shown in Fig. 1.4. It has been shown that bacterial isolates can grow on NPs as the only source of organic carbon, and also cause their mineralization.

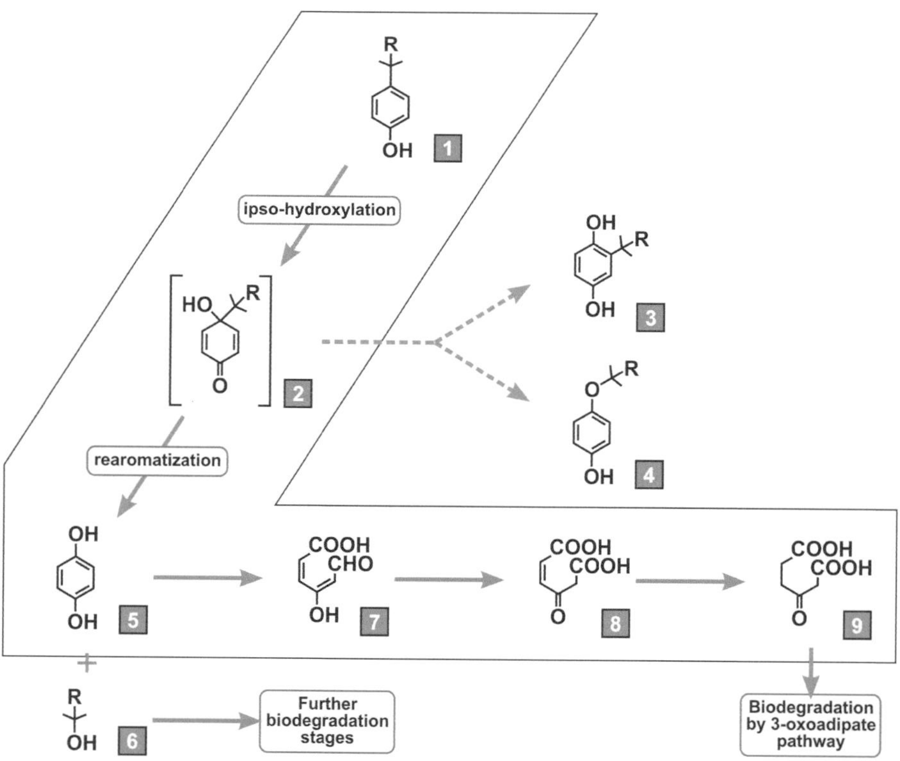

Figure 1.4. Biodegradation pathway of alkylphenols for *Sphingomonas* sp. strain TTNP3. (1) long-chain AP (NP, OP) with quaternary alpha carbon in an alkyl substituent–alkylphenolic substrate; (2) quinol intermediate; (3) alkylhydroquinone (resulting from the NIH-shift); (4) alkoxyphenol; (5) hydroquinone; (6) tertiary alcohol; (7) 4-hydroxymuconic acid semialdehyde; (8) maleylacetic acid; (9) 3-oxoadipic acid (Kolvenbach and Corvini 2012).

Bacterial degradation of long-chain APs with quaternary α-carbon (Fig. 1.4) involves several stages, the most important of which are—ipso-hydroxylation (leading to quinol intermediate), rearomatization (leading to the formation of hydroquinone and the corresponding tertiary alcohol), and further biodegradation of hydroquinone to open the aromatic ring. Full ring mineralization is obtained in the biodegradation by the 3-oxoadipate pathway (Kolvenbach and Corvini 2012). It is worth noting that at the rearomatization stage, competitive reactions can occur to create alkylhydroquinone (by the NIH-shift), and to form alkoxyphenol. These are two undesirable reactions, the products of which practically ended the biodegradation cycle of the benzene ring (Kolvenbach and Corvini 2012). It is also an interesting observation that the tertiary alcohol formed at the rearomatization stage, can also be further biodegradable.

Preferential elimination pathways of long chain branched APs by fungi can be both those that lead to oxidation of the alkyl chain and the benzene ring, and also to the formation of phenolic polymers. In all of these pathways, both intracellular and extracellular oxidative enzymes can be involved. Biodegradation of APs by fungi is not yet thoroughly investigated, and therefore it is difficult to generalize the possible paths of biodegradation. The latest, interesting publication in this field concerns biodegradation of the 4-*tert*-octylphenol by the non-ligninolytic fungus *Fusarium falciforme* RRK20. The authors propose a mechanism of biodegradation processes and make an attempt to generalize the mechanisms of biodegradation of octylphenols by bacteria and fungi (Rajendran et al. 2020).

Alkylphenols and their ethoxylates in surface waters

Research on toxicity and endocrine disrupting properties of alkylphenols and their short-chain ethoxylates resulted in increased interest in these compounds. Results published in 90s have shown that these widely used compounds can be found at very high concentrations in both surface water and sediments (Naylor et al. 1992, Bennie et al. 1997).

Studies performed during 90s in North America (Canada and the USA), Europe (Portugal and Spain), as well as in Asia (Japan and Taiwan) have shown that concentrations of alkylphenols in surface waters were high–NP was reaching µg/L liter or even hundreds µg/L, and NPEs were determined both as single ethoxylates or as the sum of all detectable ethoxylates at similar levels (Table 1.1). Moreover, in the following years, these compounds were found at even higher concentrations—up to 22 mg/L OP and 37 mg/L NP were reported in Spanish rivers in the year 2001, which is roughly 1,000 times more than in previous studies (Cespedes et al. 2005).

Growing concerns about environmental pollution led to a worldwide debate on alkylphenols and influenced the state or international authorities. As a consequence, various steps have been undertaken to reduce levels of alkylphenols polluting the environment. However, the approach to solve the problem in Europe and North America was different. The European Union issued the 2003/53/EC directive amending the 76/769/EEC aiming at restricting marketing and usage of nonylphenol and its ethoxylates. These compounds may not be used in Europe at concentrations higher than 0.1% in products intended for—industrial, institutional and domestic

Table 1.1. Concentration of APs and APEs in surface waters in ng/L.

Country	Year	OP	NP	OPEs	NPEs	Literature
USA	1989	-	< 110–640	-	< 70–1200[a] < 1600–14.9 · 10[3b]	Naylor et al. 1992
	1997	< 2–81	< 11–1190	-	< 52–17.8 · 10[3b]	Snyder et al. 1999
	≤ 2003	< 9	140–200	< 14[a]	< 14–67[a]	Loyo-Rosales et al. 2003
Canada	1994/1995	< 5–84	< 10–920	-	< 20–10 · 10[3a]	Bennie et al. 1997
	2002	< 5–12	< 10–258	-	< 20–4040[a] < 600–91.7 · 10[3b]	Mayer et al. 2007
Brazil	2015–2017	-	< 200–2320	-	-	de Araujo et al. 2018
Spain	1999	-	< 150–644 · 10[3]	-	< 200–100 · 10[3b]	Sole et al. 2000
	2001	< 20	1300	600[b]	4400[b]	Petrovic et al. 2001
	2001	< 90 · 10[3]–21900 · 10[3]	< 15 · 10[3]–37300 · 10[3]	-	< 60 · 10[3]–6870 · 10[3a]	Cespedes et al. 2005
	< 2012	< 8–110	< 30–140	-	-	Salgueiro-González et al. 2012
Portugal	1999	-	< 10–30 · 10[3]	-	-	de A. Azevedo et al. 2001
	2010	4.3–41	12.2–547	6.9–182[a]	41–780[a]	Rocha et al. 2013
	2010	30–27500	81–1000	11–2340[a]	95–18.3 · 10[3a]	Rocha et al. 2014
	2011	4.5–23.7	54.6–222.4	5.3–50.7[a]	43.2–1530[a]	Rocha et al. 2016
Poland	2008	< 50–560	1400–3400	< 50–540[a]	100–3800[a]	Zgoła-Grześkowiak and Grześkowiak 2009
	2009	< 20	< 70–560	< 20[a]	< 70–540[a]	Zgoła-Grześkowiak et al. 2010
	2009	100	< 330	100–300[a]	300–500[a]	Zgoła-Grześkowiak 2010
	2009	< 10–40	180–530	< 30–170[a]	< 30–420[a]	Zgoła-Grześkowiak and Grześkowiak 2011
	2011/2012	< 1.0–834.5	< 4.0–228.6	-	-	Staniszewska et al. 2015

Table 1.1 Contd. ...

...Table 1.1 Contd.

Country	Year	OP	NP	OPEs	NPEs	Literature
Germany	2000	< 0.5–5	13–87	< 0.5–9.6[b]	< 0.5–124[b]	Stachel et al. 2003
Netherlands	2000	nd	nd–140	nd	nd	de Voogt et al. 2000
Belgium	2003	< 5	320–2500	< 10–26[a]	nd–1200[a]	Loos et al. 2007
Italy	2003	12–111	460–700	< 10–93[a]	nd–3600[a]	
France	2009	1–81	58–426	< 1–14[a]	10–414[a]	Cladière et al. 2010
Japan	1997/1998	10–180	40–810	-	50–1080[a]	Isobe et al. 2001
	2000	< 3–118	< 10–2870	-	< 40–3380[a]	Isobe and Takada 2004
China	2005	1.2–389	20.2–28.6 · 10^3	-	-	Fu et al. 2007
	2009	1.54–45.8	106–344	-	8.92–385[a]	Zhang et al. 2014
	2010/2011	< 1.0–292.8	9.0–1687.8	-	-	Zhang et al. 2012
	2012	2–73	5–57	-	54–2074[a]	Wang et al. 2013
	< 2013	3.85–71.24	-	-	-	Yang et al. 2013
	2013	2.8–15700	36–16.2 · 10^3			Zhong et al. 2017
	2014	nd–185	nd–104	-	-	Yang et al. 2015
	2015	-	174–3411	-	-	Jie et al. 2017
Taiwan	≤ 1998	-	600–3000	-	< 10–10.3 · 10^{3a}	Ding and Tzing 1998
	2000	-	< 10–5100	-	< 10–500[a]	Cheng et al. 2006
	2004/2005	-	nd–310 · 10^3	-	nd–27.5 · 10^{3a}	Chen et al. 2010
Korea	2002	-	17.4–1533	-	-	Li et al. 2004
India	2010	< 1.1–16.3	< 0.3–2200	-	-	Selvaraj et al. 2014
	2014	-	1220–7240	-	-	Raju et al. 2018
South Africa	2005	310–6010	250–9350	720–92.7 · 10^{3a}	-	Sibali et al. 2010
Nigeria	≤ 2013	43.9–79.4	57.1–68.6	-	-	Oketola and Fagbemigun 2013
	2017	nd–1.81	nd–0.48	nd	nd–9.19[a]	Tongu et al. 2018

nd – not detected, a – concentrations apply to individual ethoxylates, b – concentrations apply to the sum of ethoxylates.

cleaning, textile and leather processing, emulsification in agriculture, metal working, manufacturing of pulp and paper, cosmetic products and other personal care products, pesticides and biocides (Directive 2003). There are only a few exceptions to the above-mentioned restrictions, e.g., textile and leather processing without release into wastewater. What is important, the European restrictions are not aimed at OP and OPEs. On the contrary, Swiss regulations based on this directive included OP and OPEs with the same limits as for NP and NPEs (Swiss Ordinance 2009).

Further, European legislation initiatives were related directly to the environmental levels of NP. The 2008/105/EC directive listed nonylphenol as the priority hazardous substance, and set the environmental quality standards for water. The limits have been set to—0.3 and 2.0 μg/L for annual average in surface water and maximum allowable concentration in surface water, respectively (Directive 2008).

Both Canada and the USA set only environmental quality criteria without any restrictions on the usage of alkylphenols. According to the US Environmental Protection Agency, in freshwater the one-hour average concentration of NP should not exceed 28 μg/L, and the four-day average concentration of NP should not exceed 6.6 μg/L. For saltwater, analogous criteria were set to 7.0 μg/L and 1.7 μg/L, respectively (US EPA 2005). Canada set freshwater criterion for NP to 1.0 μg/L, and marine criterion for NP to 0.7 μg/L, and included toxic equivalency factors to account for OP as well as NPEs and OPEs (Canadian Water Act 2002).

Rising concerns about environmental pollution with alkylphenols led to increase of worldwide interest in these compounds. Within the European Union, the widest research was performed in Spain, Portugal, and Poland. The results show lowering concentrations of alkylphenols in surface waters, although there are places where still a lot needs to be done (Rocha et al. 2014). Studies conducted in Portugal on samples gathered between 2010 and 2011 show that concentrations of AP and APEs in less urbanized areas are relatively low, while the amounts of these compounds in highly urbanized regions are very high, reaching 27.5, 1.0, 2.3, and 18.3 μg/L of OP, NP, OPE, and NPE, respectively (Table 1.1) (Rocha et al. 2013, 2014, 2016).

Concentrations of alkylphenols and their ethoxylates in natural waters of the USA were measured between 1992 and 2003. The detected amounts were below the value 6.6 μg/L—the limit of the National Recommended Aquatic Life Criteria, even though there are no restrictions on the marketing and use of these compounds in the USA (Naylor et al. 1992, Snyder et al. 1999, Loyo-Rosales et al. 2003, US EPA 2005).

In Canada, monitoring of alkylphenols and their ethoxylates had been intensively conducted between 1994 and 2002, before the legal regulations entered into force. The measured concentrations appeared far too high in relation to the limit of 1.0 μg/L of the Canadian Water Quality Guidelines for the Protection of Aquatic Life, and considering in calculations the toxic equivalency factors for nonylophenol ethoxylates in Canada (Bennie et al. 1997, Mayer et al. 2007, Canadian Water Act 2002).

In Asian countries such as China, Taiwan, and India, where there is a lack of legal regulations on marketing and use of alkylphenols and their derivatives, studies on environmental presence of these compounds have also been carried out. Research done in China between 2005 and 2015 revealed significant environmental pollution. The maximum concentration of nonylphenol in surface water noted in 2005 reached 28.6 μg/L (Fu et al. 2007). In the next decade, more studies were performed, in which maximum NP concentration was reported from 0.104 μg/L determined in Honghu Lake in 2014 (Yang et al. 2015) to higher values in more polluted rivers, where NP concentration reached μg/L levels, such as Xiangjiang River, where NP was found up to 3.4 μg/L in 2015 (Jie et al. 2017), and Perl River estuary, where maximum NP level was 16.2 μg/L in 2013 (Zhong et al. 2017). Moreover, OP was found in Perl River estuary at a concentration of 15.7 μg/L (Zhong et al. 2017).

The results of studies performed in Taiwan between 1998 and 2015 also show significant pollution of the aquatic environment. The highest concentrations of NP and its ethoxylates found in surface waters regularly reach μg/L levels (Ding and Tzing 1998, Cheng et al. 2006, Chen and Yeh 2010). Similarly, studies done in India in 2010 and 2014 show NP at μg/L level (Selvaraj et al. 2014, Raju et al. 2018).

Studies in Africa were conducted only in two countries—the Republic of South Africa (one of the most developed African countries) and Nigeria (the most populated African country). The study from the Republic of South Africa shows high contamination of surface waters with OP, NP, an OPEs (Table 1.1), which is characteristic of countries with developed industry and no regulation on APs (Sibali et al. 2010). On the other hand, the concentrations of APs and APEs in Nigeria are much lower, at the ng/L level (Oketola and Fagbemigun 2013, Tongu et al. 2018).

The presented concentrations of alkylphenols and their ethoxylates in natural surface waters of selected countries show that their levels are not always satisfactory, even in places where some legal regulations on their production and use exist. It is worth noticing that the existence of restrictions for endocrine disrupting chemicals is not sufficient, if these legal regulations are not implemented. However, in case of adhering to the contamination limits, the natural surface waters are rarely highly polluted. On the other hand, concentrations of alkylphenols and their ethoxylates in surface waters of countries with no regulations and highly developed industry are very high, and regularly reach even a few dozen μg/L.

Alkylphenols and their ethoxylates in sediments

Alkylphenols and their short-chain ethoxylates can be easily accumulated in sediments due to their hydrophobic properties. Numerous studies show that these compounds can be found in both river and marine sediments (Table 1.2). The highest concentrations of NP and NPEs in sediments have been reported so far for Dutch channels, where NP was found at 600 to 1700 mg/kg, and NPEs (determined as the sum of ethoxymers) were found at 2,600 to 5,700 mg/kg (de Voogt et al. 2000). Concentrations of NP in sediments reported worldwide usually vary between a few μg/kg and a few mg/kg. Compared to NP occurring in surface waters (Table 1.1), it is about 1,000 times more per kg. Moreover, the maximum concentrations reported in the literature often exceed the limits proposed for sediments. The interim sediment quality guideline for nonylphenol and its ethoxylates issued by the Canadian Council of Ministers and the Environment includes the limit 1.4 mg/kg for freshwater sediment and 1.0 mg/kg for marine and estuarine water sediment. The limits are

Table 1.2. Concentration of APs and APEs in water sediments in µg/kg.

Country	Year	OP	NP	OPEs	NPEs	Literature
USA	1989	-	< 2.9–2960	-	< 2.3–175[a]	Naylor et al. 1992
	≤ 2003	< 40–410	410–6700	< 40–110[a]	24–1800[a]	Loyo-Rosales et al. 2003
Canada	1994/1995	< 10–1800	170–72 · 10^3	-	< 15–38 · 10^3[a]	Bennie et al. 1997
	≤ 1998	< 1–23700	< 46–368 · 10^3	-	-	Bennett and Metcalfe 1998
	2001/2002	< 5–52	< 10–1750	-	< 25–1250[a] < 80–1320[b]	Mayer et al. 2007
Spain	2001	15	235	20[b]	350[b]	Petrovic et al. 2001
	2003	30.2–103	89.2–2331	-	-	Lacorte et al. 2006
Poland	1998	< 1–9.75	< 1–762	-	-	Kannan et al. 2003
	2009	< 2	24–97	< 7–69[a]	< 6–168[a]	Zgoła-Grześkowiak et al. 2010
Germany	2000	4–62	27–428	25–813[b]	24–3667[b]	Stachel et al. 2003
Netherlands	2000	nd	630 · 10^3–1700 · 10^3	nd	2600 · 10^3–5700 · 10^3[b]	de Voogt et al. 2000
Japan	1997/1998	3–670	30–13 · 10^3	-	10–3470[a]	Isobe et al. 2001
	2000	< 3–40	30–1820	-	< 40–460[a]	Isobe and Takada 2004
China	2005	nd–146	3.6–299	-	-	Fu et al. 2007
	2009	0.14–1.70	11.2–117	-	0.10–169[a]	Zhang et al. 2014
	2010/2011	0.8–9.3	349.5–1642.8	-	-	Duan et al. 2014
	2013	0.3–42	8.7–92	-	-	Zhong et al. 2017
	2014	2.59–95.97	nd–242.85	-	-	Yang et al. 2015
Taiwan	2013	1.1–1150	18–27.9 · 10^3	-	-	Dong et al. 2015
	2013/2014	0.36–55.81	1.16–324	< 10–18.6[a]	< 9–45.0[a]	Shiu et al. 2019

Table 1.2 Contd. ...

...Table 1.2 Contd.

Country	Year	OP	NP	OPEs	NPEs	Literature
South Korea	2002	-	10.4–5054	-	-	Li et al. 2004
	2007	nd–1.8	0.7–159.3	-	nd–105.7[a]	Hong and Shin 2009
India	2014	-	3.31–30.96	-	-	Raju et al. 2018
South Africa	2005	1.94–76.1	2.76–134	1.72–941[a]	-	Sibali et al. 2010
	2013	-	-	-	3.44–55.4[a]	Chokwe et al. 2016
Nigeria	≤ 2013	2.2–24.5	1.1–79.4	-	-	Oketola and Fagbemigun 2013
	2017	nd–2.16	0.01–0.84	0.01–0.23[a]	nd–10.5[a]	Tongu et al. 2018

nd – not detected, a – concentrations apply to individual ethoxylates, b – concentrations apply to the sum of ethoxylates.

based on the toxic equivalency approach, so that the toxic equivalency factors for OP, OPEs, and NPEs are also taken into account (Canadian Sediment Act 2002). Therefore, the simultaneous presence of different APs and APEs is additive, and the limits are exceeded to a greater extent. Furthermore, as biotransformation of short-chained NPEs to NP was observed in marine sediments (Hong and Shin 2009), the toxicity of sediments slowly grows also because of this process.

The accumulation of APs and short-chain APEs in sediments indicates the importance of considering their impact on the aquatic environment. It must be taken into account that these compounds easily bioaccumulate in many organisms from mussels (Vidal-Liñán et al. 2015, Staniszewska et al. 2017) to fish (Lacorte et al. 2006). Therefore, appropriate action should be taken to prevent further devastation of the environment.

Summary

Alkylphenols and their ethoxylates had been used widely for a very long time without restrictions, before scientists associated them to disruption of hormonal systems. This led to restriction laws introduced in the European and North American countries, which limited usage of APs and APEs, and their presence in the environment. Studies conducted on APEs showed their satisfactory primary biodegradation. However, it was found, that the biodegradation of these compounds leads to APs, short-chain APEs, and their carboxylated derivatives. All these compounds are endocrine disruptors, posing real threats to the environment, because their further biodegradation is poor and they are hydrophobic, which leads to their accumulation in sediments. As the biodegradation studies show, there is limited biodegradation of APs and their short-chain ethoxylates, and simultaneously usage of APs and APEs is rising in highly populated countries like China and India. Thus, environmental pollution will not cease until proper law regulations are introduced. Moreover, in many regions, sediments accumulate large amounts of these compounds, and taking into account the poor biodegradation rate of APs and short-chain APEs, their removal from the environment can take many years.

Acknowledgments

This work was supported by the Polish Government grants 03/31/SBAD/0382 and 502-01-3314429-03439.

References

Ahel, M. and W. Giger. 1985. Determination of alkylphenols and alkylphenol mono- and diethoxylates in environmental samples by high performance liquid chromatography. Anal Chem 57: 1577–1583.

Ahel, M., W. Giger and M. Koch. 1986. Behaviour of nonionic surfactants in biological waste water treatment. pp. 414–428. *In*: Bjørseth, A. and G. Angeletti (eds.). Organic Micropollutants in the Aquatic Environment. Springer, Dordrecht.

Ahel, M., T. Conrad and W. Giger. 1987. Persistent organic chemicals in sewage effluents. 3. Determination of nonylphenoxy carboxylic acids by high-resolution gas chromatography/mass spectrometry and high-performance liquid chromatography. Environ Sci Technol 21: 697–703.

Ahel, M., W. Giger and M. Koch. 1994. Behaviour of alkylphenol polyethoxylate surfactants in the aquatic environment—I. Occurrence and transformation in sewage treatment. Water Res 28: 1131–1142.

An, B.-S., H.-J. Ahn, H.-S. Kang, E.-M. Jung, H. Yang, E.-J. Hong and E.-B. Jeung. 2013. Effects of estrogen and estrogenic compounds, 4-tert-octylphenol, and bisphenol A on the uterine contraction and contraction-associated proteins in rats. Mol Cell Endocrin 375: 27–34.

Aydogan, M., A. Korkmaz, N. Barlas and D. Kolankaya. 2008. The effect of vitamin C on bisphenol A, nonylphenol and octylphenol induced brain damages of male rats. Toxicology 249: 35–39.

Ball, H.A., M. Reinhard and P.L. McCarty. 1989. Biotransformation of halogenated and nonhalogenated octylphenol polyethoxylate residues under aerobic and anaerobic conditions. Environ Sci Technol 23: 951–961.

Bennett, E.R. and C.D. Metcalfe. 1998. Distribution of alkylphenol compounds in great lakes sediments, United States and Canada. Env Toxicol Chem 17: 1230–1235.

Bennie, D.T., C.A. Sullivan, H.B. Lee, T.E. Peart and R.J. Maguire. 1997. Occurrence of alkylphenols and alkylphenol mono- and diethoxylates in natural waters of the Laurentian Great Lakes basin and the upper St. Lawrence river. Sci Total Environ 193: 263–275.

Bergé, A., J. Gasperi, V. Rocher, A. Coursimault and R. Moilleron. 2012. Phthalate and alkylphenol removal within wastewater treatment plants using physicochemical lamellar clarification and biofiltration. pp. 357–368. *In*: Brebbia, C.A. (ed.). Water Pollution XI, WIT Transactions on Ecology and the Environment vol. 164, WIT Press, Wessex, UK.

BGBl. 1977. Verordnung über die Abbaubarkeit anionischer und nichtionischer grenzflächenaktiver Stoffe in Wasch- und Reinigungsmitteln. BGBl (Bundesgesetzblatt), Jahrgang 1977 (vom 30.01.1977), Teil I, 244.

Birch, R.R. 1984. Biodegradation of nonionic surfactants. J Am Oil Chem Soc 61: 340–343.

Brook, J., I. Johnson, R. Mitchell and C. Watts. 2005. Environmental risk evaluation report: 4-tert-octylphenol. Environment Agency, Bristol, UK.

Brüschweiler, H., H. Gämperle and F. Schwager. 1983. Primary biodegradation, ultimate biodegradation, and biodegradation intermediates of alkylphenol ethoxylates. Tenside Deterg 20: 317–324.

Canadian Act. 2001. Canadian Environmental Protection Act, 1999. Priority substances list assessment report. Nonylphenol and its ethoxylates. Environment Canada, Health Canada.

Canadian Water Act. 2002. Canadian water quality guidelines for the protection of aquatic life—nonylphenol and its ethoxylates. Canadian Council of Ministers of the Environment.

Canadian Sediment Act. 2002. Canadian sediment quality guidelines for the protection of aquatic life—nonylphenol and its ethoxylates. Canadian Council of Ministers of the Environment.

Cespedes, R., S. Lacorte, D. Raldua, A. Ginebreda, D. Barcelo and B. Pina. 2005. Distribution of endocrine disruptors in the Llobregat River basin (Catalonia, NE Spain). Chemosphere 61: 1710–1719.

Chen, M.L., H.Y. Lee, H.-Y. Chuang, B.-R. Guo and I.-F. Mao. 2009. Association between nonylphenol exposure and development of secondary sexual characteristics. Chemosphere 76: 927–931.

Chen, T.-Ch. and Y.-L. Yeh. 2010. Ecological risk, mass loading, and occurrence of nonylphenol (NP), NP mono-, and diethoxylate in Kaoping River and its tributaries, Taiwan. Water Air Soil Pollut 208: 209–220.

Cheng, Ch.-Y., Ch.-Y. Wu, Ch.-H. Wang and W.-H. Ding. 2006. Determination and distribution characteristics of degradation products of nonylphenol polyethoxylates in the rivers of Taiwan. Chemosphere 65: 2275–2281.

Cheng, M.-C., H.-I. Chiang, J.-W. Liao, C.-M. Hung, M.-Y. Tsai, Y.-H. Chen, J.-C. Ju, M.-P. Cheng, K.-H. Tso and Y.-K. Fan. 2016. Nonylphenol reduces sperm viability and fertility of mature male breeders in Brown Tsaiya ducks (*Anas platyrhynchos*). Anim Reprod Sci 174: 114–122.

Chokwe, T.B., O.J. Okonkwo, L.L. Sibali and S.M. Mporetji. 2016. Occurrence and distribution pattern of alkylphenol ethoxylates and brominated flame retardants in sediment samples from Vaal River, South Africa. Bull Environ Contam Toxicol 97: 353–358.

Cladière, M., J. Gasperi, S. Gilbert, C. Lorgeoux and B. Tassin. 2010. Alkylphenol ethoxylates and bisphenol A in surface water within a heavily urbanized area, such as Paris. pp. 131–142. *In*: Marinov, A.M. and C.A. Brebbia (eds.). Water Pollution X, WIT Transactions on Ecology and the Environment vol. 135, WIT Press, Wessex, UK.

Corti, A., S. Frassinetti, G. Vallini, S. D'Antone, C. Fichi and R. Solaro. 1995. Biodegradation of nonionic surfactants. I. Biotransformation of 4-(1-nonyl)phenol by a Candida maltosa isolate. Environ Pollut 90: 83–87.

Darbre, P.D. 2015. What are endocrine disrupters and where are they found? pp. 3–26. *In*: Darbre, P.D. (ed.). Endocrine Disruption and Human Health. Academic Press, San Diego and Oxford.

de Araujo, F.G., G.F. Bauerfeldt and Y.P. Cid. 2018. Determination of 4-nonylphenol in surface waters of the Guandu River Basin by high performance liquid chromatography with ultraviolet detection. J Braz Chem Soc 29: 2046–2053.

de A. Azevedo, D., Lacorte, S., Viana, P. and Barceló, D. 2001. Occurrence of nonylphenol and bisphenol-A in surface waters from Portugal. J Braz Chem Soc 12: 532–537.

de Voogt, P., O. Kwast, R. Hendriks and N. Jonkers. 2000. Alkylphenol ethoxylates and their degradation products in abiotic and biological samples from the environment. Analusis 28: 776–782.

de Vries, Y.P., Y. Takahara, Y. Ikunaga, Y. Ushiba, M. Hagesawa, Y. Kasahara, H. Shimomura, S. Hayashi, Y. Hirai and H. Ohta. 2001. Organic nutrient-dependent degradation of branched nonylphenol by Sphingomonas sp. YT isolated from a river sediment sample. Microb Environ 16: 240–249.

Di Corcia, A., R. Samperi and A. Marcomini. 1994. Monitoring aromatic surfactants and their biodegradation intermediates in raw and treated sewages by solid-phase extraction and liquid chromatography. Environ Sci Technol 28: 850–858.

Di Corcia, A., A. Constantino, C. Crescenzi, E. Marinoni and R. Samperi. 1998. Characterization of recalcitrant intermediates from biotransformation of the branched alkyl side chain of nonylphenol ethoxylate surfactants. Environ Sci Technol 32: 2401–2409.

Ding, W.-H. and S.-H. Tzing. 1998. Analysis of nonylphenol polyethoxylates and their degradation products in river water and sewage effluent by gas chromatography–ion trap (tandem) mass spectrometry with electron impact and chemical ionization. J Chromatogr A 824: 79–90.

Directive. 2003. Directive 2003/53/EC of the European Parliament and of the Council of 18 June 2003 amending for the 26th time Council Directive 76/769/EEC relating to restrictions on the marketing and use of certain dangerous substances and preparations (nonylphenol, nonylphenol ethoxylate and cement).

Directive. 2008. Directive 2008/105/EC of the European Parliament and of the Council of 16 December 2008 on environmental quality standards in the field of water policy, amending and subsequently repealing Council Directives 82/176/EEC, 83/513/EEC, 84/156/EEC, 84/491/EEC, 86/280/EEC and amending Directive 2000/60/EC of the European Parliament and of the Council.

Dobrzyńska, M.M. 2014. DNA damage in organs of female and male mice exposed to nonylphenol, as a single agent or in combination with ionizing irradiation: A comet assay study. Mutation Res 772: 14–19.

Dodds, E.C. and W. Lawson. 1938. Molecular structure in relation to oestrogenic activity. Compounds without a phenanthrene nucleus. Proc Roy Soc Lond B 125: 222–232.

Dong, C.-D., C.-W. Chen and C.-F. Chen. 2015. Seasonal and spatial distribution of 4-nonylphenol and 4-tert-octylphenol in the sediment of Kaohsiung Harbor, Taiwan. Chemosphere 134: 588–597.

Duan, X.-Y., Y.-X. Li, X.-G. Li, D.-H. Zhang and Y. Gao. 2014. Alkylphenols in surface sediments of the Yellow Sea and East China Sea inner shelf: Occurrence, distribution and fate. Chemosphere 107: 265–273.

Ekelund, R., A. Granmo, K. Magnusson and M. Berggren. 1993. Biodegradation of 4-nonylphenol in seawater and sediment. Environ Pollut 79: 59–61.

Frańska, M., R. Frański, A. Szymański and Z. Łukaszewski. 2003. A central fission pathway in alkylphenol ethoxylate biodegradation. Water Res 37: 1005–1014.

Frańska, M., D. Ginter-Kramarczyk, A. Szymański, T. Kozik and R. Frański. 2009. Resistance of alkylphenol ethoxylate containing six ethoxylene units to biodegradation under the conditions of OECD (Organization for Economic Cooperation and Development) screening test. Intern Biodet Biodegr 63: 1066–1069.

Fu, M., Z. Li and H. Gao. 2007. Distribution characteristics of nonylphenol in Jiaozhou Bay of Qingdao and its adjacent rivers. Chemosphere 69: 1009–1016.

Fujii, K., N. Urano, H. Ushio, M. Satomi and S. Kimura. 2001. Sphingomonas cloacae sp nov., a nonylphenol-degrading bacterium isolated from wastewater of a sewage-treatment plant in Tokyo. Int J Syst Evol Microbiol 51: 603–610.

Gabriel, F.L.P., W. Giger, K. Günther and H.-P.E. Kohler. 2005. Differential degradation of nonylphenol isomers by Sphingomonas xenophaga Bayram. Appl Environ Microbiol 71: 1123–1129.

Gabriel, F.L.P., E.J. Routledge, A. Heidlberger, D. Rentsch, K. Guenther, W. Giger, J.P. Sumpter and H.-P.E. Kohler. 2008. Isomer-specific degradation and endocrine disrupting activity on nonylphenols. Environ Sci Technol 42: 6399–6408.

Giger, W., P.H. Brunner and C. Schaffner. 1984. 4-Nonylphenol in sewage sludge: accumulation of toxic metabolites from nonionic surfactants. Science 225: 623–625.

Giger, W., M. Ahel, M. Koch, H.U. Laubscher, C. Schaffner and J. Schneider. 1987. Behaviour of alkylphenol polyethoxylate surfactants and of nitrilotriacetate in sewage treatment. Wal Sci Tech 19: 449–460.

Gimeno, S., H. Komen, S. Jobling, J. Sumpter and T. Bowmer. 1998. Demasculinisation of sexually mature male common carp, *Cyprinus carpio*, exposed to 4-tert-pentylphenol during spermatogenesis. Aquat Toxicol 43: 93–109.

González, S., M. Petrovic and D. Barceló. 2007. Removal of a broad range of surfactants from municipal wastewater—Comparison between membrane bioreactor and conventional activated sludge treatment. Chemosphere 67: 335–343.

González, S., M. Petrovic and D. Barceló. 2008. Evaluation of two pilot scale membrane bioreactors for the elimination of selected surfactants from municipal wastewaters. J Hydrol 356: 46–55.

Gray, M.A. and Ch.D. Metcalfe. 1997. Induction of testis-ova in Japanese medaka (*Oryzias latipes*) exposed to p-nonylophenol. Environ Toxicol Chem 16: 1082–1086.

Groshart, C.P., P.C. Okkerman and W.B.A. Wassenberg. 2001. Chemical study on alkylphenols. Report: RIKZ/2001.029. BKH Consulting Engineers (Delft) and RIKZ (Den Haag).

Grześkowiak, T., B. Czarczyńska-Goślińska and A. Zgoła-Grześkowiak. 2018. Biodegradation of selected endocrine disrupting compounds. pp. 1–27. *In*: Bidoia, E.D. and R.N. Montagnolli (eds.). Toxicity and Biodegradation Testing. Methods in Pharmacology and Toxicology. Humana Press, New York, NY.

Guart, A., F. Bono-Blay, A. Borrell and S. Lacorte. 2011. Migration of plasticizers phthalates, bisphenol A and alkylphenols from plastic containers and evaluation of risk. Food Addit Contam Part A: Chem Anal Control Expo Risk Assess 28: 676–85.

Hawrelak, M., M. Bennett and C. Metcalfe. 1999. The Environmental fate of the primary degradation products of alkylphenol ethoxylate surfactants in recycled paper sludge. Chemosphere 39: 745–752.

Heemken, O.P., H. Reincke, B. Stachel and N. Theobald. 2001. The occurrence of xenoestrogens in the Elbe river and the North Sea. Chemosphere 45: 245–259.

Hejmej, A., M. Kotula-Balak, J. Galas and B. Bilińska. 2011. Effects of 4-tert-octylphenol on the testes and seminal vesicles in adult male bank voles. Reprod Toxicol 31: 95–105.

Hernandez-Raquet, G. 2012. Fate of emerging contaminants during aerobic and anaerobic sludge treatment. pp. 73–112. *In*: Vicent, T., G. Caminal, E. Eljarrat and D. Barceló (eds.). Emerging Organic Contaminants in Sludges. The Handbook of Environmental Chemistry, Vol 24. Springer, Berlin, Heidelberg.

Holt, M.S., G.C. Mitchell and R.J. Watkinson. 1992. The surfactants chemistry, fate and effects of non-ionic surfactants. pp. 89–144. *In*: Hutzinger, O. and N.T. de Oude (eds.). The Handbook of Environmental Chemistry, Volume 3 Part F: Anthropogenic Compounds: Detergents. Springer-Verlag, Berlin.

Hong, S. and K.-H. Shin. 2009. Alkylphenols in the core sediment of a waste dumpsite in the East Sea (Sea of Japan), Korea. Marine Pollut Bull 58: 1566–1571.

Isobe, T., H. Nishiyama, A. Nakashima and H. Takada. 2001. Distribution and behaviour of nonylphenol, octylphenol, and nonylphenol monoethoxylate in Tokyo metropolitan area: their association with aquatic particles and sedimentary distributions. Environ Sci Technol 35: 1041–1049.

Isobe, T. and H. Takada. 2004. Determination of degradation products of alkylphenol polyethoxylates in municipal wastewaters and rivers in Tokyo, Japan. Environ Toxicol Chem 23: 599–605.

Jie, Y., Z. Jie, L. Ya, Y. Xuesong, Y. Jing, Y. Yu, Y. Jiaqi and X. Jie. 2017. Pollution by nonylphenol in river, tap water, and aquatic in an acid rain-plagued city in southwest China. Int J Environ Health Res 27: 179–190.

Jobling, S. and J.P. Sumpter. 1993. Detergent components in sewage effluent are weakly oestrogenic to fish: an *in vitro* study using rainbow trout (*Oncorhynchus mykiss*) hepatocytes. Aquat Toxicol 27: 361–372.

Jobling, S., D. Sheahan, J.A. Osborne, P. Matthiessen and J.P. Sumpter. 1996. Inhibition of testicular growth in raibow trout (*Oncorhynchus mykiss*) exposed to estrogenic alkylphenolic chemicals. Environ Toxicol Chem 15: 194–202.

Jonkers, N., T.P. Knepper and P. De Voogt. 2001. Aerobic biodegradation studies of nonylphenol ethoxylates in river water using liquid chromatography-electrospray tandem mass spectrometry. Environ Sci Technol 35: 335–340.

Kannan, K., J.L. Kober, J.S. Khim, K. Szymczyk, J. Falandysz and J.P. Giesy. 2003. Polychlorinated biphenyls, polycyclic aromatic hydrocarbons and alkylphenols in sediments from the Odra river and its tributaries, Poland. Toxicol Environ Chem 85: 51–60.

Kim, J., E.-J. Kang, M.-N. Park, J.-E. Kim, S.-Ch. Kim, E.-B. Jeung, G.-S. Lee, D.-Y. Hwang and B.-S. An. 2015. The adverse effect of 4-tert-octylphenol on fat metabolism in pregnant rats via regulation of lipogenic proteins. Environ Toxicol Pharmacol 40: 284–291.

Kinnberg, K., B. Korsgaard, P. Bjerregaard and Å. Jespersen. 2000. Effects of nonylphenol and 17beta estradiol on vitellogenin synthesis and testis morphology in male platyfish *Xiphophorus maculatus*. J Exp Biol 203: 171–181.

Kolvenbach, B.A. and P.F.-X. Corvini. 2012. The degradation of alkylphenols by Sphingomonas sp. strain TTNP3—a review on seven years of research. New Biotechnol 30: 88–95.

Kravetz, L., H. Chung, K.F. Guin, W.T. Shebs and L.S. Smith. 1984. Primary and ultimate biodegradation of an alcohol elhoxylate and an alkylphenol ethoxylate under average winter conditions in the USA. Tenside Deterg 21: 1–6.

Kurihara, R., E. Watanabe, Y. Ueda, A. Kakuno, K. Fujii, F. Shiraishi and S. Hashimoto. 2007. Estrogenic activity in sediments contaminated by nonylphenol in Tokyo Bay (Japan) evaluated by vitellogenin induction in male mummichogs (*Fundulus heteroclitus*). Marine Poll Bull 54: 1315–1320.

Lacorte, S., D. Raldúa, E. Martínez, A. Navarro, S. Diez, J.M. Bayona and D. Barceló. 2006. Pilot survey of a broad range of priority pollutants in sediment and fish from the Ebro river basin (NE Spain). Environ Pollut 140: 471–482.

Lee, J.-H., A. Chanhwan, K. Kim and E.-B. Jeung. 2016. Effect of octylphenol and bisphenol A on cation transfer genes in mouse placenta during pregnancy. Reprod Toxicol 64: 29–49.

Li, H.Q., F. Jiku and H.F. Schroder. 2000. Assessment of the pollutant elimination efficiency by gas chromatography/mass spectrometry, liquid chromatography-mass spectrometry and -tandem mass spectrometry—comparison of conventional and membrane-assisted biological wastewater treatment processes. J Chromatogr A 889: 155–176.

Li, Z., D. Li, J.-R. Oh and J.-G. Je. 2004. Seasonal and spatial distribution of nonylphenol in Shihwa Lake, Korea. Chemosphere 56: 611–618.

Loos, R., G. Hanke, G. Umlauf and S. Eisenreich. 2007. LC–MS–MS analysis and occurrence of octyl- and nonylphenol, their ethoxylates and their carboxylates in Belgian and Italian textile industry, waste water treatment plant effluents and surface waters. Chemosphere 66: 690–699.

Loyo-Rosales, J.E., I. Schmitz-Afonso, C.P. Rice and A. Torrents. 2003. Analysis of octyl- and nonylphenol and their ethoxylates in water and sediments by liquid chromatography/tandem mass spectrometry. Anal Chem 75: 4811–4817.

Loyo-Rosales, J.E., G.C. Rosales-Rivera, A.M. Lynch, C.P. Rice and A. Torrents. 2004. Migration of nonylphenol from plastic containers to water and a milk surrogate. J Agric Food Chem 52: 2016–2020.

Lubello, C. and R. Gori. 2005. Membrane bio-reactor for textile wastewater treatment plant upgrading. Water Sci Technol 52: 91–98.

Lv, Y.-Z., L. Yao, L. Wang, W.-R. Liu, J.-L. Zhao, L.-Y. He and G.-G. Ying. 2019. Bioaccumulation, metabolism, and risk assessment of phenolic endocrine disrupting chemicals in specific tissues of wild fish. Chemosphere 226: 607–615.

Mann, R.M. and M.R. Boddy. 2000. Biodegradation of a nonylphenol ethoxylate by the autochthonous microflora in lake water with observations on the influence of light. Chemosphere 41: 1361–1369.

Marcomini, A., P.O. Capel, T. Lichtensteiger, P.H. Brunner and W. Giger. 1989. Behavior of aromatic surfactants and PCBs in sludge-treated soil and landfills. J Environ Qual 18: 523–528.

Mayer, T., F. Rosa, G. Rekas, V. Palabrica and J. Schachtschneider. 2007. Occurrence of alkylphenolic substances in a Great Lakes coastal marsh, Cootes Paradise, ON, Canada. Environ Pollut 147: 683–690.

Meier, S., T.E. Andersen, B. Norberg, A. Thorsen, G.L. Taranger, O.S. Kjesbu, R. Dale, H.C. Morton, J. Klungsøyr and A. Svardal. 2007. Effects of alkylphenols on the reproductive system of Atlantic cod (*Gadus morhua*). Aquat Toxicol 81: 207–218.

Mueller, G.C. and U.H. Kim. 1978. Displacement of estradiol from estrogen receptors by simple alkyl phenols. Endocrinology 102: 14291435.

Murono, E.P., R.C. Derk and J.H. de León. 1999. Biphasic effects of octylphenol on testosterone biosynthesis by cultured Leydig cells from neonatal rats. Reprod Toxicol 13: 451–462.

Naylor, C.G., J.P. Mieure, W.J. Adams, J.A. Weeks, F.J. Castaldi, L.D. Ogle and R.R. Romano. 1992. Alkylphenol ethoxylates in the environment. J Am Oil Chem Soc 69: 695–703.

Nice, H.E., D. Morritt, M. Crane and M. Thorndyke. 2003. Long-term and transgenerational effects of nonylphenol exposure at a key stage in the development of *Crassostrea gigas*. Possible endocrine disruption? Mar Ecol Prog Ser 256: 293–300.

Oketola, A.A. and T.K. Fagbemigun. 2013. Determination of nonylphenol, octylphenol and bisphenol-A in water and sediments of two major rivers in Lagos, Nigeria. J Environ Protect 4: 38–45.

Olkowska, E., M. Ruman, A. Kowalska and Ż. Polkowska. 2013. Determination of surfactants in environmental samples. Part III. Non-ionic compounds. Ecol Chem Eng S 20: 449–461.

OSPAR. 2006. Octylphenol. OSPAR Commission 2003 (2006 Update). Publication 273/2006.

Patoczka, J. and G.W. Pulliam. 1990. Biodegradation and secondary effluent toxicity of ethoxylated surfactants. Water Res 24: 965–972.

Petrovic, M., A. Diaz, F. Ventura and D. Barcelo. 2001. Simultaneous determination of halogenated derivatives of alkylphenol ethoxylates and their metabolites in sludges, river sediments, and surface, drinking, and wastewaters by liquid chromatography-mass spectrometry. Anal Chem 73: 5886–5895.

Potter, T.L., K. Simmons, J. Wu, M. Sanchez-Olvera, P. Kostecki and E. Calabrese. 1999. Static die-away of a nonylphenol ethoxylate surfactant in estuarine water samples. Environ Sci Technol 33: 113–118.

Pitter, P. and T. Fuka. 1979. Biodegradation of non-sulfated and sulfated nonylphenol ethoxylate surfactants. Environ Prot Eng 5: 47–56.

Rajendran, R.K., Y.-W. Lee, P.-H. Chou, S.-L. Huang, R. Kirschner and Ch.-Ch. Lin. 2020. Biodegradation of the endocrine disrupter 4-t-octylphenol by the nonligninolytic fungus Fusarium falciforme RRK20: Process optimization, estrogenicity assessment, metabolite identification and proposed pathways. Chemosphere 240: 124876.

Raju, S., M. Sivamurugan, K. Gunasagaran, T. Subramani and M. Natesan. 2018. Preliminary studies on the occurrence of nonylphenol in the marine environments, Chennai—a case study. J Basic App Zoology 79: 52.

Reed, H.W.B. 1978. Alkylphenols. pp. 72–96. *In*: Grayson, M. and D. Eckroth (eds.). Kirk-Othmer Encyclopedia of Chemical Technology. 3rd ed. Vol 2. John Wiley & Sons, New York, NY, USA.

Reinhard, M., N. Goodman and K.E. Mortelmans. 1982. Occurrence of brominated alkylphenol polyethoxy carboxylates in mutagenic wastewater concentrates. Environ Sci Technol 16: 351–362.

Renner, R. 1997. European bans on surfactant trigger transatlantic debate. Env Sci Technol 31: 316A–320A.

Rocha, M.J., C. Cruzeiro., M. Reis, E. Rocha and M. Pardal. 2013. Determination of seventeen endocrine disruptor compounds and their spatial and seasonal distribution in Ria Formosa Lagoon (Portugal). Environ Monit Assess 185: 8215–8226.

Rocha, M.J., C. Cruzeiro, M. Reis, M.Â. Pardal and E. Rocha. 2014. Spatial and seasonal distribution of 17 endocrine disruptor compounds in an urban estuary (Mondego River, Portugal): evaluation of the estrogenic load of the area. Environ Monit Assess 186: 3337–3350.

Rocha, M.J., C. Cruzeiro, M. Reis, M.Â. Pardal and E. Rocha. 2016. Pollution by oestrogenic endocrine disruptors and β-sitosterol in a south-western European river (Mira, Portugal). Environ Monit Assess 188: 240.

Routledge, E.J. and J.P. Sumpter. 1997. Structural features of alkylphenolic chemicals associated with estrogenic activity. J Biolog Chem 272: 3280–3288.

Sabbieti, M.G., D. Agas, F. Palermo, G. Mosconi, G. Santoni, C. Amantini, V. Farfariello and L. Marchetti. 2011. 4-Nonylphenol triggers apoptosis and affects 17-estradiol receptors in calvarial osteoblasts. Toxicology 290: 334–341.

Sadakane, K., T. Ichinose, H. Takano, R. Yanagisawa, E. Koike and K.-I. Inoue. 2014. The alkylphenols 4-nonylphenol, 4-tertoctylphenol and 4-tert-butylphenol aggravate atopic dermatitis-like skin lesions in NC/Nga mice. J Appl Toxicol 34: 893–902.

Salgueiro-González, N., E. Concha-Grana, I. Turnes-Carou, S. Muniategui-Lorenzo, P. López-Mahía and D. Prada-Rodríguez. 2012. Determination of alkylphenols and bisphenol A in seawater samples by dispersive liquid-liquid microextraction and liquid chromatography tandem mass spectrometry for compliance with environmental quality standards (Directive 2008/105/EC). J Chromatogr A 1223: 1–8.

Sato, H., A. Shibata, Y. Wang, H. Yoshikawa and H. Tamura. 2001. Characterization of biodegradation intermediates of non-ionic surfactants by matrix-assisted laser desorption/ionization–mass spectrometry. 1. Bacterial biodegradation of octylphenol polyethoxylate under aerobic conditions. Polym Degrad Stab 74: 69–75.

Sciubba, L., L. Bertin, D. Todaro, C. Bettini, F. Fava and D. Di Gioia. 2014. Biodegradation of low-ethoxylated nonylphenols in a bioreactor packed with a new ceramic support (Vukopor®S10). Environ Sci Pollut Res 21: 3241–3253.

Selvaraj, K.K., G. Shanmugam, S. Sampath, D.G.J. Larsson and B.R. Ramaswamy. 2014. GC–MS determination of bisphenol A and alkylphenol ethoxylates in river water from India and their ecotoxicological risk assessment. Ecotox Environ Safe 99: 13–20.

Servos, M.R. 1999. Review of the aquatic toxicity, estrogenic responses and bioaccumulation of alkylphenols and alkylphenol polyethoxylates. Water Qual Res J Canada 34: 123–177.

Shang, D.Y., M.G. Ikonomou and R.W. Macdonald. 1999. Quantitative determination of nonylphenol polyethoxylate surfactants in marine sediment using normal-phase liquid chromatography-electrospray mass spectrometry. J Chromatogr A 849: 467–482.

Shiu, R.-F., J.-J. Jiang, H.-Y. Kao, M.-D. Fang, Y.-J. Liang, C.-C. Tang and C.-L. Lee. 2019. Alkylphenol ethoxylate metabolites in coastal sediments off southwestern Taiwan: Spatiotemporal variations, possible sources, and ecological risk. Chemosphere 225: 9–18.

Sibali, L.L., J.O. Okwonkwo and R.I. McCrindle. 2010. Levels of selected alkylphenol ethoxylates (APEs) in water and sediment samples from the Jukskei River catchment area in Gauteng, South Africa. Water SA 36: 229–238.

Snyder, S.A., T.L. Keith, D.A. Verbrugge, E.M. Snyder, T.S. Gross, K. Kannan and J.P. Giesy. 1999. Analytical methods for detection of selected estrogenic compounds in aqueous mixtures. Environ Sci Technol 33: 2814–2820.

Sole, M., M.J. Lopez de Alda, M. Castillo, C. Porte, K. Ladegaard-Pedersen and D. Barcelo. 2000. Estrogenicity determination in sewage treatment plants and surface waters from the Catalonian area (NE Spain). Environ Sci Technol 34: 5076–5083.

Soto, A.M., C. Sonnenschein, K.L. Chung, M.F. Fernandez, N. Olea and F.O. Serrano. 1995. The E-SCREEN assay as a tool to identify estrogens: an update on estrogenic environmental pollutants. Environ Health Persp 103: 113–122.

Stachel, B., U. Ehrhorn, O.-P. Heemken, P. Lepom, H. Reincke, G. Sawal and N. Theobald. 2003. Xenoestrogens in the River Elbe and its tributaries. Environ Pollut 124: 497–507.

Staniszewska, M., I. Koniecko, L. Falkowska and E. Krzymyk. 2015. Occurrence and distribution of bisphenol A and alkylphenols in the water of the gulf of Gdansk (Southern Baltic). Marine Pollut Bull 91: 372–379.

Staniszewska, M., B. Graca, A. Sokołowski, I. Nehring, A. Wasik and A. Jendzula. 2017. Factors determining accumulation of bisphenol A and alkylphenols at a low trophic level as exemplified by mussels *Mytilus trossulus*. Environ Pollut 220/Part B: 1147–1159.

Staples, Ch.A., J. Weeks, J.F. Hall and C.G. Naylor. 1998. Evaluation of aquatic toxicity and bioaccumulation of C8- and C9-alkylphenol ethoxylates. Environ Toxicol Chem 17: 2470–2480.

Stephanou, E. and W. Giger. 1982. Persistent organic chemicals in sewage effluents. 2. Quantitative determinations of nonylphenols and nonylphenol ethoxylates by glass capillary gas chromatography. Environ Sci Technol 16: 800–805.

Suen, J.-L., C.-H. Hung, H.-S. Yu and S.-K. Huang. 2012. Alkylphenols—potential modulators of the allergic response. Kaohsiung J Med Sci 28: S43–S48.

SUBSPORT. 2013. SUBSPORT Specific substances alternatives assessment—nonylphenols and nonylphenol ethoxylates.

Swisher, R.D. 1987. Surfactant biodegradation (Second Edition – Revised and Expanded). Surfactant Science Series, Vol. 18, Marcel Dekker Inc., New York.

Swiss Ordinance. 2009. Ordinance on the Reduction of Risks Relating to the Use of Certain Particularly Dangerous Substances, Preparations and Articles (Chemical Risk Reduction Ordinance, ORRChem) of 18 May 2005 (Status as of 7 May 2019). Ordinance number 814.81.

Tabak, H.H. and R.L. Bunch. 1981. Measurement of nonionic surfactants in aqueous environments. Proceedings of the 36th Industrial Waste Conference, May 12–14: Purdue University, Lafayette, Indiana. 36: 888–907.

Talmage, S.S. 1994. Environmental and Human Safety of Major Surfactants: Alcohol Ethoxylates and Alkylphenol Ethoxylates (1st Edition). CRC Press (Lewis Publishers), Boca Raton.

Tanghe, T., W. Dhooge and W. Verstraete. 1999. Isolation of a bacterial strain able to degrade branched nonylphenol. Appl Environ Microbiol 65: 746–751.

Tongu, S.M., R. Sha'Ato, O.J. Okonkwo, I.S. Eneji, T.B. Chokwe and T.A. Tor-Anyiin. 2018. Determination of alkylphenol ethoxylates (APEs) and alkylphenols (APs) in water and sediment from River Benue, North Central Nigeria. J Chem Soc Nigeria 43: 156–174.

Terzic, S., M. Matosic, M. Ahel and I. Mijatovic. 2005. Elimination of aromatic surfactants from municipal wastewaters: Comparison of conventional activated sludge treatment and membranę biological reactor. Water Sci Technol 51: 447–453.

Topp, E. and A. Starratt. 2000. Rapid mineralization of the endocrine-disrupting chemical 4-nonylphenol in soil. Environ Toxicol Chem 19: 313–318.

US EPA. 2005. Aquatic life ambient water quality criteria-nonylphenol. United States Environmental Protection Agency, Office of Water, Office of Science and Technology, Washington, DC, EPA-822-R-05-005, US.

Vallini, G., S. Frassinetti and G Scorzetti. 1997. *Candida aquaetextoris* sp. nov., a new species of yeast occurring in sludge from a textile industry wastewater treatment plant in Tuscany, Italy. Int J Syst Bacteriol 47: 336–340.

Van Ginkel, C.G. 1996. Complete degradation of xenobiotic surfactants by consortia of aerobic microorganisms. Biodegradation 7: 151–164.

Venkatesan, A.K. and R.U. Halden. 2013. National inventory of alkylphenol ethoxylate compounds in U.S. sewage sludges and chemical fate in outdoor soil mesocosms. Environ Pollut 174: 189–193.

Vidal-Liñán, L., J. Bellas, N. Salgueiro-González, S. Muniategui and R. Beiras. 2015. Bioaccumulation of 4-nonylphenol and effects on biomarkers, acetylcholinesterase, glutathione-S-transferase and glutathione peroxidase, in Mytilus galloprovincialis mussel gills. Environmental Pollution, 200/Part B, 133–139.

Wang, B., B. Huang, W. Jin, S. Zhao, F. Li, P. Hu and X. Pan. 2013. Occurrence, distribution, and sources of six phenolic endocrine disrupting chemicals in the 22 river estuaries around Dianchi Lake in China. Environ Sci Pollut Res 20: 3185–3194.

Warhurst, A.M. 1995. An environmental assessment of alkylphenol ethoxylates and alkylphenols. Friends of the Earth Scotland and Friends of the Earth, Edinburgh and London.

Wheeler, T.F., J.R. Heim, M.R. LaTorre and A. Blair Janes. 1997. Mass spectral characterization of p-nonylphenol isomers using high-resolution capillary GC-MS. J Chromatogr Sci 35: 19–30.

White, R., S. Jobling, S.S. Hoare, J.P. Sumpter and M.G. Parker. 1994. Environmentally persistent alkylphenolic compounds are estrogenic. Endocrinology 135: 175–182.

Whitehouse, P. 2002. Environmental impacts of alkylphenol ethoxylates and carboxylates. Part 1: Proposals for the development of environmental quality standards. R&D Technical Report P2-115/TR3. Environment Agency, Bristol, UK.

Wild, D. and M. Reinhard. 1999. Biodegradation residual of 4-octylphenoxyacetic acid in laboratory columns under groundwater recharge conditions. Environ Sci Technol 33: 4422–4426.

Wu, Q., J.C.W. Lam, K.Y. Kwok, M.M.P. Tsui and P.K.S. Lam. 2017. Occurrence and fate of endogenous steroid hormones, alkylphenol ethoxylates, bisphenol A and phthalates in municipal sewage treatment systems. J Environ Sci 61: 49–58.

Wyrwas, B., Ł. Chrzanowski, Ł. Ławniczak, A. Szulc, P. Cyplik, W. Białas, A. Szymański and A. Hołderna-Odachowska. 2011. Utilization of Triton X-100 and polyethylene glycols during surfactant-mediated biodegradation of diesel fuel. J Hazard Mater 197: 97–103.

Yang, X., M. Liu, Z. Wang, Q. Li and Z. Zhang. 2013. Determination of 4-tert-octylphenol in surface water samples of Jinan in China by solid phase extraction coupled with GC-MS. J Environ Sci 25: 1712–1717.

Yang, Y., X. Cao, M. Zhang and J. Wang. 2015. Occurrence and distribution of endocrine-disrupting compounds in the Honghu Lake and East Dongting Lake along the Central Yangtze River, China. Environ Sci Pollut Res 22: 17644–17652.

Ying, G.-G., B. Williams and R. Kookana. 2002. Environmental fate of alkylphenols and alkylphenol ethoxylates—a review. Environ Int 283: 215–226.

Ying, G.-G. 2006. Fate, behavior and effects of surfactants and their degradation products in the environment. Environ Int 32: 417–431.

Yoshimura, K. 1986. Biodegradation and fish toxicity of nonionic surfactants. J Am Oil Chem Soc 63: 1590–1596.

Zgoła-Grześkowiak, A. and T. Grześkowiak. 2009. Liquid chromatography with fluorescence detection as a tool for separation of endocrine disrupting alkylphenols and their mono- and diethoxylates in analysis of river water samples. Tenside Surf Det 46: 200–204.

Zgoła-Grześkowiak, A., T. Grześkowiak, R. Rydlichowski and Z. Łukaszewski. 2010. Concentrations of endocrine disrupting alkylphenols and their mono- and diethoxylates in sediments and water from artificial Lake Malta in Poland. Tenside Surf Det 47: 222–227.

Zgoła-Grześkowiak, A. 2010. Dispersive liquid-liquid microextraction applied to isolation and concentration of alkylphenols and their short-chained ethoxylates in water samples. J Chromatogr A 1217: 1761–1766.

Zgoła-Grześkowiak, A. and T. Grześkowiak. 2011. Determination of alkylphenols and their short-chained ethoxylates in Polish river waters. Intern J Environ Anal Chem 91: 576–584.

Zgoła-Grześkowiak, A., T. Grześkowiak and A. Szymański. 2014. Comparison of biodegradation of nonylphenol propoxylates with usage of two different sources of activated sludge. J Surfact Deterg 17: 121–132.

Zgoła-Grześkowiak, A., T. Grześkowiak and A. Szymański. 2015. Biodegradation of nonylphenol monopropoxyethoxylates. J Surfact Deterg 18: 355–364.

Zhang, X., D. Zhang, H. Zhang, Z. Luo and C. Yan. 2012. Occurrence, distribution, and seasonal variation of estrogenic compounds and antibiotic residues in Jiulongjiang River, South China. Environ Sci Pollut Res 19: 1392–1404.

Zhang, Z., N. Ren, K. Kannan, J. Nan, L. Liu, W. Ma, H. Qi and Y. Li. 2014. Occurrence of endocrine-disrupting phenols and estrogens in water and sediment of the Songhua River, Northeastern China. Arch Environ Contam Toxicol 66: 361–369.

Zhang, Q., F. Wang, C. Xue, C. Wang, S. Chi and J. Zhang. 2016. Comparative toxicity of nonylphenol, nonylphenol-4-ethoxylateand nonylphenol-10 ethoxylate to wheat seedlings (*Triticum aestivum* L.). Ecotoxicol Environ Saf 131: 7–13.

Zhong, M., P. Yin and L. Zhao. 2017. Nonylphenol and octylphenol in riverine waters and surface sediments of the Pearl River Estuaries, South China: occurrence, ecological and human health risks. Wat Supp 17: 1070–7079.

Zumbado, M., L.D. Boada, S. Torres, J.G. Monterde, B.N. Díaz-Chico and J.L. Afonso. 2002. Evaluation of acute hepatotoxic effects exerted by environmental estrogens nonylphenol and 4-octylphenol in immature male rats. Toxicology 175: 49–62.

2

Selective-enrichment as a Tool to Obtain Microbial Degrading Consortia for the Remediation of Pesticide Residues

Carlos E. Rodríguez-Rodríguez,[1,*] *Juan Carlos Cambronero-Heinrichs,*[2] *Edward Fuller*[3] and *Víctor Castro-Gutiérrez*[3]

Introduction—What is selective enrichment?

Selective enrichment methods are, in general, phenotypical bottlenecks where a microbial community is exposed to an induced selection force that drives the growth of a desired group of microorganisms. That selection force is a certain ingredient in the culture media that favors a specific metabolic capacity. Bile salts, for example, are employed in culture media to isolate enteric bacteria to which they are naturally exposed (Madigan and Martinko 2005). It is also common to employ isolation media enriched with a certain xenobiotic to obtain isolates or microbial consortia that degrade such compounds.

Many times, hydrocarbons have been used as carbon sources to select bacteria with the capacity to remove these compounds, to be then used in bioaugmentation strategies for the remediation of oil spills, for example (Von Wedel et al. 1988, Aislabie et al. 2006). Moreover, *in vitro* selective enrichment using pesticides as a nutrient source, in particular as a carbon and less frequently as a nitrogen source, is also commonly employed to search for degrading consortia (Table 2.1).

[1] Research Center of Environmental Contamination (CICA), Universidad de Costa Rica, San Pedro, Montes de Oca, 2060 San José, Costa Rica.
[2] Faculty of Microbiology, Universidad de Costa Rica, San Pedro, Montes de Oca, 2060 San José, Costa Rica.
[3] Department of Biology, University of York, Heslington Road, YO10 5DD, York, United Kingdom.
Emails: juan.cambroneroheinrichs@ucr.ac.cr; edf504@york.ac.uk; victormanuel.castro@ucr.ac.cr
* Corresponding author: carlos.rodriguezrodriguez@ucr.ac.cr

Table 2.1. Selected examples of selective enrichment processes applied for the isolation of pesticide degrading microorganisms.

Pesticide(s) used	Origin of consortium	Use of the pesticide	Incubation conditions	Consortium or isolated organism	Reference
Organophosphates					
Tetrachlorvinphos	pre-exposed agricultural soils	carbon source	one passage in aerobiosis at 25°C for 7 d	*Stenotrophomonas malthophilia*, *Proteus vulgaris*, *Vibrio metschnikovii*, *Serratia ficaria*, *Serratia* spp. and *Yersinia enterocolitica*	Ortiz-Hernández and Sánchez-Salinas 2010
Methyl-parathion Chlorpyrifos	soil from a waste disposal site	carbon source	one passage in aerobiosis at 30°C for 5 d	*Acinetobacter* sp., *Pseudomonas putida*, *Bacillus* sp., *Pseudomonas aeruginosa*, *Citrobacter freundii*, *Stenotrophomonas* sp., *Flavobacterium* sp., *Proteus vulgaris*, *Pseudomonas* sp., *Acinetobacter* sp., *Klebsiella* sp. and *Proteus* sp.	Pino and Peñuela 2011
Chlorpyrifos	pre-exposed agricultural soils	carbon source	two passages in aerobiosis at 37°C every 7 d	*Staphylococcus* sp., *Micrococcus* sp., *Rhizobium* sp., *Comamonas aquatica*, *Staphylococcus hominis*, *Klebsiella* sp., *Pseudomonas aeruginosa*, *Pseudomonas stutzeri*, *Pseudomonas putida*, *Streptomyces radiopugnans*. Fungal strains: *Aspergillus niger*, *Tricophyton* spp.	Sasikala et al. 2012
Chlorpyrifos	pre-exposed agricultural soils	carbon source	four passages in aerobiosis at 30°C every 7 d	*Pseudomonas putida*, *Acinetobacter* sp., *Stenotrophomonas* sp., *Paracoccus* sp., *Pseudomonas mendocina*, *Pseudomonas aeruginosa*, *Sphingomonas* sp. and *Burkholderia* sp.	Akbar et al. 2014
Chlorpyrifos	pre-exposed agricultural soils	carbon source	five passages in aerobiosis at 30°C every 7 d	*Sphingobacterium* sp.	Abraham and Silambarasan 2013
Chlorpyrifos	pre-exposed soils	carbon source	one passage in aerobiosis at 30°C for 3 d	*Bacillus cereus*	Liu et al. 2011
Chlorpyrifos	pesticide contaminated soil	carbon source	five passages in aerobiosis every 2–4 weeks	*Bacillus pumilus*	Anwar et al. 2009

Diazinon	pre-exposed soils	carbon source	one passage in aerobiosis at 30°C for 72 h	*Serratia marcescens, S. liquefaciens, Pseudomonas* sp.	Cycoń et al. 2009
Fenamiphos	pre-exposed agricultural soils	carbon source	five passages	*Pseudomonas putida* and *Acinetobacter rhizosphaerae*	Chanika et al. 2011
Cadusafos Ethoprophos	pre-exposed agricultural soils	carbon source	two passages	*Flavobacterium* sp. and *Sphingomonas* sp.	Karpouzas et al. 2005
Profenofos	pre-exposed agricultural soils	carbon source	five passages in aerobiosis at 37°C every two weeks	*Achromobacter xylosoxidans, P. aeruginosa, Bacillus* sp. and *Citrobacter koseri*	Jabeen et al. 2015
Neonicotinoids					
Acetamiprid	pre-exposed agricultural soils	carbon and nitrogen source	several passages in aerobiosis every 5 d	*Rhodococcus* sp.	Phugare and Jadhav 2015
Imidacloprid	pre-exposed agricultural soils	carbon, nitrogen or carbon and nitrogen source	four passages in aerobiosis at 28°C every 7 d	*Kocuria rhizophila, Paraburkholderia phymatum, Paenibacillus odorifer, Rhodococcus aetherovorans, Microbacterium binotii, Pseudoacidovorax intermedius, Rhodotorula toruloides*	Rodríguez-Castillo et al. 2019
Imidacloprid	pre-exposed agricultural soils	carbon and nitrogen source	first passage in aerobiosis at 27°C for 3 weeks; five additional passages every two weeks	*Mycobacterium* sp.	Kandil et al. 2015
Imidacloprid	pre-exposed agricultural soils	Carbon and nitrogen source; cometabolic	several passages in aerobiosis at 27°C	*Leifsonia*	Anhalt et al. 2007

Table 2.1 Contd. ...

...Table 2.1 Contd.

Pesticide(s) used	Origin of consortium	Use of the pesticide	Incubation conditions	Consortium or isolated organism	Reference
Imidacloprid Thiamethoxam	pre-exposed agricultural soils	cometabolic	Five passages in aerobic, anaerobic, and microaerophilic conditions at 28°C every 14 d	*Pseudomonas* sp.	Pandey et al. 2009
Carbamates					
Carbofuran	pre-exposed agricultural soils	carbon source	first passage in aerobiosis at 28°C for 3 d; three additional passages every 2 d	*Cupriavidus*, *Achromobacter* and *Pseudomonas*	Castro-Gutiérrez et al. 2016
Carbofuran	pre-exposed agricultural soils	carbon source	one passage in aerobiosis at 28°C for 4 d	*Enterobacter* sp.	Mohanta et al. 2012
Carbofuran	pre-exposed agricultural soils	carbon source	two passages in aerobiosis at 25°C	*Pseudomonas*, Flexibacter/Cytophaga/Bacteroides group, *Chrysobacterium*, *Flavobacterium*, *Paenibacillus*, *Staphylococcus*, *Bacillus*, *Bordetella*, *Microbacterium*	Karpouzas et al. 2000
Organochlorines					
2,4,6-Trichlorophenol	river sediment	substrate for dechlorination	Stable dechlorination activity maintained over more than 25 transfers in anaerobiosis	*Clostridium celerecrescens*, *C. indolis*, *C. hydroxybenzoicum*, *C. sporosphaeroides*, *C. glycolicum*, *Desulfitobacterium frappieri*	Breitenstein et al. 2001

Pentachlorophenol	anaerobic granular sludge	dechlorination and carbon source	Enrichment in a UASB reactor at 35°C, with a hydraulic residence time of 28 h	*Desulfitobacterium frappieri* and an undescribed consortium	Guiot et al. 2002
Endosulfan	sludge from an endosulfan producing facility	carbon source	five passages in aerobiosis at 30°C every 7 d	*Alcaligenes faecalis*	Kong et al. 2013
Pentachloronitrobenzene	green bean coffee	carbon source	five passages in aerobiosis every 7 days	*Pseudomonas aeruginosa*, *Pseudomonas putida*, *Stenotrophomonas maltophilia*, *Flavimonas oryzihabitans* and *Morganella morganii*	Barragán-Huerta et al. 2007
Dieldrin Endrin	pre-exposed agricultural soils	structural analogs as carbon source	passages in aerobiosis at 25°C every 7–10 days	*Burkholderia* sp., *Alcaligenes* sp., *Cupriavidus* sp.	Matsumoto et al. 2008
Triazines					
Atrazine	pre-exposed soils	nitrogen source	continuous selective enrichment in a chemostat	*Agrobacterium tumefaciens*, *Caulobacter crescentus*, *Pseudomonas putida*, *Sphingomonas yaniokuyae*, *Nocardia* sp., *Rhizobium* sp., *Flavobacterium oryzihabitans*, and *Variovorax paradoxus*	Smith et al. 2005
Atrazine	enriched culture	carbon source	7–10 passages in aerobiosis at 30°C every 72 h	*Alcaligenes xylosoxidans* ssp. *denitrificans*, *Sphingobacterium* sp., and *Pseudomonas* sp.	Kontchou and Gschwind 1999
Other Pesticides					
Isoproturon	pre-exposed agricultural soils	carbon and nitrogen source	> 15 passages monitored by mineralization	*Sphingomonas* sp.	Sørensen et al. 2001

Table 2.1 Contd. ...

...Table 2.1 Contd.

Pesticide(s) used	**Origin of consortium**	**Use of the pesticide**	**Incubation conditions**	**Consortium or isolated organism**	**Reference**
Mesotrione	lake and stream water near pre-exposed agricultural fields	carbon source	one passage in aerobiosis at 18–37°C for 24 h	*Pantoea ananatis*	Pileggi et al. 2012
Fipronil	pre-exposed agricultural soils	carbon source	two passages in aerobiosis at 18–28°C every 4 d	*Bacillus thuringiensis*	Mandal et al. 2013
Nicosulfuron	sludge from industrial wastewater treatment pond	nitrogen source	weekly passages in aerobiosis at 30°C for 4 weeks	*Serratia marcescens*	Zhang et al. 2012
Metaldehyde	agricultural soils	carbon source	4 passages at 25°C, 150 rpm, every 72 h	*Acinetobacter bohemicus, Sphingobium* sp., *Rhodococcus globerulus, Pseudomonas vancouverensis, Acinetobacter lwoffii, Caballeronia jiangsuensis*	Castro-Gutiérrez et al. 2020
Pesticide Mixtures					
Chlorpyrifos Monocrotophos Endosulfan	pre-exposed agricultural soils	carbon source	not available	*Alcaligenes* sp., *Ochrobactrum* sp., and *Sphingobacterium* sp. (chlorpyrifos contaminated soil); *Enterobacter ludwigii*, *Pseudomonas moraviensis* and *Serratia marcescens* (monocrotophos spiked soil); *Klebsiella pneumoniae*, *Enterobacter cloacae*, *Halophilic bacterium* and *Enterobacter asburiae* (endosulfan polluted soil)	Abraham et al. 2014

Methyl-parathion Lindane Carbofuran	pre-exposed soils from the vicinity of pesticide production facility	carbon source (simultaneous enrichment)	several passages in aerobiosis at 28°C	*Pseudomonas aeruginosa*, *Bacillus* sp. and *Chryseobacterium joostei*	Krishna and Philip 2008
Carbaryl	soils from pesticide disposal site	carbon or nitrogen source	several passages in aerobiosis at 24°C every 2–4 weeks	*Pseudomonas* sp.	Chapalamadugu and Chaudhry 1991

Pesticides comprise a wide range of agrochemicals employed to control pests, and their use in agriculture greatly increases the yields of crops (Ware 2000). However, many of these compounds are persistent in soil and their leaching can pollute surface and ground waters. Furthermore, some of them are toxic to non-target organisms, even at low doses (Van der Werf 1996, Margni et al. 2002). That is why many are the strategies developed using pesticide degrading isolates and consortia obtained from selective enrichment that could be employed in bioremediation approaches. The description of such strategies and the general aspects of performing a selective enrichment comprise the aims of this chapter.

How to do a selective enrichment?

The general procedure to obtain degrading bacteria by selective enrichment starts with the growth of microorganisms at the expense of the compound to break down, as the sole source of carbon, other nutrients, such as nitrogen, or energy. That is how the enrichment procedure usually starts—by inoculating a culture medium, supplemented with the xenobiotic, with a natural source of degrading microorganisms. Consequently, source samples are generally taken from the area supposedly contaminated with the given compound.

Then, the cultures are incubated, and when sufficient microbial growth has been generated, successive transfers are made into fresh medium. These passages are made in order to remove nutrient debris from the environmental sample and to avoid the isolation of microorganisms that are just metabolizing the debris and not actually removing the xenobiotic. Nothing is written in stone regarding the number of passages or the incubation time between each successive transfer. For instance, there are reports of selective enrichments in which a single passage is conducted (Ortiz-Hernández and Sánchez-Salinas 2010, Pino and Peñuela 2011), and others in which more than ten are applied (Kontchou and Gschwind 1999, Sørensen et al. 2001). It is likely that including more passages is related with the isolation of more stable degrading communities, which is translated into longer incubation periods. The incubation time between each passage is generally days (Table 2.1), in order to allow the growth of these often slow-growing organisms.

Finally, after successive passages/transfers, a degrading consortium is obtained; the subsequent isolation of pure cultures (from the consortium members) is then conducted—generally by plating on solid media. Even though selective enrichment processes mostly result in the isolation of bacteria, the isolation of mixed consortia, also including fungi or yeasts is not uncommon (Peng et al. 2012, Rodríguez-Castillo et al. 2019, Silambarasan and Abraham 2013).

Methodological recommendations

Choose a proper isolation source

In general, within the microbial diversity of any environment, microorganisms with the capacity to naturally attenuate most organic pollutants are likely to exist. For example, wood-decay fungi produce ligninolytic extracellular enzymes that are very promiscuous and have the capacity to remove many pesticides (Yang et al. 2013).

However, the best strategy is not to randomly choose an isolation source, but to use one from an environment previously exposed to the pollutant of interest. Certainly, enrichment occurs in environments that are chronically exposed to contaminants, and this is an important mechanism for the attenuation of agrochemicals in nature (Mulligan and Yong 2004).

The most common isolation source employed in enrichment cultures for pesticide removal is pre-exposed agricultural soil (Table 2.1). However, samples from pesticide manufacturing wastewater treatment systems (Chen et al. 2014) and soil from pesticide disposal sites (Chapalamadugu and Chaudhry 1991), have been successfully used. The goal of using such pre-exposed matrices is to take advantage of selective forces driven by the pollutants even before the enrichment culture procedure is conducted. Also, it is common to do a pre-incubation of the matrix with the target pollutant, in order to stimulate communities that are resistant, or able to metabolize the molecule of interest.

Know your molecule of interest

Even when pesticides are usually removed by bacteria and fungi by oxidative pathways, knowing the structure of the molecule of interest is very helpful to select isolation conditions during enrichment culture. For example, a common strategy is to use a defined selective media spiked with the pesticide as a carbon or nitrogen source (or both) (Table 2.1), although in some cases the enrichment is done relying on a cometabolic transformation of the target compound (Pandey et al. 2009). However, some pesticide molecules lack nitrogen atoms in their structure (e.g., the organophospate malathion or the organochlorine aldrin), and a uniform media choice could result in the failure of the selective enrichment.

Furthermore, poly-halogenated molecules, such as organochlorine pesticides and other substances included in the Stockholm Convention on Persistent Organic Pollutants, are very slowly removed by aerobic oxidation. However, they could be used as electron acceptors in reductive anaerobic processes (Tartakovsky et al. 2001). For example, molecules as hexachlorobenzene have been reported to be slowly removed by bacteria such as *Dehalococcoides* spp. (Taş et al. 2010), and pesticides as trichlorophenol and pentachlorophenol have been successfully used in anaerobic selective enrichments (Breitenstein et al. 2001, Guiot et al. 2002).

The simultaneous enrichment using two or more different pesticides at once, has resulted in the successful isolation of degrading consortia with the capacity to remove the target compounds (Krishna and Philip 2008, Pino and Peñuela 2011), even though they were from different chemical families in some cases (Table 2.1).

Use high purity standards for the xenobiotic of interest

Standards may be expensive, however, using analytical or high purity molecules is the best way to do a selective enrichment. Commercial products used in farming contain excipients that can hinder the selective enrichment process. Excipients, for example, may be used by the microorganisms in the enrichment culture as a nutrient source, thus likely resulting in the isolation of a consortium lacking the capacity to remove the target pollutant.

How to select the enrichment culture medium

In general, a selective enrichment process proceeds using a defined medium spiked with the pesticide or xenobiotic to be studied. Different formulations have been used (Table 2.2), and a precise composition does not seem to be critical. However, it is clear that if the xenobiotic is intended to be employed in the enrichment as a carbon, nitrogen, sulfur, or phosphorus source, those elements should be omitted in the preparation of basal media.

Otherwise, during the preparation of the defined medium, inorganic macronutrients, such as sulfur, nitrogen, and phosphorus should be supplied as sulfate, ammonium or nitrate, and phosphate, respectively. Other cations, such as Mg^{2+}, Ca^{2+}, Na^{+}, and K^{+} should always be added, and other metals, such as Fe, Mn, Zn, Cu, Mo, and W are supplemented in lower concentrations. They are necessary in the formulation, as required for enzymatic functionality, acting as cofactors, for example. Trace element requirements for the more fastidious anaerobic bacteria include selenium as selenite and tungsten as tungstate (Neilson and Allard 2012).

Defined minimum media are regularly designed using a buffer system and salts that are employed to maintain an adequate osmotic concentration. Media are often buffered using phosphate, but high concentrations should be avoided because precipitation with cations may occur during autoclaving. Media may be sterilized by autoclaving, but in order to avoid degradation of the xenobiotics to be used as nutrient sources, a separate solution containing the compound should be prepared as a stock solution and sterilized by filtration, using 0.2 μm cellulose nitrate or cellulose acetate filters.

Commonly used media include different formulations of minimal salt medium (Cycoń et al. 2013, Sasikala et al. 2012), Bushnell-Haas (Castro-Gutiérrez et al. 2016, Rodríguez-Castillo et al. 2019) and Dorn´s broth (Mandal et al. 2013). To include nutrients from the original sample source, supplements including soil extract medium have also been incorporated (Chanika et al. 2011, Karpouzas et al. 2005).

Incubation conditions

Incubation is normally carried out in the dark, in order to avoid photodegradation of the supplemented xenobiotic and avoid the growth of autotrophic organisms. In general, selective enrichments are mostly carried out in aerobiosis, at incubation temperatures close to environmental conditions (24–30°C, Table 2.1). Such conditions are usually employed to emulate the environmental parameters from where the sample used as isolation source was withdrawn.

Isolation and storage

During isolation, the use of complex media may introduce isolation bias, since overgrowth of unwanted, rapidly growing organisms may occur. That is why, a solid version of the mineral medium employed in the enrichment procedure is usually employed for the isolation of the degrading consortium members. High quality agar should be employed to avoid the unintended supplementation of additional sources of carbon or other nutrients.

Table 2.2. Culture media (and composition) employed in selective enrichment processes for the isolation of pesticide degrading microorganisms.

Culture media (media were supplemented with the target pesticide)	Reference
Mineral medium (MM) (g/L): 0.2 KH_2PO_4; 0.5 K_2HPO_4; 1 $(NH_4)_2SO_4$; 0.2 $MgSO_4 \cdot 7H_2O$; 0.2 NaCl; 0.05 $CaCl_2 \cdot 2H_2O$; 0.025 $FeSO_4 \cdot 7H_2O$; 0.005 Na_2MoO_4; 0.0005 $MnSO_4$	Ortiz-Hernández and Sánchez-Salinas 2010
(g/L) 4.8 KH_2PO_4; 1.2 K_2HPO_4; 1.0 NH_4NO_3; 0.2 $MgSO_4 \cdot 7H_2O$, 0.04 $Ca(NO_3)_2 \cdot 4H_2O$; 0.001 $Fe_2(SO_4)_3$	Pino and Peñuela 2011
Minimal salt medium (MSM)(g/L): 1.5 K_2HPO_4; 0.5 KH_2PO_4; 0.5 $(NH_4)_2SO_4$; 0.5 NaCl; 0.2 $MgSO_4$; 0.05 $CaCl_2$; 0.02 $FeSO_4$	Sasikala et al. 2012
Minimal salt medium (MSM)(g/L): 1.5 K_2HPO_4; 0.5 KH_2PO_4; 0.5 NaCl; 0.5 $(NH_4)_2SO_4$; 0.2 $MgSO_4 \cdot 7H_2O$, 0.2. Trace solution (10 mL; mg/L): 500 $Na_2EDTA \cdot 2H_20$; 143 $FeCl_2 \cdot 4H_2O$; 4.7 $ZnCl_2$; 3.0 $MnCl_2 \cdot 4H_2O$; 30 H_3BO_3; 20 $CoCl_2 \cdot 6H_2O$; 1.0 $CuCl_2 \cdot 2H_2O$; 2.0 $NiCl_2 \cdot 6H_2O$; 3.0 $Na_2MoO_4 \cdot 2H_2O$; 100 $CaCl_2 \cdot 2H_2O$	Akbar et al. 2014
Minimal salt medium (MSM)(g/L): 5.8 Na_2HPO_4; 3.0 K_2HPO_4; 0.5 NaCl; 1 NH_4Cl; 0.25 $MgSO_4$	Abraham and Silambarasan 2013, Anwar et al. 2009
Minimal salt medium (MSM)(g/L): 1.5 K_2HPO_4; 0.5 KH_2PO_4; 1.0 $(NH_4)_2SO_4$; 0.5 NaCl; 0.2 $MgSO_4$; 0.02 $FeSO_4$	Liu et al. 2011
Minimal salt medium (MSM) (g/L): 2.0 $(NH_4)_2SO_4$; 0.2 $MgSO_4 \cdot 7H_2O$; 0.01 $CaCl_2 \cdot 2H_2O$; 0.001 $FeSO_4 \cdot 7H_2O$; 1.5 $Na_2HPO_4 \cdot 12H_2O$; 1.5 KH_2PO_4	Cycoń et al. 2009
Mineral salts medium supplemented with nitrogen (MSMN) and soil extract medium (SEM)	Chanika et al. 2011, Karpouzas et al. 2005
Minimal salt medium (MSM) (g/L): 6 Na_2HPO_4; 5 NaCl; 3 KH_2PO_4; 0.1 $MgSO_4$; 2 NH_4Cl; 5 glucose	Phugare and Jadhav 2015
Bushnell-Haas broth (g/L): 0.2 $MgSO_4$; 0.02 $CaCl_2$; 1.0 K_2HPO_4; 1.0 KH_2PO_4; 1.0 NH_4NO_3; 0.05 $FeCl_3$	Castro-Gutiérrez et al. 2016, Rodríguez-Castillo et al. 2019
R-salts minimal medium (RSM): 67.0 mL 1 M KH_2PO_4, 5 mL R-salt mixture (16.0 g $MgSO_4 \cdot 7H_2O$, 0.4 g $FeSO_4 \cdot 7H_2O$, and 0.8 mL HCl per 200 mL H_2O), 200 μL 1 M $CaCl_2 \cdot 2H_2O$, and 1.0 mL trace element solution (10.0 mg $ZnSO_4 \cdot 7H_2O$, 3.0 mg $MnCl_2 \cdot 4H_2O$, 30 mg H_3BO_4, 20.0 mg $CoCl_2 \cdot 6H_2O$, 1.0 mg $CuCl_2 \cdot 2H_2O$, 2.0 mg $NiCl_2 \cdot 6H_2O$, and 3.0 mg $Na_2MoO_4 \cdot 2H_2O$ per liter)	Kandil et al. 2015
Kaufman and Kearney's minimal salts media	Anhalt et al. 2007
Phosphate, 0,95 mM; trace elements; 16 nM Na_2SeO_3; 12 nM Na_2WoO_4; 62 nM 1,4-naphthoquinone; 0.1 g/L yeast extract; 10 mM lactate; 10 mM pyruvate	Breitenstein et al. 2001
Basal medium (g/L): 5.8 K_2HPO_4; 4.5 KH_2PO_4; 2.0 $(NH_4)_2SO_4$; 5 peptone; 0.16 $MgSO_4$; 0.02 $CaCl_2$; 0.002 Na_2MoO_4; 0.001 $FeSO_4$; 0.001 $MnCl_2$	Kong et al. 2013

Table 2.2 Contd. ...

...Table 2.2 Contd.

Culture media (media were supplemented with the target pesticide)	Reference
Minimal salt medium (MSM) (g/L): 2.4 Na_2HPO_4; 2.0 KH_2PO_4; 0.1 NH_4NO_3; 0.01 $MgSO_4 \cdot 7H_2O$; 0.01 $CaCl_2$	Barragán-Huerta et al. 2007
RM2 supplemented with 0.05% yeast extract Minimal salt medium (MSM) (g/L): 1.0 $(NH_4)_2SO_4$; 0.8 K_2HPO_4; 0.2 KH_2PO_4; 0.2 $MgSO_4 \cdot 7H_2O$; 0.1 $CaCl_2 \cdot 2H_2O$; 0.05 $FeCl_3 \cdot 6H_2O$; 0.01 $(NH_4)_6Mo_4O_{24} \cdot 4H_2O$	Matsumoto et al. 2008
Nitrogen- and chloride-free minimal salts medium (MS) (mM): 10 K_2HPO_4, 3 NaH_2PO_4, 1 $MgSO_4$ and trace minerals	Smith et al. 2005
Basal medium: 10 mM disodium potassium phosphate buffer (pH 7.5); 20 mg $MgSO_4 \cdot 7H_2O$ per liter, and 10 mL trace element solution. Trace element solution (mg/L): 20 $MgSO_4 \cdot 7H_2O$; 500 EDTA; 200 $FeSO_4 \cdot 7H_2O$; 10 $ZnSO_4 \cdot 7H_2O$; 3 $MnSO_4 \cdot H_2O$; 30 H_3BO_3; 24 $CoSO_4 \cdot 7H_2O$; 1 $CuSO_4 \cdot 5H_2O$; 2 $NiSO_4 \cdot 7H_2O$; 3 $Na_2MoO_4 \cdot 2H_2O$; 50 $Ca(OH)_2$	Kontchou and Gschwind 1999
Dorn's broth: 3.0 $Na_2HPO_4 \cdot 12H_2O$; 1.0 KH_2PO_4; 1.0 $(NH_4)_2SO_4$; 10.0 $MgSO_4 \cdot 7H_2O$; 2.0 $CaCl_2 \cdot 2H_2O$; 3.0 $MnSO_4 \cdot H_2O$; 0.2 $FeSO_4 \cdot 7H_2O$; 0.01 ammonium ferric citrate; 0.1 yeast extract	Mandal et al. 2013
Minimal salt medium (MSM) (g/L): 5 glucose; 0.5 KH_2PO_4; 0.2 $MgSO_4 \cdot 7H_2O$; 0.5 K_2HPO_4; 0.2 NaCl	Zhang et al. 2012
Minimal salts medium (MSM) plus 2.0 mL trace elements	Castro-Gutiérrez et al. 2020
Minimal salt medium (MSM) (g/L): 3 KH_2SO_4; 0.5 NaCl; 5.8 Na_2SO_4; 1 NH_4Cl; 0.25 $MgSO_4 \cdot 7H_2O$; supplemented with Focht trace element solution	Jabeen et al. 2015
Nutrient medium (NM) (g/L): 1.0 KH_2PO_4; 1.0 K_2HPO_4; 1.0 NH_4NO_3; 1.0 NaCl; 0.2 $MgSO_4 \cdot 7H_2O$; 0.02 $CaCl_2$; 0.02 $Fe(SO_4)_3$; 1 mL trace metal solution	Krishna and Philip 2008
Minimal medium (MM) (g/L): 4.8 K_2HPO_4; 1.2 KH_2PO_4; 1.0 NH_4NO_3; 0.2 $MgSO_4 \cdot 7H_2O$; 0.4 $Ca(NO_3)_2 \cdot 4H_2O$; $Fe_2(SO_4)_3$, 0.001 MMG: MM with glucose and lacking nitrogen sources	Chapalamadugu and Chaudhry 1991

Sometimes it is impossible to isolate a single bacteria that grows on the pollutant supplemented because more than one organism may cooperate in the degradation of the substrate. Also, lack of repeatability due to loss of degrading activity using laboratory strains or consortia that have been maintained by repeated transfers for long periods under nonselective conditions, may be encountered in metabolic studies. For these reasons, strains should be maintained in the presence of a cryoprotectant such as glycerol at low temperatures (−80°C) as soon as possible after isolation.

During isolation, a practice that helps to distinguish between isolates that remove the compound of interest and those lacking this capacity, is cultivating the isolates at the same time in a supplemented mineral medium without supplementing with the xenobiotic. If the microorganism grows in both media, it is likely to grow from some trace in the medium and not from the pollutant of interest.

Although potential removal of the target compound is carried out by several individual strains within the consortium, several works report better removal capacity when the consortium is employed as a whole, rather than using individual isolates (Castro-Gutiérrez et al. 2016, Jariyal et al. 2018).

Analytical evaluation of removal capacity

Once the degrading consortium is isolated, a removal test should be performed to properly corroborate the transformation capacity of the target compound. The test should measure the residual concentration of the target compound by analytical approach, ideally methods based on chromatography coupled with mass spectrometry. Other approaches may use radiolabeled target compounds to estimate their mineralization (Sørensen et al. 2001). Controls should include an abiotic control, to determine losses of the target compound by abiotic phenomena, such as hydrolysis or photodegradation, and a heat-killed control to estimate its adsorption to the consortium biomass (Castro-Gutiérrez et al. 2016). If possible, analytical determination of the pesticide removal should be performed during the selective enrichment process at each passage. Although expensive and time consuming, this practice could help to better estimate the incubation period between transfers, and to perform modifications of the enrichment itself.

Of great value is also the practice to evaluate the cross-degradation of other compounds of structural similarity to the target molecule, as several consortia have shown successful removal of other pesticides from the same chemical family of the target compound, like in the case of carbamates (Castro-Gutiérrez et al. 2016), organophosphates (Karpouzas et al. 2005), neonicotinoids (Rodríguez-Castillo et al. 2019) and triazines (Kontchou and Gschwind 1999), or even from different chemical families (organophosphates/carbamates) (Chanika et al. 2011).

Use of consortia for the removal of pesticides: applications

Consortia or individual strains isolated by selective enrichment means have been employed as biocatalytic agents for the removal of pesticides in different matrices, including several bioreactor configurations for the treatment of polluted water or soil, or as amendment in soil microcosms.

The use of such consortia is quite a flexible approach- in some cases, other than the whole consortium, individual isolates of high removal capacity are selected and employed in the treatment of the matrix (Ahmad et al. 2012, Cycoń et al. 2013). On the other hand, several works have created "artificial" degrading-consortia, by mixing organisms isolated from individual selective enrichments. For instance, several combinations of *Streptomyces* sp. isolates have been used to treat different organochlorine pesticides (Fuentes et al. 2014, 2017, Saez et al. 2014). Similarly, Bazot and Lebeau (2009) took advantage of two complementary metabolic abilities to create a consortium formed by *Arthrobacter* N4, capable of transforming diuron in 3,4-dichloroaniline (Tixier et al. 2002), and *Delftia acidovorans* WDL34, able to degrade 3,4-dichloroaniline (Dejonghe et al. 2003), to remove the herbicide diuron.

This section presents some of the approaches described in scientific literature to apply pesticide degrading consortia (Table 2.3).

Immobilization of degrading-consortia

The application of the consortium or individual isolates has been performed in most cases as a direct inoculation on the polluted matrix. However, given that cellular viability can be increased by cell immobilization by providing cell protection against potentially toxic compounds (Cassidy et al. 1996), several works have employed different immobilization approaches to apply their microbial consortia (Table 2.3). One of the most commonly employed is the use of alginate beads, which has systematically shown better results than free cells, at least when applied in liquid matrices. Fuentes et al. (2013) employed a mixed culture of six *Streptomyces* spp. isolates to remove chlorpyrifos and pentachlorophenol from a defined medium. Free cell systems removed 40.2% chlorpyrifos and 5.2% pentachlorophenol after 72 hours, while elimination by cells immobilized in alginate beads reached 71.1% and 14.7%, respectively. The dissipation of diuron by a two member consortium (*Arthrobacter* N4 and *Delftia acidovorans* WDL34) using different combinations of cell-free or alginate-immobilized cells, demonstrated an optimal removal when free-cells of *Arthrobacter* N4 were co-applied with immobilized cells of *Delftia acidovorans* WDL34 (Bazot and Lebeau 2009). Nonetheless, from the operational point of view, the authors recommended the co-immobilization of both strains. Consortium immobilization in alginate beads was also applied by Yañez-Ocampo et al. (2009) for the simultaneous treatment of two organophosphates, methyl-parathion and tetrachlorvinphos. The immobilization improved the removal of the former, but the effect was not significantly different in the latter. The same authors applied immobilization by means of biofilm formation in tezontle, a highly porous volcanic stone, and achieved similar results, i.e., enhancement in the removal of methyl-parathion, but not in the case of tetrachlorvinphos. The use of this consortium also proved to be useful to remove these pesticides in a packed-bed reactor, using the biofilm on tezontle as the packing material (Yañez-Ocampo et al. 2011). Another support material, polyurethane foam pellets, permitted the immobilization of an atrazine degrading consortium to achieve the simultaneous elimination of atrazine, simazine, terbuthylazine, ametryn, cyromazine, and the transformation products

Table 2.3. Application strategies of pesticide degrading consortia isolated by selective enrichment for the removal of pesticides.

Pesticide	Polluted matrix	Application	Consortium	Removal	Inoculum	Reference
Organophosphates						
Cadusafos **Ethoprophos**	Soil microcosm	Direct inoculation	*Sphingomonas paucimobilis*	Cadusaphos 85% in 24 d Ethoprophos 100% in 6 d	4.3×10^8 cells/g soil	Karpouzas et al. 2005
Methyl-parathion **Tetrachlorvinphos**	Mineral salt medium supplemented with the pesticides	Immobilized cells in alginate beads and in biofilm on tezontle	Bacterial consortium	Methyl-parathion: 41% (suspended consortium) in 6 d 72% (alginate immobilized consortium) in 11 d 66% (tezontle immobilized consortium) in 13 d Tetrachlorvinphos: 53% (suspended consortium) in 6 d 65% (alginate immobilized consortium) in 11 d 47% (tezontle immobilized consortium) in 13 d	Alginate beads, 1.23×10^6 CFU/bead Tezontle, 2.8×10^7 CFU/tezontle unit	Yañez-Ocampo et al. 2009
Methyl-parathion **Tetrachlorvinphos**	Tezontle-packed up-flow reactor	Immobilized cells	Bacterial consortium	Operating time of 8 h and hydraulic residence time 0.313 h: 73% (methyl-parathion) 75% (Tetrachlorvinphos)	Not available	Yañez-Ocampo et al. 2011
Chlorpyrifos (CP) **Fenitrothion (FT)** **Parathion (PT)**	Sterile and non-sterile soil microcosms	Direct inoculation	*Serratia marcescens*	Sterile soil at 42 d: 1. Sandy loam soil PT: 75.7% 2. Silty soil CP: 78.4% 3. Silty soil FT: 89.9% Non-sterile soil at 42 d: 1. Sandy loam soil PT: 89.7% 2. Silty soil CP: 87.7% FT: 96.7%	3×10^6 cells $\times$ g^{-1} of soil	Cycoń et al. 2013

Table 2.3 Contd. ...

...Table 2.3 Contd.

Pesticide	Polluted matrix	Application	Consortium	Removal	Inoculum	Reference
Methyl-parathion Coumaphos	Mineral Salts Medium (MSM) containing mixture of pesticides	Immobilized cells on *Luffa cylindrica* fibers	Bacterial consortium	Immobilized cells in 72 h 98% (methyl-parathion) 100% (coumaphos)	10^5 CFU/mL	Moreno-Medina et al. 2014
Fenamiphos	Soil microcosm	Direct inoculation	*Pseudomonas putida*	100% in 14 d	10^6 cells/g soil	Chanika et al. 2011
Chlorpyrifos (including its major degradation product 3,5,6-trichloro-2-pyridinol (TCP))	Soil microcosm	Direct inoculation	*Aspergillus terreus*	Chlorpyrifos 100% in 12 h TCP 100% in 24 h	3×10^7 spores/g soil	Silambarasan and Abraham 2013
Chlorpyrifos	Slurry with and without surfactant	Direct inoculation	*Pseudomonas, Klebsiella, Stenotrophomonas, Ochrobactrum* and *Bacillus*	Slurry without surfactant in 10 d, 81% Slurry with surfactant in 10 d, 96%	$O.D_{540} = 0.75$	Singh et al. 2016
Chlorpyrifos	Soil microcosm with plants	Direct inoculation	*Bacillus pumilus* C2A1	97% in 45 d	3.3×10^7 CFU/g soil	Ahmad et al. 2012
Phorate	Soil microcosm	Direct inoculation	*Brevibacterium frigoritolerans, Bacillus aerophilus* and *Pseudomonas fulva*	98% in 42 d	$\sim 4.5 \times 10^6$ cells/g soil	Jariyal et al. 2018

Methyl-parathion **Chlorpyrifos**	Soil microcosm	Direct inoculation	*Acinetobacter* sp., *Pseudomonas putida, Bacillus* sp., *Pseudomonas aeruginosa, Citrobacter freundii, Stenotrophomonas* sp., *Flavobacterium* sp., *Proteus vulgaris, Pseudomonas* sp., *Acinetobacter* sp., *Klebsiella* sp. and *Proteus* sp.	Methyl-parathion 98% in 120 h Chlorpyrifos 97% in 120 h	3×10^{6} CFU $\times$ g^{-1} of soil	Pino and Peñuela 2011
Organochlorides						
Lindane **Chlordane Methoxychlor**	Sterile and non-sterile soil microcosms	Direct inoculation	*Streptomyces* consortium	Sterile soil in 16 d: 30% (lindane), 23% (chlordane) 39% (methoxychlor) Non-sterile soil in 16 d: 25% (lindane) 5% (chlordane) 21% (methoxychlor)	2.0×10^{-3} g biomass/g soil	Fuentes et al. 2017
Lindane	Soil slurry	Consortium immobilized in cloth sachets	*Streptomyces*	70.6 in 7 d	10^{7} CFU/g soil	Saez et al. 2014
Pentachlorophenol **Chlorpyrifos** **(organophosphate)**	Minimal medium containing the mixture of pesticides	Immobilized cells	Mixed culture of *Streptomyces* spp.	Pentachlorophenol 14.7% in 72 h Chlorpyrifos 71.1% in 72 h	Not available	Fuentes et al. 2013

Table 2.3 Contd. ...

...Table 2.3 Contd.

Pesticide	Polluted matrix	Application	Consortium	Removal	Inoculum	Reference
Endosulfan (organochlorine) Chlorpyrifos Monocrotophos (organophosphates)	2.5 L stirred tank bioreactor	Direct inoculation	*Alcaligenes* sp., *Ochrobactrum* sp., *Sphingobacterium* sp., *Enterobacter ludwigii, Pseudomonas moraviensis, Serratia marcescens, Klebsiella pneumoniae, Enterobacter cloacae, Enterobacter asburiae. Halophilic* bacteria	100% in 24 h	3×10^6 cells/mL	Abraham et al. 2014
Methoxychlor	Slurry bioreactor, batch operation: soil/water (SB-water) soil/trypteine soy broth (SB-TSB)	Direct inoculation	*Streptomyces* spp.	SB-TSB in 7 d: 56.2% SB-water at 7 d: 45.6%	SB-TSB 10^7 CFU/mL SB-water 10^4 CFU/mL	Fuentes et al. 2014
Triazines						
Atrazine Ametryne Cyromazine Desethylatrazine and cyanuric acid (transformation products)	Aerobic sequence batch reactors	Immobilized cells on polyurethane foam pellets	*Alcaligenes xyloxoxydans* ssp. *denitrificans*, *Sphingobacterium* sp., and *Pseudomonas* sp.	Atrazine 90% in 16 d 60–80% in 16 d (ametryne, cyromazine, desethylatrazine, and cyanuric acid)	Not available	Kontchou and Gschwind 1999

Atrazine	Soil microcosm	Direct inoculation	Mixed microbial culture	100% removed in 30 d; 86% mineralized after 145 d	Not available	Assaf and Turco 1994
Atrazine	Soil microcosm: corn planted and non planted	Direct inoculation	*Clavibacter michiganese, Pseudomonas* sp., and *Cytophaga* sp.	Non-planted soil: 71% (mineralized) after 4 weeks Planted soil: 84% (mineralized) after 4 weeks	4.5×10^6 CFU/g soil	Alvey and Crowley 1996
Atrazine	Soil microcosm	Direct inoculation	*Pseudomonas* sp.	Sterilised soil in 24 h 79.9% (mineralized) Non-sterilised soil in 72 h 66.3% (mineralized)	2×10^7 cells/g soil	Silva et al. 2004
Atrazine	Soil microcosm	Direct inoculation	*Pseudomonas* sp.	70% in 3 weeks	2×10^6 cells/g soil	Mandelbaum et al. 1995
Atrazine	Soils with and without microcosm surfactants	Direct inoculation	*Acinetobacter* sp.	Soils without surfactants 55% in 6 d Soils with surfactants > 70% in 6 d	5×10^4 CFU/g soil	Singh and Cameotra 2014
Phenylureas						
Diuron	Sediment culture medium inoculated with different formulations of the bacterial co-culture	Direct inoculation Immobilized cells in Ca-alginate beads of both strains or one strain with free cells of the another strain	*Delftia acidovorans* and *Arthrobacter* sp.	100% dissipation in 144 h when *Delftia acidovorans* was immobilized with *Arthrobacter* sp. as free cells. 100% dissipation in 192 h when both strains were immobilized	10^6 cells/ mL	Bazot and Lebeau 2009

Table 2.3 Contd. ...

...*Table 2.3 Contd.*

Pesticide	Polluted matrix	Application	Consortium	Removal	Inoculum	Reference
Isoproturon	Soil microcosm	Applied via clay particles	Bacterial community of "Calcaric Regosol" soil	58.7% (mineralized) in 20 d	5.5×10^7 CFU/g (content in clay particles)	Grundmann et al. 2007
Diuron	Soil slurries inoculated with the bacterial consortium, with and without HPBCD solution	Direct inoculation	*Arthrobacter sulfonivorans, Variovorax soli*	System with bacterial consortium in 120 d 78.3 (removed) 45.25% (mineralized) System with bacterial consortium and HPBCD solution in 120 d 93.2% (removed) 98.67% (mineralized)	10^7 CFU/mL	Villaverde et al. 2012
Diuron	Soil microcosm	Direct inoculation	*Arthrobacter* sp.	35% in 30 d	10^7–10^9 cells/g soil	Widehem et al. 2002
Diuron	Soil microcosm	Direct inoculation	*Arthrobacter globiformis* and *Variovorax* sp.	59.9% in 26 d	10^8 cells/g soil	Sørensen et al. 2008
Other Pesticides						
Fenpropathrin	Soil microcosm	Direct inoculation	*Bacillus* sp.	93.3% in 72 h	1.0×10^8 CFU/g soil	Chen et al. 2014
Bensulphuron-methyl	Soil microcosm	Direct inoculation	*Penicillium pinophilum*	Sterilised soil in 60 h 72.8–80.7% Non-sterilised soil in 60 h 79.2–85.7%	Not available	Peng et al. 2012

Carbofuran Aldicarb Methiocarb Methomyl	Fluidized-bed reactor and flask scale	Direct inoculation	*Cupriavidus*, *Achromobacter* and *Pseudomonas*	Carbofuran: Batch reactor 100% in 16–17 h Continuous reactor 100% at HRT of 23 h Methiocarb: 100% in 15 h, flask scale Methomyl: 100% in 7 d, flask scale Aldicarb: 90.9% in 14 d, flask scale	Not available	Castro-Gutiérrez et al. 2016
Carbendazim 2,4-Dichlorophenoxyacetic acid (2,4-D)	Culture medium at flask scale	Immobilized cells on *Luffa cylindrica*	Consortium obtained from soil samples in Japanese paddy fields	Carbendazim 100% in 5.5 d 2,4-D 100% in 1,5 d	10^7 CFU/ support	Pattanasupong et al. 2004
2,4-Dichlorophenoxyacetic acid (2,4-D)	Soil microcosm	Direct inoculation	*Novosphingobium* spp.	Sterilised soil in 7 d: 96.3% Non-sterilised soil in 7 d: 96.5%	2.3×10^6 CFU/g soil	Dai et al. 2015
Imidacloprid + thiamethoxam Imidacloprid + thiamethoxam + acetamiprid	Stirred tank reactors operating in batch mode	Direct inoculation	*Kocuria rhizophila*, *Paraburkholderia phymatum*, *Paenibacillus odorifer*, *Rhodotorula toruloides*	95,8% (Imidacloprid + thiamethoxam) in 30 d 94,4% (imidacloprid + thiamethoxam + acetamiprid) in 30 d	2.5×10^2 CFU × mL^{-1}	Rodríguez-Castillo et al. 2019

hydroxyatrazine and deethylatrazine in sequence batch reactors (Kontchou and Gschwind 1999).

Immobilization of a consortium isolated from rice paddy fields on loofa sponge (*Luffa cyindrica*), showed the best removal efficiency for the fungicide carbendazim and the herbicide 2,4-D, among several supports—coconut fiber, activated carbon, nonwoven polyester fiber and Fabios® (Pattanasupong et al. 2004). Besides being a biodegradable support, the loofa sponge system was also able to detoxify the matrix polluted with pesticides, as demonstrated with tests on *Daphnia magna*. The removal of the organophosphates methyl-parathion and coumaphos with a consortium immobilized in the same support material revealed that elimination was enhanced when using the loofa sponge, basically due to the sorption of the pesticides to this matrix, and not necessarily due to microbial activity (Moreno-Medina et al. 2014). Such studies should include adsorption assays, to properly differentiate the input given by the microbial community and that from the adsorption properties of the support material.

Slurry systems

Slurry systems have emerged as a useful approach for the *ex situ* treatment of soils, particularly when small polluted areas contain high concentrations of xenobiotics. They allow the treatment of the polluted matrix under controlled conditions, and promote the biological depuration activity as the main removal process. In a slurry, soil is treated in an aqueous suspension at concentrations ranging from 10 to 30% w/v, with the mechanical mixing necessary to sustain a homogeneous suspension, which leads to advantages, such as increased mass transfer rates and close contact between degrading microorganisms and the pollutant, thus resulting in higher rates of pollutant biodegradation compared to *in situ* or other solid-phase treatments (Robles-González et al. 2008).

Slurries are considered as good systems to support bioaugmentation, and hence, they have also been tested with pesticide degrading consortia (Table 2.3). Although slurries are usually performed in agitated systems, such as stirred tank bioreactors, those described here for the removal of pesticides were set up at flask scale. Singh et al. (2016) created a consortium by combining two individual degrading isolates (*Klebsiella* sp. and *Stenotrophomonas maltophilia*) that removed 82% chlorpyrifos (50 mg/kg) in 10 days in a slurry system. When the system was amended with a rhamnolipid biosurfactant produced by another degrading isolate (*Pseudomonas* sp.), the elimination was enhanced to more than 95% within the same period, thanks to the increase in the aqueous phase solubility of chlorpyrifos. Even though it was not tested, the rhamnolipid could be potentially produced *in situ* by the addition of the surfactant-producing isolate.

Streptomyces sp.-based defined consortia created by mixing isolates from soil and sediments polluted with organochlorine insecticides have been used as bioaugmentation agents in slurries designed for the treatment of these agrochemicals. Saez et al. (2014) achieved the elimination of 35.3 mg/kg of lindane in 7 days when the *Streptomyces* sp.-based consortium immobilized in cloth bags was inoculated in a high concentration slurry (2:3 soil-water ratio). Similarly, another non-immobilized *Streptomyces* sp.-based consortium removed 46–56% methoxychlor in 7 days (Fuentes et al. 2014). When the same consortium employed for lindane removal was applied for the treatment of a mixture of organochlorines, the efficiency of the slurry decreased, reaching eliminations of 26% (methoxychlor), 12.5% (lindane), and 10% (chlordane) after 16 days. Interestingly, removal values were much higher (40% lindane; > 99% methoxychlor and chlordane) when the elimination was assayed in liquid phase (Fuentes et al. 2017). Such a difference could be ascribed to the strong sorption of the insecticides to soil, which was a problem that could be overcome by the use of surfactants, as previously discussed (Singh et al. 2016). In this respect, Villaverde et al. 2012 improved the mineralization of diuron (from 45.3% to 98.7%) by a consortium of *Arthrobacter sulfonivorans* (diuron degrading strain) and *Variovorax soli* (linuron mineralizing strain) in a soil slurry, after the addition of hydroxypropyl-β-cyclodextrin, which is capable of enhancing the herbicide bioavailability.

The use of bioaugmented slurry systems at pilot scale is still necessary to determine their efficiency in the removal of pesticides.

Other liquid-phase bioreactors

Several liquid-phase approaches have explored different configurations at reactor scale (Table 2.3). Castro-Gutiérrez et al. (2016) employed a fluidized-bed bioreactor to treat the carbamate carbofuran in water, using a carbofuran-degrading consortium that was also capable of cross-degrading other carbamates (aldicarb, methiocarb, and methomyl). The bioreactor showed fast removal of carbofuran in batch systems, eliminating concentrations between 50 and 200 mg/L in less than 20 hours. During continuous operation, the process was optimized and achieved the complete elimination of 100 mg/L at a hydraulic residence time of 23 hours. As the consortium employs carbofuran as a carbon source, the system did not require the addition of other carbon sources, which represents an advantage for the treatment of pesticide wastewater of agricultural origin, whose nutritional content is usually poor.

A work by Rodríguez-Castillo et al. (2019) described the use of a stirred tank bioreactor for the batch treatment of several neonicotinoid insecticides (mixtures of imidacloprid, thiamethoxam, and acetamiprid) employing an imidacloprid-degrading consortium. The isolation of the consortium was motivated by the extremely slow removal of neonicotinoids in solid-phase biopurification systems aimed at pesticide removal, which made them unfeasible for application in agricultural fields (Huete-Soto et al. 2017, Rodríguez-Castillo et al. 2019). The reactor achieved over 99% elimination of imidacloprid in 6 days or 17 days (depending on the mixture of neonicotinoids to be treated), acetamiprid was removed in 28 days, while thiamethoxam reached 87% elimination after 30 days of operation. The ecotoxicological evaluation of the treated water with Microtox®, bees, and seed germination tests revealed partial, although significant, detoxification of the system. A similar configuration (stirred tank bioreactor) was employed to successfully and simultaneously remove a mixture of three pesticides (chlorpyrifos, monocrotophos, and endusulfan) in just 24 hours; the consortium employed consisted of a mixture of three individual consortia, individually obtained for each pesticide (Abraham et al. 2014). This work proved that the mixture of pesticides did not hinder the elimination capacity of the individual consortia.

The use of aerobic sequence batch reactors (SBR), three in parallel, loaded with an atrazine-degrading consortium immobilized on polyurethane foam pellets was able to remove not only atrazine, but other triazines and some of their transformation products, at more than 90% in the effluent. Moreover, the elimination of atrazine was higher in wastewater, than in defined medium (Kontchou and Gschwind 1999).

A packed-bed up-flow configuration described by Yañez-Ocampo et al. (2011) employed a biofilm of the degrading consortium supported on volcanic rocks as packing material. The continuous system achieved the maximum removal, 72.7% and 75.5% for methyl-parathion and tetrachlorvinphos, respectively, at a hydraulic residence time of 0.313 hours. Although incomplete, the removal correlated with the detoxification of the effluent, as demonstrated with an acute toxicity test on *Eisenia fetida*.

Solid-phase approaches: soil microcosms and biopurification systems

Most of the solid-phase systems employed in bioaugmentation processes with degrading consortia comprise soil microcosms (Table 2.3). Organophosphates encompass the most commonly explored group of pesticides, usually with high removal efficiencies in soil microcosms, of over 80% in most cases. Chlorpyrifos elimination has been assayed in soil microcosms bioaugmented with diverse degrading isolates. The use of the Gram-negative bacteria *Serratia marcescens* achieved removals between 51–88%, depending on the type of soil. Nonetheless, the effect was enhanced when the isolate was applied to non-sterile soil—a finding also observed for the other organophosphates assayed, fenitrothion and parathion, which marks the desired combined effect of natural attenuation and bioaugmentation (Cycoń et al. 2013). The use of the Gram-positive *Bacillus pumilus* in a system containing plants achieved 97% chlorpyrifos removal (Ahmad et al. 2012), while the addition of the fungus *Aspergillus terreus* completely removed the insecticide and its toxic major transformation product, 3,5,6-trichloro-2-pyridinol (Silambarasan and Abraham 2013).

Other organophosphates removed in soil microcosms included fenamiphos, completely eliminated after 14 days with the addition of *Pseudomonas putida* (Chanika et al. 2011), and simultaneously cadusafos (~ 80% after 24 days) and ethoprophos (100% after 3 days), although the *Sphingomonas paucimobilis* strain employed was isolated in an enrichment with cadusafos alone (Karpouzas et al. 2005). Similarly, Jariyal et al. (2018) employed a consortium formed by *Brevibacterium frigotolerans*, *Bacillus aerophilus*, and *Pseudomonas fulva*, strains previously isolated from sugarcane fields (Jariyal et al. 2014), to reach 98% removal of phorate.

The elimination of the herbicide atrazine in soil microcosms has been demonstrated in several works; most of them evaluated mineralization rather than merely elimination of the pesticide. Mineralization in aerobic conditions refers to the complete oxidation of the molecule to CO_2 and water, thus reducing the possibility of accumulation of potentially toxic transformation products (Rodríguez-Rodríguez et al. 2018). Hence this process takes longer periods than simple transformation of the parent compound. Assaf and Turco (1994) employed a mixed culture that achieved 86% mineralization after 146 days, and proved to enhance by 20-fold the mineralization with respect to the non-bioaugmented soil. The joint biostimulation with citrate and succinate and bioaugmentation with *Pseudomonas* sp., resulted in 54% atrazine mineralization after 67 days (Silva et al. 2004). This combined approach not only enhanced the process with respect to the bioaugmentation alone (30.6%), but also increased the survival of the degrading strain. The use of the same isolate resulted in the removal of 70% atrazine when biostimulation with citrate was co-applied (Mandelbaum et al. 1995). The combined use of plants (corn) and a bacterial consortium (*Clavibacter michiganese*, *Pseudomonas* sp., and *Cytophaga* sp.) described by Alvey and Crowley (1996), increased atrazine mineralization in soil (84% after 28 days), versus the non-planted soil (71%), although mineralization rates were not significantly different. Moreover, the planted system enhanced long-term survival of the consortium and the production of hydroxyatrazine. Ecotoxicological evaluation with a bioassay of germination and survival of wheat revealed the key role of corn plants in the detoxification, as only planted systems (and not only bioaugmented) supported the survival of wheat. As in slurry systems, the use of rhamnolipids also enhanced the removal of pesticides in soil microcosms, as reported by Singh and Cameotra (2014) in the case of atrazine elimination with *Acinetobacter* sp.

The removal of the highly persistent organochlorines lindane, chlordane, and methoxychlor has been rather poor, below 30%, in microcosms inoculated with mixtures of *Streptomyces* sp. isolates. To improve the removal of isoproturon in soil, Grundmann et al. (2007) applied a microbial community (extracted from soil with high ability to mineralize this herbicide), via clay particles, and reached mineralization values of 58.7% after 20 days. This so-called "hot spots" approach seemed promising, although its success depends on physicochemical properties of soil and the pesticide. The removal of pesticides from other chemical groups can be found in Table 2.3. Overall, as is the case with other bioremediation approaches, the application of consortia in soils should be scaled-up for the treatment of polluted soil, for instance, at field conditions.

The elimination of pesticides has been also assayed in biopurification systems, solid-phase devices designed for the treatment of pesticide-containing wastewater of agricultural origin (Rodríguez-Rodríguez et al. 2013). These systems employ the degrading microbiota contained within the materials that constitute the biomixture (biological active core of biopurification systems), mainly from the use of primed soil (Karanasios et al. 2012). A couple of works describe the bioaugmentation of biomixtures with a microbial degrading consortium isolated by selective enrichment with carbofuran. The removal performance of the biomixture was quite high for the pesticides evaluated (carbofuran, plus atrazine, carbendazim, and metalaxyl), with half-lives ranging from 2.7 to 9.9 days, which is why, the addition of the consortium did not exert any improvement in the system (Castro-Gutiérrez et al. 2018), not even for carbofuran or in the case of the co-bioaugmentation with a degrading fungus, *Trametes versicolor* (Castro-Gutiérrez et al. 2019). In the end, the authors recommended employing the non-bioaugmented biopurification system, and applying the consortium in biomixtures with low depuration capabilities of the tested pesticides.

Future perspectives

Even if time consuming and expensive, future assays for the selective enrichment of pesticide removal consortia should include not only pesticide removal monitoring, but also description of the microbial community by culture independent methods. In this way, one could determine not only the adequate time to make passages in the enrichment, but also to determine whether the composition of the consortium is stable or not.

Despite being applied since several decades, the isolation of pesticide-degrading consortia has not evolved to a systematic application on polluted sites, contrary to observations in the bioremediation of hydrocarbons. In this respect, the lack of field studies involving bioaugmentation with isolated consortia is highly remarkable.

The standardization of proper inocula is still a topic to be addressed. Most works employ concentrations of the degrading consortia or individual isolates ranging from 10^4 to 10^8 CFU per gram or milliliter of the matrix to be treated when conformed by bacteria. In the case of fungal treatments, there are just a few reports, and indicate around 10^7 spores per gram or milliliter (Silambarasan and Abraham 2013). However, no systematic optimization of inoculum size has been developed for the treatment of pesticides, and future research should cover this topic. Alternative ways of adding the inoculum include different immobilization techniques, as described above. The amount of inocula employed in different treatment approaches can be seen in Table 2.3.

For bioremediation purposes, the transformation of the pollutants by the degrading-consortia or isolates should result in the detoxification of the polluted matrix. Degradation processes may result in the formation and accumulation of transformation products of higher toxicity than the parent compound, thus representing an environmental risk. Table 2.4 summarizes the ecotoxicological findings identified during the biological treatment of pesticides with microbial consortia. As ecotoxicological assessment of the removal process is not commonly included in most works, such analysis should be implemented to provide a wider panorama of the environmental feasibility of the proposed process.

Acknowledgments

The authors acknowledge Gabriel Acuña, Ariel Loaiza, Adriana López, and Annette Vaglio for their technical support.

Table 2.4. Ecotoxicological evaluation during pesticide treatment in approaches that employ microbial degrading consortia.

Pesticide	Treatment	Bioindicator	Ecotoxicological findings	Reference
Neonicotinoid insecticides (imidacloprid + thiamethoxam; and imidacloprid + thiamethoxam + acetamiprid)	Consortium obtained from enrichment with imidacloprid and cross-degrading ability to remove other neonicotinoids, applied in 5 L-stirred tank reactors during batch operation	Lettuce seeds (*Lactuca sativa*) *Vibrio fischeri* Honeybees	Ecotoxicity of the wastewater was partially decreased by the reactor treatment for every ecotoxicological assay	Rodríguez-Castillo et al. 2019
Lindane	A defined consortium of actinobacteria consisting of *Streptomyces* strains immobilized in cloth sachets applied in slurry systems	Lettuce seeds (*L. sativa*)	Ecotoxicity was decreased in the bioaugmented systems with respect to non-bioaugmented systems	Saez et al. 2014
Methyl-parathion Tetrachlorvinphos	A tezontle-packed up-flow reactor (TPUFR) with an immobilized bacterial consortium	*Eisenia foetida*	Detoxification of the effluent is suggested, as the percentage of the affected individuals diminished gradually in the TPUFR effluent after biological treatment times of 0, 4, and 8 h	Yañez-Ocampo et al. 2011
Carbendazim 2,4-Dichlorophenoxyacetic acid (2,4-D)	Microbial consortium immobilized on loofa (*Luffa cylindrica*) sponge or free-living consortium at flask scale	*Daphnia magna*	The immobilized consortium decreased the toxicity towards *D. magna*	Pattanasupong et al. 2004
Atrazine	Combined application of an atrazine-mineralizing bacterial consortium (*Clavibacter michiganese*, *Pseudomonas* sp., and *Cytophaga* sp.) and corn plants in soil	Wheat seeds	Germination rates were slightly lower in atrazine-amended treatments that had not been previously planted with corn than in the previously planted soils. The importance of corn plants for detoxification of soil was evident by the 0% survival of the germinated seedlings in non-planted soils	Alvey and Crowley 1996
Methyl-parathion Chlorpyrifos	Bacterial consortium (*Acinetobacter* sp., *Pseudomonas putida*, *Bacillus* sp., *Pseudomonas aeruginosa*, *Citrobacter freundii*, *Stenotrophomonas* sp., *Flavobacterium* sp., *Proteus vulgaris*, *Pseudomonas* sp., *Acinetobacter* sp., *Klebsiella* sp. and *Proteus* sp.) applied in soil	*Vibrio fischeri*	Toxicity increased in every treatment applied to soil	Pino and Peñuela 2011

References

Abraham, J. and S. Silambarasan. 2013. Biodegradation of chlorpyrifos and its hydrolyzing metabolite 3,5,6-trichloro-2-pyridinol by *Sphingobacterium* sp. JAS3. Process Biochem 48: 1559–1564.

Abraham, J., S. Silambarasan and P. Logeswari. 2014. Simultaneous degradation of organophosphorus and organochlorine pesticides by bacterial consortium. J Taiwan Inst Chem E 45: 2590–2596.

Aislabie, J., D.J. Saul and J.M. Foght. 2006. Bioremediation of hydrocarbon-contaminated polar soils. Extremophiles 10: 171–179.

Akbar, S., S. Sultan and M. Kertesz. 2014. Bacterial community analysis in chlorpyrifos enrichment cultures via DGGE and use of bacterial consortium for CP biodegradation. World J Microbiol Biotechnol 30: 2755–2766.

Ahmad, F., S. Iqbal, S. Anwar, M. Afzal, E. Islam, T. Mustafa and Q.M. Khan. 2012. Enhanced remediation of chlorpyrifos from soil using ryegrass (*Lollium multiflorum*) and chlorpyrifos-degrading bacterium *Bacillus pumilus* C2A1. J Hazard Mater 237-238: 110–115.

Alvey, S. and D.E. Crowley. 1996. Survival and activity of an atrazine-mineralizing bacterial consortium in rhizosphere soil. Environ Sci Technol 30: 1596–1603.

Anhalt, J.C., T.B. Moorman and W.C. Koskinen. 2007. Biodegradation of imidacloprid by an isolated soil microorganism. J Environ Sci Heal B 42: 509–514.

Anwar, S., F. Liaquat, Q.M. Khan, Z.M. Khalid and S. Iqbal. 2009. Biodegradation of chlorpyrifos and its hydrolysis product 3, 5, 6-trichloro-2-pyridinol by *Bacillus pumilus* strain C2A1. J Hazard Mater 168: 400–405.

Assaf, N.A. and R.F. Turco. 1994. Accelerated biodegradation of atrazine by a microbial consortium is possible in culture and soil. Biodegradation 5: 29–35.

Barragán-Huerta, B.E., C. Costa-Pérez, J. Peralta-Cruz, J. Barrera-Cortés, F. Esparza-García and R. Rodríguez-Vázquez. 2007. Biodegradation of organochlorine pesticides by bacteria grown in microniches of the porous structure of green bean coffee. Int Biodeter Biodegr 59: 239–244.

Bazot, S. and T. Lebeau. 2009. Effect of immobilization of a bacterial consortium on diuron dissipation and community dynamics. Bioresour Technol 100: 4257–4261.

Breitenstein, A., A. Saano, M. Salkinoja-Salonen, J.R. Andreesen and U. Lechner. 2001. Analysis of a 2, 4, 6-trichlorophenol-dehalogenating enrichment culture and isolation of the dehalogenating member *Desulfitobacterium frappieri* strain TCP-A. Arch Microbiol 175: 133–142.

Cassidy, M.B., H. Lee and J.T. Trevors. 1996. Environmental applications of immobilized microbial cells: a review. J Ind Microbiol Biotechnol 16: 79–101.

Castro-Gutiérrez, V., M. Masís-Mora, G. Caminal, T. Vicent, E. Carazo-Rojas, M. Mora-López and C.E. Rodríguez-Rodríguez. 2016. A microbial consortium from a biomixture swiftly degrades high concentrations of carbofuran in fluidized-bed reactors. Process Biochem 51: 1585–1593.

Castro-Gutiérrez, V., M. Masís-Mora, E. Carazo-Rojas, M. Mora-López and C.E. Rodríguez-Rodríguez. 2018. Impact of oxytetracycline and bacterial bioaugmentation on the efficiency and microbial community structure of a pesticide-degrading biomixture. Environ Sci Pollut Res Int 25: 11787–11799.

Castro-Gutiérrez, V., M. Masís-Mora, E. Carazo-Rojas, M. Mora-López and C.E. Rodríguez-Rodríguez. 2019. Fungal and bacterial co-bioaugmentation of a pesticide-degrading biomixture: pesticide removal and community structure variations during different treatments. Water Air Soil Pollut 230: 247.

Castro-Gutiérrez, V., E. Fuller, J.C. Thomas, C.J. Sinclair, S. Johnson, T. Helgason and J.W.B. Moir. 2020. Genomic basis for pesticide degradation revealed by selection, isolation and characterisation of a library of metaldehyde-degrading strains from soil. Soil Biol Biochem 107702.

Chanika, E., D. Georgiadou, E. Soueref, P. Karas, E. Karanasios, N.G. Tsiropoulos, E.A. Tzortzakakis and D.G. Karpouzas. 2011. Isolation of soil bacteria able to hydrolyze both organophosphate and carbamate pesticides. Bioresour Technol 102: 3184–3192.

Chapalamadugu, S. and G.R. Chaudhry. 1991. Hydrolysis of carbaryl by a *Pseudomonas* sp. and construction of a microbial consortium that completely metabolizes carbaryl. Applied and Environ Microbiol 57: 744–750.

Chen, S., C. Chang, Y. Deng, S. An, Y.H. Dong, J. Zhou, M. Hu, G. Zhong and L.H. Zhang. 2014. Fenpropathrin biodegradation pathway in *Bacillus* sp. DG-02 and its potential for bioremediation of pyrethroid-contaminated soils. J Agr Food Chem 62: 2147–2157.

Cycoń, M., M. Wójcik and Z. Piotrowska-Seget. 2009. Biodegradation of the organophosphorus insecticide diazinon by *Serratia* sp. and *Pseudomonas* sp. and their use in bioremediation of contaminated soil. Chemosphere 76: 494–501.

Cycoń, M., A. Żmijowsk, M. Wójcik and Z. Piotrowska-Seget. 2013. Biodegradation and bioremediation potential of diazinon-degrading *Serratia marcescens* to remove other organophosphorus pesticides from soils. J Environ Manage 117: 7–16.

Dai, Y., N. Li, Q. Zhao and S. Xie. 2015. Bioremediation using *Novosphingobium* strain DY4 for 2, 4-dichlorophenoxyacetic acid-contaminated soil and impact on microbial community structure. Biodegradation 26: 161–170.

Dejonghe, W., E. Berteloot, J. Goris, N. Boon, K. Crul, S. Maertens, M. Höfte, P. De Vos, W. Verstraete and E.M. Top. 2003. Synergistic degradation of linuron by a bacterial consortium and isolation of a single linuron-degrading *Variovorax* strain. Appl Environ Microbiol 69: 1532–1541.

Fuentes, M.S., G.E. Briceño, J.M. Saez, C.S. Benimeli, M.C. Diez and M.J. Amoroso. 2013. Enhanced removal of a pesticides mixture by single cultures and consortia of free and immobilized *Streptomyces* strains. Biomed Res Int 2013: 9.

Fuentes, M.S., A. Alvarez, J.M. Saez, C.S. Benimeli and M.J. Amoroso. 2014. Use of actinobacteria consortia to improve methoxychlor bioremediation in different contaminated matrices. pp. 267–277. *In*: Carmichael, W.W. (ed.). Bioremediation in Latin America. Springer, Cham.

Fuentes, M.S., E.E. Raimondo, M.J. Amoroso and C.S. Benimeli. 2017. Removal of a mixture of pesticides by a *Streptomyces* consortium: Influence of different soil systems. Chemosphere 173: 359–367.

Guiot, S.R., B. Tartakovsky, M. Lanthier, M.J. Lévesque, M.F. Manuel, R. Beaudet, C.W. Greer and R. Villemur. 2002. Strategies for augmenting the pentachlorophenol degradation potential of UASB anaerobic granules. Water Sci Technol 45: 35–41.

Grundmann, S., R. Fuß, M. Schmid, M. Laschinger, B. Ruth, R. Schulin, J.C. Munch and R. Schroll. 2007. Application of microbial hot spots enhances pesticide degradation in soils. Chemosphere 68: 511–517.

Huete-Soto, A., M. Masís-Mora, V. Lizano-Fallas, J.S. Chin-Pampillo, E. Carazo-Rojas and C.E. Rodríguez-Rodríguez. 2017. Simultaneous removal of structurally different pesticides in a biomixture: detoxification and effect of oxytetracycline. Chemosphere 169: 558–567.

Jabeen, H., S. Iqbal, S. Anwar and R.E. Parales. 2015. Optimization of profenofos degradation by a novel bacterial consortium PBAC using response surface methodology. Int Biodeter Biodegr 100: 89–97.

Jariyal, M., V.K. Gupta, K. Mandal, V. Jindal, G. Banta and B. Singh. 2014. Isolation and characterization of novel phorate-degrading bacterial species from agricultural soil. Environ Sci Pollut Res Int 21: 2214–2222.

Jariyal, M., V. Jindal, K. Mandal, V.K. Gupta and B. Singh. 2018. Bioremediation of organophosphorus pesticide phorate in soil by microbial consortia. Ecotoxicol Environ Saf 159: 310–316.

Kandil, M.M., C. Trigo, W.C. Koskinen and M.J. Sadowsky. 2015. Isolation and characterization of a novel imidacloprid-degrading *Mycobacterium* sp. strain MK6 from an Egyptian soil. J Agric Food Chem 63: 4721–4727.

Karanasios, E., N.G. Tsiropoulos and D.G. Karpouzas. 2012. On-farm biopurification systems for the depuration of pesticide wastewaters: recent biotechnological advances and future perspectives. Biodegradation 23: 787–802.

Karpouzas, D.G., J.A.W. Morgan and A. Walker. 2000. Isolation and characterization of 23 carbofuran-degrading bacteria from soils from distant geographical areas. Lett Appl Microbiol 31: 353–358.

Karpouzas, D.G., A. Fotopoulou, U. Menkissoglu-Spiroudi and B.K. Singh. 2005. Non-specific biodegradation of the organophosphorus pesticides, cadusafos and ethoprophos, by two bacterial isolates. FEMS Microbiol Ecol 53: 369–378.

Kong, L., S. Zhu, L. Zhu, H. Xie, K. Su, T. Yan, J. Wang, J. Wang, F. Wang and F. Sun. 2013. Biodegradation of organochlorine pesticide endosulfan by bacterial strain *Alcaligenes faecalis* JBW4. Int J Environ Sci 25: 2257–2264.

Kontchou, C.Y. and N. Gschwind. 1999. Biodegradation of s-Triazine compounds by a stable mixed bacterial community. Ecotox Environ Safe 43: 47–56.

Krishna, K.R. and L. Philip. 2008. Biodegradation of mixed pesticides by mixed pesticide enriched cultures. J Environ Sci Health B 44: 18–30.

Liu, Z.Y., X. Chen, Y. Shi and Z.C. Su. 2011. Bacterial degradation of chlorpyrifos by *Bacillus cereus*. Open J Adv Mater Res 356-360: 676–680.

Mandal, K., B. Singh, M. Jariyal and V.K. Gupta. 2013. Microbial degradation of fipronil by *Bacillus thuringiensis*. Ecotoxicol Environ Saf 93: 87–92.

Mandelbaum, R.T., D.L. Allan and L.P. Wackett. 1995. Isolation and characterization of a *Pseudomonas* sp. that mineralizes the s-triazine herbicide atrazine. Appl Environ Microbiol 61: 1451–1457.

Madigan, M.T. and J. Martinko. 2005. Brock Biology of Microorganisms. Prentice Hall, New Jersey.

Margni, M., D. Rossier, P. Crettaz and O. Jolliet. 2002. Life cycle impact assessment of pesticides on human health and ecosystems. Agric Ecosyst Environ 93: 379–392.

Matsumoto, E., Y. Kawanaka, S.J. Yun and H. Oyaizu. 2008. Isolation of dieldrin- and endrin-degrading bacteria using 1,2-epoxycyclohexane as a structural analog of both compounds. Appl Microbiol Biot 80: 1095–1103.

Mohanta, M.K., A.K. Saha, M.T. Zamman, A.E. Ekram, A.S. Khan, S.B. Mannan and M. Fakruddin. 2012. Isolation and characterization of carbofuran degrading bacteria from cultivated soil. Biochem Cell Arch 12: 313–320.

Moreno-Medina, D.A., E. Sánchez-Salinas and L. Ortiz-Hernández. 2014. Removal of methyl parathion and coumaphos pesticides by a bacterial consortium immobilized in *Luffa cylindrica*. Rev Int Contam Ambient 30: 51–63.

Mulligan, C.N. and R.N. Yong. 2004. Natural attenuation of contaminated soils. Environ Int 30: 587–601.

Neilson, A.H. and A.S. Allard. 2012. Organic Chemicals in the Environment: Mechanisms of Degradation and Transformation. CRC press.

Ortiz-Hernández, L. and E. Sánchez-Salinas. 2010. Biodegradation of the organophosphate pesticide tetrachlorvinphos by bacteria isolated from agricultural soil in Mexico. Rev Int Contam Ambient 26: 27–38.

Pandey, G., S.J. Dorrian, R.J. Russell and J.G. Oakeshott. 2009. Biotransformation of the neonicotinoid insecticides imidacloprid and thiamethoxam by *Pseudomonas* sp. 1G. Biochem Biophys Res Commun 380: 710–714.

Pattanasupong, A., H. Nagase, E. Sugimoto, Y. Hori, K. Hirata, K. Tani, M. Nasu and K. Miyamoto. 2004. Degradation of carbendazim and 2,4-dichlorophenoxyacetic acid by immobilized consortium on loofa sponge. J Biosci Bioeng 98: 28–33.

Peng, X., J. Huang, C. Liu, Z. Xiang, J. Zhou and G. Zhong. 2012. Biodegradation of bensulphuron-methyl by a novel *Penicillium pinophilum* strain, BP-H-02. J Hazard Mater 213: 216–221.

Phugare, S.S. and J.P. Jadhav. 2015. Biodegradation of acetamiprid by isolated bacterial strain *Rhodococcus* sp. BCH2 and toxicological analysis of its metabolites in silkworm (*Bombax mori*). Clean 43: 296–304.

Pileggi, M., S.A.V. Pileggi, L.R. Olchanheski, P.A.G. da Silva, A.M. Munoz-Gonzalez, W.C. Koskinen, B. Barber and M.J. Sadowsky. 2012. Isolation of mesotrione-degrading bacteria from aquatic environments in Brazil. Chemosphere 86: 1127–1132.

Pino, N. and G. Peñuela. 2011. Simultaneous degradation of the pesticides methyl parathion and chlorpyrifos by an isolated bacterial consortium from a contaminated site. Int Biodeter Biodegr 65: 827–831.

Robles-González, I.V., F. Fava and H.M. Poggi-Varaldo. 2008. A review on slurry bioreactors for bioremediation of soils and sediments. Microb Cell Fact 7: 5.

Rodríguez-Castillo, G., M. Molina-Rodríguez, J.C. Cambronero-Heinrichs, J.P. Quirós-Fournier, V. Lizano-Fallas, C. Jiménez-Rojas, M. Masís-Mora, V. Castro-Gutierrez and C.E. Rodriguez-Rodriguez. 2019. Simultaneous removal of neonicotinoid insecticides by a microbial degrading consortium: Detoxification at reactor scale. Chemosphere 235: 1097–1106.

Rodríguez-Rodríguez, C.E., V. Castro-Gutiérrez, J.S Chin-Pampillo and K. Ruiz-Hidalgo. 2013. On-farm biopurification systems: role of white rot fungi in depuration of pesticide-containing wastewaters. FEMS Microbiol Lett 345: 1–12.

Rodríguez-Rodríguez, C.E., V. Castro-Gutiérrez and V. Lizano-Fallas. 2018. Alternative approaches to determine the efficiency of biomixtures used for pesticide degradation in biopurification systems. pp. 57–73. *In*: Bidoia, E. and R. Montagnolli (eds.). Springer Protocols. Toxicity and Biodegradation Testing. Humana Press, New York, NY.

Saez, J.M., A. Álvarez, C.S. Benimeli and M.J. Amoroso. 2014. Enhanced lindane removal from soil slurry by immobilized *Streptomyces* consortium. Int Biodeter Biodegr 93: 63–69.

Sasikala, C., S. Jiwal, P. Rout and M. Ramya. 2012. Biodegradation of chlorpyrifos by bacterial consortium isolated from agriculture soil. World J Microbiol Biotechnol 28: 1301–1308.

Silambarasan, S. and J. Abraham. 2013. Ecofriendly method for bioremediation of chlorpyrifos from agricultural soil by novel fungus *Aspergillus terreus* JAS1. Water Air Soil Poll 224: 1369.

Silva, E., A.M. Fialho, I. Sá-Correia, R.G. Burns and L.J. Shaw. 2004. Combined bioaugmentation and biostimulation to cleanup soil contaminated with high concentrations of atrazine. Environ Sci Technol 38: 632–637.

Singh, A.K. and S.S. Cameotra. 2014. Influence of microbial and synthetic surfactant on the biodegradation of atrazine. Environ Sci Pollut Res Int 2088–2097.

Singh, P., H.S. Saini and M. Raj. 2016. Rhamnolipid mediated enhanced degradation of chlorpyrifos by bacterial consortium in soil-water system. Ecotoxicol Environ Saf 134: 156–162.

Smith, D., S. Alvey and D.E. Crowley. 2005. Cooperative catabolic pathways within an atrazine-degrading enrichment culture isolated from soil. FEMS Microbiol Ecol 53: 265–273.

Sørensen, S.R., Z. Ronen and J. Aamand. 2001. Isolation from agricultural soil and characterization of a *Sphingomonas* sp. able to mineralize the phenylurea herbicide isoproturon. Appl Environ Microbiol 67: 5403–5409.

Sørensen, S.R., C.N. Albers and J. Aamand. 2008. Rapid mineralization of the phenylurea herbicide diuron by *Variovorax* sp. strain SRS16 in pure culture and within a two-member consortium. Appl Environ Microbiol 74: 2332–2340.

Tartakovsky, B., M.F. Manuel, D. Beaumier, C.W. Greer and S.R. Guiot. 2001. Enhanced selection of an anaerobic pentachlorophenol-degrading consortium. Biotechnol Bioeng 73: 476–483.

Taş, N., M.H. Van Eekert, W.M. De Vos and H. Smidt. 2010. The little bacteria that can–diversity, genomics and ecophysiology of *'Dehalococcoides'* spp. in contaminated environments. Microb Biotechnol 3: 389–402.

Tixier, C., M. Sancelme, S. Aït-Aïssa, P. Widehem, F. Bonnemoy, A. Cuer, N. Truffaut and H. Veschambre. 2002. Biotransformation of phenylurea herbicides by a soil bacterial strain, *Arthrobacter* sp. N2: structure, ecotoxicity and fate of diuron metabolite with soil fungi. Chemosphere 46: 519–526.

Van der Werf, H.M. 1996. Assessing the impact of pesticides on the environment. Agric Ecosyst Environ 60: 81–96.

Villaverde, J., R. Posada-Baquero, M. Rubio-Bellido, L. Laiz, C. Saiz-Jimenez, M.A. Sanchez-Trujillo and E. Morillo. 2012. Enhanced mineralization of diuron using a cyclodextrin-based bioremediation technology. J Agr Food Chem 60: 9941–9947.

Von Wedel, R.J., J.F. Mosquera, C.D. Goldsmith, G.R. Hater, A. Wong, T.A. Fox, W.T. Hunt, M.S. Paules, J.M. Quiros and J.W. Wiegand. 1988. Bacterial biodegradation of petroleum hydrocarbons in groundwater: in situ augmented bioreclamation with enrichment isolates in California. Water Sci Technol 20: 501–503.

Ware, G.W. 2000. The Pesticide Book. Thomson Publications.

Widehem, P., S. Aït-Aïssa, C. Tixier, M. Sancelme, H. Veschambre and N. Truffaut. 2002. Isolation, characterization and diuron transformation capacities of a bacterial strain *Arthrobacter* sp. N2. Chemosphere 46: 527–534.

Yang, S., F.I. Hai, L.D. Nghiem, W.E. Price, F. Roddick, M.T. Moreira and S.F. Magram. 2013. Understanding the factors controlling the removal of trace organic contaminants by white-rot fungi and their lignin modifying enzymes: a critical review. Bioresour Technol 141: 97–108.

Yañez-Ocampo, G., E. Sánchez-Salinas, G.A. Jiménez-Tobon, M. Penninckx and M.L. Ortiz-Hernández. 2009. Removal of two organophosphate pesticides by a bacterial consortium immobilized in alginate or tezontle. J Hazard Mater 168: 1554–1561.

Yáñez-Ocampo, G., E. Sánchez-Salinas and M.L. Ortiz-Hernández. 2011. Removal of methyl parathion and tetrachlorvinphos by a bacterial consortium immobilized on tezontle-packed up-flow reactor. Biodegradation 22: 1203–1213.

Zhang, H., W. Mu, Z. Hou, X. Wu, W. Zhao, X. Zhang, H. Pan and S. Zhang. 2012. Biodegradation of nicosulfuron by the bacterium *Serratia marcescens* N80. J Environ Sci Heal B 47: 153–160.

3

Using Molecular Methods to Identify and Monitor Xenobiotic-Degrading Genes for Bioremediation

Edward Fuller,[1] *Víctor Castro-Gutiérrez,*[1] *Juan Carlos Cambronero-Heinrichs*[2] and *Carlos E. Rodríguez-Rodríguez*[3,*]

Introduction

Increases in farming and manufacturing has been linked to a rise in xenobiotic compounds found in both the soil and water (Haiser and Turnbaugh 2013). As many of these compounds can be harmful, they pose a great cause for concern. Currently, bioremediation provides the most cost effective and least environmentally damaging approach in the removal of these recalcitrant compounds. Using bioremediation methods that include bioaugmentation, biostimulation, and bioreactors, a wide range of xenobiotic compounds have been effectively removed from the environment (Boon et al. 2003, Robles-González et al. 2008, Sakultantimetha et al. 2011). As microbial communities can possess a diverse set of degradative mechanisms and pathways, they provide an invaluable tool in the removal of contaminants (Haiser and Turnbaugh 2013). Through the identification of the corresponding catabolic genes and the monitoring of their presence and abundance throughout the environment, the

[1] Department of Biology, University of York, Heslington Road, YO10 5DD, York, United Kingdom.
[2] Faculty of Microbiology, Universidad de Costa Rica, San Pedro, Montes de Oca, 2060 San José, Costa Rica.
[3] Research Center of Environmental Contamination (CICA), Universidad de Costa Rica, San Pedro, Montes de Oca, 2060 San José, Costa Rica.
Emails: edf504@york.ac.uk; victormanuel.castro@ucr.ac.cr; juan.cambroneroheinrichs@gmail.com
* Corresponding author: carlos.rodriguezrodriguez@ucr.ac.cr

degradative ability of the community can be assessed. Recent technological advances and improvements to molecular methods allow for numerous novel approaches to be undertaken in the characterization and monitoring of these catabolic communities. Through developing our understanding and knowledge of these processes, the degradative rates and abilities of these microbial communities can potentially be enhanced (Haiser and Turnbaugh 2013).

Methods for identification of xenobiotic-degrading genes

The identification of novel xenobiotic catabolic genes provides not only insight into the degradative mechanisms, but also provides both a larger arsenal of degradative tools and allows for a greater ability to detect and monitor degradative genes throughout the environment (Gedalanga et al. 2014, Huang et al. 2018, Thakur et al. 2019). Traditional gene identification methods, such as genomic libraries, mutagenesis analysis, and induction assays rely on the ability to isolate, culture, and characterize the xenobiotic-degrading microorganism (Ruan et al. 2013, Sadauskas et al. 2017). With reduced cost in nucleotide sequencing, increased accessibility of bioinformatic tools, and improved molecular methods, additional strategies are available. With the majority of environmental microorganisms considered to be unculturable, a significant percent of catabolic genes may potentially have been overlooked till now (Wade 2002). By optimizing the identification methods and combining traditional approaches with modern techniques, a greater number of catabolic genes can be identified. Various methods and techniques currently used in the identification of novel xenobiotic-degrading genes will now be described.

Clone libraries

Genomic libraries are a traditional technique used in molecular biology, and when coupled with the ability to isolate and culture xenobiotic-degrading microorganisms, they can provide a powerful tool in the identification of catabolic genes. Following the extraction and then either shearing or digesting of the genomic DNA of the xenobiotic-degrading microorganism, the DNA fragments are cloned into a suitable vector and then transformed into an appropriate host organism (Borden and Papoutsakis 2007). Following this transformation, either a selection step, such as growing the colonies on the xenobiotic compound, or a screen approach, ranging from a colorimetric assay to observing zones of clearing surrounding the clones, is used to determine the presence of the catabolic gene(s) (Ruan et al. 2013, Sadauskas et al. 2017). Once the catabolic activity has been verified, the degradative colonies can be subjected to nucleotide sequencing and the relevant gene(s) identified. This approach has the advantage of requiring little or no information regarding the metabolic pathway or the degradative microorganism.

This strategy has shown to be very effective in the identification of numerous novel catabolic genes (Goyal and Zylstra 1996, Zhang et al. 2012, Ruan et al. 2013). One example is the identification of an esterase capable of hydrolysing a pyrethroid pesticide—by coupling the genomic library with a fast blue RR colorometric screen, the degradative clone was successfully identified and verified (Ruan et al. 2013).

In situations where the xenobiotic-degrading microorganism cannot be isolated or cultured, a metagenomic library approach can be undertaken. This method involves extracting the DNA directly from the environmental samples capable of xenobiotic degradation and then cloning these fragments into an appropriate vector (Meier et al. 2016, Popovic et al. 2017). This approach has been used to identify novel catabolic genes for numerous pollutants, such as pyrethroids, carbamates, and catechols, amongst many others (Li et al. 2008, Gong et al. 2012, Silva et al. 2013, Terrón-González et al. 2016, Soares Bragança et al. 2017, Ufarté et al. 2017). As less than 1% of bacteria are believed to be culturable in a laboratory environment, this approach is invaluable in allowing the study of degradative mechanisms that may be overlooked when using culture-dependent methods (Vartoukian et al. 2010).

Despite the fact that numerous novel degradative genes have been identified using the library approaches, several limitations do exist. As the utilisation of complex molecules can require multiple novel degradative steps, several novel genes may be required simultaneously to convey this ability to another organism (Nordin et al. 2005, Cámara et al. 2007). Therefore, clones containing the degradative genes would not be able to grow on the xenobiotic compound and go undetected. A further limitation is based on the screening approach to gene identification. The screening reagent is typically a sophisticated substrate with chromogenic properties, and as such may require information regarding either the metabolic pathway or the degradative mechanism in order to select the correct detection method (Rabausch et al. 2013).

Mutagenesis

Unlike the genomic library methods, which seek to insert the catabolic gene(s) into a non-native host, the mutagenesis approach aims to disrupt these genes in the degradative microorganism. As spontaneous mutations are a rare occurrence, active approaches to cause gene disruptions are undertaken (Vartoukian et al. 2010). For the identification of novel genes, these approaches can be simply divided into insertional mutagenesis and random mutagenesis.

Transposons are mobile elements capable of random insertion within a host's genome. These mobile elements can therefore be used as powerful molecular tools to disrupt genes throughout the genome, whilst simultaneously inserting selectable markers into the host genome to identify mutated genes. The advantage of this insertional mutagenesis approach, relative to random mutagenesis, is a reduced lethality rate and a potentially higher mutation rate (Seifert et al. 1986). Transposon mutagenesis has been used successfully to identify numerous catabolic genes capable of degrading dicholprop, long chain alkanes, and rubber, amongst other compounds (Schleinitz et al. 2004, Throne-Holst et al. 2007, Kasai et al. 2017, Qiu et al. 2018). As transposons are typically transformed into the host cell through a suicide vector, the main limitation regarding this approach is that it is largely restricted to naturally competent microorganisms (Rabausch et al. 2013).

Random mutagenesis is the second mutagenic approach used to identify novel degradative genes. The aim of this strategy is to create either point mutations and/or deletions throughout the genome. This can be achieved typically through either radiation, such as ultraviolet (UV), or through the use of mutagenic compounds, such as ethyl methanesulfonate (EMS), methyl methanesulfonate (MMS), and acridine, amongst others (Todd et al. 1979, Ferguson and Denny 1991, Geissdorfer et al. 1995, Pan et al. 2012, Shibai et al. 2017). Chemical mutagens, which are typically favored in random mutagenesis, exist in several classes with numerous modes of actions, ranging from nucleotide base analogs, whereby they resemble purine and pyridines, to chemical modifiers, where they are capable of modifying a single pair to create faulty base pairing (Goncharova and Kuzhir 1989, Holroyd and Van Mourik 2014).

Using random chemical mutagenesis with the base modifying compound EMS, researchers obtained mutants deficient in the ability to degrade the molluscicide metaldehyde, and by coupling this with whole genome sequencing of the strains and comparative genomics, the initial gene in metaldehyde degradation was then identified (Castro-Gutiérrez et al. 2020). With the reduced costs to whole genome sequencing and the increased accessibility of bioinformatic packages, random mutagenesis is a gene identification method that can increasingly be applied to a wide variety of microorganisms (Ruan et al. 2013).

The main limitation of both mutagenesis methods are the non-targeted disruptions, whereby downstream metabolic genes can be mutated and therefore, may give misleading results in the selection procedure (Ruan et al. 2013). Although random mutagenesis provides an advantage over insertional mutagenesis by requiring no complex molecular techniques and therefore needing no information regarding the microorganism, it does suffer with the shortcoming of the handling of highly carcinogenic compounds.

Plasmid curing

Catabolic plasmids are typically relatively large and highly mobile genetic elements that confer the ability to degrade and utilize both naturally occurring and xenobiotic compounds as either energy, carbon, or nitrogen sources (Chen et al. 2016). As such, catabolic plasmids are found widely in the environment and are valuable assets in their ability to aid in the formation of new catabolic pathways (Sayler et al. 1990, Schmidt et al. 2011). As such, the ability to remove or 'cure' the degradative microorganism of these plasmids is a valuable tool in the identification of catabolic genes. Loss of the degrading ability upon curing will point out that degrading genes are located on a plasmid.

Plasmid curing can occur spontaneously through segregation and deletion; however, most plasmids are very stable. Therefore, treatments either with elevated temperatures or chemical curing agents, such as mitomycin C, acridine orange, and ethidium bromide are required (Salisbury et al. 1972, Letchumanan et al. 2015). Using curing methods has been shown to be an effective approach in the identification of novel xenobiotic-degrading genes. For example, through repeated sub-culturing of a carbamate-degrading bacterium, in the presence of mitomycin C and the absence of selective pressures, researchers were able to use curing to identify a carbaryl hydrolase gene present on a plasmid (Hashimoto et al. 2002). Additionally, the purification of the catabolic plasmid coupled with its transformation into a non-degrading strain can be useful in both the verification of the curing assay or potentially as an alternative gene identification tool. The conveyance of xenobiotic degradation through transformation of a catabolic plasmid has been shown to be a simple and useful method in catabolic gene identification (Wang et al. 2009).

Although the curing method is simple and effective, there are certain limitations that should be considered. The curing assay relies on the catabolic genes being situated on a plasmid, and as such, catabolic genes located within the genome will not be identified. A further consideration relating to the curing assay is that many catabolic plasmids possess a plasmid addiction system, whereby loss of the plasmid leads to either an arrest in cell growth or cell death (Kroll et al. 2010, Tsang 2017). In such situations, the ability to cure the host from the plasmid may not be an appropriate gene identification tool.

Induction methods

Gene expression regulation prevents the synthesis of unnecessary enzymes, thereby saving energy and nutrients. Where expression is induced by the substrate, differences in either gene or protein expression levels may be used to identify the catabolic genes. By comparing either transcriptome or the proteome with and without substrate addition, inducible genes can be identified, and information regarding the metabolic pathway can potentially be obtained (Pankaj et al. 2016, Wang et al. 2018).

With the reduced cost of next generation sequencing, RNA-Seq provides a very powerful method in the detection and quantification of RNA within a given sample. The method relies on the isolation of RNA, followed by the conversion into the more stable cDNA. Once converted, the sequences can be subjected to nucleotide sequencing (Stark et al. 2019). By quantifying the change in expression levels in different conditions, such as with and without the xenobiotic substrate, the expression profile can determine inducible genes and potential catabolic operons. This approach can be used on single bacterial strains, where the ability to culture them is available (Wang et al. 2018, Levy-Booth et al. 2019). However, a metagenomic approach using RNA-Seq can also be undertaken. By detecting and quantifying gene expression in the environment, inducible genes can be identified without the need to isolate and culture the xenobiotic-degrading microorganism(s). This is one of RNA-Seq's greatest strengths and allows for the identification of inducible novel catabolic genes that may have been overlooked using other methods (Culligan et al. 2014).

When the xenobiotic-degrading microorganism is isolated and culturable, protein expression analysis provides a simple and effective method to identify inducible catabolic genes. Comparison of the proteomic induction profiles using traditional gel separation methods, such as sodium dodecyl sulfate polyacrylamide gel electrophoresis (SDS-PAGE) or two-dimensional electrophoresis (2D-GE), has proved to be an effective gene identification method (Seung et al. 2007, Pankaj et al. 2016). As such, this approach was used in the identification of a novel hydrolase capable of mineralizing the herbicide linuron (Sangwan et al. 2014). Extraction of the protein band/spot identified in the analysis, followed by mass spectrometry can reveal the sequence of the protein and ultimately the inducible gene (Pankaj et al. 2016).

It should be noted, however, that evolutionarily, newer degradative pathways typically demonstrate constitutive expression, and therefore are not induced upon addition of the substrate. Where this is the case, the above induction methods would be unable to identify the catabolic genes (Cases and De Lorenzo 2001).

Bioinformatic analysis

In cases where the catabolic mechanism or a catabolic gene has been identified previously for a xenobiotic compound, a similarity analysis provides a simple and powerful tool. As high sequence similarity infers functional similarity, programs such as BLAST can allow the xenobiotic-degrading microorganism's genome to be appropriately searched and analyzed. By using this approach, numerous novel xenobiotic-degrading genes have been identified requiring only whole genome sequence information and a similarity program (Heiss et al. 2003, Kube et al. 2013, Nanthini et al. 2017). Despite the simplicity and effectiveness of the similarity analysis, this approach requires past information regarding the mechanism of action and/or the catabolic genes, and as such will not be effective in the identification of uncharacterized or completely novel xenobiotic compounds.

Where multiple xenobiotic-degrading strains have been isolated, whole-genome sequencing and comparative genomics can be very powerful in the identification of novel genes. By identifying genes shared among three strains capable of mineralizing isoproturon, Yan et al. (2016) were able to narrow down the number of candidates to 84 gene sequences. From this number, manual curation was possible, and as such were able to identify and verify the isoproturon-mineralizing genes. This method has also been used to identify genes responsible for the degradation of other xenobiotic compounds, such as metaldehyde and chloroacetanilide (Cheng et al. 2019, Castro-Gutiérrez et al. 2020). Since this strategy relies on similar degradative mechanisms and sequences to identify the novel gene(s), it is most powerful with a large dataset of degrading microorganisms. One drawback is that as xenobiotic compounds may be degraded using different degrading pathways in different microorganisms, degrading genes may not necessarily be identified using this approach.

As mentioned above, catabolic plasmids are very important in the analysis of xenobiotic degradation. However, due to the limitations of the curing procedure, alternative approaches for their identification and analysis have been developed. Traditional plasmid analysis involved the individual extraction and sequencing of the plasmid. With the development of new bioinformatic tools, whole-genome sequencing of the xenobiotic-degrading microorganism, coupled with *in silico* plasmid detection is becoming increasingly common (Kudirkiene et al. 2018).

Using conserved replicon sites, variation in GC content and copy number analysis, bioinformatic tools, and databases, such as PlasmidFinder, PLSDB, and HyAsP can allow for the detection of catabolic plasmids (Carattoli et al. 2014, Galata et al. 2019, Müller and Chauve 2019). Even though *in silico* plasmid analysis is not being directly used to identify novel degradative genes, bioinformatic analysis has been used in combination with various other methods to aid in their identification, and it can provide a greater insight into the degradative pathways (Lykidis et al. 2010, Yan et al. 2016). By obtaining the sequences present in the catabolic plasmid(s), the potential xenobiotic sequences can be manually curated, or experimental analysis becomes a practical and realistic possibility.

Detection and quantification of pesticide-degrading genes

The assessment and prediction of the fate of xenobiotics in the environment during bioremediation or natural attenuation can be carried out through different strategies—analyzing the disappearance of the parent compound, detecting its transformation products, or through evidence of transformation potential in a given setting (Fenner et al. 2013). It is this latter approach, in which knowledge regarding degrading genes can be used to detect and quantify their copy numbers or their expression, which provides a measure of degradation activity.

Several approaches for detection and quantification of specific xenobiotic-degrading strains in environmental samples through marker (non-degrading) genes have been undertaken in the past (Widada et al. 2002). However, since pesticide-degrading genes can be acquired by previously non-degrading microbes (DiGiovanni et al. 1996) or lost from degrading organisms (Changey et al. 2011), detecting and quantifying the degrading genes themselves carries the obvious advantage of more accurately reflecting the status of the degrading community.

Different methods have been used for this purpose. Techniques involving DNA hybridization (using labelled DNA probes) were used in the past few decades to detect pesticide-degrading genes in soil microbial populations (Walia et al. 1990, Holben et al. 1992, Parekh et al. 1995). However, due to limitations mainly regarding their sensitivity and reliance on cultivation, they were gradually replaced by methods based on PCR amplification (Widada et al. 2002).

Initially, methods such as most probable number PCR (MPN-PCR) (Picard et al. 1992) were developed to provide a quantitative component to PCR amplifications, and they have been used for assessment of contaminant-degrading genes (Salminen et al. 2008).

Improved quantitative PCR methods providing higher sensitivity and accuracy, such as quantitative competitive PCR (QC-PCR) and replicate limited dilution PCR (RLD-PCR) were also developed. QC-PCR relies on the inclusion of an internal control competitor, to which each reaction is normalized (Piatak et al. 1993). This technique has been used to monitor the size of the atrazine-degrading bacterial community in soil, finding a transitory increase in the amount of *atzC* genes in soils pre-treated with atrazine (Martin-Laurent et al. 2003). However, the design of the competing molecule and the validation of the amplification efficiencies can be a laborious process (Heid et al. 1996). RLD-PCR relies on replicative dilution to extinction (Chandler 1998), and has been used to quantify hydrocarbon-degrading gene copies in soil (Tuomi et al. 2004). However, it can be time and resource consuming.

Quantitative PCR (qPCR)-based studies

Since its introduction and dissemination, most studies dealing with pesticide catabolic gene quantification have relied on real-time PCR, also known as quantitative PCR (qPCR) (Klein 2002). qPCR allows the monitoring of the PCR amplification and quantification of its products in real time through fluorescence measurements. This technique can provide a sensitive, robust, and fast estimation of catabolic gene copy numbers in samples taken from pesticide bioremediation settings. Despite the obvious advantages of the method, the adoption and transfer of inadequate and varied protocols amongst researchers has led to problems when trying to reproduce data (Hayden et al. 2008, Jaworski et al. 2018, Taylor et al. 2019). Therefore, the minimum information for publication of quantitative real-time PCR experiments (MIQE) guidelines have been proposed to increase the uniformity and reliability of published research using this method (Bustin et al. 2009).

Numerous studies have used this technique to quantify copy numbers of pesticide-degrading genes to estimate the biodegradation potential and the response of the microbial community to these contaminants (Table 3.1).

The gene quantification approach is usually complemented with other strategies to maximize the information obtained from the system under study—either disappearance of the parent compound, detection of transformation products, or both. Many studies also carry out microbial community analysis through fingerprinting or amplicon sequencing techniques to observe modifications in the microbial community structure of the soil.

Most of these studies have found a correlation between degrading gene copy number and pesticide degradation/mineralization activity (Bælum et al. 2012, Sagarkar et al. 2013, Rousidou et al. 2017), highlighting the value of this approach for biodegradation monitoring. Pre-exposed soils usually harbored higher degrading gene copies than non-exposed soils (Lal et al. 2015, Castillo et al. 2016). Very low or undetectable numbers of degrading genes have been observed in instances where elimination was negligible (Nousiainen et al. 2014), others in which degradation was mainly through abiotic processes (Ben Salem et al. 2018), or where degradation was presumably occurring through alternative biological pathways for pesticides, such as diuron (Pesce et al. 2013) and metaldehyde. Notably, soils with similar genetic degrading potential can exhibit different pesticide-degrading activities, depending on factors, such as soil type (Martin-Laurent et al. 2004) and pH (Yale et al. 2017). This emphasizes the value of assessing biodegradation potential/activity using several different strategies simultaneously.

Even though qPCR has long been the workhorse for quantification of functional genes in bioremediation, it is not without limitations. Most notably, DNA extracts from environmental samples often contain diverse contaminants and different backgrounds that can variably affect the activity of the Taq polymerase, producing misleading results (Carvalhais et al. 2013). This can sometimes be remediated by sample dilution, which will likely lead to undetectable target levels or by extensive sample purification, which is not always feasible (Rački et al. 2014).

Digital PCR (dPCR)-based studies

Newer technologies with distinctive advantages have emerged. One of such technologies is digital PCR (dPCR), in which the sample is partitioned in droplets or distributed in nanolitre chambers, and then a standard Taq polymerase PCR reaction is used to amplify target DNA. However, different to qPCR, data is acquired at the reaction endpoint (Taylor et al. 2017). This minimizes the influence of sample contaminants on the amplification, and leads to more accurate and reproducible

Table 3.1. Studies using qPCR to quantify pesticide-degrading gene copies in environmental samples.

Pesticide class	Main use	Specific pesticide	Gene(s)	Enzyme(s)	Matrix	Complementary techniques	Main findings	Reference
Carbamates	Insecticides/ nematicides	oxamyl, carbofuran	*cehA*	carbamate hydrolase	Agricultural soil	Parent compound quantification (HPLC/UV)	*ceh*A gene abundance and pH were negatively correlated with oxamyl half-life.	(Rousidou et al. 2017)
			mcd	carbamate hydrolase			Carbofuran stimulated the abundance of *ceh*A and *mcd* genes.	
		oxamyl	*cehA*	carbamate hydrolase	Agricultural soil	Pesticide mineralization	*ceh*A copy numbers increased concomitantly with oxamyl mineralization.	(Gallego et al. 2019)
						16S rRNA gene amplicon sequencing	The diversity of the bacterial community was modified by oxamyl addition.	
						16S rRNA reverse transcription and amplicon sequencing		
Organochlorines	Insecticides/ acaricides	hexachlorocyclohexane	*linA*	Gamma-hexachlorocyclohexane dehydrochlorinase	Dumpsite soil, agricultural soil	Parent compound quantification (GC-ECD)	An indigenous community capable of HCH degradation was detected in the contaminated soils.	(Lal et al. 2015)

Table 3.1 Contd. ...

...Table 3.1 Contd.

Pesticide class	Main use	Specific pesticide	Gene(s)	Enzyme(s)	Matrix	Complementary techniques	Main findings	Reference
			*lin*B	haloalkane dehalogenase		16S rRNA gene T-RFLP		
Organophosphates	Insecticides/ nematicides/ fungicides	chlorpyrifos	*mpd*	methyl parathion hydrolase	Agricultural soil	Parent compound quantification (GC-MS)	Chlorpyrifos was degraded mainly through abiotic processes.	(Ben Salem et al. 2018)
			opd	parathion hydrolase		Pesticide mineralization	Neither *mpd* nor *opd* sequences were detected in the soil.	
		tolclofos-methyl	*opd*	parathion hydrolase	Agricultural soil	Parent compound quantification (HPLC/UV)	*opd* gene copy number was significantly higher at week 4 than at week 0.	(Kwak et al. 2012)
Phenoxyalkanoic acids	Herbicides	2,4-D, MCPA, MCPP	*tfd*A	Alpha-ketoglutarate-dependent dioxygenase	Arable soil, forest soil, other soil	Pesticide mineralization	Class III *tfd*A genes were most involved in mineralization of the pesticides.	(Bælum and Jacobsen 2009)
		2,4-D, MCPA	*tfd*A	Alpha-ketoglutarate-dependent dioxygenase	Agricultural soil, forest soil	*tfd*A gene diversity analysis	Unknown, diverse, *tfd*A-like genes were abundant in soil samples.	(Zaprasis et al. 2010)
		MCPP	*tfd*A	Alpha-ketoglutarate-dependent dioxygenase	Agricultural soil	Parent compound quantification (HPLC-UV)	MCPP half-life increased progressively with soil depth, with a considerable lag phase in subsoil.	(Rodríguez-Cruz et al. 2010)

		MCPA	*tfd*A	Alpha-ketoglutarate-dependent dioxygenase	Urban soil, agricultural soil	Pesticide mineralization	A correlation was found between class III *tfdA* gene copy numbers and the rate of mineralization.	(Nielsen et al. 2011)
		2,4-D, MCPA, MCPP	*tfd*A	Alpha-ketoglutarate-dependent dioxygenase	Arable soil, forest soil, other soil	Pesticide mineralization	*tfd*A gene copy numbers involved in growth linked mineralization kinetics.	(Bælum et al. 2012)
							Degradability of pesticides was 2,4-D > MCPA > MCPP.	
Phenylureas/ Sulphonylureas	Herbicides	chlorimuron-ethyl	*sul*E	sulfonylurea herbicide de-esterification esterase	Soil from abandoned land	Parent compound quantification (HPLC/UV)	An inoculated strain effectively contributed to pesticide removal.	(Yang et al. 2014)
						Quantification of N-cycling functional genes	Copy numbers of *sul*E genes decreased with time after inoculation.	
		diuron	*puh*A	phenylurea hydrolase A	River sediment	Pesticide mineralization	Mineralization rates in soil were linked to *puh*B gene copies.	(Pesce et al. 2013)
			*puh*B	phenylurea hydrolase B	Agricultural soil		*puh*A gene copies were not detected in soil or sediment samples.	

Table 3.1 Contd. ...

...Table 3.1 Contd.

Pesticide class	Main use	Specific pesticide	Gene(s)	Enzyme(s)	Matrix	Complementary techniques	Main findings	Reference
		diuron	*puh*A	phenylurea hydrolase A	Agricultural soil with organic amendments	Parent compound quantification (HPLC/UV)	Olive vermicompost increased enzymatic activities, as well as bacterial and *puh*B gene abundance.	(Castillo et al. 2016)
			*puh*B	phenylurea hydrolase B		A-RISA fingerprinting	Diuron pre-treatment correlated with high abundance of *puh*B gene copies.	
						Enzymatic activity		
Triazines	Herbicides	atrazine	*atzA*	atrazine chlorohydrolase	Agricultural soil	Pesticide mineralization	Organic amendment did not modify atrazine mineralization and genetic potential.	(Martin-Laurent et al. 2004)
			atzB	hydroxydechloroatrazine ethylaminohydrolase		RISA fingerprinting	Structure of the bacterial community was significantly affected.	
			atzC	N-isopropylammelide isopropyl amidohydrolase			Soils with similar genetic potential can exhibit different atrazine-degrading activities, which depended on soil type.	

		atrazine	*atzA*	atrazine chlorohydrolase	Rhizosphere soil	Pesticide mineralization	Addition of a bioaugmentation strain leads to more efficient mineralization of atrazine.	(Thompson et al. 2010)
						Parent compound quantification (GC/MS)		
		atrazine	*trzN*	triazine hydrolase	Agricultural soil	Parent compound quantification (HPLC/UV)	Bioaugmentation could achieve a quick and significant reduction in atrazine concentrations.	(Wang et al. 2013)
			atzB	Hydroxydechloroatrazine ethylaminohydrolase				
			atzC	N-isopropylammelide isopropyl amidohydrolase				
		atrazine	*trzN*	triazine hydrolase	Agricultural soil	Parent compound quantification (GC/FID)	Degrading gene quantification correlated with degradation efficiencies.	(Sagarkar et al. 2013)
		atrazine	*atzA*	atrazine chlorohydrolase	Boreal soil (various depths), agricultural soil	Pesticide mineralization	Low levels of atrazine-degrading genes in boreal soils; no mineralization detected.	(Nousiainen et al. 2014)

Table 3.1 Contd. ...

...Table 3.1 Contd.

Pesticide class	Main use	Specific pesticide	Gene(s)	Enzyme(s)	Matrix	Complementary techniques	Main findings	Reference
			atzB	hydroxydechloroatrazine ethylaminohydrolase				
			trzN	triazine hydrolase				
		atrazine	*trzN*	triazine hydrolase	Agricultural soil	Parent compound quantification (HPLC/UV)	In most soils, a single dose of the pesticide induced accelerated degradation.	(Yale et al. 2017)
						16S rRNA gene amplicon sequencing	Neutralization of a low pH soil restored accelerated degradation.	

results when compared with qPCR, and allows for very low target quantitation (Rački et al. 2014, Taylor et al. 2015). dPCR is yet to be used for pesticide-degrading gene quantification. However, it has been used for the determination of alkane hydrocarbon-degrading genes (*alk*B1) in a pilot scale biopile field experiment at freeze-thaw temperatures (Kim et al. 2018). The authors performed nutrient amendment in one of the biopiles and found that degrading gene copy numbers were higher in the treated biopile versus the untreated one during the seasonal freezing and thawing phases. More studies using this technology for xenobiotic-degrading gene quantification will undoubtedly emerge in the coming years as it becomes increasingly available.

Loop mediated isothermal amplification (LAMP)-based studies

Loop mediated isothermal amplification (LAMP) has been used as an alternative to qPCR as a point-of-care, lower cost, diagnostic tool. LAMP is performed at constant temperature, so it removes the need for thermal cycling equipment. It is faster than PCR due to the high amplicon yield and can amplify target DNA even in complex matrices that can inhibit PCR (Notomi et al. 2000); detection, and can be visual or via fluorescence. Visual-based LAMP was combined with most probable number for the quantification of reductive dehalogenase genes in groundwater samples (Kanitkar et al. 2017) using centrifuged cells instead of DNA extracts, and a strong correlation was found with qPCR results.

Reverse transcription quantitative PCR (RT-qPCR)-based studies

All the techniques discussed so far suffer from an important limitation—they reveal a genetic potential which is only expressed under favorable conditions. A more powerful way to monitor biodegradation involves the measurement of degrading gene expression, routinely carried out by reverse transcription-quantitative polymerase chain reaction (RT-qPCR). Here, RNA instead of DNA is used as a target for amplification. Since RNA cannot be used as a template for PCR directly, reverse transcription of the RNA template into cDNA is performed, and then followed by the exponential amplification, as done in qPCR (Bustin 2000). Studies using RT-qPCR to quantify expression of pesticide-degrading genes in environmental samples are shown in Table 3.2.

Several studies have found good correlations between degrading gene expression and pesticide mineralization activity (Nicolaisen et al. 2008, Monard et al. 2013). Moreover, while studying MCPA mineralization in agricultural soil samples, researchers found that *tfd*A mRNA levels had a much better correlation with pesticide mineralization than *tfd*A gene copy numbers per gram of soil (Bælum et al. 2008). Even so, care should be taken in cases where the degrading gene is constitutively expressed, because the quantification of its expression would merely reflect the amount of active degraders, rather than active biodegradation (Monard et al. 2013). In these instances, if expression is normalized to gene copy numbers, this ratio has been shown to remain stable (Albers et al. 2015).

Table 3.2. Studies using RT-qPCR to quantify pesticide-degrading gene expression in environmental samples.

Pesticide Class	Main use	Specific Pesticide	Gene(s)	Enzyme(s)	Matrix	Complementary techniques	Main findings	Reference
Benzonitriles	Herbicides	dichlobenil	*bbd*A	Amidase	Sand filter material, groundwater	BAM quantification (HPLC-UV)	Bioaugmentation produced significant BAM degradation for 2–3 weeks.	(Albers et al. 2015)
						Total bacteria, *bbd*A gene copy and MSH1 strain quantification (qPCR)	Ratio of *bbd*A mRNA per *bbd*A gene (amidase expression) was stable over time.	
						Protozoan quantification (MPN)	Protozoa were observed to grow in the sand filters.	
Phenoxyalkanoic acids	Herbicides	2,4-D, MCPA	*tfd*A	Alpha-ketoglutarate-dependent dioxygenase	Agricultural soil	Pesticide mineralization	Relatively high degree of correlation was found between functional gene expression and mineralization rates, not so with gene copy numbers.	(Bælum et al. 2008)
						*tfd*A gene copy quantification (qPCR)		
						*tfd*A gene DGGE		
		MCPA	*tfd*A	Alpha-ketoglutarate-dependent dioxygenase	Agricultural soil	Pesticide mineralization	Transient maximums of gene expression were observed only during active mineralization.	(Nicolaisen et al. 2008)
						*tfd*A gene copy quantification (qPCR)	Microbial degrader activity correlated with *tfd*A mRNA presence.	

Triazines	Herbicides	atrazine	*atzA*	atrazine chlorohydrolase	Agricultural soil	Pesticide mineralization	No *atz*A expression was detected in the tested soils.	(Monard et al. 2010)
			*atz*D	cyanuric acid amidohydrolase		16S rRNA gene copy quantification (qPCR)	Expression of *atz*D was greater in the soil with the highest atrazine mineralization activity.	
		atrazine	*atzD*	cyanuric acid amidohydrolase	Agricultural soil	Pesticide mineralization	Relative gene expression was positively correlated with the maximum rate of pesticide mineralization.	(Monard et al. 2013)
						16S rRNA gene copy quantification (qPCR)		

Metagenomic and metatranscriptomic-based studies

Even though the amplification-based approaches discussed so far greatly increase the sensitivity to allow the quantification of degrading genes in environmental samples, they may introduce specific biases, and are subject to limitations. For instance, protein-coding genes in the environment usually have high variability in their nucleotide sequences (when compared with small subunit rRNA genes) (Gaby and Buckley 2017), and therefore, primers that target functional genes are sometimes designed to degenerate in sequence to be able to encompass that variability (Jin and Mattes 2011, Morales et al. 2020). It has been shown that degenerate qPCR primers are subject to PCR bias (i.e., certain templates are favored in their amplification, and therefore their quantification will not reflect their true original template ratios), which can lead to dramatic mis-estimation of real gene copy numbers—even more than 10,000-fold in some instances (Gaby and Buckley 2017). Furthermore, amplicon-targeted methods are usually low throughout technologies, in which a small number of targets can be detected per run and *a priori* knowledge of the degrading genes is essential for primer design (Forbes et al. 2017). Newer methodologies have found a way to circumvent these issues.

The fact that DNA sequencing technologies are becoming increasingly affordable and accessible has led to a growing number of studies using metagenomic strategies to evaluate pesticide contaminated environments (Sangwan et al. 2014, Regar et al. 2019). Metagenomics involves the direct sequencing of extracted DNA from a sample (Gorski et al. 2019), and can be used to assess and quantify functional genes in environmental matrices without targeted amplification and without the use of sequence-specific primers. Essentially, it provides a detailed survey of all the genes that exist within a particular community. Nonetheless, some of its limitations include that it heavily samples the dominant microbes in the community, and members with low abundance are sparsely covered (Techtmann and Hazen 2016), or not at all, depending on sequencing power. Furthermore, data analysis can be very complicated, and may involve genes with no homologs in bioinformatic databases (Delmont et al. 2012).

Metagenomic analysis has been used to evaluate the genetic pool present in pesticide contaminated environments and quantify its functional potential. In a study focusing on hexachlorocyclohexane, researchers characterized the metagenome of the soil microbial community of three sites with different contamination levels (Sangwan et al. 2012). Among different analyses, they selected twelve genes known to be involved in the degradation pathways for this pesticide, and quantified the matching reads. The two sites with the highest contamination levels had a higher metabolic potential to degrade hexachlorocyclohexane and higher relative abundance of degrading genes (*lin* genes) when compared with the least polluted one. Another study analyzing activated sludge samples from pesticide wastewater treatment plants (Fang et al. 2018) detected 68 subtypes of pesticide-degrading genes, which together showed a relative abundance from 2.08 to 7.14%, which varied depending on the seasons and pesticide wastewater properties. They also found that activated sludge from pesticide wastewater treatment plants had a higher abundance of pesticide-degrading genes compared to others in which saline or freshwater was treated.

Researchers have also evaluated the metagenome of agricultural soils with a history of chlorpyrifos exposure, nevertheless at different times and intensity (Jeffries et al. 2018), while also undertaking chlorpyrifos degradation assays in the laboratory. Relative abundance of metabolic pathways resulted in clustering of samples in two distinct groups, which coincided with effective and ineffective control of pests in the field, and also with slow and rapid experimental degradation of the pesticide in the samples. Soils with faster degradation harbored a distinct community with increased nutrient cycling and transport pathways, as well as enzymes putatively involved in the metabolism of phosphorus.

Another methodology that has been gaining increasing momentum is metatranscriptomic analysis. It has the advantage over metagenomics in that it indicates the specific microbial genes being expressed at the moment of sampling. It involves RNA extraction from an environmental sample, conversion into cDNA (akin to RT-qPCR), and sequencing in a similar manner to metagenomics (Techtmann and Hazen 2016). Metatranscriptomic analysis has been used to assess gene transcripts linked with the degradation of aromatic compounds and pesticides from wheat rhizosphere sample sequences obtained from the EBI metagenomics database (Singh et al. 2018). The researchers found abundant transcripts associated with the degradation of xenobiotics (including pesticides), aromatic amines, carbazoles, benzoates, naphthalene, ketoadipate pathway, phenols, and biphenyls.

Conclusions

Newer, more informative, and accurate methods are being applied by researchers to understand the role of microorganisms in polluted environmental systems. Lower costs in sequencing and advances in molecular methods have allowed scientists to gain insight into xenobiotic-degrading genes and pathways, not only related to culturable microorganisms, but increasingly to genes associated with the yet to be cultured microbes.

However, the most informative genetic and biochemical analyses can still only be performed on culturable organisms. In order to be able to obtain the most information out of environmental systems, efforts must be made to effectively isolate and culture these environmentally relevant microbes. This would allow a great amount of information to be added to genomic and proteomic bioinformatic databases which are, ultimately, the ones used in metagenomic studies to assign gene functions.

Furthermore, monitoring studies that aim to assess not only the genetic potential, but also microbial taxonomic shifts, parent compound and metabolite concentrations, physical and chemical parameters of the environment have and will continue to prove most informative, especially if done in the field, and should be encouraged.

Acknowledgments

The authors would like to thank Claire E. Brown for her helpful suggestions that significantly improved this chapter.

References

Albers, C.N., L. Feld, L. Ellegaard-Jensen and J. Aamand. 2015. Degradation of trace concentrations of the persistent groundwater pollutant 2,6-dichlorobenzamide (BAM) in bioaugmented rapid sand filters. Water Res 83: 61–70.

Bælum, J., M.H. Nicolaisen, W.E. Holben, B.W. Strobel, J. Sørensen and C.S. Jacobsen. 2008. Direct analysis of tfdA gene expression by indigenous bacteria in phenoxy acid amended agricultural soil. ISME J 2: 677–687.

Bælum, J. and C.S. Jacobsen. 2009. TaqMan probe-based real-time PCR assay for detection and discrimination of class i, ii, and III tfdA genes in soils treated with phenoxy acid herbicides. Appl Environ Microbiol 75: 2969–2972.

Bælum, J., E. Prestat, M.M. David, B.W. Strobel and C.S. Jacobsen. 2012. Modeling of phenoxy acid herbicide mineralization and growth of microbial degraders in 15 soils monitored by quantitative real-time PCR of the functional tfdA gene. Appl Environ Microbiol 78: 5305–5312.

Ben Salem, A., N. Rouard, M. Devers, J. Béguet, F. Martin-Laurent, P. Caboni, H. Chaabane and S. Fattouch. 2018. Environmental fate of the insecticide chlorpyrifos in soil microcosms and its impact on soil microbial communities. pp. 387–389. *In*: Kallel, A., M. Ksibi, H. Ben Dhia and N. Khélifi (eds.). Recent Advances in Environmental Science from the Euro-Mediterranean and Surrounding Regions Proceedings of Euro-Mediterranean Conference for Environmental Integration (EMCEI-1), Tunisia 2017, Advances in Science, Technology & Innovation, Springer. doi:10.1177/1753193417746055.

Boon, N., E.M. Top, W. Verstraete and S.D. Siciliano. 2003. Bioaugmentation as a tool to protect the structure and function of an activated-sludge microbial community against a 3-chloroaniline shock load. Appl Environ Microbiol 69: 1511–1520.

Borden, J.R. and E.T. Papoutsakis. 2007. Dynamics of genomic-library enrichment and identification of solvent tolerance genes for Clostridium acetobutylicum. Appl Environ Microbiol 73: 3061–3068.

Bustin, S. 2000. Absolute quantification of mRNA using real-time reverse transcription polymerase chain reaction assays. J Mol Endocrinol 169–193.

Bustin, S.A., V. Benes, J.A. Garson, J. Hellemans, J. Huggett, M. Kubista, R. Mueller, T. Nolan, M.W. Pfaffl, G.L. Shipley, J. Vandesompele and C.T. Wittwer. 2009. The MIQE guidelines: Minimum information for publication of quantitative real-time PCR experiments. Clin Chem 55: 611–622.

Cámara, B., P. Bielecki, F. Kaminski, V.M. Dos Santos, I. Plumeier, P. Nikodem and D.H. Pieper. 2007. A gene cluster involved in degradation of substituted salicylates via ortho cleavage in Pseudomonas sp. strain MT1 encodes enzymes specifically adapted for transformation of 4-methylcatechol and 3-methylmuconate. J Bacteriol 189: 1664–1674.

Carattoli, A., E. Zankari, A. Garciá-Fernández, M.V. Larsen, O. Lund, L. Villa, F.M. Aarestrup and H. Hasman. 2014. *In silico* detection and typing of plasmids using plasmidfinder and plasmid multilocus sequence typing. Antimicrob Agents Chemother 58: 3895–3903.

Carvalhais, V., M. Delgado-Rastrollo, L.D.R. Melo and N. Cerca. 2013. Controlled RNA contamination and degradation and its impact on qPCR gene expression in *S. epidermidis* biofilms. J Microbiol Methods 95: 195–200.

Cases, I. and V. De Lorenzo. 2001. The black cat/white cat principle of signal integration in bacterial promoters. EMBO J 20: 1–11.

Castillo, J.M., J. Beguet, F. Martin-Laurent and E. Romero. 2016. Multidisciplinary assessment of pesticide mitigation in soil amended with vermicomposted agroindustrial wastes. J Hazard Mater 304: 379–387.

Castro-Gutiérrez, V., E. Fuller, J.C. Thomas, C.J. Sinclair, S. Johnson, T. Helgason and J.W.B. Moir. 2020. Genomic basis for pesticide degradation revealed by selection, isolation and characterisation of a library of metaldehyde-degrading strains from soil. Soil Biol Biochem 142.

Chandler, D.P. 1998. Redefining relativity: Quantitative PCR at low template concentrations for industrial and environmental microbiology. J Ind Microbiol Biotechnol 21: 128–140.

Changey, F., M. Devers-Lamrani, N. Rouard and F. Martin-Laurent. 2011. *In vitro* evolution of an atrazine-degrading population under cyanuric acid selection pressure: Evidence for the selective loss of a 47 kb region on the plasmid ADP1 containing the atzA, B and C genes. Gene 490: 18–25.

Chen, K., X. Xu, L. Zhang, Z. Gou, S. Li, S. Freilich and J. Jiang. 2016. Comparison of four Comamonas catabolic plasmids reveals the evolution of pBHB to catabolize haloaromatics. Appl Environ Microbiol 82: 1401–1411.

Cheng, M., X. Yan, J. He, J. Qiu and Q. Chen. 2019. Comparative genome analysis reveals the evolution of chloroacetanilide herbicide mineralization in Sphingomonas wittichii DC-6. Arch Microbiol 201: 907–918.

Culligan, E.P., R.D. Sleator, J.R. Marchesi and C. Hill. 2014. Metagenomics and novel gene discovery: Promise and potential for novel therapeutics. Virulence 5.

Delmont, T.O., P. Simonet and T.M. Vogel. 2012. Describing microbial communities and performing global comparisons in the omic era. ISME J 6: 1625–1628.

DiGiovanni, G.D., J.W. Neilson, I.L. Pepper and N.A. Sinclair. 1996. Gene transfer of Alcaligenes eutrophus JMP134 plasmid pJP4 to indigenous soil recipients. Appl Environ Microbiol 62: 2521–6.

Fang, H., H. Zhang, L. Han, J. Mei, Q. Ge, Z. Long and Y. Yu. 2018. Exploring bacterial communities and biodegradation genes in activated sludge from pesticide wastewater treatment plants via metagenomic analysis. Environ Pollut 243: 1206–1216.

Fenner, K., S. Canonica, L.P. Wackett and M. Elsner. 2013. Evaluating pesticide degradation in the environment: Blind spots and emerging opportunities. Science (80-) 341: 752–758.

Ferguson, L.R. and W.A. Denny. 1991. The genetic toxicology of acridines. Mutat Res 258: 123–60.

Forbes, J.D., N.C. Knox, J. Ronholm, F. Pagotto and A. Reimer. 2017. Metagenomics: The next culture-independent game changer. Front Microbiol 8.

Gaby, J.C. and D.H. Buckley. 2017. The use of degenerate primers in qPCR analysis of functional genes can cause dramatic quantification bias as revealed by investigation of nifH primer performance. Microb Ecol 74: 701–708.

Galata, V., T. Fehlmann, C. Backes and A. Keller. 2019. PLSDB: A resource of complete bacterial plasmids. Nucleic Acids Res 47: D195–D202.

Gallego, S., M. Devers-Lamrani, K. Rousidou, D.G. Karpouzas and F. Martin-Laurent. 2019. Assessment of the effects of oxamyl on the bacterial community of an agricultural soil exhibiting enhanced biodegradation. Sci Total Environ 651: 1189–1198.

Gedalanga, P.B., P. Pornwongthong, R. Mora, S.Y.D. Chiang, B. Baldwin, D. Ogles and S. Mahendraa. 2014. Identification of biomarker genes to predict biodegradation of 1,4-dioxane. Appl Environ Microbiol 80: 3209–3218.

Geissdorfer, W., S.C. Frosch, G. Haspel, S. Ehrt and W. Hillen. 1995. Two genes encoding proteins with similarities to rubredoxin and rubredoxin reductase are required for conversion of dodecane to lauric acid in Acinetobacter calcoaceticus ADP1. Microbiology 141: 1425–1432.

Goncharova, R.I. and T.D. Kuzhir. 1989. A comparative study of the antimitagenic effects of antioxidants on chemical mutagenesis in Drosophila melanogaster. Mutat Res—Fundam Mol Mech Mutagen 214: 257–265.

Gong, X., R.J. Gruninger, M. Qi, L. Paterson, R.J. Forster, R.M. Teather and T.A. McAllister. 2012. Cloning and identification of novel hydrolase genes from a dairy cow rumen metagenomic library and characterization of a cellulase gene. BMC Res Notes 5: 566.

Gorski, L., P. Rivadeneira and M.B. Cooley. 2019. New strategies for the enumeration of enteric pathogens in water. Environ Microbiol Rep 00.

Goyal, A.K. and G.J. Zylstra. 1996. Molecular cloning of novel genes for polycyclic aromatic hydrocarbon degradation from Comamonas testosteroni GZ39. Appl Environ Microbiol 62: 230–6.

Haiser, H.J. and P.J. Turnbaugh. 2013. Developing a metagenomic view of xenobiotic metabolism. Pharmacol Res 69: 21–31.

Hashimoto, M., M. Fukui, K. Hayano and M. Hayatsu. 2002. Nucleotide sequence and genetic structure of a novel carbaryl hydrolase gene (cehA) from Rhizobium sp. strain AC100. Appl Environ Microbiol 68: 1220–1227.

Hayden, R.T., K.M. Hokanson, S.B. Pounds, M.J. Bankowski, S.W. Belzer, J. Carr, D. Diorio, M.S. Forman, Y. Joshi, D. Hillyard, R.L. Hodinka, M.N. Nikiforova, C.A. Romain, J. Stevenson, A. Valsamakis and H.H. Balfour. 2008. Multicenter comparison of different real-time PCR assays for quantitative detection of Epstein-Barr virus. J Clin Microbiol 46: 157–163.

Heid, C.A., J. Stevens, K.J. Livak and P.M. Williams. 1996. Real time quantitative PCR. Genome Res 6: 986–994.

Heiss, G., N. Trachtmann, Y. Abe, M. Takeo and H.J. Knackmuss. 2003. Homologous npdGI genes in 2,4-dinitrophenol- and 4-nitrophenol-degrading Rhodococcus spp. Appl Environ Microbiol 69: 2748–2754.

Holben, W.E., B.M. Schroeter, V.G.M. Calabrese, R.H. Olsen, J.K. Kukor, V.O. Biederbeck, A.E. Smith and J.M. Tiedje. 1992. Gene probe analysis of soil microbial populations selected by amendment with 2,4-dichlorophenoxyacetic acid. Appl Environ Microbiol 58: 3941–3948.

Holroyd, L.F. and T. Van Mourik. 2014. Stacking of the mutagenic DNA base analog 5-bromouracil. Theor Chem Acc 133: 1–13.

Huang, Y., L. Xiao, F. Li, M. Xiao, D. Lin, X. Long and Z. Wu. 2018. Microbial degradation of pesticide residues and an emphasis on the degradation of cypermethrin and 3-phenoxy benzoic acid: A review. Molecules 23.

Jaworski, J.P., A. Pluta, M. Rola-Luszczak, S.L. McGowan, C. Finnegan, K. Heenemann, H.A. Carignano, I. Alvarez, K. Murakami, L. Willems, T.W. Vahlenkamp, K.G. Trono, B. Choudhury and J. Kuźmak. 2018. Interlaboratory comparison of six real-time PCR assays for detection of bovine leukemia virus proviral DNA. J Clin Microbiol 56.

Jeffries, T.C., S. Rayu, U.N. Nielsen, K. Lai, A. Ijaz, L. Nazaries and B.K. Singh. 2018. Metagenomic functional potential predicts degradation rates of a model organophosphorus xenobiotic in pesticide contaminated soils. Front Microbiol 9: 1–12.

Jin, Y.O. and T.E. Mattes. 2011. Assessment and modification of degenerate qPCR primers that amplify functional genes from etheneotrophs and vinyl chloride-assimilators. Lett Appl Microbiol 53: 576–580.

Kanitkar, Y.H., R.D. Stedtfeld, P.B. Hatzinger, S.A. Hashsham and A.M. Cupples. 2017. Most probable number with visual based LAMP for the quantification of reductive dehalogenase genes in groundwater samples. J Microbiol Methods 143: 44–49.

Kasai, D., S. Imai, S. Asano, M. Tabata, S. Iijima, N. Kamimura, E. Masai and M. Fukuda. 2017. Identification of natural rubber degradation gene in Rhizobacter gummiphilus NS21. Biosci Biotechnol Biochem 81: 614–620.

Kim, J., A.H. Lee and W. Chang. 2018. Enhanced bioremediation of nutrient-amended, petroleum hydrocarbon-contaminated soils over a cold-climate winter: The rate and extent of hydrocarbon biodegradation and microbial response in a pilot-scale biopile subjected to natural seasonal freeze-thaw t. Sci Total Environ 612: 903–913.

Klein, D. 2002. Quantification using real-time PCR technology: Applications and limitations. Trends Mol Med 8: 257–260.

Kroll, J., S. Klinter, C. Schneider, I. Voß and A. Steinbüchel. 2010. Plasmid addiction systems: Perspectives and applications in biotechnology. Microb Biotechnol 3: 634–657.

Kube, M., T.N. Chernikova, Y. Al-Ramahi, A. Beloqui, N. Lopez-Cortez, M.E. Guazzaroni, H.J. Heipieper, S. Klages, O.R. Kotsyurbenko, I. Langer, T.Y. Nechitaylo, H. Lünsdorf, M. Fernández, S. Juárez, S. Ciordia, A. Singer, O. Kagan, O. Egorova, P.A. Petit, P. Stogios, Y. Kim, A. Tchigvintsev, R. Flick, R. Denaro, M. Genovese, J.P. Albar, O.N. Reva, M. Martínez-Gomariz, H. Tran, M. Ferrer, A. Savchenko, A.F. Yakunin, M.M. Yakimov, O.V. Golyshina, R. Reinhardt and P.N. Golyshin. 2013. Genome sequence and functional genomic analysis of the oil-degrading bacterium Oleispira antarctica. Nat Commun 4: 2156.

Kudirkiene, E., L.A. Andoh, S. Ahmed, A. Herrero-Fresno, A. Dalsgaard, K. Obiri-Danso and J.E. Olsen. 2018. The use of a combined bioinformatics approach to locate antibiotic resistance genes on plasmids from whole genome sequences of Salmonella enterica Serovars from humans in ghana. Front Microbiol 9.

Kwak, Y., S.J. Kim, I.K. Rhee and J.H. Shin. 2012. Application of quantitative real-time polymerase chain reaction on the assessment of organophosphorus compound degradation in *in situ* soil. J Korean Soc Appl Biol Chem 55: 757–763.

Lal, D., S. Jindal, H. Kumari, S. Jit, A. Nigam, P. Sharma, K. Kumari and R. Lal. 2015. Bacterial diversity and real-time PCR based assessment of linA and linB gene distribution at hexachlorocyclohexane contaminated sites. J Basic Microbiol 55: 363–373.

Letchumanan, V., K.-G. Chan and L.-H. Lee. 2015. An insight of traditional plasmid curing in Vibrio species. Front Microbiol 6.

Levy-Booth, D.J., M.M. Fetherolf, G.R. Stewart, J. Liu, L.D. Eltis and W.W. Mohn. 2019. Catabolism of alkylphenols in Rhodococcus via a meta-cleavage pathway associated with genomic Islands. Front Microbiol 10.

Li, G., K. Wang and Y. Liu. 2008. Molecular cloning and characterization of a novel pyrethroid-hydrolyzing esterase originating from the Metagenome. Microb Cell Fact 7: 1–10.

Lykidis, A., D. Pérez-Pantoja, T. Ledger, K. Mavromatis, I.J. Anderson, N.N. Ivanova, S.D. Hooper, A. Lapidus, S. Lucas, B. González and N.C. Kyrpides. 2010. The complete multipartite genome sequence of Cupriavidus necator JMP134, a versatile pollutant degrader. PLoS One 5.

Martin-Laurent, F., S. Piutti, S. Hallet, I. Wagschal, L. Philippot, G. Catroux and G. Soulas. 2003. Monitoring of atrazine treatment on soil bacterial, fungal and atrazine-degrading communities by quantitative competitive PCR. Pest Manag Sci 59: 259–268.

Martin-Laurent, F., L. Cornet, L. Ranjard, J.C. López-Gutiérrez, L. Philippot, C. Schwartz, R. Chaussod, G. Catroux and G. Soulas. 2004. Estimation of atrazine-degrading genetic potential and activity in three French agricultural soils. FEMS Microbiol Ecol 48: 425–435.

Meier, M.J., E.S. Paterson and I.B. Lambert. 2016. Use of substrate-induced gene expression in metagenomic analysis of an aromatic hydrocarbon-contaminated soil. Appl Environ Microbiol 82: 897–909.

Monard, C., F. Martin-Laurent, M. Devers-Lamrani, O. Lima, P. Vandenkoornhuyse and F. Binet. 2010. Atz gene expressions during atrazine degradation in the soil drilosphere. Mol Ecol 19: 749–759.

Monard, C., F. Martin-Laurent, O. Lima, M. Devers-Lamrani and F. Binet. 2013. Estimating the biodegradation of pesticide in soils by monitoring pesticide-degrading gene expression. Biodegradation 24: 203–213.

Morales, M.E., M. Allegrini, J. Basualdo, M.B. Villamil and M.C. Zabaloy. 2020. Primer design to assess bacterial degradation of glyphosate and other phosphonates. J Microbiol Methods 169.

Müller, R. and C. Chauve. 2019. HyAsP, a greedy tool for plasmids identification. Bioinformatics 35: 4436–4439.

Nanthini, J., S.Y. Ong and K. Sudesh. 2017. Identification of three homologous latex-clearing protein (lcp) genes from the genome of Streptomyces sp. strain CFMR 7. Gene 628: 146–155.

Nicolaisen, M.H., J. Bælum, C.S. Jacobsen and J. Sørensen. 2008. Transcription dynamics of the functional tfdA gene during MCPA herbicide degradation by Cupriavidus necator AEO106 (pRO101) in agricultural soil. Environ Microbiol 10: 571–579.

Nielsen, M.S., J. Bælum, M.B. Jensen and C.S. Jacobsen. 2011. Mineralization of the herbicide MCPA in urban soils is linked to presence and growth of class III tfdA genes. Soil Biol Biochem 43: 984–990.

Nordin, K., M. Unell and J.K. Jansson. 2005. Novel 4-chlorophenol degradation gene cluster and degradation route via hydroxyquinol in Arthrobacter chlorophenolicus A6. Appl Environ Microbiol 71: 6538–6544.

Notomi, T., H. Okayama, H. Masubuchi, T. Yonekawa, K. Watanabe, N. Amino and T. Hase. 2000. Loop-mediated isothermal amplification. Nucleic Acids Res 28: e63.

Nousiainen, A.O., K. Björklöf, S. Sagarkar, S. Mukherjee, H.J. Purohit, A. Kapley and K.S. Jørgensen. 2014. Atrazine degradation in boreal nonagricultural subsoil and tropical agricultural soil. J Soils Sediments 14: 1179–1188.

Pan, X., B. Lei, N. Zhou, B. Feng, W. Yao, X. Zhao, Y. Yu and H. Lu. 2012. Identification of novel genes involved in DNA damage response by screening a genome-wide Schizosaccharomyces pombe deletion library. BMC Genomics 13: 662.

Pankaj, Negi, G., S. Gangola, P. Khati, G. Kumar, A. Srivastava and A. Sharma. 2016. Differential expression and characterization of cypermethrin-degrading potential proteins in Bacillus thuringiensis strain, SG4. 3 Biotech 6.

Parekh, N.R., A. Hartmann, M.P. Charnay and J.C. Fournier. 1995. Diversity of carbofuran-degrading soil bacteria and detection of plasmid-encoded sequences homologous to the mcd gene. FEMS Microbiol Ecol 17: 149–160.

Pesce, S., J. Beguet, N. Rouard, M. Devers-Lamrani and F. Martin-Laurent. 2013. Response of a diuron-degrading community to diuron exposure assessed by real-time quantitative PCR monitoring of phenylurea hydrolase A and B encoding genes. Appl Microbiol Biotechnol 97: 1661–1668.

Piatak, M., K.C. Luk, B. Williams and J.D. Lifson. 1993. Quantitative competitive polymerase chain reaction for accurate quantitation of HIV DNA and RNA species. Biotechniques 14: 70–76+78.

Picard, C., C. Ponsonnet, E. Paget, X. Nesme and P. Simonet. 1992. Detection and enumeration of bacteria in soil by direct DNA extraction and polymerase chain reaction. Appl Environ Microbiol 58: 2717–2722.

Popovic, A., T. Hai, A. Tchigvintsev, M. Hajighasemi, B. Nocek, A.N. Khusnutdinova, G. Brown, J. Glinos, R. Flick, T. Skarina, T.N. Chernikova, V. Yim, T. Brüls, D. Le Paslier, M.M. Yakimov, A. Joachimiak, M. Ferrer, O.V. Golyshina, A. Savchenko, P.N. Golyshin and A.F. Yakunin. 2017. Activity screening of environmental metagenomic libraries reveals novel carboxylesterase families. Sci Rep 7.

Qiu, J., B. Liu, L. Zhao, Y. Zhang, D. Cheng, X. Yan, J. Jiang, Q. Hong, J. He and M. Kivisaar. 2018. A novel degradation mechanism for pyridine derivatives in Alcaligenes faecalis JQ135. Appl Environ Microbiol 84: e00910-18.

Rabausch, U., J. Juergensen, N. Ilmberger, S. Böhnke, S. Fischer, B. Schubach, M. Schulte and W.R. Streit. 2013. Functional screening of metagenome and genome libraries for detection of novel flavonoid-modifying enzymes. Appl Environ Microbiol 79: 4551–4563.

Rački, N., T. Dreo, I. Gutierrez-Aguirre, A. Blejec and M. Ravnikar. 2014. Reverse transcriptase droplet digital PCR shows high resilience to PCR inhibitors from plant, soil and water samples. Plant Methods 10: 42.

Regar, R.K., V.K. Gaur, A. Bajaj, S. Tambat and N. Manickam. 2019. Comparative microbiome analysis of two different long-term pesticide contaminated soils revealed the anthropogenic influence on functional potential of microbial communities. Sci Total Environ 681: 413–423.

Robles-González, I.V., F. Fava and H.M. Poggi-Varaldo. 2008. A review on slurry bioreactors for bioremediation of soils and sediments. Microb Cell Fact 7.

Rodríguez-Cruz, M.S., J. Bælum, L.J. Shaw, S.R. Sørensen, S. Shi, T. Aspray, C.S. Jacobsen and G.D. Bending. 2010. Biodegradation of the herbicide mecoprop-p with soil depth and its relationship with class III tfdA genes. Soil Biol Biochem 42: 32–39.

Rousidou, C., D. Karaiskos, D. Myti, E. Karanasios, P.A. Karas, M. Tourna, E.A. Tzortzakakis and D.G. Karpouzas. 2017. Distribution and function of carbamate hydrolase genes cehA and mcd in soils: the distinct role of soil pH. FEMS Microbiol Ecol 93: 1–12.

Ruan, Z., Y. Zhai, J. Song, Y. Shi, K. Li, B. Zhao and Y. Yan. 2013. Molecular cloning and characterization of a newly isolated pyrethroid-degrading esterase gene from a genomic library of Ochrobactrum anthropi YZ-1. PLoS One 8.

Sadauskas, M., J. Vaitekunas, R. Gasparavičiute and R. Meškys. 2017. Indole biodegradation in Acinetobacter sp. strain O153: Genetic and biochemical characterization. Appl Environ Microbiol 83.

Sagarkar, S., S. Mukherjee, A. Nousiainen, K. Björklöf, H.J. Purohit, K.S. Jorgensen and A. Kapley. 2013. Monitoring bioremediation of atrazine in soil microcosms using molecular tools. Environ Pollut 172: 108–115.

Sakultantimetha, A., H.E. Keenan, T.K. Beattie, S. Bangkedphol and O. Cavoura. 2011. Effects of organic nutrients and growth factors on biostimulation of tributyltin removal by sediment microorganisms and Enterobacter cloacae. Appl Microbiol Biotechnol 90: 353–360.

Salisbury, V., R.W. Hedges and N. Datta. 1972. Two modes of "curing" transmissible bacterial plasmids. J Gen Microbiol 70: 443–452.

Salminen, J.M., P.M. Tuomi and K.S. Jørgensen. 2008. Functional gene abundances (nahAc, alkB, xylE) in the assessment of the efficacy of bioremediation. Appl Biochem Biotechnol 151: 638–652.

Sangwan, N., P. Lata, V. Dwivedi, A. Singh, N. Niharika, K. Jasvinder, S. Anand, J. Malhotra, S. Jindal, A. Nigam, D. Lal, A. Dua, S. Saxena, N. Garg, M. Verma, J. Kaur, U. Mukherjee, J.A. Gilbert, S.E. Dowd, R. Raman, P. Khurana, J.P. Khurana and R. Lal. 2012. Comparative metagenomic analysis of soil microbial communities across three hexachlorocyclohexane contamination levels. PLoS One 7: 1–12.

Sangwan, N., H. Verma, R. Kumar, V. Negi, S. Lax, P. Khurana, J.A. Gilbert and R. Lal. 2014. Reconstructing an ancestral genotype of two hexachlorocyclohexane-degrading Sphingobium species using metagenomic sequence data. ISME J 8: 398–408.

Sayler, G.S., S.W. Hooper, A.C. Layton and J.M.H. King. 1990. Catabolic plasmids of environmental and ecological significance. Microb Ecol 19: 1–20.

Schleinitz, K.M., S. Kleinsteuber, T. Vallaeys and W. Babel. 2004. Localization and characterization of two novel genes encoding stereospecific dioxygenases catalyzing 2(2,4-dichlorophenoxy)propionate cleavage in Delftia acidovorans MC1. Appl Environ Microbiol 70: 5357–5365.

Schmidt, R., A. Ahmetagic, D.S. Philip and J.M. Pemberton. 2011. Catabolic Plasmids. In: eLS. Chichester, UK: John Wiley & Sons, Ltd.

Seifert, H.S., E.Y. Chen, M. So and F. Heffron. 1986. Shuttle mutagenesis: A method of transposon mutagenesis for Saccharomyces cerevisiae. Proc Natl Acad Sci USA 83: 735–739.

Seung, I.K., J.S. Choi and H.Y. Kahng. 2007. A proteomics strategy for the analysis of bacterial biodegradation pathways. Omi A J Integr Biol 11: 280–294.

Shibai, A., Y. Takahashi, Y. Ishizawa, D. Motooka, S. Nakamura, B.W. Ying and S. Tsuru. 2017. Mutation accumulation under UV radiation in *Escherichia coli*. Sci Rep 7.

Silva, C.C., H. Hayden, T. Sawbridge, P. Mele, S.O. De Paula, L.C.F. Silva and P.M.P. Vidigal. 2013. Identification of genes and pathways related to phenol degradation in metagenomic libraries from petroleum refinery wastewater. PLoS One 8(4): e61811.

Singh, D.P., R. Prabha, V.K. Gupta and M.K. Verma. 2018. Metatranscriptome analysis deciphers multifunctional genes and enzymes linked with the degradation of aromatic compounds and pesticides in the wheat rhizosphere. Front Microbiol 9: 1–15.

Soares Bragança, C.R., T.-M. Dooley-Cullinane, C. O'Reilly and L. Coffey. 2017. Applying functional metagenomics to search for novel nitrile-hydrolyzing enzymes using environmental samples. Biomater Tissue Technol 1: 1000108.

Stark, R., M. Grzelak and J. Hadfield. 2019. RNA sequencing: the teenage years. Nat Rev Genet 20: 631–656.

Taylor, S.C., J. Carbonneau, D.N. Shelton and G. Boivin. 2015. Optimization of droplet digital PCR from RNA and DNA extracts with direct comparison to RT-qPCR: Clinical implications for quantification of Oseltamivir-resistant subpopulations. J Virol Methods 224: 58–66.

Taylor, S.C., G. Laperriere and H. Germain. 2017. Droplet digital PCR versus qPCR for gene expression analysis with low abundant targets: From variable nonsense to publication quality data. Sci Rep 7: 1–8.

Taylor, S.C., K. Nadeau, M. Abbasi, C. Lachance, M. Nguyen and J. Fenrich. 2019. The ultimate qPCR experiment: producing publication quality, reproducible data the first time. Trends Biotechnol 37: 761–774.

Techtmann, S.M. and T.C. Hazen. 2016. Metagenomic applications in environmental monitoring and bioremediation. J Ind Microbiol Biotechnol 43: 1345–1354.

Terrón-González, L., G. Martín-Cabello, M. Ferrer and E. Santero. 2016. Functional metagenomics of a biostimulated petroleum-contaminated soil reveals an extraordinary diversity of extradiol dioxygenases. Appl Environ Microbiol 82: 2467–2478.

Thakur, M., I.L. Medintz and S.A. Walper. 2019. Enzymatic bioremediation of organophosphate compounds—progress and remaining challenges. Front Bioeng Biotechnol 7: 1–21.

Thompson, B.M., C.H. Lin, R.J. Kremer, R.N. Lerc, H.Y. Hsieh and H.E. Garrett. 2010. Evaluation of PCR-based quantification techniques to estimate the abundance of atrazine chlorohydrolase gene atzA in rhizosphere soils. J Environ Qual 39: 1999–2005.

Throne-Holst, M., A. Wentzel, T.E. Ellingsen, H.K. Kotlar and S.B. Zotchev. 2007. Identification of novel genes involved in long-chain n-alkane degradation by Acinetobacter sp. strain DSM 17874. Appl Environ Microbiol 73: 3327–3332.

Todd, P.A., C. Monti-Bragadin and B.W. Glickman. 1979. MMS mutagenesis in strains of *Escherichia coli* carrying the R46 mutagenic enhancing plasmid: Phenotypic analysis of Arg+ revertants. Mutat Res - Fundam Mol Mech Mutagen 62: 227–237.

Tsang, J. 2017. Bacterial plasmid addiction systems and their implications for antibiotic drug development. Postdoc J 5(5): 3–9.

Tuomi, P.M., J.M. Salminen and K.S. Jørgensen. 2004. The abundance of nahAc genes correlates with the 14C-naphthalene mineralization potential in petroleum hydrocarbon-contaminated oxic soil layers. FEMS Microbiol Ecol 51: 99–107.

Ufarté, L., E. Laville, S. Duquesne, D. Morgavi, P. Robe, C. Klopp, A. Rizzo, S. Pizzut-Serin and G. Potocki-Veronese. 2017. Discovery of carbamate degrading enzymes by functional metagenomics. PLoS One 12: e0189201.

Vartoukian, S.R., R.M. Palmer and W.G. Wade. 2010. Strategies for culture of 'unculturable' bacteria. FEMS Microbiol Lett 309(1): 1–7.

Wade, W. 2002. Unculturable bacteria—The uncharacterized organisms that cause oral infections. In, J R Soc Med 95(2): 81–83.

Walia, S., A. Khan and N. Rosenthal. 1990. Construction and applications of DNA probes for detection of polychlorinated biphenyl-degrading genotypes in toxic organic-contaminated soil environments. Appl Environ Microbiol 56: 254–259.

Wang, M., G. Yang, H. Min and Z. Lv. 2009. A novel nicotine catabolic plasmid pMH1 in Pseudomonas sp. strain HF-1. Can J Microbiol 55: 228–233.

Wang, Q., S. Xie and R. Hu. 2013. Bioaugmentation with Arthrobacter sp. strain DAT1 for remediation of heavily atrazine-contaminated soil. Int Biodeterior Biodegrad 77: 63–67.

Wang, W., L. Wang and Z. Shao. 2018. Polycyclic aromatic hydrocarbon (PAH) degradation pathways of the obligate marine PAH degrader Cycloclasticus sp. strain P1. Appl Environ Microbiol 84.

Widada, J., H. Nojiri and T. Omori. 2002. Recent developments in molecular techniques for identification and monitoring of xenobiotic-degrading bacteria and their catabolic genes in bioremediation. Appl Microbiol Biotechnol 60: 45–59.

Yale, R.L., M. Sapp, C.J. Sinclair and J.W.B. Moir. 2017. Microbial changes linked to the accelerated degradation of the herbicide atrazine in a range of temperate soils. Environ Sci Pollut Res 24: 7359–7374.

Yan, X., T. Gu, Z. Yi, J. Huang, X. Liu, J. Zhang, X. Xu, Z. Xin, Q. Hong, J. He, J.C. Spain, S. Li and J. Jiang. 2016. Comparative genomic analysis of isoproturon-mineralizing sphingomonads reveals the isoproturon catabolic mechanism. Environ Microbiol 18: 4888–4906.

Yang, L., Li, Xinyu, Li, Xu, Z. Su, C. Zhang and H. Zhang. 2014. Bioremediation of chlorimuron-ethyl-contaminated soil by Hansschlegelia sp. strain CHL1 and the changes of indigenous microbial population and N-cycling function genes during the bioremediation process. J Hazard Mater 274: 314–321.

Zaprasis, A., Y.J. Liu, S.J. Liu, H.L. Drake and M.A. Horn. 2010. Abundance of novel and diverse tfdA-like genes, encoding putative phenoxyalkanoic acid herbicide-degrading dioxygenases, in soil. Appl Environ Microbiol 76: 119–128.

Zhang, J., J.G. Yin, B.J. Hang, S. Cai, J. He, S.G. Zhou and S.P. Li. 2012. Cloning of a novel arylamidase gene from paracoccus sp. Strain FLN-7 that hydrolyzes amide pesticides. Appl Environ Microbiol 78: 4848–4855.

4

Bioprospecting Contaminated Soil for Degradation of the Drimaren X-BN Azo Dye

Carolina Rosai Mendes,[1,*] *Guilherme Dilarri,*[1]
Paulo Renato Matos Lopes,[2] *Ederio Dino Bidoia*[1] and
Renato Nallin Montagnolli[3]

Introduction

The existence of dyes in human history is dated from the Paleolithic period to about 40,000 B.C., when man used naturally colored substances to create rock drawings. The availability of colors in this period was limited, and the main pigments used were blood, charcoal, and plant and mineral extracts. Many minerals have shades due to their chemical nature, such as metal oxides, sulfides, carbonates, chromates, sulfates, phosphates, and silicates (Adedayo et al. 2004). The most diverse types of natural dyes are reported throughout history in India, Persia, Phenicia, Egypt, Greece, and Rome. In colonial Brazil, the red dye extracted from *Paubrasilia echinata* was intensely commercialized in Europe (Pinto 2006), as well as the indigo dye extracted from the vegetable *Indigofera tinctoria*. Natural dyes had a high monetary value due to the complex extraction. In addition, large quantities of raw materials were needed to produce the dye. The availability of shades was limited when compared to

[1] Department of Biochemistry and Microbiology, Sao Paulo State University (UNESP), 24-A Avenue, 1515, Postal Code: 13506-900, Rio Claro-SP, Brazil.
[2] College of Technology and Agricultural Sciences – Sao Paulo State University (UNESP), SP-294, km 651 – Dracena-SP, Brazil.
[3] Department of Natural Sciences, Mathematics and Education, Agricultural Sciences Centre – Federal University of Sao Carlos (UFSCar), SP-330, km 174 – Araras-SP, Brazil.
Emails: gui_dila@hotmail.com; prm.lopes@unesp.br; ederio.bidoia@unesp.br; renatonm@ufscar.br
* Corresponding author: carolina_rosai@hotmail.com

the availability of shades of synthetic dyes. As cotton became a commodity of great economic value in England, so did the demand for dyes and the need for cheaper dyes. The changes from an agrarian to an industrial economy forced the demand for chemicals to whiten and dye. It was in 1856 that the English chemistry William Perkin synthesized the first organic molecule of synthetic dye, from tar (Mills 1987). Perkin's discovery prompted a race among chemistry to produce other molecules of different shades. In a short time, the production of synthetic dyes met the needs of the fabric industries in Europe. Synthetic dyes guarantee low cost and intense color.

Currently, more than 100,000 different dyes are used industrially, with a worldwide production of 700,000 tons per year (Raval et al. 2016). In the last decades, new classes of dyes were created to meet the current needs of the market. Textile dyes are subdivided into classes according to their molecular structure, and among them are acidic, alkaline, dispersive, azoic, diazoic, metallic, reactive, and bleaching dyes (Ratnamala et al. 2012). In general, the color of a dye is the result of the interaction of the functional chromophor groups and aromatic groups (Bhatia et al. 2017). The functional groups, called auxochromes, in dyes can be chlorine, hydroxyl, bromine, nitro, sulfonic, amino, methoxy, ethoxy, and ethyl (Kimura et al. 2000). In this way, they allow to reach different wave frequencies in the visible spectrum, and therefore, different shades of colors. Some dyes have an electrophilic group in their molecule capable of forming a covalent bond with anionic and cationic interactions (Mendes et al. 2015). These bonds guarantee greater stability between the dye molecules and the fibers after the dyeing processes. Despite the advances made in the dye production sector, there are still problems that must be overcome in the future in order to make full use of the dye molecules. The main problem is in the dyeing stages, where a large part of the dye used does not favorably fix the fibers of the fabric and remain in the industrial wastewater (Corso et al. 2012). As a result, the excess dye remains in the form of effluent in the industry and reaches water bodies. It is estimated that about 280,000 tons of dyes enter the environment annually (Brillas and Martínez-Huttle 2015). The released organic load generates impacts for the aquatic ecosystem, and the increase in the turbidity in the water can generate a toxic environment due to the reduction of photosynthetic activities (Gupta et al. 2012). In addition, dyes are toxic and have stable molecules. In this way, the contaminant becomes persistent and accumulative, and can remain for more than 50 years in the soil and aquatic environment (Dilarri et al. 2018). The absence of color in the effluent does not result in low toxicity. This is because some processes degrade only the chromophore group of the dye molecule, which is characteristic of reaching a wavelength in the visible spectrum. Therefore, it is necessary to properly treat industrial effluents in order to mineralize the organic load.

The main treatments are degradation, filtration, flocculation, ozonation, precipitation, and adsorption. Some effluent treatments have disadvantages, such as excessive use of chemicals and high costs (Mendes et al. 2015). Biological treatment methods include microbial biodegradation under aerobic, anaerobic, anoxic, or anaerobic/aerobic conditions combined (Sarayu and Sandhya 2012). Biodegradation is an energy-dependent process and involves breaking the dyes through the action of various enzymes (Hameed and Ismail 2020). The complete breakdown of organic molecules with the production of water, carbon dioxide, or other inorganic products is called mineralization. The use of biological treatment consists of agitating the effluents in the presence of microorganisms and oxygen, for the time necessary to metabolize much of the organic matter. It is evident that biological effluent treatment processes have more advantages due to their low cost and complete mineralization of the dye molecule (Bouraie and Din 2016). One of the factors that accelerate the discoloration is the conditions of pure culture of strains that already have an enzymatic apparatus capable of degrading dye molecules. Maximum dye discoloration can be achieved by optimizing the conditions of pH, temperature, incubation time, dye concentration, and availability of suitable sources of carbon or nitrogen.

The use of bacteria, such as *Pseudomonas* sp. has been described in the literature for biodegradation of textile dyes. These microorganisms are particularly useful for degradation of azo dyes, as they have the ability to perform reductive cleavage in the azo bonds of this type of compound. Most of the time the degradation occurs through the reducible pathway of the enzyme azoreductase. Azoreductases are the diverse group of enzymes widely present in the enzymatic apparatus of various microorganisms. Although there is diversity in structure and function associated with the azoreductase pathways, they have a common potential for reducing the azo N=N bonds, common in dye molecules of the azo groups.

Thus, the aim of this work was to evaluate the potential of gram-negative *P. aeruginosa* bacteria selected from the soil in an area contaminated with effluent from the textile industry to degrade the Drimaren X-6BN Red azo-dye molecule. Therefore, investigating new microbial strains by bioprospecting allows selecting strains with enzymatic apparatus better adapted to the target contaminant. Thus, it is a biotechnological alternative for the treatment of areas contaminated by textile effluents.

Materials and methods

Drimaren X-6BN azo dye

The Drimaren X-6BN azo dye from Sigma Aldrich®, contains 40% purity as indicated by the manufacturer, and belongs to the group of Azo dyes and molar mass 416.38 g mol^{-1} (Fig. 4.1).

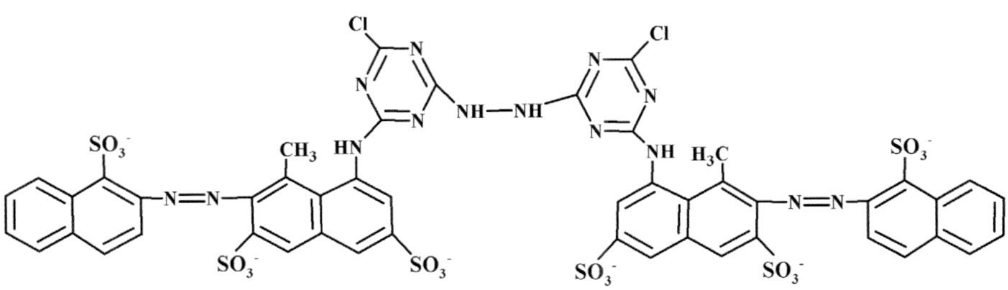

Figure 4.1. Chemical structure of Drimaren X-6BN azo dye.

Microorganism identification

Molecular microbial ecology has made progress through the construction of metagenomic libraries, which is a powerful approach to exploring soil microbial diversity, including providing data on non-cultivable microorganisms in the laboratory.

The microorganism was selected from a sample of soil contaminated with textile industrial effluent from the region of Americana-SP in Brazil. The sample followed a serial dilution of 10^5 in sterile 0.85% saline. An aliquot of the sample was placed on Nutrient Agar medium and grown in an oven at 25°C. The microorganism selected from the colony-forming unit (UFC) was identified at the species level by the molecular biology technique proposed by Chen et al. (1993).

The 16S rDNA sequences were evaluated by the GenBank database using the MegAlign software (DNASTAR Inc., Madison, Wis.), which included 136 sequences from 42 species of *Pseudomonas* as well as several other phylogenetically related γ-Proteobacteria. The microorganism was identified by the 16S region through the Polymerization Chain Reaction (PCR), using the following primers–27F (5'AGAGTTTGATCCTGGCTCAG 3') and 1401R (5'CGGTGTGTACAAGACCC 3'). The DNA amplification of the selected microorganism was performed in volumes of 25 μL, each containing 2 mM of MgCl2, 50 mM of Trizma (pH 8.30; Sigma, St. Louis, Mo.), 250 μM deoxynucleoside triphosphates (Promega, Madison, Wis.), 0.40 μM of primer, 1 U of Taq polymerase (Invitrogen, Carlsbad, Califórnia), and 2 μL of lysed cells. Amplification was performed on a RapidCycler (Idaho Technology Inc., Salt Lake City, Utah) thermocycler. After an initial 2 min, denaturation at 95°C with 25 cycles was completed. Each cycle was adjusted from 20 seconds at 94°C, 20 seconds to the pairing temperature, and 40 seconds at 72°C. A final extension of 1 min at 72°C was applied. With this program, the total time to amplify the target DNA was approximately 45 minutes. Then, the amplifications were analyzed by electrophoresis on 1% agarose gel (Sambrook et al. 1989). The amplicons were sequenced and compared with others already known and available in the GenBank database.

Preparation of the inoculum for biodegradation

The selected CFU was grown in liquid nutrient medium with agitation for 48 hours and controlled temperature of 28°C. The cells were centrifuged at 10,000 × *g* for 8 min, and resuspended in sterile 0.85% saline. The supernatant was measured on the UV-Vis spectrophotometer (Spectrophotometer Thermo–Mod Biomate 3) to quantify the number of cells present in the liquid.

Treatment in liquid medium with different concentrations

The experiment analyzed the biodegradation of Drimaren X-6BN azo dye from the strain selected by bioprospecting. Biodegradation was carried out in triplicates, varying the concentrations of the dye in solution, with 90 μg mL^{-1}, 80 $μg^{-1}$, 70 μg mL^{-1}, and 60 μg mL^{-1}. In each treatment, 10^8 cells were inoculated and the dye solutions were made in minimal mineral medium (10 mL of $FeSO_4$ at 1 M, 1 mL of $MgSO_4$ at 1 M, 0.5 mL of $CaCl_2$ at 1 M, and 10 mL of NH_4Cl_2 M, Na_2HPO_4 17.4 g, KH_2PO_4 10.6 g). This medium was chosen due to the minimal concentrations of nutrients essential for the growth of the microorganism, however, the dye molecule is the only available carbon source. The flasks were analyzed on a UV-Vis spectrophotometer $\lambda = 520$ nm every 24 hours. Before each UV-Vis analysis, the samples were centrifuged at 10,000 × *g*, for 8 min to separate the cell suspension from the liquid. The remaining dye concentration was measured from the calibration curve that correlates the dye concentration in solution with the absorbance values at a maximum wavelength in a UV-Vis spectrophotometer (Shimadzu®, model 2401 PC). From the absorbance results, the model proposed by Chan and Chu (2003), showed in Equation 1, was applied.

$$\frac{C_r}{C_i} = \frac{(1-t)}{(\rho + \sigma \cdot t)} \tag{1}$$

where, C_r is the concentration (absorbance) of the product remaining after time t (hours), and C_i is the initial concentration (absorbance) of the dye. The kinetic parameter ρ represents the reaction kinetics (hours) and σ the oxidative capacity of the system.

In this study, the percentage in the decolorization efficiency (DE %) (Xiao et al. 2017) shown in Equation 2 was calculated.

$$DE(\%) = \left[\frac{C_i - C_r}{C_i}\right] \times 100 \tag{2}$$

The confirmation of all data obtained in the present study was done with the adequate fit to the standard deviation (*SD*) shown in Equation 3.

$$SD = \sqrt{\frac{1}{N-1}\sum_{i=1}^{N}\left(\frac{Q_{ie} - Q_{ic}}{Q_{ie}}\right)^2} \tag{3}$$

in which Q_{ic} and Q_{ie} (μg mL^{-1}) are calculated and experimental mass of dye adsorbed by the adsorbent, respectively, and N is the number of measurements made.

Results and discussion

The study confirmed the identification of the *P. aeruginosa* strain isolated from the contaminated soil based on the 98% similarity in the GenBank database. The *P. aeruginosa* is a gram-negative, heterotrophic bacterium with metabolic versatility that guarantees advantages over other microorganisms. Therefore, it is common to find this species among the organisms that colonize contaminated substrates, such as soil and water. The advantage of using a microbial strain selected from a contaminated site is to ensure that the microorganism is able, through metabolic apparatus, to degrade the contaminating compound.

In this study, a reduction in the absorbance value was observed for the four different concentrations used after the biodegradation processes. However, the C_r of the Drimaren X-6BN dye in solution was different for each treatment. A reduction in the dye concentration was observed in all treatments in the first 24 hours after the contaminant molecule entered to be available to the *P. aeruginosa* cells. It is possible to state that there was no time for microbial stress to trigger an alternative metabolic pathway for degradation of the molecule. Therefore, the metabolic pathway of degradation is expressed under normal conditions. This characteristic is very advantageous, as it speeds up the degradation time of the molecule of interest.

Figure 4.2 shows the graph of the degradation kinetics of the Drimaren X-6BN dye molecule at different concentrations as a function of time.

The graph shows the decrease every 24 hours in the concentration values of the dye in solution. This type of graph indicates that the microorganism is using the dye molecule as a carbon source, so the C_r of the dye is reduced at the end of the experiment. In addition, none of the concentrations used showed toxicity to the microorganism, because even at higher concentrations of dye, it is possible to see the decrease. Therefore, these microbial activities demonstrate that *P. aeruginosa* has a high tolerance to the contaminating compound.

The study showed better results of reducing the dye in solution at C_i of 60 μg mL^{-1}. After 96 hours, the C_r was 10.30 μg mL^{-1}, which means a reduction of 83% of the coloration in solution. However, there was an approximation in the results of color reduction to C_i of 70 μg mL^{-1}, which showed very close values. After the

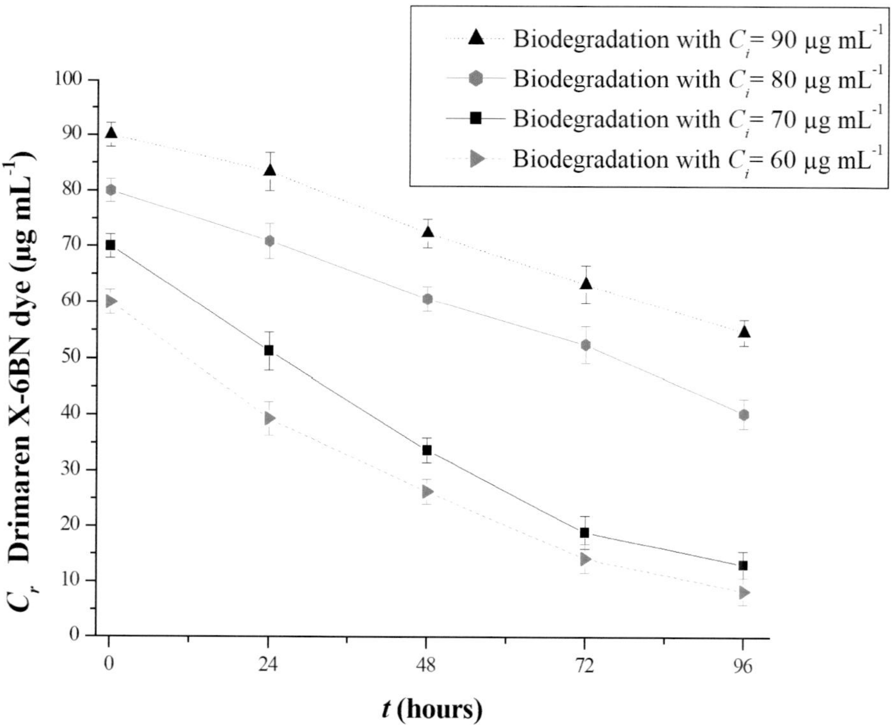

Figure 4.2. Biodegradation kinetics of Drimaren X-6BN dye by *P. aeruginosa.*

same period of degradation, the C_r was 13.10 µg mL^{-1}, which was a reduction of 81 percent. Thus, the values did not show significant variations in the results, when compared.

The results obtained at C_i of 90 and 80 µg mL^{-1}, had a reduction in the rate of degradation when compared with studies of degradation in smaller C_i. The values obtained were C_r 57.70 µg mL^{-1}, with a reduction of 64.10% to C_i of 90 µg mL^{-1}, and C_r 40.30 µg mL^{-1} with a reduction of 50.40% to C_i of 80 µg mL^{-1}.

The results also prove the tolerance of *P. aeruginosa* in relation to the various concentrations of the contaminating compound tested, presenting potential for use in bioremediation processes. The results indicate a continuous decrease in C_r, so it is predicted that the increase in the time factor would bring better results in the biodegradation processes.

Biodegradation is a viable alternative in the treatment for textile dye molecules. The clearing and breaking of the Drimaren X-6BN azo dye molecules showed promising results, with reduction values of more than 80 percent. Biodegradation is an operation that consists of the decomposition of the dye molecules into smaller organic molecules, whose impact factor on the environment is lower. The azo dye molecules have amine groups as the main composition and may contain one or more characteristic azo bonds (–N = N–) in their aromatic structure. This semi-covalent azo bond makes dyes more recalcitrant to microbial degradation, and is not easily decomposed under environmental conditions (Zimmermann et al. 1982). However, there are many microorganisms used in the degradation processes of aromatic compounds and have many advantages for industrial treatment. Biological treatment is not expensive, and some microorganisms are able to break down aromatic chains until mineralization, transforming a contaminating molecule into compounds such as H_2O and CO_2. Several gram-negative bacteria have been described as resistant to certain contaminants. The structure of their outer membranes, composed of lipopolysaccharides, lipoproteins, and phospholipids, is one of the main factors that make these microorganisms resistant to these toxic compounds (Zimmermann et al. 1984). The outer membrane of gram-negative bacteria provides a barrier to certain compounds, detergents, heavy metals, salts, and certain dyes. In addition, these microorganisms have an enzymatic apparatus capable of degrading the aromatic ring and other chains of the contaminant molecule for their own use as a carbon source.

In this work, the degradation mechanism and the way in which *P. aeruginosa* decomposes the molecule of the dye Drimaren X-6BN is proposed. Figure 4.3 shows the process of initial decomposition of the dye molecule and the reduction in coloration as a result.

The model shows the possibility of the *P. aeruginosa* microorganism to use the degradation pathway from different microbial oxidases and reductases. According to the studies by Koppel et al. (2017), some isolates had high activity of azo nitroreductase in the presence of flavins and NAD(P)H$^+$ when in contact with contaminating molecules of the phenolic chain. Azoreductase is involved in the reductive cleavage of the azo bond as an initial stage in the degradation of azo dyes. Azoreductases are part of a diverse group of microbial enzymes widely present in different species of microorganisms (Zimmermann et al. 1982). However,

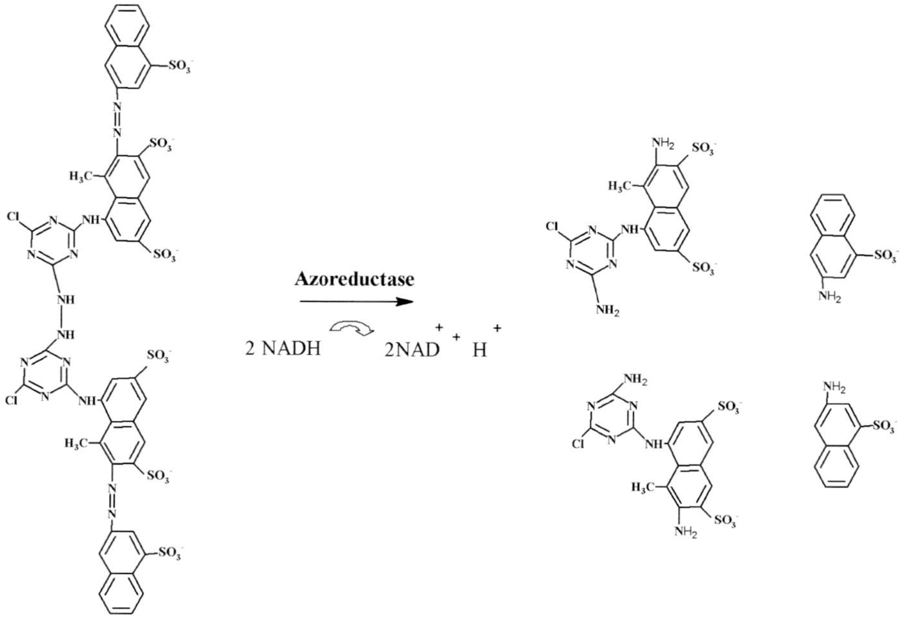

Figure 4.3. Decomposition model of the Drimaren X-6BN azo dye molecule after biodegradation with *P. aeruginosa*.

each microorganism has a different production path with variations in its structures and functions. Although there is diversity in structure and function, they have a common potential to break down the azo dye molecules. In the studies proposed by Morrison and John (2015), several types of azoreductases were isolated, and after the purification had their coded genes identified. In addition, azoreductases were biochemically characterized and classified based on the division of aerobic and anaerobic microorganisms. Azoreductases from different sources are part of the enzymatic apparatus of microorganisms resistant to certain contaminants. These microorganisms use the contaminated decomposed chains as a carbon source, essential cofactor elements, catalytic activity, and synthesis of organic compounds.

In this study, the variation of the degradation from the wavelength in analysis in spectrophotometer UV-Vis was observed, as shown in Fig. 4.4.

When there is biodegradation, there is a decrease in the peaks in different proportions, forming new peaks in the wavelengths in the UV-Vis spectrum. This is due to the result of absorbances of new metabolites or fragments of the dye molecule after degradation (Mitter and Corso 2013). Figure 4.4 shows the spectra of the dye biodegradation in the intervals of 48, 72, and 96 hours, through the formation of new peaks in the wavelength. The changes in plasmonic bands occurred after 48 hours, and this shows that the degradation process occurs as soon as *P. aeruginosa* comes in contact with the dye molecule. In addition, the degradation of the dye molecule alters the chromophore groups, and therefore changes the characteristic of wavelength. Therefore, after the degradation processes, a clearing of the solution and a change to a color close to the yellow wavelength was observed. The change in color is shown in

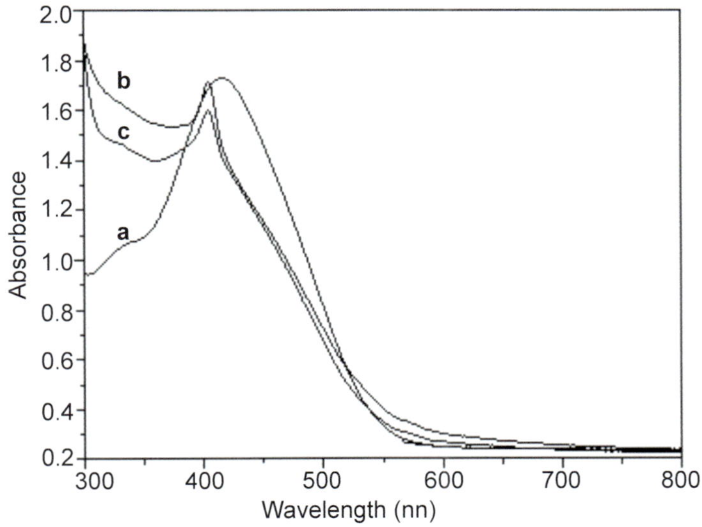

Figure 4.4. Biodegradation spectrum of Drimaren X-6BN azo dye after 48 hours (a), 72 hours (b) and 96 hours (c).

Figure 4.5. Comparison between the Drimaren X-6BN azo dye solution before and after biodegradation with *P. aeruginosa.*

Figure 4.5, which compares the coloring of the Drimaren X-6BN azo in the solution used as a control, and the color of the solution after the degradation processes with *P. aeruginosa.*

Conclusion

The bacteria *Pseudomonas aeruginosa* showed a relevant resistance to the various concentrations of the red dye Drimaren X-6BN, in addition to demonstrating a good rate of degradation of the dye tested *in vitro*. The experiments also proved the efficiency of biological treatment for effluents with dye residues, and the biological treatment showed a reduction of more than 80% of the azo dye concentration. Therefore, the use of the *P. aeruginosa* microorganism as an agent for the biodegradation of dye molecules could be considered a low cost alternative to reduce the incorrect disposal of these effluents in the soil and rivers. It was also concluded that the *P. aeruginosa* bacteria have the potential for use in bioremediation processes, using it in processes of allogeneic bio-increase in effluents contaminated with textile dyes, thus reducing the impacts caused by the textile effluents discarded incorrectly in the environment.

Acknowledgments

This study received funding from the Brazilian fostering agency: Conselho Nacional de Desenvolvimento Científico e Tecnológico (CNPq)—Process 144912/2017-1.

References

Adedayo, O., S. Javadpour, C. Taylor, W.A. Anderson and M. Moo-Young. 2004. Decolourization and detoxification of methyl red by aerobic bacteria from a wastewater treatment plant. World J Microbiol Biotecnhnol 20: 545–550.

Bhatia, D., N.R. Sharma, J. Singh and R.S. Kanwar. 2017. Biological methods for textile dye removal from wastewater: A review. Crit Rev Env Sci Tec 47: 1836–1876.

Bouraie, M.E. and W.S.E. Din. 2016. Biodegradation of reactive Black 5 by Aeromonas hydrophila strain isolated from dye-contaminated textile wastewater. Sustain Environ Res 26(5): 209–216.

Brillas, E. and C.A. Martínez-Huttle. 2015. Decontamination of wastewaters containing synthetic organic dyes by electrochemical methods. Appl Catal-B: Environ 166: 603–613.

Chan, K.H. and W. Chu. 2003. Modeling the reaction kinetics of Fenton's process on the removal of atrazine. Chemosphere 51(4): 305–311.

Chen, Z., L. Kuo, R.R.R. Rowland, C. Even, K.S. Faaberg and P.G.W. Plagemann. 1993. Sequences of 3' end of genome and of 5' end of open reading frame 1a of lactate dehydrogenase-elevating virus and common junction motifs between 5' leader and bodies of seven subgenomic mRNAs. J GenVirol 74: 643–660.

Corso, C.R., E.J.R. Almeida, G.C. Santos, L.G. Morao, G.S.L. Fabris and E.K. Mitter. 2012. Bioremediation of direct dyes in simulated textile effluents by a paramorphogenic form of *Aspergillus oryzae*. Water Sci Technol 65: 1490–1495.

Dilarri, G., R.N. Montagnolli, E.D. Bidoia, C.R. Mendes and C.R. Corso. 2018. Kinetic, isothermal, and thermodynamic models to evaluate Acid Blue 161 dye removal using industrial chitosan powder. Desalin Water Treat 109: 261–270.

Gupta, V.K., D. Pathania, S. Agarwal and P. Singh. 2012. Adsorptional photocatalytic degradation of methylene blue onto pectin-CuS nanocomposite under solar light. J Hazar Mater 243: 179–186.

Hameed, B.B. and Z.Z. Ismail. 2020. Biodegradation of reactive yellow dye using mixed cells immobilized in different biocarriers by sequential anaerobic/aerobic biotreatment: experimental and modelling study. Environ Technol. https://doi.org/10.1080/09593330.2020.1720306.

Kimura, I.Y., V.T. Fávere, M.C.M. Laranjeira, A. Josué and A. Nascimento. 2000. Evaluation of the adsorption capacity of reactive dye orange 16 by chitosan. Acta Sci 22(5): 1161–1166.

Koppel, N., V.M. Rekdal and E.P. Balskus. 2017. Chemical transform of xenobiotics by the human gut microbiota. Science 356: 2770.

Mendes, C.R., G. Dilarri and R.T. Pelegrini. 2015. Application of biomass *Saccharomyces cerevisiae* as adsorption agent of dye Direct Orange 2GL and possible interactions mechanisms adsorbate/adsorbent. Matéria (RJ) 20: 898–908.

Mills, J.S. and R. White. 1987. The Organic Chemistry of Museum Objects. Butterworth and Co, London.

Mitter, E.K. and C.R. Corso. 2013. Acid dye biodegradation using Saccharomyces cerevisiae immobilized with polyethyleneimine-treated sugarcane bagasse. Water Air Soil Pollut 224: 1391.

Morrison, J.M. and G.H. John. 2015. Non-classical azoreductase secretion in Clostridium perfringens in response to sulfonated azo dye exposure. Anaerobe 34: 34–43.

Pinto, A.C.O. 2006. Pau-brasil e um pouco da história brasileira. SBQ 6: 15–24.

Ratnamala, G.M., K.V. Shetty and G. Srinikethan. 2012. Removal of remazol brilliant blue dye from dye-contaminated water by adsorption using red mud: equilibrium, kinetic, and thermodynamic studies. Water Air Soil Poll 223: 6187–6199.

Raval, N.P., P.U. Shah and N.K. Shah. 2016. Adsorptive amputation of hazardous azo dye Congo red from wastewater: a critical review. Environ Sci Poll Res 23: 14810–14853.

Sambrook, J., E.F. Fritsch and T. Maniatis. 1989. Molecular Cloning: A Laboratory Manual. Cold Spring Harbor Laboratory Press, Nova York.

Sarayu, K. and S. Sandhya. 2012. Current technologies for biological treatment of textile wastewater—a review. Appl Biochem Biotechnol 167: 645–661.

Xiao, X., T.T. Li, X.R. Lu, X.L. Feng, X. Han, W.W. Li, Q. Li and H.Q. Yu. 2017. A simple method for assaying anaerobic biodegradation of dyes. Bioresour Technol 251: 204–209.

Zimmermann, T., H.G. Kulla and T. Leisinger. 1982. Properties of purified Orange II azoreductase, the enzyme initiating azo dye degradation by *Pseudomonas* KF46. Eur J Biochem 129: 197–203.

Zimmermann, T., F. Gasser, H.G. Kulla and T. Leisinger. 1984. Comparison of two azoreductases acquired during adaptation to growth on azo dyes. Arch Microbiol 138: 37–43.

5

Saccharomyces cerevisiae—A Platform for Delivery of Drugs and Food Ingredients

Encapsulation and Analysis

Bahman Khameneh,[1] *Bibi Sedigheh Fazly Bazzaz*[2,*] and *Maryam Nakhaee Moghadam*[3]

Introduction: a glance to encapsulation process by the yeasts

Microencapsulation is a process of covering active compounds or living microorganisms by a carrier material for various proposes, such as stabilizing the active compounds, extending the shelf-life, and protecting them from micro-environments. The optimum release of components is another benefit of this process. It is applied in many fields of sciences and different industries, including pharmaceuticals, cosmetic, and food industries (Madene et al. 2006, Schrooyen et al. 2001).

The yeast cell of *Saccharomyces cerevisiae* is one of the most important one, which has been used from about 50 years ago as an encapsulation coating material for essential oils and flavors (NA 1990). Being natural causes many advantages over the other microencapsulation carriers (Nelson 2002). It is present in human nutrition

[1] Department of Pharmaceutical Control, School of Pharmacy, Mashhad University of Medical Sciences, Vakilabad Blvd, Mashhad, +9851, Iran.

[2] Department of Pharmaceutical Control, School of Pharmacy/Biotechnology Research Center, Pharmaceutical Technology Institute, Mashhad University of Medical Sciences, Vakilabad Blvd, Mashhad, +9851, Iran.

[3] Department of Food Hygiene, Faculty of Veterinary Medicine, Ferdowsi University of Mashhad, Vakilabad Blvd, Mashhad, +9851, Iran.

Emails: khamenehbagherib@mums.ac.ir; mn.moghadam@gmail.com

* Corresponding author: fazlis@mums.ac.ir

(generally recognized as safe, GRAS) with low cost. It is shown that the fungal cell wall with a dynamic structure protects the cell from environmental stresses, including changes in osmotic pressure. However, it allows the cell to interact with its surrounding environment (Bowman and Free 2006). Nowadays, technologists have incorporated several compounds with different physicochemical (hydrophobic/hydrophilic) properties in the yeast cells (Bishop et al. 1998, Mokhtari et al. 2017b, Nelson 2002, Paramera et al. 2011a, Pham-Hoang et al. 2013).

The structure of this carrier is composed of the water soluble β-1,3 and β-1,6 D-glucans, the major polysaccharides of the yeast cell walls. These structures have a wide range of important properties, including antibacterial activities, wound-healing, antioxidant, anti-mutagenic, and anti-genotoxic properties, which make them potent agents for application in anti-infective and anti-cancer therapies (Kogan et al. 2008).

Passive diffusion processes are the more important encapsulation process mechanisms with respect to the active transport procedures. The solubility of the hydrophobic compound in the cell wall, which is inversely related to partition coefficient (log P), is the major factor for the successful encapsulation (Ciamponi et al. 2012).

The encapsulation process involves mixing the yeast cells at different conditions, such as alive, plasmolyzed, or non-plasmolyzed, wet or dried, or even the yeast walls, with the active ingredient in a solvent solution (water or water/organic). This is followed by controlled speed stirring for some hours with controlled temperature of usually 20–60°C to obtain the desired droplet size for optimum diffusion. To remove the non-encapsulated active compounds, the loaded non-viable cells are washed with water or an organic solvent. Finally, spray drying, fluidized bed, or freeze drying is used for drying the loaded yeasts (Paramera et al. 2014).

The method of vacuum infusion has also shown to have a rapid and low energy with lower cost to increase the loading bioactive materials, either multiple or complex mixtures, in yeast cells, used in functional food and beverage (Young et al. 2017).

It is important to notice that the molecular structure of active ingredients is another factor affecting the encapsulation process. The penetration yield is directly correlated to active compound size (molecular weight), shape, and even polarity (hydrophobic properties) (Paramera et al. 2014).

Although some researchers report that the molecules with molecular weight lower than 760, diffuse freely through the cell wall, but others showed that molecules with MW as big as 200,000 to 400,000 can diffuse quite freely through the cell wall of the yeast. However, one should not forget that the major permeability barrier to the molecules is the cell membrane (De Nobel et al. 1990).

Amphiphilic molecules with similar character to the cell lipid bilayers can enter into the cells freely. Molecules having strong polar area and high log P can encapsulate easily, because the polarity extends the localization of the polar part and the general hydrophobicity (Pham-hoang et al. 2016).

Polar groups, such as hydroxyl when present in the structure of the molecules, can facilitate the encapsulation process. For these molecules, the better interactions with the polar head groups of the membrane phospholipids' bilayer occur, and led to higher percentage encapsulation yield (EY) values (Paramera et al. 2014).

The chain saturation or ramification, cis or trans isomerism, are other parameters playing key roles in the shape and volume of the molecule which affect encapsulation (Pham-hoang et al. 2016).

Structure of the yeast cell envelope

Yeast cells with about 25 μm diameter, have spherical or ellipsoidal shape. The main cell parts, responsible for encapsulation properties, are rigid cell wall and the inner cell membrane (Nakhaee Moghadam et al. 2019a).

The yeast cell walls act as a protecting envelope to control the osmotic pressure and the exchanges of materials between the cell and its environment (Pham-Hoang et al. 2013).

The wall is a thick capsule (100 to 200 nm), which is about 15% to 25% of the dry weight of the cell. The polysaccharides, which are major structural constituents of the cell wall (80–90%), are mainly mannans and glucans, with a minor amount of chitin. A high amount of proteins is entrapped into the network of these polysaccharides and made into carbohydrate polymers (Kapteyn et al. 1999).

The mechanical rigidity of the cell wall is due to the presence of glucans and chitin, while the porosity, and therefore permeability of the cell wall, is because of mannoproteins (Lipke and Ovalle 1998, Paramera et al. 2014).

The outer of either sides of the cell wall (10 nm) are glycosylated mannoprotein layer (Kapteyn et al. 1999). The mannoproteins supply a negative charge at physiological pH and a hydrophilic environment (Rodriguez-Pena et al. 2000).

The inner layer, plasma membrane, is limited by the cell wall and the cytoplasm. The bilayer membrane is mainly composed of phospholipids, sterols, and neutral lipids (including triacylglycerols and sterol esters) (Suomalainen and Nurminen 1970).

Selective permeability is a major role for the plasma membrane. It is responsible for the components that enter or leave the cell, while it has an attractive role for encapsulation. This permeability of membrane is highly linked with its fluidity, known as phase transition temperature (Paramera et al. 2014).

All together, these factors made the Baker's yeast, *S. cerevisiae*, a suitable host for the drug delivery system (Blanquet et al. 2005, Lipke and Ovalle 1998).

Transfer across the cell envelope

Cell wall

S. cerevisiae cell wall, which provides a mechanical support for the cell, allows molecules with molecular weight up to 620–1,100 to diffuse freely. However, plasma membrane is significantly related to permeability barrier and controls the diffusion of components (De Nobel et al. 1990). Notably, it has been shown that the external β-glucan and mannoprotein layers of cell wall are permeable to small molecules, both polar and non-polar, in aqueous solution (Zlotnik et al. 1984). In another words, the reticulated β-glucan network acts as skeleton and the external protein layer as a sieve with a mesh-size evolving with the water content (Normand et al. 2005).

The cell wall is mainly hydrophilic and has limiting effects in the diffusion of hydrophobic molecules into the cells. The knowledge about the cell wall properties, which is involved in the uptake of either hydrophobic or hydrophilic compounds or its mechanism, is little and also insufficient (Ciamponi et al. 2012, PC 1999).

It was demonstrated that certain components of the cell wall play an important role in the hydrophobicity of the cell surface (Pham-Hoang et al. 2013). However, the cell wall is able to adapt itself to morphological and physiological changes during the cell cycle (Klis et al. 2006, 2002), also in response to conditions of growth and cultivation mode (Aguilar-Uscanga and Francois 2003).

Cell membrane

After the cell wall, the cell membrane is the next barrier for crossing the compounds. Dissolving in the lipid part and diffusing through the cell membrane is the way that fat-soluble molecules can cross the cell, and this occurs mainly by passive diffusion (Bishop et al. 1998, Ciamponi et al. 2012), or active transport in rare events (Kohlwein and Paltauf 1984). These hydrophobic molecules have an impact on the membrane's physical properties, and modify its fluidity. Diffusion occurs by a concentration gradient from high concentration to low concentration. Major factors affecting their ability to cross the membrane are molecular size and lipid solubility, i.e., small and more lipid soluble molecules will diffuse easily (Nelson et al. 2006). The diffusion rate also depends on the interaction of the molecule with the selective action of the membrane. Amphiphilic compounds, such as methyl oleate, when present in the culture medium, induce a cell adaptation at the membrane level (Ta et al. 2010, 2012). The membrane becomes very fluid and acquires buffer capacity, which makes the cells resistant to "fluidizing stresses". This fluidizing stress can be triggered by heat, and it is even believed that higher encapsulation will occur at higher temperatures (Bishop et al. 1998, Paramera et al. 2011b).

Organelles for encapsulation

There is not a clear report on the passage of foreign molecules through the yeast cell envelope, and it is believed that storage occurs inside the cytoplasm organelles, such as vacuoles or lipid bodies (Pham-Hoang et al. 2013).

Vacuole is a storage organelle of fungal cells with degradative potential (Li and Kane 2009), therefore, degradation of the compound in the vacuole is usually a problem which has to be checked. Loading of active components in the vacuole is an active process, therefore using inactivated cells cannot be a solution (Pham-Hoang et al. 2013). In a report to evaluate the vacuolar pH (Cao-Hoang et al. 2008), fluorescent probe was used to assess the intracellular pH extruded by the cells from the cytoplasm. By this observation, they showed the vacuole can act as a storage compartment.

Lipid bodies in the form of intracellular lipid droplets are another organelle of fungi capable of foreign materials storage. This process involves the synthesis of lipids in the endoplasmic reticulum and assembling between the two layer of membrane phospholipids. Finally, when these lipids reach the critical size, mature, and bud off in the form of a micro-droplet surrounded by a membrane monolayer (Czabany et al. 2007).

In the beginning, the lipid bodies were considered a reservoir for encapsulation of hydrophobic molecules. A technique was used in which encapsulating molecules were localized in the intracellular lipid droplets (JL 1977). Many researchers based on cell physiology mechanism tried to increase the encapsulation yield through enhancing the lipid droplets' size. This was achieved by the use of oleaginous species with oil contents excess of 40% to 60% biomass dry weight in a culture medium with nutrient limitations (Beopoulos et al. 2009). Similarly, a method based on the solvent usage, known as "lipid-extending substances", has been proposed. In this method, cells with lipid content less than 40% of dry weight are mixed with the lipid content of microorganisms to more than 40% of dry weight by passive diffusion of the solvent/co-solvent into the cell (NA 1994).

Yeast cell encapsulation properties

The superior properties of yeast cells as carrier are the nutritional value and health benefits. These properties make them more useful for the drug and food industries. The yeast cell wall is composed of reticulated network of fibrous β-1,3-glucans and mannoproteins surrounding the membrane lipid bilayer. This mechanically strong structure prevents destructive effects of drastic environmental condition on the active encapsulate. Other properties of yeast cells which make it superior are being a cost-effective technology, the ease of encapsulation which could be accomplished with no additive, high loading capacity (either in the form of intact or plasmolyzed cell), long-lasting controlled release, and enhanced stability of the active ingredient (Paramera et al. 2014). In addition, yeast cells being resistant to environmental conditions, particularly to high temperatures (Normand et al. 2005) and light (Shi et al. 2007), gives rise to a high protection of encapsulants during production process. Reports indicate the stability of yeast cells at temperatures as high as 250°C. It seems that the network of the cell wall containing fibrous β-1,3-glucans and mannoproteins, along with the lipid bilayer, provides additional protection to the encapsulated molecules (Bishop et al. 1998, Paramera et al. 2011a).

Applications

Hydrophobic molecules

Encapsulation procedure in empty yeast cells was applied for coating dyes. The results suggested that this procedure can be used for numerous fat-soluble substances, such as drugs, condiments, flavors, fragrances, chemicals, vitamins, and adhesives (Pham-Hoang et al. 2013).

Aroma is one of the important factors in the sensorial value of food products for the consumer. It has a complex composition and is composed of volatile odorous organic molecules having different physicochemical properties. Therefore, it is unstable and its storage and delivery in food products undergoes many difficulties. Microencapsulation of aroma improves the functionality and stability, and it also releases in food products at the right time (Zuidam and Heinrich 2010).

Empty *S. cerevisiae* yeast cells were used as a thermo-stable delivery system to encapsulate flavors. Limonene with highly hydrophobic character, which is not considered a very volatile molecule, has been encapsulated in the yeast cells (Dardelle et al. 2007).

Another compound used in the food processing industry is curcumin, with poor water solubility. This compound is susceptible to different conditions, such as alkaline, light, oxidation, and heat, which all limit its clinical efficacy. Encapsulation of curcumin in the yeast cells of *S. cerevisiae* was a success to overcome most of its drawbacks (de Medeiros et al. 2019, Paramera et al. 2011b).

Berberin, an effective component of *Berberis vulgaris,* with many pharmacological properties, was effectively encapsulated in plasmolized and intact cells of *S. cerevisiae* (Salari et al. 2013, 2015).

Drug

Yeast cells have also been used to target delivery of different drug molecules, such as ibuprofen, propranolol hydrochloride, and itraconazole. Yeast cells can adhere to the human epithelial cells and enhance permeability of the drug molecules (Fuller et al. 2005).

Additionally, the drug remains stable within the encapsulated yeast cell until release is initiated. This will occur by addition of a surfactant or by contact with a mucous membrane. When the encapsulated drug is administered directly into the duodenum, lipophilic drug is released from the cell and enters the blood stream, which has a reduced burst effect and prolonged release profile (Nelson et al. 2006).

Essential oils

Phytochemicals with bioactivity and stability can act as preservative, and provide antimicrobial and antioxidant potencies. However, most essential oils (EOs) are sensitive to environmental factors (oxygen, light, and temperature) with poor water solubility. To overcome these drawbacks, *Zataria multiflora* Bioss. EO was successfully encapsulated in *S. cerevisiae* cells (Nakhaee Moghadam et al. 2019b).

Yeast cells, when used as microcapsules, were able to non-specifically sequester high concentrations of a variety of EOs. Meanwhile, the EO becomes stabilized within the cell as an oil droplet(s). It is reported that variations in encapsulation rates were also due to the solubility of the permeating molecules in the cell membrane, and their molecular size and the shape. In this situation, *S. cerevisiae* yeast cell and its membrane bilayer acts as a liposome during the process of microencapsulation, and allows the stabilization of oil droplets within the yeast cell (Bishop et al. 1998).

Natural compounds

In commercial applications, encapsulated enzymes in permeabilized *S. cerevisiae* cells offer an economical method of harvesting and reusing recombinant proteins. This process protects the enzyme from shear stresses and maintains it in the solution, while allowing easy separation of starting materials and products (Chow and Palecek 2004).

In the food industry, the use of flavors composed of volatile molecules, subject them to the large losses. To minimize flavor loss during such harsh industrial processes, protecting the volatile molecules within an empty *S. cerevisiae* yeast cell with thermo-stable property is desirable (Bishop et al. 1998, Dardelle et al. 2007, Normand et al. 2005). To this end, spray drying of residual yeast cells, from which β-glucans have been partially extracted, is a promising approach for successful encapsulation of flavors (Sultana et al. 2017).

Fish oil is valuable for human diet and contains high amounts of polyunsaturated fatty acids. Fish oil, due to the strong undesirable odor and subject to oxidation, is a good candidate for microencapsulation. Consequently, encapsulation in the yeast cells is a possible solution which provides great results (Czerniak et al. 2015).

Resveratrol is a photosensitive natural phenol with high antioxidant activities. To achieve the sustained release and improved solubility and bioavailability, resveratrol was successfully encapsulated in *S. cerevisiae* cells. Furthermore, *in vitro* releasing property was shown in the potential applicability of yeast cell as an effective delivery carrier for this compound (Shi et al. 2008).

Chlorogenic acid (CGA) is a phenolic compound that widely exists in fruits and vegetables with many beneficial properties. This compound can directly interact with reactive oxygen species, which is the most important concern for its applications. Stabilization of CGA has been demonstrated successfully in yeast-encapsulation, which results in a water-soluble anti-oxidant (Shi et al. 2007).

Carvacrol, a natural monoterpene derivative of cymene, was encapsulated in yeast cell walls of *S. cerevisiae*. The encapsulated form of this molecule showed good activity, decreasing volatility, and also increasing the action time. In addition, it was used in acaricidal formulations for the control of Asian blue tick *Rhipicephalus microplus* (da Silva Lima et al. 2017).

Vitamin D with profitable effects to help maintain human health has poor water solubility and susceptibility to environmental conditions. An applied strategy to overcome these limitations is incorporation of vitamin D3 into microcapsules. Microencapsulation in yeast was used for encapsulation of cholecalciferol (Dadkhodazade et al. 2018).

Hydrophilic molecules

FR2179528 patent demonstrated that hydrophilic molecules could be surrounded by yeast cells. According to the patent, plasmolyzed yeast cells were used for onion juice (Laboratoire 1973). However, there are a few reports in hydrophilic compound encapsulation. Recently, the technique of encapsulation in the yeast cells has been adopted for water-soluble flavors (Benczedi et al. 2007). Modification was achieved by mixing yeast cells directly with encapsulable flavor of tetrahydro methyl furanthiol or S-(2-methyl-3-furyl)-ethanethioate. The novelty of this patent was the use of matrix components of maltodextrin. After flavor absorbed in the yeast, and prior to spray-drying, maltodextrin was added to the mixture to form the final capsule. Maltodextrin matrix helps increase the loading of the flavor in the capsule through increasing hydrophilicity (Pham-Hoang et al. 2013). This is again a reminder that flavor encapsulation, like other molecules, by the yeast is governed by the effect of molecular size and hydrophilicity, and the growth phase of the cell, i.e., small lipophilic molecules are encapsulated in an easier manner (Pham-hoang et al. 2016).

Probiotics

It was reported that on the addition of encapsulated *Lactobacillus acidophilus* and *Bifidobacterium bifidum* in calcium alginate microbeads, and coating with the cell wall of *S. cerevisiae* to grape juice shows that the survival of probiotics at the end of storage time was higher than recommended minimum value (10^7 cfu/mL). However, this process did not affect the survival of *B. bifidum* ($P > 0.05$) (Mokhtari et al. 2017a, b). This demonstrated the suitable protection of the cell wall of *S. cerevisiae* to improve the survival of probiotics within the food products until their release at the appropriate location(s) in the gastrointestinal tract (Mokhtari et al. 2017a).

Affecting factors

Growth phase

S. cerevisiae cell wall has different porosity at different stages of growth, and it is higher in the logarithmic phase in comparison with stationary phase. However, when the cells at stationary phase are treated with alpha mannosidase or tunicamycin (an inhibitor of N-glycosylation) that eliminate glucanase-soluble mannoproteins (the main factor limiting the cell wall porosity), the porosity may revert and even increase (De Nobel et al. 1990). Notably, in *S. cerevisiae*, the encapsulation of terpenes, hydrophobic model compounds, was not affected by the growth phase, indicating that it was not affected by the porosity, while it was affected by the solubility of the hydrophobic agent in the cell wall, and the solubility is inversely related to the hydrophobic partition coefficient (logP) (Ciamponi et al. 2012).

Viability of yeast cell

It is shown by the plate count method that yeast cells quickly lost their viability during the encapsulation process. The estimated viability when cultured on MYGP (containing maltose, yeast extract, glucose, and peptone) media was under 10% and 0.1% after 1 hour and 2 hours, respectively. Therefore, it seems that culturability is not essential for the encapsulation process (Bishop et al. 1998).

Using freeze drying or spray drying which was applied for pretreatment of the cells or in the final stage of the encapsulation process, is lethal for nearly all the cells (Paramera et al. 2011b).

Mostly, when the yeast cells are applied for the encapsulation, the inactivated form is used, and can still keep the interesting encapsulation properties (Pham-Hoang et al. 2013).

The structure of dead cells is similar to the living cells and composed of a thin protein layer that covers on a β(1–3) glucan layer. The protein layer is 15 ± 2 nm thick and the β(1–3) glucan layer is 130 ± 17 nm. The average diameter of the dry cells is 2 μm, and the external protein phase occupies about 10% of the cell wall volume as compared to the living cells (31%), and the inner polysaccharide phase occupies the remaining (90%) when compared with the living cells (69%) (Zlotnik et al. 1984).

Transport of materials in live yeast occurs either by passive or facilitated diffusion and active transport. However, as encapsulation processes take place in both dead and live cells, therefore it takes place by simple diffusion. Simple diffusion follows Overton's Rule, indicating the correlation of permeability coefficients with oil/water partition coefficients (Nelson et al. 2006).

Yeast cell pretreatment

Culture conditions is the most important factor affecting the yeast cells permeability (Shi et al. 2010, Zlotnik et al. 1984), and subsequently chemical treatment increases permeability (De Nobel et al. 1989). Since culture medium and the condition of growth all affect the intermolecular structures of the cell wall constituents, the mannoproteins, hydrophobic linkages, and disulfide bonds, therefore all are responsible in the porosity and permeability (De Nobel and Barnett 1991). If a treatment can destroy the membrane and intermolecular network or rupture the disulfide bonds, it will increase the permeability of *S. cerevisiae* yeast cells greatly. However, for an efficient encapsulation yield of the active core materials, it is necessary to have a balance between the permeability of the cell wall and the preservation. Shi et al. (2010) reported that all the chemical treatments, such as adding surfactant (Trition X100), salt (Na Cl), and organic solvent (ethanol) have similar effects on the FT-IR spectra of *S. cerevisiae* with increased absorption bands to lipids and fatty acids and decreased ones to the proteins and nucleic acids (Shi et al. 2010).

Chemical treatments for encapsulation in the yeast cells are different. They are based on osmosis (immersion in NaCl concentrations) (Paramera et al. 2011b, Shi et al. 2010, 2007, 2008), or on membrane-disrupting compounds (use of surfactants such as Triton X100) (Chow and Palecek 2004), or on solvents (use of ethanol or acetone) (Inoue et al. 1991, Paramera et al. 2011b, Shi et al. 2008). To empty the yeast cells, various techniques have been used, including plasmolysis, autolysis, or hydrolysis with exogenous enzymes (Tanguler and Erten 2008).

The benefits of plasmolysis on the encapsulation of hydrophilic compounds are obvious, however, they can also remove the cell materials that could interact with the encapsulated substance or could decrease the unpleasant taste of the yeast. The important point is the relation between the taste and the culture medium, and therefore it is possible to avoid the plasmolysis step (Pham-Hoang et al. 2013). It was postulated that the improvement of the encapsulation of both fisetin and curcumin could be achieved in the presence of intercellular organelles through binding (Young et al. 2017).

Proteases and nucleases cause cell autolysis, i.e., degradation of insoluble cellular components by the cell's own native hydrolytic enzymes. This process is accompanied by the release of hydrolysis products, such as amino acids, nucleotides, and peptides, to the environment. It should be noted that large scale autolysis is used in the production of yeast extract for applications in food industry and laboratory (Dziezak 1987). The influence of incubation temperature on the extent of autolysis indicated that optimum activity of the native yeast hydrolases occurs at temperature in the range between 55–65°C. Below or above these temperatures, adverse effects on yeast extract yield were seen (Czerniak et al. 2015). In another study, the optimum temperature between 50–60°C for the yeast extract production by autolysis was reported (Peppler 1982). Therefore, 50°C, with a slight difference from 50–60°C, is the optimum temperature for the autolysis of spent brewery yeast (Tanguler and Erten 2008). The differences between the optimum temperature may be resulted from differences in the yeast strain, culture medium and conditions, or even phase of growth (Czerniak et al. 2015).

Pretreatment with dithiothreitol (DTT) increased the cell wall porosity because of an increased loss of cell wall proteins (mannoproteins) and reduced disulphide bridges (De Nobel et al. 1989).

All of the mentioned parameters affected the encapsulation efficiency (EE) of final preparations. The EE, when treatment occurs, depends on different factors, including treatments diversity, encapsulation protocols, and the hydrophobicity of the encapsulated compound (Pham-Hoang et al. 2013).

Shi et al. (Shi et al. 2007) reported the increase of EE values after treatment with a plasmolyzer, while Salari et al. reports indicate structural changes in the yeast cell of *S. cerevisiae*, with no change in loading properties. They demonstrated disorganization of the cell plasma membrane and change of cell wall thickness with no additional disruptions at higher concentrations of plasmolizer, sodium chloride solutions (Salari et al. 2013). The increased flowing and encapsulation yield of curcumin by plasmolysis has been reported (Paramera et al. 2011b). Dadkhodazade et al. (Dadkhodazade et al. 2018) demonstrated that employing plasmolysis before encapsulation causes higher EE values. Czerniak et al. (Czerniak et al. 2015) showed ethyl acetate pretreatment of the cells, facilitated the cell autolysis, and gave rise to maximum EE of fish oil.

Temperature

One of the important factors which affect encapsulation of active material in the yeast is temperature (Paramera et al. 2014). Yeast cell encapsulation is not increased linearly with the temperature, and depends on the phase transition between the gel and liquid crystalline states of the cell. It was shown that there were almost no detectable changes in encapsulation below the membrane phase transition, but a remarkable increase was likely to include the phase transition temperature, with no further increase between 35°C and 45°C (Paramera et al. 2011b).

Another report by Bishop et al. showed that encapsulation of orange peel oil in fresh yeast occurs almost independent of changes in the temperature (from 6°C to 40°C). However, they showed a significant increase in the rate of encapsulation between 40°C and 50°C, which shows the process appears more temperature dependent (Bishop et al. 1998). In other words, temperature increases membranes' fluidity and facilitates the diffusion of orange peel oil across the membrane barriers (Czerniak et al. 2015).

Characterization of capsules

Characterization of micro and nanocapsules is necessary for the optimization and validation of the yeast-cell-based microencapsulation process, as well as encapsulated agent. Moreover, in food and pharmaceutical applications, characteristics of the carriers are one of the major concerns in safety and efficacy regulatory aspects (Gaonkar et al. 2014). The most frequently used techniques for characterization of capsules are reviewed below. Figures 5.1 and 5.2 show the mainly used characterization methods of microcapsules in food and pharmaceutical industries.

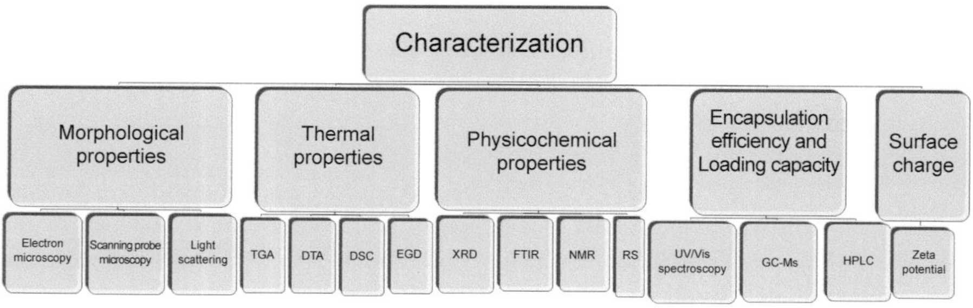

Figure 5.1. Frequently used methods for characterization of microcapsules in food and pharmaceutical industries.

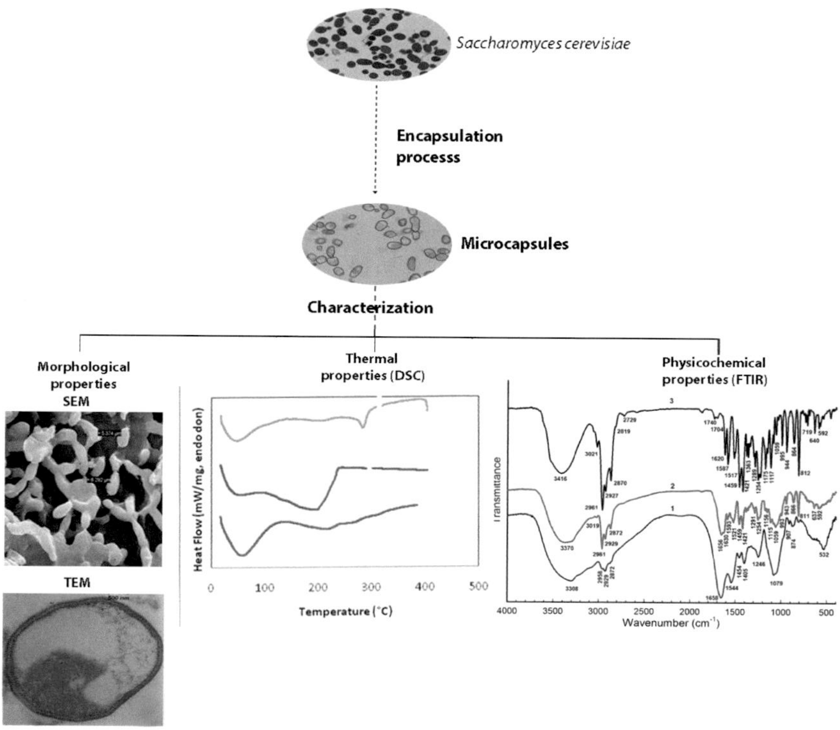

Figure 5.2. The features of *Saccharomyces cerevisiae* capsules analyzed by different methods.

Morphological properties

Morphology and particle size analyses provide important information about shape, structure, size, and range of particle, particle size distribution, and state of agglomeration. This information is essential for understanding the physicochemical and functional characteristics of the encapsulated particle (Jafari and Esfanjani 2017). The microscopic studies are easy and common methods for this purpose, and could be classified into various methods, such as electron microscopy techniques (scanning electron microscopy (SEM) and transmission electron microscopy (TEM)), scanning probe microscopy technique (atomic force microscopy (AFM) and confocal laser scanning microscopy (CLSM)), and light scattering techniques (dynamic light scattering (DLS), static light scattering (SLS), and laser diffraction (LD)). Instruments based on light scattering method are the most widely used instruments to characterize microcapsules particle size. Electron microscopy techniques are common methods which have been used to study the structure of nanocapsules (< 1000 nm). These techniques are similar to light microscopy methods, except that using electron beams instead of light for creating images of particles provides as much as a thousand-fold increase in resolving power. SEM provides information about the surface of the sample, whereas TEM is useful for analyzing the micro-structural properties, such as thickness of wall material (Shah et al. 2018, Tiede et al. 2008, Vishwakarma et al. 2016). In TEM instruments, the beams of electron interact with the particles by passing through an ultrathin specimen. High energy electron beams (100–400 keV) are emitted from the cathode and accelerated into anode by electron field. After that, these electron beams are concentrated by the electromagnetic and electrostatic lenses and are transmitted through a particle. In the case of SEM, electrons are emitted through the electron gun with the energy of 0.1–30 keV. Thus, TEM provides superior resolution than that of SEM (Jafari and Esfanjani 2017).

These electron microscopic imaging techniques have been used extensively to study the interaction of encapsulated agents with yeast cell. In a study, the morphological analysis by SEM revealed size differences between the yeast cell walls and the encapsulated agents. The results indicated that upon encapsulation of carvacrol with yeast cell walls, the size of the particles enhanced remarkably (da Silva Lima et al. 2017). In another study, SEM technique was used to evaluate the effect of encapsulation process on the surface structure of yeast cells. The results showed that no significant differences were observed in cell wall morphological properties of the yeast cells during the encapsulation procedures, and the cell wall organization was intact (Salari et al. 2015). Bishop et al. used *S. cerevisiae* to encapsulate high concentrations of essential oils. TEM results demonstrated that the cell wall and membrane remain intact during the process (Bishop et al. 1998). The objective of another study was to encapsulate a synthetic compound in glucan-rich particles, mainly composed by the cell wall of *S. cerevisiae* and TEM method was used to analysis morphological properties. It was found that by TEM, the structural difference between the samples could not easily be distinguished (Volpato et al. 2018).

Scanning probe microscopy methods use a physical probe to study the probe-specimen interaction. In the case of CLSM, the images show advantages, such as high-resolution, selectivity in depth, and facilitate construction of 3D structures (Jafari and Esfanjani 2017, Vimala Bharathi et al. 2018). AFM is a method, with dimensional resolutions from 0.1 up to 10 nm, which provides a unique opportunity for visualizing particles (Drozdek and Bazylinska 2016). In a study, *S. cerevisiae* was used as model to investigate the interaction of nanoparticles (NPs) with cell surface (Nomura et al. 2018). In another work, a type of carrier was constructed by loading polymer-lipid hybrid nanoparticles into yeast cell wall microparticles for macrophage-targeted oral delivery of cabazitaxel. The AFM results revealed that after loading, the holes of the yeast cell wall were filled with the NPs. CLSM method was used, and the results indicated that the NPs were efficiently and successfully packaged into the yeast cell wall (Ren et al. 2018). The morphology and viability of encapsulated *S. cerevisiae* in polymeric shells were studied by different microscopic methods, such as AFM and CLSM. The CLSM images showed that upon encapsulation of living cells in polymeric shells, the cells could preserve their sub-cellular structure, and duplication capability was maintained. Additionally, coupling AFM and CLSM data, provided a correlation between local stiffness and duplication rate (Svaldo-Lanero et al. 2007).

Light scattering methods, such as DLS and SLS, are widely used to investigate the particle size analysis. The DLS method was used for analyzing the particle size of yeast-encapsulated doxorubicin microparticles. The results indicated that the size of the yeast cell was not significantly changed after encapsulation of the drug. This study showed that *S. cerevisiae* shows potential in loading and delivering of anti-cancer drugs (Wu et al. 2018). In the latest research, this method was also employed for evaluating the particle size of carriers (Kavetsou et al. 2019).

Thermal properties

The development of biodegradable films based on *S. cerevisiae* biomass carriers shows lots of advantages to the food industry. As thermal processing is one of the most important processes in the food, knowing about the thermal properties of the *S. cerevisiae* carriers is very important, which will help to be scaled-up to an industrial level. These properties give us more important information about thermal stability, melting and freezing temperatures, heat energy storage, decomposition, and glass transition temperature characterization of the carriers. Different thermo-analytical techniques which have been generally used to investigate these properties are thermo-gravimetric analysis (TGA), differential scanning calorimetry (DSC), differential thermal analysis (DTA), and evolved gas detection (EGA) (Jafari and Esfanjani 2017, Vishwakarma et al. 2016).

By TGA method, the changes in weight of the samples during the application of controlled heat have been determined and used for evaluation of thermal stability of the carriers (Anandharamakrishnan 2015). The films made of *S. cerevisiae* biomass were employed extensively in the food industry. The results indicated that the yeast biodegradable films showed potential feature for use in packaging industry (Delgado et al. 2018). EGD is used for detection of volatile substances from the sample during thermal degradation. Additionally, this technique could be coupled with techniques such as FTIR or mass spectrometry to study structural characterization of carriers during thermal degradation (Xie and Pan 2001).

In the case of DTA, both the sample and a reference material are exposed to similar heat treatment, and the difference in temperature is recorded. In DSC the heat required to enhance the temperature of sample and the reference material is measured. It should be noted that DSC provides different information, such as melting point, heat capacity, and recrystallization times of carrier and glass transition temperature, so that this technique is more common than DTA (Jafari and Esfanjani 2017). The thermal properties of the yeast cell micro-carrier of *Mentha pulegium* essential oil were analyzed by using DSC and TGA methods, and the results confirmed the protection of the essential oil (Kavetsou et al. 2019).

Purslane seed oil, as a potential nutritious source of ω-3, is susceptible to oxidation, so that encapsulation in *S. cerevisiae* is a possible approach to overcome this problem. Purslane seed oil was encapsulated in *S. cerevisiae* cells, and thermal behavior was analyzed. Based on the results, the presence of purslane seed oil in yeast microcapsules was confirmed by DSC analyses (Kavosi et al. 2018). The other studies used DSC method for investigation of encapsulated agents with *S. cerevisiae* cell in coupling with other techniques (Paramera et al. 2011b, Salari et al. 2013).

Physicochemical properties

Physicochemical properties of *S. cerevisiae* capsules can be characterized by different techniques, such as X ray diffraction (XRD), FTIR, nuclear magnetic resonance (NMR), and Raman spectroscopy (RS). XRD and NMR are widely used to characterize the degree of crystallinity of the capsules, especially wall materials (Anandharamakrishnan 2015). The application of NMR for characterization of the crystalline degree is based on the different observations for crystalline domains (broad component) and amorphous ones (narrow component) of nuclei of the same functional group (de Oca et al. 2004, Hronský et al. 2014). The fingerprint region of IR spectrum helps in measuring the changes of the sample during the encapsulation process (Li et al. 2016). This method is also used to evaluate the chemical interactions between capsules and the molecules (Ahmad et al. 2019).

Plasmolysis procedure is an important process in controlling an ingredient's permeation into the yeast cell. By IR technique, we can estimate the effect of the plasmolysis procedure on the structure of the cell wall and the membrane of the yeast cell. For plasmolyzed cells, some characteristic IR peaks were changed, which provided information about the effect of plasmolysis on other components of the yeast cell, such as proteins, lipids, nucleic acids, and carbohydrates of the whole yeast cell (Paramera et al. 2011b).

FTIR method was used to evaluate the physicochemical properties of yeast cells incorporation with purslane oil. The results indicated that increase of lipid concentration in the cell wall was due to the loss of other molecular cell constituents, and also that the vibration bands of amides zone were changed and their intensity was decreased. The changes in these regions are due to the protein degradation and transition of the degraded protein to an unfolded state (Kavosi et al. 2018).

De Medeiros et al. used FTIR method for investigation of yeast cell interaction with curcumin. The spectrum indicates that curcumin was encapsulated inside the yeast cell structure so that the signals for these characteristic bands are hidden (de Medeiros et al. 2019).

Jafari et al. reported that the chemical interactions between the encapsulated molecule and the microorganism cell components lead to weak absorption bands of the IR spectrum (Jafari et al. 2016).

In another study, FT-IR was also used to study changes in the yeast cells associated plasmolysis. It seems that upon plasmolysis, the yeast cell was disorganized to some degree (Paramera et al. 2011b). The results of other studies revealed the presence of cholecalciferol in the yeast cell microcapsules (Dadkhodazade et al. 2018).

Beside the FT-IR, the XRD method also was used to evaluate the physicochemical properties of both encapsulated material and carrier, such as the state of the material. The amorphous state shows a wide broad peak in XRD patterns. The *S. cerevisiae* capsules usually have the amorphous state (Guler and Sarioglu 2014, Volpato et al. 2018). The XRD patterns of encapsulated carriers were different form unloaded ones. For example, by encapsulation of cholecalciferol in the *S. cerevisiae* microcapsules, the XRD pattern was changed, which slightly moved and showed more intensity. Additionally, the characteristic peaks of cholecalciferol were not observed, suggesting that the vitamin was not in a crystalline state (Dadkhodazade et al. 2018).

Encapsulation efficiency and loading capacity

As definition, encapsulation yield (%EY) is the percentage of the encapsulated material with respect to the total used material for the preparation of the formulation, whereas loading capacity (LC) is the percentage of mass ratio between the encapsulated material and the total mass of the capsules (Hosseini et al. 2013). These parameters are usually characterized by spectroscopic and chromatographic techniques. UV/Vis spectroscopy is a simple and common method for evaluation of %EY, LC, and also releasing profile of encapsulated agents. Chromatographic techniques, including high performance liquid chromatography (HPLC) and gas chromatography (GC), are useful methods for both identification and quantification of the components. It should be noted that all of these methods require extraction of encapsulated molecule from the microcapsule by using solvents, and then preparation of calibration curve (Vishwakarma et al. 2016).

By encapsulation of berberine in *S. cerevisiae* capsules, the amount of this compound was determined by using indirect UV/Visible spectroscopy method (Salari et al. 2013).

Carvacrol and curcumin were encapsulated in *S. cerevisiae* microcapsules, and %EY of them was evaluated with the HPLC method (da Silva Lima et al. 2017, Paramera et al. 2011b). Shi et al. quantified the %EY of encapsulated resveratrol by the validated HPLC method (Shi et al. 2008).

Surface charge

The overall charge of a carrier in a given medium is mainly described in terms of zeta potential value. This value describes the potential stability, electrostatic interaction, and also mobility of the suspension or colloids (Anandharamakrishnan 2015, Tiede et al. 2008). The values ranging from −10 to +10 mV, indicate a neutral system and the formulation is not stable, whereas values lesser than −30 mV and more than +30 mV represent strongly anionic and cationic systems, respectively, and the formulation is stable and has lesser or no tendency to aggregate (McNeil 2011, Peres 2011).

References

Aguilar-Uscanga, B. and J.M. Francois. 2003. A study of the yeast cell wall composition and structure in response to growth conditions and mode of cultivation. Lett Appl Microbiol 37: 268–274.

Ahmad, M., P. Mudgil, A. Gani, F. Hamed, F.A. Masoodi and S. Maqsood. 2019. Nano-encapsulation of catechin in starch nanoparticles: Characterization, release behavior and bioactivity retention during simulated *in-vitro* digestion. Food Chem 270: 95–104.

Anandharamakrishnan, C. 2015. Spray drying techniques for food ingredient encapsulation. John Wiley & Sons, Ltd: Chichester.

Benczedi, D., A. Hahn, G. Trophardy, E. Cantergiani and W.R. 2007. Encapsulated hydrophilic compounds. In US0122398A1.

Beopoulos, A., T. Chardot and J.M. Nicaud. 2009. Yarrowia lipolytica: A model and a tool to understand the mechanisms implicated in lipid accumulation. Biochimie 91: 692–696.

Bishop, J.R.P., G. Nelson and J. Lamb. 1998. Microencapsulation in yeast cells. J Microencapsul 15: 761–773.

Blanquet, S., G. Garrait, E. Beyssac, C. Perrier, S. Denis, G. Hebrard and M. Alric. 2005. Effects of cryoprotectants on the viability and activity of freeze dried recombinant yeasts as novel oral drug delivery systems assessed by an artificial digestive system. European Journal of Pharmaceutics and Biopharmaceutics: Official Journal of Arbeitsgemeinschaft fur Pharmazeutische Verfahrenstechnik e.V, 61: 32–39.

Bowman, S.M. and S.J. Free. 2006. The structure and synthesis of the fungal cell wall. BioEssays: news and reviews. In Molecular, Cellular Develop Biol 28: 799–808.

Braun, P.C. 1999. Nutrient uptake by Candida albicans: the influence of cell surface mannoproteins. Can J Microbiol 45: 353–359.

Cao-Hoang, L., P.-A. Marechal, M. Lê-Thanh, P. Gervais and Y. Waché. 2008. Fluorescent probes to evaluate the physiological state and activity of microbial biocatalysts: A guide for prokaryotic and eukaryotic investigation. Biotechnol J 3: 890–903.

Chow, C.K. and S.P. Palecek. 2004. Enzyme encapsulation in permeabilized *Saccharomyces cerevisiae* cells. Biotechnol Prog 20: 449–456.

Ciamponi, F., C. Duckham and N. Tirelli. 2012. Yeast cells as microcapsules. Analytical tools and process variables in the encapsulation of hydrophobes in *S. cerevisiae*. Appl Microbiol Biotechnol 95: 1445–1456.

Czabany, T., K. Athenstaedt and G. Daum. 2007. Synthesis, storage and degradation of neutral lipids in yeast. Biochimica et Biophysica Acta 1771: 299–309.

Czerniak, A., P. Kubiak, W. Białas and T. Jankowski. 2015. Improvement of oxidative stability of menhaden fish oil by microencapsulation within biocapsules formed of yeast cells. J Food Eng 167: 2–11.

da Silva Lima, A., A.P. Maciel, C.D.J.S. Mendonça and L.M. Costa Junior. 2017. Use of encapsulated carvacrol with yeast cell walls to control resistant strains of Rhipicephalus microplus (Acari: Ixodidae). Ind Crops Prod 108: 190–194.

Dadkhodazade, E., A. Mohammadi, S. Shojaee-Aliabadi, A.M. Mortazavian, L. Mirmoghtadaie and S.M. Hosseini. 2018. Yeast cell microcapsules as a novel carrier for cholecalciferol encapsulation: development, characterization and release properties. Food Biophys 13: 404–411.

Dardelle, G., V. Normand, M. Steenhoudt, P.-E. Bouquerand, M. Chevalier and P. Baumgartner. 2007. Flavour-encapsulation and flavour-release performances of a commercial yeast-based delivery system. Food Hydrocoll 21: 953–960.

de Medeiros, F.G.M., S. Dupont, L. Beney, G. Roudaut, R.T. Hoskin and M.R. da Silva Pedrini. 2019. Efficient stabilisation of curcumin microencapsulated into yeast cells via osmoporation. Appl Microbiol Biotechnol.

De Nobel, J.G., C. Dijkers, E. Hooijberg and F.M. Klis. 1989. Increased cell wall porosity in *Saccharomyces cerevisiae* after treatment with dithiothreitol or EDTA. Microbiology 135: 2077–2084.

De Nobel, J.G., F.M. Klis, T. Munnik, J. Priem and H. van den Ende. 1990. An assay of relative cell wall porosity in *Saccharomyces cerevisiae*, *Kluyveromyces lactis* and Schizosaccharomyces pombe. Yeast (Chichester, England) 6: 483–490.

De Nobel, J.G. and J.A. Barnett. 1991. Passage of molecules through yeast cell walls: a brief essay-review. Yeast (Chichester, England) 7: 313–323.

de Oca, H.M., I. Ward, P. Klein, M. Ries, J. Rose and D. Farrar. 2004. Solid state nuclear magnetic resonance study of highly oriented poly (glycolic acid). Polymer 45: 7261–7272.

Delgado, J.F., M.A. Peltzer, A.G. Salvay, O. de la Osa and J.R. Wagner. 2018. Characterization of thermal, mechanical and hydration properties of novel films based on *Saccharomyces cerevisiae* biomass. Innov Food Sci Emerg Technol 48: 240–247.

Drozdek, S. and U. Bazylinska. 2016. Biocompatible oil core nanocapsules as potential co-carriers of paclitaxel and fluorescent markers: preparation, characterization, and bioimaging. Colloid Polym Sci 294: 225–237.

Dziezak, J.D. 1987. Yeast and yeast derivatives: applications. Food Technol 41: 122–125.

Fuller, E.J., C. Duckham and E.J. Wood. 2005. Using yeast to enhance the delivery of insulin through an epithelial monolayer. J Pharm Phramacol 57: S05.

Gaonkar, A.G., N. Vasisht, A.R. Khare and R. Sobel. 2014. Microencapsulation in the food industry: a practical implementation guide, Amsterdam: Elsevier.

Guler, U.A. and M. Sarioglu. 2014. Mono and binary component biosorption of Cu (II), Ni (II), and Methylene Blue onto raw and pretreated S. cerevisiae: equilibrium and kinetics. Desalination and Water Treat 52: 4871–4888.

Hosseini, S.F., M. Zandi, M. Rezaei and F. Farahmandghavi. 2013. Two-step method for encapsulation of oregano essential oil in chitosan nanoparticles: preparation, characterization and *in vitro* release study. Carbohydr Polym 95: 50–56.

Hronský, V., M. Koval'aková, P. Vrábel, M. Uhrínová and D. Olčák. 2014. Estimation of the degree of crystallinity of partially crystalline polypropylenes using 13 C NMR. Acta Physica Polonica, A 125.

Inoue, C., M. Ishiguro, N. Ishiwaki and Y. K. 1991. Process for preparation of microcapsules. In EP0453316B1.

Jafari, S.M. and A.F. Esfanjani. 2017. Instrumental analysis and characterization of nanocapsules. In: Nanoencapsulation Technologies for the Food and Nutraceutical Industries, pp. 524–544. Elsevier.

Jafari, Y., H. Sabahi and M. Rahaie. 2016. Stability and loading properties of curcumin encapsulated in *Chlorella vulgaris*. Food Chem 211: 700–706.

Kapteyn, J.C., H. Van Den Ende and F.M. Klis. 1999. The contribution of cell wall proteins to the organization of the yeast cell wall. Biochimica et Biophysica Acta 1426: 373–383.

Kavetsou, E., S. Koutsoukos, D. Daferera, M.G. Polissiou, D. Karagiannis, D.C. Perdikis and A. Detsi. 2019. Encapsulation of *Mentha pulegium* essential oil in yeast cell microcarriers: an approach to environmentally friendly pesticides. J Agric Food Chem 67: 4746–4753.

Kavosi, M., A. Mohammadi, S. Shojaee-Aliabadi, R. Khaksar and S.M. Hosseini. 2018. Characterization and oxidative stability of purslane seed oil microencapsulated in yeast cells biocapsules. J Sci Food Agric 98: 2490–2497.

Klis, F.M., P. Mol, K. Hellingwerf and S. Brul. 2002. Dynamics of cell wall structure in *Saccharomyces cerevisiae*. FEMS Microbiol Rev 26: 239–256.

Klis, F.M., A. Boorsma and P.W. De Groot. 2006. Cell wall construction in *Saccharomyces cerevisiae*. Yeast (Chichester, England) 23: 185–202.

Kogan, G., M. Pajtinka, M. Babincova, E. Miadokova, P. Rauko, D. Slamenova and T.A. Korolenko. 2008. Yeast cell wall polysaccharides as antioxidants and antimutagens: can they fight cancer? Neoplasma 55: 387–393.

Kohlwein, S.D. and F. Paltauf. 1984. Uptake of fatty acids by the yeasts, Saccharomyces uvarum and Saccharomycopsis lipolytica. Biochimica et Biophysica Acta 792: 310–317.

Laboratoire, S. 1973. Procédé pour faire pénétrer, absorber et/ou fixer, par des microorganismes des substances diverses. In FR2179528.

Li, J., G.H. Shin, I.W. Lee, X. Chen and H.J. Park. 2016. Soluble starch formulated nanocomposite increases water solubility and stability of curcumin. Food Hydrocoll 56: 41–49.

Li, S.C. and P.M. Kane. 2009. The yeast lysosome-like vacuole: endpoint and crossroads. Biochimica et Biophysica Acta 1793: 650–663.

Lipke, P.N. and R. Ovalle. 1998. Cell wall architecture in yeast: new structure and new challenges. J Bacteriol 180: 3735–3740.

Madene, A., M. Jacquot, J. Scher and S. Desobry. 2006. Flavour encapsulation and controlled release—a review. Int J Food Sci 41: 1–21.

McNeil, S.E. 2011. Characterization of Nanoparticles Intended for Drug Delivery. New York: Springer.

Mokhtari, S., S.M. Jafari, M. Khomeiri, Y. Maghsoudlou and M. Ghorbani. 2017a. The cell wall compound of *Saccharomyces cerevisiae* as a novel wall material for encapsulation of probiotics. Food Research International (Ottawa, Ont.) 96: 19–26.

Mokhtari, S., M. Khomeiri, S.M. Jafari, Y. Maghsoudlou and M. Ghorbani. 2017b. Descriptive analysis of bacterial profile, physicochemical and sensory characteristics of grape juice containing *Saccharomyces cerevisiae* cell wall-coated probiotic microcapsules during storage. Inter J Food Sci Technol 52: 1042–1048.

Nakhaee Moghadam, M., B. Khameneh and B.S. Fazly Bazzaz. 2019a. *Saccharomyces cervisiae* as an efficient carrier for delivery of bioactives: a review. Food Biophys 14: 346–353.

Nakhaee Moghadam, M., J. Movaffagh, B.S. Fazli Bazzaz, M. azizzadeh and A. Jamshidi. 2019b. Encapsulation of Zataria multiflora essential oil in *Saccharomyces cerevisiae*: sensory evaluation and antibacterial activity in commercial soup. Iranian Journal of Chemistry and Chemical Engineering (IJCCE).

Nelson, G. 2002. Application of microencapsulation in textiles. Inter J Pharmaceutics 242: 55–62.

Nelson, G., S.C. Duckham and M.E.D. Crothers. 2006. Microencapsulation in yeast cells and applications in drug delivery. pp. 268–281. *In*: Sonkë, S. (ed.). Polymeric Drug Delivery I: Particulate Drug Carriers, Vol. 923. American Chemical Society.

Nomura, T., Y. Kuriyama, S. Toyoda and Y. Konishi. 2018. Direct measurements of colloidal behavior of polystyrene nanoparticles into budding yeast cells using atomic force microscopy and confocal microscopy. Colloids Surf A Physicochem Eng Asp 555: 653–659.

Normand, V., G. Dardelle, P.E. Bouquerand, L. Nicolas and D.J. Johnston. 2005. Flavor encapsulation in yeasts: limonene used as a model system for characterization of the release mechanism. J Agric Food Chem 53: 7532–7543.

Pannell, N.A. 1990. Microencapsulation in microorganisms. European. Patent, EP 0242 135, B1.

Pannell, N.A. 1994. Encapsulation of material in microbial cells. In US005288632A.

Paramera, E.I., S.J. Konteles and V.T. Karathanos. 2011a. Stability and release properties of curcumin encapsulated in *Saccharomyces cerevisiae*, β-cyclodextrin and modified starch. Food Chem 125: 913–922.

Paramera, E.I., S.J. Konteles and V.T. Karathanos. 2011b. Microencapsulation of curcumin in cells of *Saccharomyces cerevisiae*. Food Chem 125: 892–902.

Paramera, E.I., V.T. Karathanos and S.J. Konteles. 2014. Chapter 23—Yeast cells and yeast-based materials for microencapsulation. pp. 267–281. *In*: Gaonkar, A.G., N. Vasisht, A.R. Khare and R. Sobel (eds.). Microencapsulation in the Food Industry. San Diego: Academic Press.

Peppler, H.J. 1982. Yeast extracts. pp. 293–312. *In*: Peppler, H.J. (ed.). Economic Microbiology. New York: Academic Press.

Peres, I.M.N.F.V. 2011. Encapsulation of active compounds: particle characterization, loading efficiency and stability. Doctor of Philosophy in Chemical and Biological Engineering, University of Porto Franc.

Pham-Hoang, B.N., C. Romero-Guido, H. Phan-Thi and Y. Wache. 2013. Encapsulation in a natural, preformed, multi-component and complex capsule: yeast cells. Appl Microbiol Biotechnol 97: 6635–6645.

Pham-hoang, B.N., A. Voilley and Y. Waché. 2016. Molecule structural factors influencing the loading of flavoring compounds in a natural-preformed capsule: Yeast cells. Colloids and Surfaces B: Biointer 148: 220–228.

Ren, T., J. Gou, W. Sun, X. Tao, X. Tan, P. Wang, Y. Zhang, H. He, T. Yin and X. Tang. 2018. Entrapping of nanoparticles in yeast cell wall microparticles for macrophage-targeted oral delivery of cabazitaxel. Mol Pharm 15: 2870–2882.

Rodriguez-Pena, J.M., V.J. Cid, J. Arroyo and C. Nombela. 2000. A novel family of cell wall-related proteins regulated differently during the yeast life cycle. Mol Cellular Biol 20: 3245–3255.

Salari, R., B.S. Bazzaz, O. Rajabi and Z. Khashyarmanesh. 2013. New aspects of *Saccharomyces cerevisiae* as a novel carrier for berberine. Daru 21: 73.

Salari, R., O. Rajabi, Z. Khashyarmanesh, M. Fathi Najafi and B.S. Fazly Bazzaz. 2015. Characterization of encapsulated berberine in yeast cells of *Saccharomyces cerevisiae*. Iran J Pharm Res 14: 1247–1256.

Schrooyen, P.M., R. van der Meer and C.G. De Kruif. 2001. Microencapsulation: its application in nutrition. The Proceedings of the Nutrition Soc 60: 475–479.

Shah, M.A., S.A. Mir and M. Bashir. 2018. Nanoencapsulation of food ingredients. pp. 218–234. *In*: Food Science and Nutrition: Breakthroughs in Research and Practice. IGI Global.

Shank, J.L. 1977. Encapsulation process utilizing microorganisms and products produced thereby. In US4001480.

Shi, G., L. Rao, H. Yu, H. Xiang, G. Pen, S. Long and C. Yang. 2007. Yeast-cell-based microencapsulation of chlorogenic acid as a water-soluble antioxidant. J Food Eng 80: 1060–1067.

Shi, G., L. Rao, H. Yu, H. Xiang, H. Yang and R. Ji. 2008. Stabilization and encapsulation of photosensitive resveratrol within yeast cell. Int J Pharm 349: 83–93.

Shi, G., L. Rao, Q. Xie, J. Li, B. Li and X. Xiong. 2010. Characterization of yeast cells as a microencapsulation wall material by Fourier-transform infrared spectroscopy. Vibrational Spectro 53: 289–295.

Sultana, A., A. Miyamoto, Q. Lan Hy, Y. Tanaka, Y. Fushimi and H. Yoshii. 2017. Microencapsulation of flavors by spray drying using *Saccharomyces cerevisiae*. J Food Eng 199: 36–41.

Suomalainen, H. and T. Nurminen. 1970. The lipid composition of cell wall and plasma membrane of baker's yeast. Chem Phys Lipids 4: 247–256.

Svaldo-Lanero, T., S. Krol, R. Magrassi, A. Diaspro, R. Rolandi, A. Gliozzi and O. Cavalleri. 2007. Morphology, mechanical properties and viability of encapsulated cells. Ultramicroscopy 107: 913–921.

Ta, T.M., L. Cao-Hoang, H. Phan-Thi, H.D. Tran, N. Souffou, J. Gresti, P.A. Marechal, J.F. Cavin and Y. Wache. 2010. New insights into the effect of medium-chain-length lactones on yeast membranes. Importance of the culture medium. Appl Microbiol Biotechnol 87: 1089–1099.

Ta, T.M., L. Cao-Hoang, C. Romero-Guido, M. Lourdin, H. Phan-Thi, S. Goudot, P.A. Marechal and Y. Wache. 2012. A shift to 50 degrees C provokes death in distinct ways for glucose- and oleate-grown cells of Yarrowia lipolytica. Appl Microbiol Biotechnol 93: 2125–2134.

Tanguler, H. and H. Erten. 2008. Utilisation of spent brewer's yeast for yeast extract production by autolysis: The effect of temperature. Food and Bioproducts Processing 86: 317–321.

Tiede, K., A.B. Boxall, S.P. Tear, J. Lewis, H. David and M. Hassellov. 2008. Detection and characterization of engineered nanoparticles in food and the environment. Food Addit Contam Part A Chem Anal Control Expo Risk Assess 25: 795–821.

Vimala Bharathi, S.K., J.A. Moses and C. Anandharamakrishnan. 2018. Nano and microencapsulation using food grade polymers. pp. 357–400. *In*: Polymers for Food Applications. Springer International Publishing.

Vishwakarma, G.S., N. Gautam, J.N. Babu, S. Mittal and V. Jaitak. 2016. Polymeric encapsulates of essential oils and their constituents: A review of preparation techniques, characterization, and sustainable release mechanisms. Pol Rev 56: 668–701.

Volpato, H., D.B. Scariot, E.F.P. Soares, A.P. Jacomini, F.A. Rosa, M.H. Sarragiotto, T. Ueda-Nakamura, A.F. Rubira, G.M. Pereira, R. Manadas, A.J. Leitao, O. Borges, C.V. Nakamura and M.D.C. Sousa. 2018. *In vitro* anti-Leishmania activity of T6 synthetic compound encapsulated in yeast-derived beta-(1,3)-d-glucan particles. Int J Biol Macromol 119: 1264–1275.

Wu, Y., C. Zhong, T. Du, J. Qiu, M. Xiong, Y. Hu, Y. Chen, Y. Li, B. Liu, Y. Liu, B. Zou, S. Jiang and M. Gou. 2018. Preparation and characterization of yeast-encapsulated doxorubicin microparticles. J Drug Deliv Sci Technol 45: 442–448.

Xie, W. and W.-P. Pan. 2001. Thermal characterization of materials using evolved gas analysis. J Therm Anal Calorim 65: 669–685.

Young, S., S. Dea and N. Nitin. 2017. Vacuum facilitated infusion of bioactives into yeast microcarriers: Evaluation of a novel encapsulation approach. Food Research International (Ottawa, Ont.) 100: 100–112.

Zlotnik, H., M.P. Fernandez, B. Bowers and E. Cabib. 1984. *Saccharomyces cerevisiae* mannoproteins form an external cell wall layer that determines wall porosity. J Bacteriol 159: 1018–1026.

Zuidam, N.J. and E. Heinrich. 2010. Encapsulation of aroma. pp. 127–160. *In*: Zuidam, N.J. and V. Nedovic (eds.). Encapsulation Technologies for Active Food Ingredients and Food Processing. New York, NY: Springer New York.

6

Biotransformation of Toxic Thiosulfate into Merchandisable Elemental Sulfur by Indigenous SOB Consortium

Panteha Pirieh and *Fereshteh Naeimpoor**

Introduction

Reduced sulfur compounds (RSC), such as thiosulfate in wastewater or gaseous streams, can have adverse effects on human health and the environment. Thiosulfate ($S_2O_3^{2-}$), an oxyanion of sulfur is discharged as an effluent by many chemical process industries. In most cases, thiosulfate comes to industrial wastewater as a result of sulfide oxidation (Baquerizo et al. 2013, Xu et al. 2017). The sulfur present in the natural gas from offshore gas production installations is oxidized, and the generated wastewater contains a predominant amount of thiosulfate at concentrations as high as 3000 mg/L. Thiosulfate is also used as a photographic fixer in photographic processing laboratories, and is consumed for extraction of gold at 0 to 220 g-thiosulfate L^{-1} (Abdel-Monaem Zytoon et al. 2014, Ahmad et al. 2014, Xu et al. 2017).

Thiosulfate can be transformed into polythionates and sulfate under oxidizing conditions, or into toxic sulfide ion under anaerobic or reducing conditions. So, uncontrolled thiosulfate disposal can cause de-oxygenation of waterways and the potential formation of toxic compounds. Researchers, therefore, have been working on the development of suitable treatment methods to reduce the concentration of thiosulfate in aqueous solution down to a permissible limit, which is 100 ppm (Ahmad et al. 2014, Qian et al. 2015).

Biotechnology Research Laboratory, School of Chemical, Petroleum and Gas Engineering, Iran University of Science and Technology, Narmak, Tehran, 1684613114, Iran.
Email: pantea.pirie@gmail.com
* Corresponding author: fnaeim@iust.ac.ir

Both chemical and biological technologies have been applied to remove thiosulfate from wastewater. However, biological systems with sulfur-oxidizing bacteria (SOB) have been demonstrated to be a feasible and efficient way to treat RSC by avoiding the disadvantages associated with chemical processes, such as operational costs and the use of hazardous chemicals (Gholipour et al. 2018, Lin et al. 2018). In the presence of nutritional demands, SOB can play a key role in the biological elimination of S-compounds (Valdebenito-Rolack et al. 2011). SOB are metabolically divided into two groups—chemolithotrophs and photoautotrophs. The former oxidize RSC to provide energy, while the latter use light as energy and a reduced S-compound for reduction of CO_2 into various carbohydrates (Pokorna and Zabranska 2015, Luo et al. 2018). Phototrophic SOB grow slowly while chemolithotrophs grow faster. Colorless chemolithotrophs are known as the most useful SOB due to their highest rate of S-compound oxidation. Therefore, bio-oxidation is normally performed by autotrophic aerobic SOB, where elemental sulfur, sulfate, and/or intermediate products, such as thionates are generated (Equations 1–7) depending on the specific molar ratios of O_2/S (Tang et al. 2009, Vosoughi et al. 2015).

$$S_2O_3^{2-} + H_2O + 2\ O_2 \rightarrow 2\ SO_4^{2-} + 2\ H^+ \tag{1}$$

$$S_2O_3^{2-} + \tfrac{1}{2}\ O_2 + H_2O \rightarrow S^0 + SO_3^{2-} + 2\ OH^- \rightarrow S^0 + SO_4^{2-} + 2\ H^+ \tag{2}$$

$$S^0 + O_2 + H_2O \rightarrow SO_4^{2-} + 2\ H^+ \tag{3}$$

$$S_2O_3^{2-} + \tfrac{1}{4}\ O_2 + \tfrac{1}{2}\ H_2O \rightarrow \tfrac{1}{2}\ S_4O_6^- + OH^- \tag{4}$$

$$2\ S_4O_6^- \rightarrow S_3O_6^- + S_5O_6^- \tag{5}$$

$$S_5O_6^- \rightarrow S_4O_6^- + S^0 \tag{6}$$

$$S_3O_6^- + 3\ OH^- \rightarrow 3\ SO_3^{2-} + 3\ H^+ \tag{7}$$

Among thiosulfate oxidation products, biological-S^0 is a favorable product since it can be exploited as a basic chemical in various industries or as fertilizers in agriculture. Moreover, it has a hydrophilic character, which makes it more applicable (Lin et al. 2018). Since it is practically difficult to maintain a situation which is respectable for sulfur formation during the S-compound oxidation period, most of the research is just focused on how to remove the toxic thiosulfate, and little attention has been paid to oxidation products, such as elemental sulfur and their application in industry. So in this chapter, we aim to investigate effective factors in products distribution of thiosulfate bio-oxidation by focusing on thiosulfate biotransformation to elemental sulfur. In the end, oxidation of high thiosulfate concentrations was examined in a magnetic and stirred bioreactor.

Materials and methods

Microbial source

An appropriate microbial source for thiosulfate bio-oxidation was obtained from the soil, which was collected from Ramhormoz region in Ahvaz, South of Iran.

To prepare the microbial species source, 10 g soil sample was suspended in sterile physiological serum (8.5% NaCl), and the mixture was shaken at 30°C for 24 hours. Finally, the supernatant was filtered into sterile flasks and stored at 4°C as the original cell suspension (Luo et al. 2013, Pirieh and Naeimpoor 2019).

Culture medium

Minimal salt (MS) medium containing (in gl^{-1} each): KH_2PO_4, 2; K_2HPO_4, 2; $FeCl_3$, 0.01; $MgCl_2$, 0.2; and NH_4Cl, 0.4 was used in all experiments (Potivichayanon et al. 2006). This medium was supplemented with the desired amounts of thiosulfate pentahydrate ($Na_2S_2O_3.5H_2O$) as the energy source. Apart from experiments carried out on enrichment and investigation of the effect of additions of $NaHCO_3$ as co-carbon substrate (Gerrity et al. 2016) and $CaCl_2$ for better cell visualization, MS medium was also supplemented with a ratio of 1:4 (bi-carbonate/thiosulfate) and $CaCl_2$ at $0.05gL^{-1}$ in all experiments. For plate cultivations, 1.8% agarose was added to MSM containing 3000 ppm thiosulfate to solidify the medium.

Analytical methods

Experimental analyses

At each sampling time, the content of each flask was centrifuged, and the supernatant was stored at 4°C for thiosulfate and sulfate analyses, while the precipitate was used for S^0 analysis. Sulfate was measured by a turbimetric method based on precipitation of sulfate with barium chloride (Liu et al. 2017, Kalantari et al. 2018). Light absorbance of $BaSO_4$ suspension was then measured by spectrophotometer (Beam, model: UVS-2800) at 420 nm wavelength (Kolmerta et al. 2000). Iodometric method was exploited to determine thiosulfate concentration (Vogel 1996). Iodine solution, golden-brown in color, was titrated against sodium thiosulfate solution to turn into a pale yellow color. Subsequently, adding a few drops of a freshly prepared starch solution into this solution shifted its color to blue-black, followed by continued titration until turning into a colorless solution (Rodier 1998). For S^0 analysis, acetone was first added to the precipitate to dissolve it. After 24 hours of rest, followed by vortex mixing and centrifugation, the supernatant containing S^0 was reacted with cyanide to form CNS^-. A red complex was then formed by reacting CNS^- with Fe^{3+} and the absorbance of this solution was then measured at 460 nm (Troelsen and Jørgensen 1982, Pirieh and Naeimpoor 2019).

To make the results more comparable, percentages of thiosulfate oxidation, as well as sulfate and elemental sulfur formation, are calculated based on S-atom and concentration of other products (OP), which was then calculated according to Equation 8 (Pirieh and Naeimpoor 2019):

$$[OP]_t = \left(\left[S - S_2O_3^{2-}\right]_0 - \left[S - S_2O_3^{2-}\right]_t - \left[SO_4^{2-}\right]_t - \left[S^0\right]_t\right) \tag{8}$$

in which brackets denote concentration with subscripts 0 and t referring to time.

Microbial analyses

Culture optical density of fresh samples was read at 600 nm as a measure of cell concentration (Mirzaei et al. 2014). Readers are notified that for cultures having high S^0 levels, OD imprecisely implies cell concentration due to the interference of S^0 in the sample.

The identification of the strains in SOB culture was carried out by 16S rRNA sequence. Total DNA was extracted by a modified procedure of Marmur (Marmur 1961). Gained genes were sequenced by Applied Biosystems 3730/3730xL DNA Analyzers (Bioneer, Korea) using the Sanger method. The elated sequences were obtained from the Gen-Bank database using BLASTN program (Mahmood et al. 2009, Valle et al. 2018). The 16S rRNA sequences determined and reference sequences obtained from Gen-Bank databases were aligned using multiple sequence alignment software CLUSTAL.W ver. 1.81. Phylogenetic trees were constructed with Molecular Evolutionary Genetics Analysis software (Mega 3.1) based on the 16S rRNA sequences of some strains closer to isolated strains PN1 and PN2.

Experiments

Thiosulfate oxidation and the effect of various factors were investigated in shake flasks (SF) as well as bioreactors. Four sets of experiments were performed in shake flasks. This included enrichment, supplementation of MS medium, and the effect of inoculum age and source, as well as interference of elemental sulfur in inoculum on thiosulfate oxidation at high levels. All experiments were carried out in 250 mL Erlenmeyer flasks containing 100 mL medium. After inoculation, flasks were incubated at 30°C and 150 rpm for 6–8 days.

Two types of bioreactors: a magnetically stirred bioreactor (MSR) and a stirred bioreactor (SR) were exploited. MSR consisted of a 1 L glass vessel (500 cc working volume) placed in a water bath on top of a magnetic stirrer. SR consisted of 5 L borosilicate glass vessel (1000 cc working volume) with 2 stainless steel baffles and 3-blade-pitch impellers. Air was filtered and humidified before introduction into each bioreactor.

SF experiments

Cell enrichment

Thiosulfate was used to enrich the original microbial source (Graff and Stubner 2003, Yang et al. 2014, Jadhav and Jadhav 2017). Flasks containing MSM supplemented with thiosulfate sodium at 400 ppm (equivalent to 103 mg-SL^{-1}) were inoculated with 5 mL of the original cell suspension. After incubation, the enriched culture was centrifuged and cells were re-suspended in 10 mL of MSM, and this enriched cell suspension was used to inoculate flasks containing 700 ppm thiosulfate (equivalent to 180 mg-SL^{-1}). Cell growth and product formation were then investigated over time. Control experiments in the absence of cells were carried out to assess chemical oxidation.

Effect of bicarbonate and $CaCl_2$

The effect of supplementation of MS medium containing 700 ppm of thiosulfate with $CaCl_2$ and $NaHCO_3$ was investigated in a one factor at a time manner. At first, we examined the effect of presence/absence of $CaCl_2$ at 50 mgL^{-1} to investigate whether this compound could enhance visualization of our colorless SOB cells. Since MS medium contains no C-source (C is supplied through CO_2 transfer via aeration), $NaHCO_3$ has been used by some researchers as co-carbon substrate in oxidation of reduced sulfur compounds (Tang et al. 2009, Gerrity et al. 2016). Therefore, we examined the effect of addition of bicarbonate at a ratio of 1:4 bicarbonate/S-compound in the next step.

Effect of SOB inoculum source and age

Thiosulfate oxidation rate is known to be dependent on the activity of SOB, especially at high thiosulfate levels. Therefore, we attempted to investigate the effect of inoculum source (suspended cells and cells on plate) and age in the efficiency of thiosulfate oxidation at 3000 ppm (equivalent to 774 $mg\text{-}SL^{-1}$). Original cells were first cultured at 1000 ppm for 1 and 8 days, and cells were then separately centrifuged and re-suspended in 10 mL of MSM to prepare 1 and 8 day old inoculum (method A1 and A8). To inoculate from cells on plates, cells should first be cultivated on plates. To do this, enriched cells (at 1000 ppm thiosulfate) were separately cultivated at 3000 ppm thiosulfate for 1 and 8 days. Subsequently, these cultures were spread on agarose-plates, and incubated for 8 days at 30°C (method B1 and B8). A loopful from each plate was used to inoculate the medium containing 3000 ppm thiosulfate. Cell growth as well as products distribution obtained by using methods A and B were then compared.

Interference of elemental sulfur in inoculum

To assess the effect of elemental sulfur (trapped on cells) in inoculum on thiosulfate oxidation performance, we set an experiment in which inoculum contained biological elemental sulfur adhered to cells. Original cells enriched at 3000 ppm containing elemental sulfur were used to inoculate a medium containing 3000 ppm thiosulfate. Oxidation performance and products distribution were then investigated.

Experiments in bioreactor

Thiosulfate oxidation in MSR and effect of aeration rate

To investigate thiosulfate oxidation and its product distribution at the higher level of 4000 ppm (equivalent to 1032 $mg\text{-}SL^{-1}$), a MSR was exploited using two aeration rates of 1 and 2 vvm. Batch experiments were performed at 35 ± 0.5°C and initial pH of 7 ± 0.1 using 425 mL MSM containing 4000 ppm thiosulfate, 0.8 gl^{-1} $MgCl_2$ and $NaHCO_3$ at carbonate/thiosulfate of 1:4 (w/w) (appropriate conditions, unpublished data). Cells maintained on plates prepared by method B1 were first pre-cultured in a medium containing 3000 ppm, and 75 mL of the culture was taken on day one and used as inoculum for MSR. Samples were taken every 12 h, and concentrations of thiosulfate, sulfate, and S^0, as well as pH, were measured.

Thiosulfate oxidation in SR and effect of agitation speed

Thiosulfate oxidation in SR was investigated using two agitation speeds of 60 and 90 rpm under 1 vvm aeration rate. Batch experiments were performed at 30 ± 0.5°C, with initial pH of 7 ± 0.1 using 900 mL MS medium containing 3000 ppm thiosulfate (similar to SF experiments). The bioreactor was filled with 1 L medium, and inoculated with 100 mL of a 1-day-old SOB. A 10 ml sample was taken every 12/24 h, and concentrations of thiosulfate, sulfate, and sulfur were measured, as previously mentioned.

Results and discussions

Enrichment

Cultivation of the original cells in MSM containing 400 ppm thiosulfate resulted in an optical density of 0.11 after 8 days when all the thiosulfate was oxidized. This enriched culture was then used as the inoculum for oxidation of 700 ppm thiosulfate. Time courses of the optical density (OD) and pH at 700 ppm initial thiosulfate are illustrated in Fig. 6.1. It can be seen that cell growth was higher, with culture OD reaching 0.165. Although culture pH decreased from 6.8 to a final value of 5.2, it showed a maximum of 7. This can be explained by formation of OH^- (Equation 4) early in the culture followed by formation of sulfate (65%) as an end product through oxidation of intermediate sulfur products, resulting in lowering of the culture pH. As Fig. 6.1 illustrates, sulfate was the main product of oxidation, while an insignificant amount of elemental sulfur as the merchandisable product was observed. Sulfate has been previously reported as the main product of aerobic microbial oxidation of S-compounds by many researchers (Moussavi et al. 2007, González-Sánchez and Revah 2009, Moghanloo et al. 2010, de Graaff et al. 2012). SOB can gain more energy from oxidation of reduced S-compounds into sulfate because of the high Gibbs free energy (ΔG_{So4} = 818kJ) of this reaction (Equation 1) (Kalantari et al. 2018) compared to other S-products ($\Delta G_{S}{}^{0}$ = 213kJ). Maximum elemental sulfur formation (9 ppm) was seen on the third day. Formation of S^0 through Equation 2

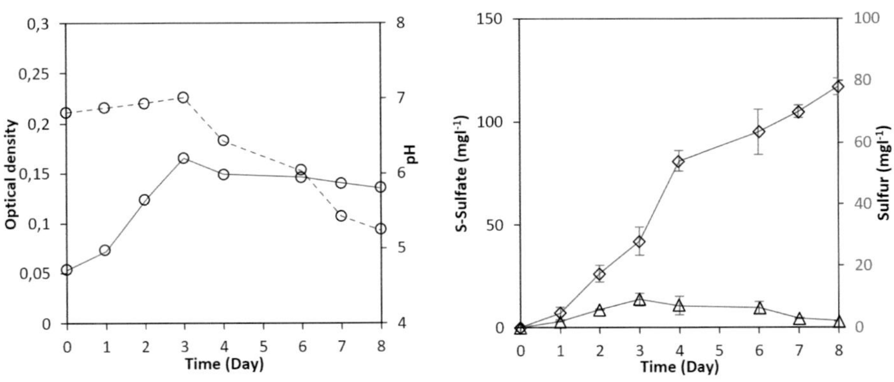

Figure 6.1. Trend of cell growth (−) and pH (- -) as well as formations of sulfate (◊) and sulfur (Δ) during oxidation of 700 ppm thiosulfate by enriched culture at 400 ppm.

or Equations 4–6 is supported by a molar ratio of O_2/(S-thiosulfate) ≤ 1.0, whereas higher ratios favor sulfate formation via Equation 1 (Bonilla-Blancas et al. 2015, Sun et al. 2017).

Despite the fact that 700 ppm thiosulfate was completely removed by this enriched culture, little elemental sulfur was formed. Therefore, to enhance sulfur formation, further investigations on the effective factors were required.

Effect of $CaCl_2$ and bicarbonate on growth and products distribution

Figure 6.2 shows the results of 700 ppm thiosulfate oxidation in the presence of $CaCl_2$. Comparison of these results with those in Fig. 6.1 in the absence of $CaCl_2$ illustrates that cell growth and products distribution was only slightly affected. However, deposition of calcium was observed in cell wall of our colorless SOB during growth. This makes visualization of colorless bacteria possible in suspended and solid state cultivations.

When bicarbonate and $CaCl_2$ were simultaneously present in medium (see Fig. 6.2), OD was always higher compared to the case without bicarbonate. The unexpected increase observed in OD on the third day could be due to the presence of elemental sulfur produced in the culture during oxidation. Increasing turbidity of the medium due to elemental sulfur production had also been previously reported by researchers (Krishnakumar and Manilal 1999).

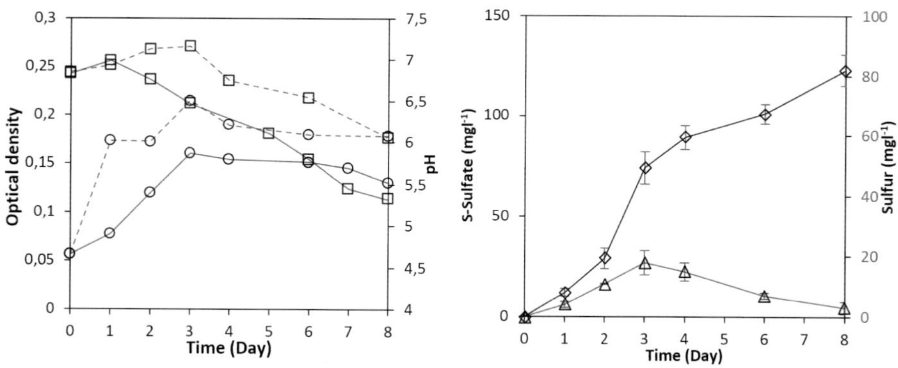

Figure 6.2. Trends of cell growth (○) and pH (□) as well as concentrations of sulfate (◊) and sulfur (Δ) during oxidation of 700 ppm thiosulfate in the presence of $CaCl_2$ (solid lines) and bicarbonate alongside $CaCl_2$ (dashed lines).

Ion analysis revealed that by the end of cultivation, thiosulfate was completely oxidized and 68% sulfate was formed. Additionally, formation of elemental sulfur of 10% (18 ppm) was enhanced two-fold. The overall decrease in culture pH was about 1 unit, while this was 1.5 in the absence of bicarbonate. This is due to the presence of bicarbonate and the amount of OH^- ions and carbon dioxide produced in the medium through the chemical interactions (Equations 9 and 10) during the oxidation.

$$H_2O + HCO_3^- \rightarrow OH^- + H_2CO_3 \quad (9)$$

$$H_2CO_3 \rightarrow H_2O + CO_2 \quad (10)$$

Bicarbonate is reported to play roles as both a carbon source for SOB and an electron acceptor in the oxidation process (Tang et al. 2009). Due to the advantages of supplementing MS medium with bicarbonate and $CalCl_2$, all subsequent experiments were performed in the presence of these two compounds.

Effect of SOB inoculum source and age on sulfur formation

As mentioned in the preceding sections, microorganisms seek to obtain more energy during the oxidation process by complete oxidation of thiosulfate into sulfate as the end product via Equation 1 (Gibbs energy of 818 KJ) (Kalantari et al. 2018). However, we tend to have sulfur as a desirable product due to its industrial applications. We investigated the effect of inoculum source and age on sulfur formation. Oxidation of 3000 ppm thiosulfate was first examined by methods A1 and A8, where method A1 was found superior with respect to cell growth and sulfur formation. The results obtained using method A1 are shown in Fig. 6.3. It can be seen that method A1 showed the maximum growth on day one.

Comparisons of the results of methods A and B are given in Table 6.1. Overall methods A1 and B1 (1 day-old inoculum) performed better than A8 and B8 (8 day-old inoculum). It can be seen that using A8 increases the time of maximum growth to day 3 compared to method A1. Similarly, the maximum sulfur formation of 20% (equivalent to 155 ppm) was higher for A1, and the time required for reaching this

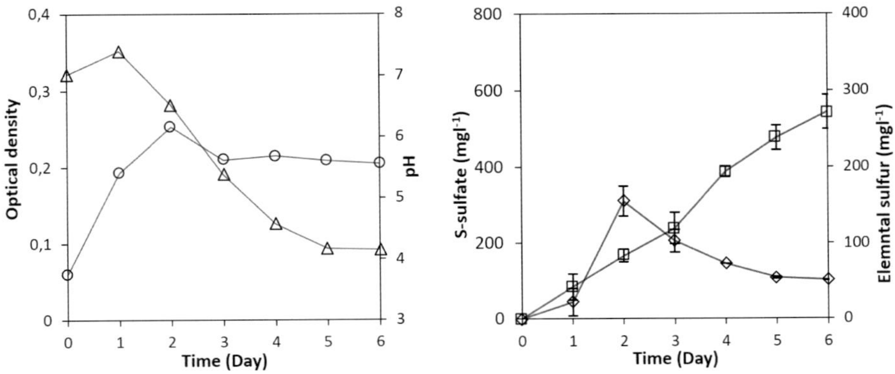

Figure 6.3. Time course of cell growth (○) and pH (Δ) as well as concentrations of sulfate (□) and sulfur (◊) during 3000 ppm thiosulfate oxidation using method A1.

Table 6.1. Comparison of 3000 mgl^{-1} thiosulfate oxidation and its products using methods A and B.

Method	Maximum growth		Final products (%)			Max. S^0	
	Time (day)	OD	Sulfate	Sulfur	OP	Time (day)	%
A1	1	0.193	70	6.5	23.5	2	20
A8	3	0.163	62	2.5	35.5	5	14
B1	1	0.481	76	0	24	1	30
B8$^+$	1	0.125	30	0	50	1	0.77

+ incomplete thiosulfate oxidation

value was lower (2 days) compared to method A8 (with 14% at day 5). Ion analyses indicated that thiosulfate was almost completely oxidized in both cases, and was transformed into 62–70% sulfate, reducing the pH of culture medium.

To check the effect of inoculum source, SOB maintained on solid plate (method B) was used. Table 6.1 shows that keeping the active microorganisms (1 day-old) on the solid plate (B1) increased the cell growth of the bacteria up to 0.481. This sharp increase of OD could also be due to the presence of elemental sulfur in the tested samples. Additionally, growth of cell by method B8 (0.125) was minor compared to the active consortium (B1), and the pH of the culture medium had not decreased significantly.

Products distribution showed that the fresh SOB (B1) have shown greater ability to remove thiosulfate and produce sulfate and sulfur. As given in Table 6.1, cells from method B8 were not capable of complete removal of thiosulfate (20% remaining), while active SOB (B1) oxidized almost all of the thiosulfates. A substantial amount of elemental sulfur, 30% (230 ppm), was produced till day one by method B1, while in the S^0 6 ppm was formed by the inactive consortium (B8). Chen et al. also indicated that using the younger inoculum enhanced S-component removal efficiency. Old inoculum potentially contains a high proportion of spores that are unable to revert to the vegetative cell cycle and have limited cell growth and metabolic pathways. This may explain why the thiosulfate oxidation ability of the old inoculum was lower (Chen et al. 2018). Therefore, younger SOB inoculum preserved on a solid plate is superior in complete thiosulfate oxidation at high concentrations and production of higher elemental sulfur.

Interference of elemental sulfur in inoculum

A certain amount of biological-elemental sulfur (22 ppm), which was produced in the previous step, was added to MS medium and the effect of elemental sulfur oxidation alongside thiosulfate oxidation (3000 ppm) was investigated. Results illustrated in Fig. 6.4 show that the cell growth rate decreased and a one-day lag phase was

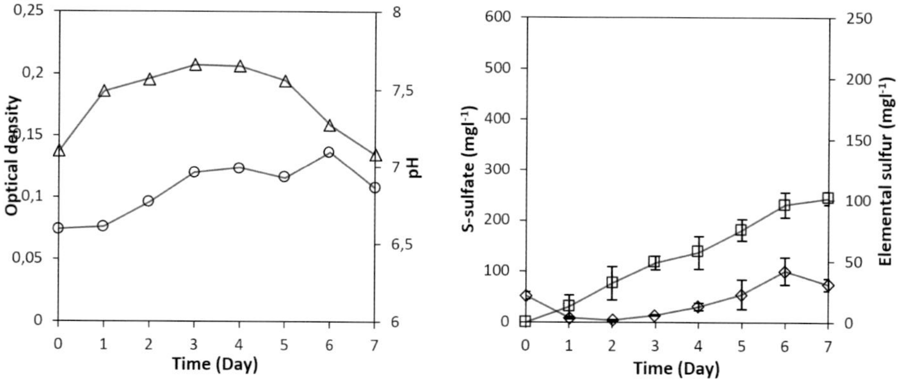

Figure 6.4. Trend of cell growth (○), pH (Δ), sulfate (□), and sulfur (◊) formation during simultaneous oxidation of 3000 ppm thiosulfate and 22 ppm biological-elemental sulfur.

observed. There was also a slight decrease in the pH of the culture, indicating low sulfate production in the medium.

Results show that sulfate production started from the first day of the oxidation process alongside the reduction of sulfur. This reveals that the primary biological-elemental sulfur has been converted to sulfate during the oxidation process by Equation 3. Gibbs energy of Equation 3 (587 KJ) is higher than the Equations 2 and 4, so the microbial consortium tends to perform this reaction for more energy, especially at the beginning of oxidation when the S/O ratio is high. Moreover, affinity constant for thiosulfate consumption ($K_{TS} = 0.0023$) is less than biological-elemental sulfur ($K_{TS} = 0.0030$–0.833) (Mora et al. 2016) and oxidation of the fresh biological-elemental sulfur occurs at a much faster rate than dried, powdered, and chemical S^0 particles (Fortuny et al. 2010). So, as the Fig. 6.4 shows, almost all the initial sulfur was oxidized to sulfate at day one. However, as in some previous experiments, the OD of culture increased on the sixth day due to the increase in sulfur content in the medium from thiosulfate oxidation by Equations 2 or 6.

Besides, titration analysis revealed that the oxidation process removed 87% of the initial thiosulfate within 7 days, but only 31% was converted to sulfate and 4% to elemental sulfur, with the remaining being intermediate products. The maximum elemental sulfur production during the process (5.4%) also decreased compared to the previous state. Thus, diminution of the ability of the microbial consortium to completely oxidize thiosulfate during 7 days and produce sulfur could possibly be due to sulfur adhesion to cells and cells agglomeration phenomenon.

Thiosulfate oxidation in MSR and effect of aeration rate

Results of thiosulfate (4000 ppm) oxidation in MSR at 1 and 2 vvm aeration are shown in Fig. 6.5. It can be seen that 80% of the initial thiosulfate was oxidized within 65 h at 1 vvm. Interestingly at this lower aeration rate, concentration of sulfur was always higher than sulfate during the oxidation process due to the limitations in aeration rate, and hence microbial activity. Of the oxidized thiosulfate, 44% and

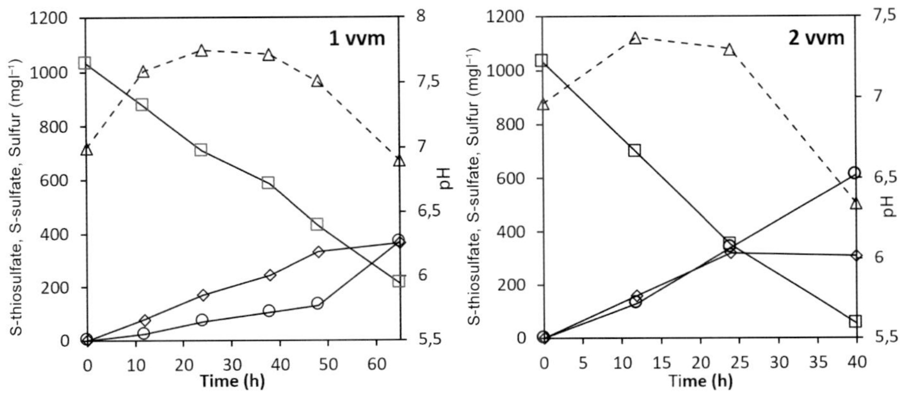

Figure 6.5. Trend of pH (Δ) and thiosulfate removal (□) as well as formations of sulfate (○) and sulfur (◊) at 4000 ppm thiosulfate at 1 and 2 vvm.

35% were transformed into elemental sulfur and sulfate, respectively, due to the low oxygen transfer rate.

Since 20% of thiosulfate remained non-oxidized at 1 vvm, aeration rate was doubled to 2 vvm in the next experiment. The results (see Fig. 6.5) show that increasing the aeration rate leads to a rise in the rate of thiosulfate oxidation from 0.41 to 0.88 mmol-SL^{-1}h^{-1} during the first 24 h, and more than 95% of the initial thiosulfate was oxidized over 40 hours. Although both rates of sulfur and sulfate production increased, the rate of sulfate production exceeded that of sulfur after 24 h due to exposure of SOB to high availability of dissolved oxygen. Sulfate production finally reached 59% at a rate of 0.48 mmol-SL^{-1}h^{-1}. Despite the increase in the rate of thiosulfate oxidation at 2 vvm, sulfur formation of 32% (310 ppm) was less than that in the lower aeration rate (370 ppm). To satisfy the original target, lower aeration rates and better mixing under modified cultivation could be used. Modification in cultivation conditions based on shake flask optimization (data not given) resulted in 98% oxidation of 3000 ppm thiosulfate after 24 h and formation of 35% elemental sulfur and 48% sulfate.

Thiosulfate oxidation in SR and effect of agitation speed

To investigate the effect of mixing at lower aeration rates on thiosulfate oxidation and products distribution, thiosulfate oxidation at 3000 ppm was examined at 1 vvm in SR at 60 and 90 rpm agitation speed. Results shown in Fig. 6.6 illustrate that complete thiosulfate removal at 60 rpm occurs at 36 hours. This long period can be explained by the low agitation speed used, which provides insufficient mixing, and hence oxygen transfer. Under this oxygen limitation condition, 40% elemental sulfur was formed at an overall rate of 0.27 mmol-SL^{-1}h^{-1}. Since the culture runs for a long time of 36 h, intermediate products (such as tetrathionate) are more exposed to aeration, and hence are further oxidized to sulfate from hour 24.

Due to insufficient mixing and hence long required period of oxidation, stirring speed was increased to 90 rpm. Figure 6.6 shows that complete oxidation occurs at

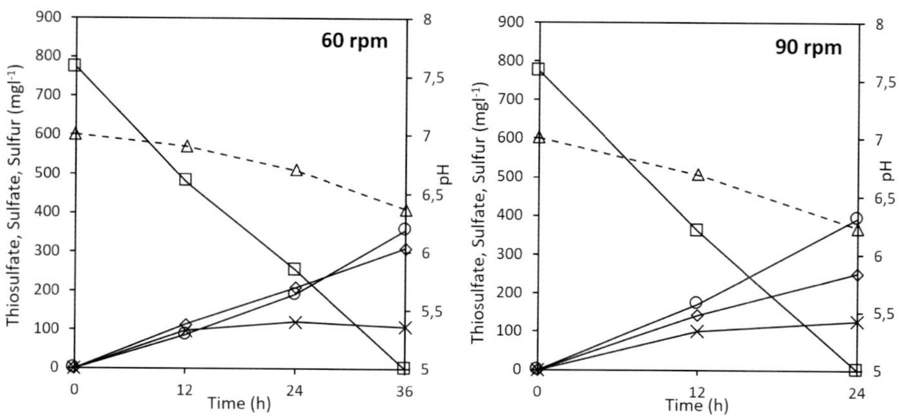

Figure 6.6. Trend of pH (Δ) and thiosulfate (□) removal sulfate (○) and sulfur (◊) formation in 3000 ppm thiosulfate at 60 and 90 rpm at 1 vvm.

24 h, which is less than for the lower speed. Formations of 32% elemental sulfur and 50% sulfate were obtained at 90 rpm.

Comparison of bioreactor results is given in Table 6.2. Thiosulfate was removed more rapidly in SR at 90 rpm (1.04 mmol-SL^{-1}h^{-1}) compared to MSR (0.98 mmol-SL^{-1}h^{-1}). The production of elemental sulfur and sulfate as the final oxidation products was almost similar to those for MSR. Elemental sulfur production decreased in SR with agitation speed due to the improved mixing.

Therefore, the results show that thiosulfate (3000 ppm) was eliminated with higher efficiency in less time and with approximately the same percentage of end-products, though with higher cost using SR. If the goal is only to produce more elemental sulfur without considering thiosulfate removal efficiency, this can be achieved by lower stirrer speed (60 rpm).

Table 6.2. Comparison of oxidation time, overall thiosulfate removal rate, and production of oxidation products (Sulfate (Sul.) and S^0) in two kinds of bioreactors at vvm = 1.

Reactor type	Agitation Speed (rpm)	Time of Oxidation (h)	Vol. rates (mmol-SL^{-1}h^{-1})			S-products (%)		Max S^0
			Thio.	Sul.	S^0	Sul.	S^0	
MSR	108 (flea)	> 24	0.98	0.49	0.36	48	35	35
SR	60 (impeller)	< 36	0.47	0.25	0.27	46	40	40
	90	< 24	1.04	0.51	0.33	50	32	32

Strains identification

Since our SOB consortium was able to transform toxic thiosulfate to S0, the dilution method was used to isolate its strains on agarose-plate. The mixture of bacteria, strain PN1 and PN2, were identified as Gram-negative and Gram-positive, respectively, being autotrophic–aerobic sulfide-oxidizing bacteria grown in MSM supplemented with thiosulfate as an energy source. The isolated strains PN1 and PN2 were phylogenetically closely related to genus *Ochrobactrum* and *Gordonia* by partial 16S rRNA method, respectively. Few previous studies exist on potential application of *Ochrobactrum* strain for the oxidation of S-compound such as thiosulfate (Pacheco Aguilar et al. 2008, Luo et al. 2013) and sulfide (Mahmood et al. 2009, Jing et al. 2010). However, very little information about S-components oxidation by *Gordonia* species has been reported (Elkin et al. 2013). A phylogenetic tree was constructed based on the 16S rRNA sequence (Fig. 6.7) for this latter species by the neighbor-joining method with the software MEGA 3. Evolutionary distances of nucleotide sequences were calculated with Kimura's 2-parameter model. The result of this phylogenetic analysis with high bootstrap value (1000 re-samplings) also identifies the bacterial isolation of PN2 as *Gordonia* sp. and its close clustering with *Gordonia polyisoprenivorans* first isolated from the effluent of rubber plants containing large amounts of sulfate and sulfide (Chaiprapat et al. 2011).

Moreover, all sequences in the phylogenetic tree were grouped into one cluster, actinobacteria, which is a major phylum of gram-positive bacteria, some of which are pathogenic to humans. For example, it has been reported that *G. polyisoprenivorans*

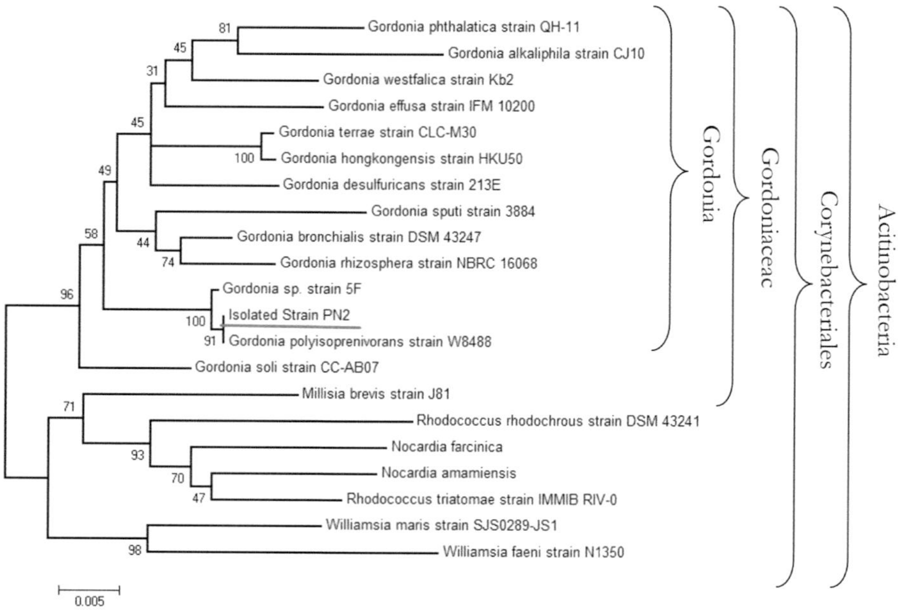

Figure 6.7. Phylogenetic tree based on the 16S rRNA gene of isolated strain and the selected sequences from formally described species. Bootstrap values (1000 replicates) above 50 are shown at each node.

can enter the body through the catheter in bone marrow transplant surgery and cause infection (Arenskötter et al. 2004).

Conclusions

The enriched mixed culture taken from the soil was found to be capable of complete oxidation of high thiosulfate concentration with the formation of a conspicuous amount of S0 alongside sulfate as an end-product. Thiosulfate oxidation along with bicarbonate as the C-source by active enriched SOB produced a considerable amount of elemental sulfur. Using bioreactor significantly enhanced the rates of thiosulfate oxidation and S0 formation. Our SOB consortium was capable of oxidizing high levels of thiosulfate—up to 4000 ppm with abundant S^0 in MSR. However, high thiosulfate removal efficiency was achieved in SR with sufficient elemental sulfur formation. Therefore, our results clarified the superiority of mixer cultivation of *Ochrobactrum* and *Gordonia* strains in SR for thiosulfate biotransformation into S0, which is the desired product rather than sulfate.

References

Abdel-Monaem Zytoon, M., A. Ahmad Alzahrani, M. Hamed Noweir and F. Ahmed El-Marakby. 2014. Bioconversion of high concentrations of hydrogen sulfide to elemental sulfur in airlift bioreactor. Sci World J.

Ahmad, N., F. Ahmad, I. Khan and A.D. Khan. 2014. Studies on the oxidative removal of sodium thiosulfate from aqueous solution. Arab J Sci Eng 40(2): 289–293.

Arenskötter, M., D. Bröker and A. Steinbüchel. 2004. Biology of the metabolically diverse genus Gordonia. Appl Environ Microbiol 70(6): 3195–3204.

Baquerizo, G., A. Chaneac, L. Arellano-García, A. González-Sánchez and S. Revah. 2013. Biological removal of high loads of thiosulfate using a trickling filter under alkaline conditions. Mine Water Environ 32(4): 278–284.

Bonilla-Blancas, W., M. Mora, S. Revah, J.A. Baeza, J. Lafuente, X. Gamisans, D. Gabriel and A. González-Sánchez. 2015. Application of a novel respirometric methodology to characterize mass transfer and activity of H2S-oxidizing biofilms in biotrickling filter beds. Biochem Eng J 99: 24–34.

Chaiprapat, S., R. Mardthing, D. Kantachote and S. Karnchanawong. 2011. Removal of hydrogen sulfide by complete aerobic oxidation in acidic biofiltration. Process Biochem 46(1): 344–352.

Chen, C.-Y., T.-H. Tsai, C.-H. Chang, C.-F. Tseng, S.-Y. Lin and Y.-C. Chung. 2018. Airlift bioreactor system for simultaneous removal of hydrogen sulfide and ammonia from synthetic and actual waste gases. J Environ Sci Health A 53(8): 694–701.

de Graaff, M., J.B. Klok, M.F. Bijmans, G. Muyzer and A.J. Janssen. 2012. Application of a 2-step process for the biological treatment of sulfidic spent caustics. Water Res 46(3): 723–730.

Elkin, A.A., T.I. Kylosova, V.V. Grishko and I.B. Ivshina. 2013. Enantioselective oxidation of sulfides to sulfoxides by Gordonia terrae IEGM 136 and Rhodococcus rhodochrous IEGM 66. J Mol Catal, B Enzym 89: 82–85.

Fortuny, M., A. Guisasola, C. Casas, X. Gamisans, J. Lafuente and D. Gabriel. 2010. Oxidation of biologically produced elemental sulfur under neutrophilic conditions. J Chem Technol Biotechnol 85(3): 378–386.

Gerrity, S., C. Kennelly, E. Clifford and G. Collins. 2016. Hydrogen sulfide oxidation in novel horizontal-flow biofilm reactors dominated by an Acidithiobacillus and a Thiobacillus species. Environ Technol 37(17): 2252–2264.

Gholipour, S., P. Mehrkesh, E. Azin, H. Nouri, A.A. Rouhollahi and H. Moghimi. 2018. Biological treatment of toxic refinery spent sulfidic caustic at low dilution by sulfur-oxidizing fungi. J Environ Chem Eng 6(2): 2762–2767.

González-Sánchez, A. and S. Revah. 2009. Biological sulfide removal under alkaline and aerobic conditions in a packed recycling reactor. Water Sci Technol 59(7).

Graff, A. and S. Stubner. 2003. Isolation and molecular characterization of thiosulfate-oxidizing bacteria from an italian rice field soil. Syst Appl Microbiol 26(3): 445–452.

Jadhav, K. and I. Jadhav. 2017. Sulfur oxidation by Achromobacter xylosoxidans strain wsp05 reveals ecological widening over which thiotrophs are distributed. World J Microbiol Biotechnol 33(10): 192.

Jing, C., Z. Ping and Q. Mahmood. 2010. Influence of various nitrogenous electron acceptors on the anaerobic sulfide oxidation. Bioresour Technol 101(9): 2931–2937.

Kalantari, H., M. Nosrati, S.A. Shojaosadati and M. Shavandi. 2018. Investigation of transient forms of sulfur during biological treatment of spent caustic. Environ Technol 39(12): 1597–1606.

Kolmerta, A.s., P. Wikstromb and K.B. Hallberga. 2000. A fast and simple turbidimetric method for the determination of sulfate in sulfate-reducing bacterial cultures. J Microbiol Methods 41: 179–184.

Krishnakumar, B. and V.B. Manilal. 1999. Bacterial oxidation of sulphide under denitrifying conditions. Biotechnol Lett 21(5): 437–440.

Lin, S., H.R. Mackey, T. Hao, G. Guo, M.C.M. van Loosdrecht and G. Chen. 2018. Biological sulfur oxidation in wastewater treatment: A review of emerging opportunities. Water Res 143: 399–415.

Liu, T., J.-H. Hou and Y.-L. Peng. 2017. Biodesulfurization from the high sulfur coal with a newly isolated native bacterium, Aspergillus sp. DP06. Environ Prog Sustain. Energy 36(2): 595–599.

Luo, J., G. Tian and W. Lin. 2013. Enrichment, isolation and identification of sulfur-oxidizing bacteria from sulfide removing bioreactor. J Environ Sci 25(7): 1393–1399.

Luo, J., X. Tan, K. Liu and W. Lin. 2018. Survey of sulfur-oxidizing bacterial community in the Pearl River water using soxB, sqr, and dsrA as molecular biomarkers. 3 Biotech 8(1): 73.

Mahmood, Q., B. Hu, J. Cai, P. Zheng, M.R. Azim, G. Jilani and E. Islam. 2009. Isolation of Ochrobactrum sp. QZ2 from sulfide and nitrite treatment system. J Hazard Mater 165(1): 558–565.

Marmur, J. 1961. A procedure for the isolation of deoxyribonucleic acid from micro-organisms. J Mol Biol 3(2): 208–IN201.

Mirzaei, M., G. Amoabediny, F. Yazdian, M. Sheikhpour, E. Ebrahimi and B.E.H. Zadeh. 2014. An immobilized Thiobacillus thioparus biosensing system for monitoring sulfide hydrogen; optimized parameters in a bioreactor. Process Biochem 49(3): 380–385.

Moghanloo, G., E. Fatehifar, S. Saedy, Z. Aghaeifar and H. Abbasnezhad. 2010. Biological oxidation of hydrogen sulfide in mineral media using a biofilm airlift suspension reactor. Bioresour Technol 101(21): 8330–8335.

Mora, M., L.R. López, J. Lafuente, J. Pérez, R. Kleerebezem, M.C.M. van Loosdrecht, X. Gamisans and D. Gabriel. 2016. Respirometric characterization of aerobic sulfide, thiosulfate and elemental sulfur oxidation by S-oxidizing biomass. Water Res 89(Supplement C): 282–292.

Moussavi, G., K. Naddafi and A. Mesdaghinia. 2007. Developing a biofilm of sulfur oxidizing bacteria, starting-up and operating a bioscrubber treating H2S. Pak J Biol Sci 10(5): 701–709.

Pacheco Aguilar, J.R., J.J. Pena Cabriales and M. Maldonado Vega. 2008. Identification and characterization of sulfur-oxidizing bacteria in an artificial wetland that treats wastewater from a tannery. Int J Phytoremediation 10(5): 359–370.

Pirieh, P. and F. Naeimpoor. 2019. Discrimination of chemical and biological sulfide oxidation in a hybrid two-phase process. J Environ Chem Eng 7(3): 103027.

Pokorna, D. and J. Zabranska. 2015. Sulfur-oxidizing bacteria in environmental technology. Biotechnol Adv 33(6, Part 2): 1246–1259.

Potivichayanon, S., P. Pokethitiyook and M. Kruatrachue. 2006. Hydrogen sulfide removal by a novel fixed-film bioscrubber system. Process Biochem 41(3): 708–715.

Qian, J., H. Lu, Y. Cui, L. Wei, R. Liu and G.-H. Chen. 2015. Investigation on thiosulfate-involved organics and nitrogen removal by a sulfur cycle-based biological wastewater treatment process. Water Res 69: 295–306.

Rodier, J. 1998. Analysis of water. Barcelona.

Sun, J., X. Dai, Y. Liu, L. Peng and B.-J. Ni. 2017. Sulfide removal and sulfur production in a membrane aerated biofilm reactor: Model evaluation. Chem Eng J 309(Supplement C): 454–462.

Tang, K., V. Baskaran and M Nemati. 2009. Bacteria of the sulphur cycle: an overview of microbiology, biokinetics and their role in petroleum and mining industries. Biochem Eng J 44: 73–94.

Troelsen, H. and B.B. Jørgensen. 1982. Seasonal dynamics of elemental sulfur in two coastal sediments. Estuar Coast Shelf Sci 15(3): 255–266.

Valdebenito-Rolack, E.H., T.C. Araya, L.E. Abarzua, N.M. Ruiz-Tagle, K.E. Sossa, G.E. Aroca and H.E. Urrutia. 2011. Thiosulphate oxidation by Thiobacillus thioparus and Halothiobacillus neapolitanus strains isolated from the petrochemical industry. Electron J Biotechnol 14(1): 7–8.

Valle, A., M. Fernández, M. Ramírez, R. Rovira, D. Gabriel and D. Cantero. 2018. A comparative study of eubacterial communities by PCR-DGGE fingerprints in anoxic and aerobic biotrickling filters used for biogas desulfurization. Bioprocess Biosyst Eng 41(8): 1165–1175.

Vogel. 1996. Qualitative Inorganic Analysis. University College Longman Singapore Publishers (Pte) Ltd.

Vosoughi, A., F. Yazdian, G. Amoabediny and M. Hakim. 2015. Investigating the effect of design parameters on the response time of a highly sensitive microbial hydrogen sulfide biosensor based on oxygen consumption. Biosens Bioelectron 70: 106–114.

Xu, B., W. Kong, Q. Li, Y. Yang, T. Jiang and X. Liu. 2017. A review of thiosulfate leaching of gold: focus on thiosulfate consumption and gold recovery from pregnant solution. Metals 7: 222.

Yang, H., K. Gao, S. Feng, L. Zhang and W. Wang. 2014. Isolation of sulfide remover strain Thermithiobacillus tepidarius JNU-2, and scale-up bioreaction for sulfur regeneration. Ann Microbiol 65(1): 553–563.

7

Removal of Oil Spills in Temperate and Cold Climates of Russia

Experience in the Creation and Use of Biopreparations Based on Effective Microbial Consortia

Filonov Andrey,[1,2] *Akhmetov Lenar*,[1] *Puntus Irina*[1,2] and *Solyanikova Inna*[3,*]

Introduction: Peculiarities of oils spills in Russia

The pollution of the environment with crude oil and petroleum products is currently a global problem (Vogt and Richnow 2014). By the extent of the adverse effects on ecosystems, petroleum hydrocarbons take the second place after radioactive contamination (Danilov-Daniljian 2007). A failure during drilling operations, transportation, processing, and storage can lead to accidental oil spillage at the rate of 60–70 million tons annually, constituting about 2% of total global oil production. Oil spills pose a serious threat to both ecosystems and human health (Xue et al. 2015), and result in environmental catastrophes worldwide (Wang et al. 2011). The self-purification of soil from oil products at the concentration of oil of 5 g/kg of the soil takes from 2 to 30 years, and up to 50 years in northern regions (Khmurchik et al. 2008). The effects caused by oil contamination may affect natural ecosystems for decades, or even centuries (Tevvors and Saier 2010).

[1] Laboratory of Plasmid Biology, Institute of biochemistry and physiology of microorganisms, the Russian Academy of Sciences, prospect Nauki 5, Pushchino, Moscow Region, 142290, Russia.

[2] Department of Biotechnology, Tula State University, prospekt Lenina 92, Tula, 300012, Russia.

[3] Laboratory of Microbial Enzymology, Institute of biochemistry and physiology of microorganisms, the Russian Academy of Sciences, prospect Nauki 5, Pushchino, Moscow Region, 142290, Russia.
Emails: filonov.andrey@rambler.ru; lenarakhmetov@yandex.ru; puntus66@mail.ru

* Corresponding author: innap_solyan@yahoo.com

In Russia, a significant number of oil producing enterprises operate in the north of the European part of the country, and in Western Siberia, in the regions with temperate and cold climates. In these regions, self-purification of soil and water reservoirs from oil pollution, driven by vital activity of indigenous hydrocarbon-oxidizing microbiota, is limited by unfavorable soil and climatic factors—low average annual temperatures, weak influence of physical and chemical factors, such as solar radiation, low evaporation rate of volatile hydrocarbon fractions, low content of nutrients, increased salt concentration, lack of aeration, and so on. In addition, a distinct feature of northern ecosystems is the presence of permafrost and wetland soils, small thickness of the humus horizon, low biological activity of soils, and low diversity of plants, microorganisms, and soil animals (Alekseev 2011, Alekseev et al. 2011).

Oil pollution leads to negative changes in soil biocenosis (Pokonova 2003, Shtina and Nekrasova 1988), and also fundamentally changes the composition of soil chemical properties, and its structure (Gainutdinov et al. 1988, Kolomytseva et al. 2005), resulting in depletion of soil fertility and reduction of crop yield. In oil-contaminated soil, the structure of the microbial community changes (Kireeva 1996, Chaillan et al. 2006, Escalante-Espinosa et al. 2005, Jirasripongpun 2002), and the photosynthetic activity of higher plants declines (Borodavkin 1981, Buzmakov and Ladygin 1993, Khabibullin and Kovalenko 1982, Shilova 1988). As a result of oil spills, soils can turn into technogenic, man-made deserts, in which the biota is almost completely suppressed.

The capability of microorganisms to transform or degrade oil hydrocarbons is well known and thus, the use of these bacteria for bioremediation of contaminated areas is actual. Bioremediation methods are based on biostimulation of indigenous oil-utilizing microorganisms or introduction of alien ones (bioaugmentation).

By the extent to which bacterial cultures participate in oil biodegradation to reduce concentration of pollutants, the components of oil and oil products are arranged in the following way: *n*-alkanes → branched alkanes → branched alkenes → low molecular *n*-alkyl aromatic compounds → monoaromatic compounds → polycyclic aromatic hydrocarbons → asphaltenes (Van Hamme et al. 2003). To date, there are many works focusing on the study of genetic control of the decomposition of various hydrocarbons, which demonstrate the increased interest of researchers in understanding the mechanisms of biodegradation.

Strategies used by microorganisms for the degradation of petroleum hydrocarbons include the synthesis of appropriate enzymes, cometabolism, transfer of catabolic plasmids, and the production of biosurfactants increasing the bioavailability of hydrophobic substrates (Balba 2003, http://home.eng.iastate.edu/~tge/ce421-521/brubaker.pdf). Bioremediation returns natural processes to their original conditions, and environmental indicators are approaching the initial state (before the technogenic impact). Bioremediation provides a cost-effective, highly specific clean-up, leading to a decrease in the concentration of both single pollutant and a mixture of pollutants (Foght and McFarlane 1999, Margesin and Schinner 2001).

The characteristic features of the soils in the West Siberian oil-producing regions are—greater amount of water, low pH, extremely low abundance or absence of indigenous oil-oxidizing microorganisms, and a short warm period when temperatures rise above 0°C. Taking into consideration the mentioned factors, the addition of active petroleum-degrading microorganisms as a component of biopreparations is the only alternative. In the process of the development of technologies for bioremediation of various hydrocarbon-contaminated sites with the use of added and/or indigenous microorganisms, it is necessary to find optimum conditions under which microbial activity increases to a maximum. Therefore, it is more efficient to combine bioremediation technologies to improve the removal of the contaminants. The bioremediation strategy will directly depend on the type of ecosystems, specific features of the pollutant, the extent and period of pollution, ambient temperature, humidity, pH, and other factors.

Main factors influencing the biodegradation process in cold climate

Temperature

Temperature plays a key role in regulating microbial metabolism, which is necessary for *in situ* bioremediation. The bioavailability and solubility of hydrophobic, poorly soluble in water substrates, such as aliphatic and aromatic hydrocarbons, are temperature dependent. At low temperatures, the viscosity of liquid hydrocarbons increases, with decreasing volatility of toxic components, and the bioavailability of nutrients and pollutants becomes more limited. The ability of oil products to spread over the surface and penetrate frozen soils depends on soil characteristics, the type of pollutant, and temperature (Chuvilin et al. 2001).

A cold environment is a habitat of different microbial species, including bacteria, archaea, yeasts, fungi, and algae (Margesin et al. 2007). Microorganisms capable of degrading oil components at low temperatures have a set of structural and adaptation mechanisms. Their adaptations are associated with growth and enzymatic activity, which makes it possible to compensate for the negative effects of low temperatures on biochemical processes (Gounot and Russell 1999, Margesin et al. 2002, Morita 1975). These microorganisms have an important role in *in situ* biodegradation of hydrocarbons in cold climate regions, where summer temperatures often coincide with their temperature growth range.

The biodegradation processes have been described for many petroleum compounds in various soil and water ecosystems at low temperatures, including arctic (Braddock et al. 1997, Mohn et al. 2001, Eriksson et al. 2001, Delille and Coulon 2007), alpine (Gounot and Russell 1999, Margesin and Schinner 2001, Margesin 2000, Margesin et al. 2007), and Antarctic soils (Aislabie et al. 1998, Ferguson et al. 2003).

Microorganisms

The representatives of the genera *Rhodococcus*, *Pseudomonas*, *Arthrobacter*, *Microbacterium* are the most common in oil-polluted sites. In addition, bacteria of the genera *Moraxella*, *Beijerinckia*, *Aeromonas*, *Flavobacterium*, *Nocardia*, *Corynebacterium*, *Acinetobacter*, *Streptomyces*, *Bacillus*, *Cyanobacterium*, and others are capable of degrading oil hydrocarbons (Thapa et al. 2012). It is shown that in the surface horizon of soil contaminated with hydrocarbons of oil under conditions of Central Siberia, the number of psychrophilic microorganisms is lower than the mesophilic ones. In soils at depths of 1–5 m, the number of psychrophilic microorganisms is comparable with the number of mesophiles, and at depths of 15 m the number of psychrophiles is an order of magnitude higher (Trusei 2018). In chronically contaminated soils, rhodococci are often the dominant bacteria. The hydrophobicity and multilayer structure of the cell wall of rhodococci allows these bacteria to mineralize a significant amount of hydrocarbons (Koronelli 1996). As a result, they have potential for accumulation of reserve hydrocarbons and lipids, and store them in the cells. This ensures high starvation survival of rhodococci rather than pseudomonads (Koronelli et al. 1988). On the other hand, rapid adaptation to a new substrate and a high growth rate on water-soluble organic compounds facilitate the survival of pseudomonads in the natural environment (Koronelli 1996). In regions with cold and temperate climates, the addition of hydrocarbon-oxidizing bacteria is vital and important, and rhodococci are highly recommended as a tool. Hydrocarbons as a substrate do not have attractant properties for motile bacteria, because they do not contain functional groups that excite cell motor receptors. Rhodococcal metabolism promotes the appearance of oxidized compounds that attract pseudomonads. Thus, the introduction of rhodococci enhances biodegradation not only directly, but also indirectly through the formation of a complex hydrocarbon-oxidizing community (Koronelli et al. 1982).

Sorbents

Cell immobilization on sorbents can facilitate the increase of the metabolism efficiency and oil-oxidizing activity of microorganisms, since immobilized cells are more resistant to the adverse effects of environmental factors (Chugunov et al. 2000). Various substances are used to immobilize microorganisms, for example, polyurethane, polyesters, activated carbon, silica, alginate, and other carriers (Koronelli 1996).

Chugunov et al. (2000) developed "Ecosorb" biopreparation based on *Mycobacterium fluorescens, Mycobacterium* sp., *Rhodococcus* sp., and *Acinetobacter* sp., immobilized on lessorb—a product of heat treatment of moss and wood. After the application of Ecosorb to the soil contaminated with petroleum, the efficiency of biodegradation of the pollutant increased by 40% within 30 days, compared with the control.

Similar results were obtained by a group of researchers headed by E.V. Karaseva, who showed that in an effort to clean oil-contaminated sand, sandy soil, and water, immobilization of *Rhodococcus* cells on a sorbent (the product of sawdust heat treatment) by absorption showed the following advantage—increased microbial activity contributed to removal of hydrocarbons (Karaseva et al. 2007, 2008a, b, 2009).

During immobilization, the rate of degradation activity of microbial cells increases due to the enhanced bioavailability of a hydrophobic substrate adsorbed on the surface of the carrier (Kovalenko et al. 2006). In this work, it was also shown that the capability of *Rhodococcus ruber* cells immobilized on a layer of catalytic filamentous carbon of ceramic foam to degrade hexadecane increased by a factor of 3.7 compared to that in a free suspended state.

A team of Russian-Finnish researchers showed (Suni et al. 2004) that a cotton grass-based plant (*Eriophorum vaginatum*, a grass of peat bogs, and a by-product of peat extraction) sorbent is capable of adsorbing various types of oil products (gasoline, diesel fuel, synthetic oil, mineral oil, heavy oil fractions) faster and more efficiently than synthetic sorbents (Teas et al. 2001). The sorbent is considered an ideal material for adsorption of oil products from water surface, because it almost does not absorb water. The low specific weight of the sorbent ensures its quick and easy removal from water surface after the end of the adsorption process of the oil and oil products; moreover, it is biodegradable.

Fertilizers

In cold climate ecosystems, with the exception of wetlands, nutrients are a limiting factor in oil biodegradation. If petroleum products permeate in wetland soil, the predominant mechanism that prevents remediation is the restriction of oxygen availability and the influence of anaerobic zones (Ron and Rosenberg 2001, Thapa et al. 2012). Nutrients and fertilizers of various types are the main tools when applying such a bioremediation approach as activation of indigenous microorganisms. Typically, water-soluble nutrients, including mineral salts, such as KNO_3, $NaNO_3$, NH_4NO_3, K_2HPO_4, $MgNH_4PO_4$ (Sendstad 1980, Lee and Levy 1989, Garcia-Blanco et al. 2001, Venosa et al. 2002), are used for biostimulation.

The introduction of Unipolymer-M urea-formaldehyde polymer (soil structure-forming agent and prolonged nitrogen fertilizer) through the upper soil horizon and the observation well system increases the number of indigenous microorganisms by 2–4 orders of magnitude, and is accompanied by an increase in the content of ammonia and nitrate nitrogen, carbon dioxide, and permanganate oxidation, which are indicators of ongoing reduction processes (Ladygina et al. 2008, Trusei et al. 2009, Trusei 2018).

In Russia, composting of soils contaminated with oil, and primarily oil sludge, is used. Previously, the soil is protected with a film, then piles with a height of up to 1.5–2 m, a width of about 2 m, a length of 5 to 20 meters, and more are built up. In the pile, the layers of contaminated soil are interspersed with peat or clean soil, organic or mineral fertilizers are applied, and vents are made for air intake and gas exit. The piles are irrigated with water. Aeration is carried out either using artificial ventilation, or by simple mixing (Chertes et al. 2010).

The use of slow-release fertilizers is one of the approaches in order to overcome the problem of fertilizers leaching and to provide a constant source of nutrients for microbial degrader in the contaminated area. They consist of relatively insoluble or sparingly soluble nutrients coated with hydrophobic materials, such as paraffin or vegetable oils. The use of oleophilic organic additives is desirable since hydrocarbons biodegradation mainly occurs in the oil-water interphase.

Researchers within the EUREKA BIOREN European program conducted field trials at river mouths for two bioremediation products—BIOREN1 and BIOREN2 (Le Floch et al. 1997, 1999). The products were obtained from fishmeal in granular form, containing urea and superphosphate as nitrogen and phosphorus sources, and a protein substance as a carbon source. Unlike BIOREN2, BIOREN1 contains a biosurfactant. The data obtained suggests that the biosurfactant in BIOREN1 was the most active component, which contributed to an increase in the rate of oil degradation.

French researchers (Societe, CECA SA, France) developed a mixture of fertilizers in the form of an emulsion consisting of urea (a source of nitrogen), lauryl sulfate (a source of phosphorus), oleic acid (a hydrophobic substance), and 2-butoxy-1-ethanol (a surfactant). Based on the invention, the French company Elf Aquitaine began to manufacture the preparation Inipol (Inipol EAP 22). Inipol is currently the most widespread biostimulating nutritional supplement used in technologies for bioremediation of oil-contaminated areas in cold climates (Delille et al. 2002, 2004, Delille and Coulon 2007, Margesin and Schinner 2001, Margesin et al. 2007).

The effectiveness of the various types of nutritional supplements described above depends on their composition and the state of contaminated ecosystems. Successful application of commercial products in bioremediation will always require appropriate experimental models, tests, and analyses based on the specific conditions of any polluted ecosystem.

Thus, the introduction of microorganisms into contaminated sites is necessary for bioremediation of oil-contaminated soils, or when oil products enter aquatic environments, where the development of the natural population of microorganisms occurs slowly, even under favorable environmental conditions. In cleaning up soils and water surface at low temperatures, the bioremediation process gives good results after introduction of preparations that include microorganisms isolated directly from the pollution sites. Plowing and breaking up of upper soil layers are important factors due to which the necessary level of oxygen supply is ensured. A good effect is observed when biopreparations with mineral fertilizers (sources of nitrogen, phosphorus, and potassium) are added to contaminated sites.

Biopreparations in Russia

To date, several dozen biopreparations have been developed in Russia to clean up the environment from pollution by crude oil and oil products (Rogozina et al. 2010, Kireeva et al. 2010, Kuznetsov et al. 2010), but not all of them have found their application. Many of them have not even been patented, and some of them are known only by advertising information.

As a result of a patent search in the Russian database of the Federal Institute of Industrial Property, information on biopreparations for bioremediation of natural ecosystems from crude oil and oil products has been summarized (Table 7.1).

Biopreparations can be divided into three groups—the first group includes biopreparations with one microbial strain isolated from a natural or industrial source, which is capable of efficiently degrading petroleum or its derivatives. The second group is formed by biopreparations which have consortia of various petroleum degrading strains. Biopreparations of the third group have a more complex composition. They include, in addition to microorganisms, sorbents (natural and non-natural), carriers, stabilizers, preservatives, enzymes, surfactants (biosurfactants), organic, and mineral substances.

Table 7.1. Russian biopreparations for bioremediation of oil-contaminated sites.

Biopreparation group	**Brand**	**Microorganisms, genus/ species**	**Field of Application**	**Inventors**
1	2	3	4	5
I one strain	Bacispecin	*Bacillus* sp.	oil contaminated grey forest soil	Andreson et al. 1997
	Destroil	*Acinetobacter* sp.	+5…+38°C pH 4,5–8,5	Ikhsanov and Ikhsanova 2000
	Diezoil-M	*Candida maltosa*	oil-contaminated soil and water, up to 41°C pH 3,0–9,0, and saline biotopes	Avchieva 1998
	Pseudomin	*Pseudomonas putida*	+0…+34°C pH 4,8–8	Stankevich 2002
	Agent for treatment	*Bacillus firmus*	the ecosystem contaminated by petroleum hydrocarbons, 24% NaCl	Soprunova et al. 2011
II some strains	Biodegrader	*Acinetobacter bicoccum* *Acinetobacter valentis* *Arthrobacter* sp. *Rhodococcus* sp.	+10… +37°C pH 4–9,5	Murzakov et al.1996
	Devoroil	yeast *Candida* sp. *Rhodococcus* sp. *Rhodococcus morus* *Rhodococcus erythropolis* *Alcaligenes* sp. *Pseudomonas stutzeri*	+5 + 45°C pH 4,5–9 up to 15 g/l NaCl	Borzenkov et al. 1994
	Consortium of Strains	*Pseudomonas aeruginosa* *Pseudomonas fluorescens*	petroleum products, oil-contaminated groundwater	Maksimovich and Khmurchik 2007
	Naphtox	*Mycobacterium phlei* *Pseudomonas aeruginosa* *Rhodococcus* sp.	cleanup of soils contaminated by petroleum and its products	Belonin et al. 1995
	Bacterial consortium	*Rhodococcus* sp. *Acinetobacter* sp.	cleanup of petroleum-contaminated soils +10°C	Tretiakova 2017
	Lenoil®-Mycostat	*Pseudomonas nitroreducens* *Rhodococcus* sp.	biorestoration of petroleum-hydrocarbons-contaminated soils	Smolova 2015

Table 7.1 Contd. …

...Table 7.1 Contd.

Biopreparation group	Brand	Microorganisms, genus/ species	Field of Application	Inventors
	Lenoil®, SHP	*Acinetobacter calcoaceticus* *Ochrobactrum intermedium*	cleanup of water surface and industrial wastewater treatment after pollution with crude oil and oil products	Korshunova 2019
	Lenoil®– Super, SHP	*Acinetobacter calcoaceticus* *Ochrobactrum intermedium* *Pseudomonas koreensis*	cleanup of soils contaminated by crude oil and oil products	Korshunova 2019
	Lenoil®– Grand, SHP	*Acinetobacter calcoaceticus* *Ochrobactrum intermedium* *Pseudomonas koreensis* *Paenibacillus ehimensis*	cleanup of oil- and oil products-contaminated soils and plant growth stimulation	Korshunova 2019
	Lenoil®– NORD, SHP	*Pseudomonas turukhanskensis*	cleanup of crude oil- and oil products-contaminated soils at low temperature, but above freezing in West Siberia	Korshunova 2019
	Rhoder	*Rhodococcus rubber* *Rhodococcus erythropolis*	+8 + 35°C up to 100 g/l NaCl	Murygina et al. 2002
	Bacterial consortium	*Rhodococcus* sp. *Arthrobacter* sp. *Microbacterium* sp. *Thalassospira* sp. *Halomonas* sp. *Salinicola socius*	crude oil, PAHs, up to 7% NaCl	Ananjina et al. 2010
	Ecoil	*Mycobacterium flavescens* *Pseudomonas putida* *Acinetobacter* sp.	+5 +37°C pH 5–8,5	Ermolenko et al. 1997
	Consortium of oil degrading strains	*Acinetobacter* sp. *Pseudomonas* sp. *Bacillus* sp.	a wide spectrum of petroleum compounds in soils 10–15ºC	Ilyicheva et al. 2014
	Consortium of strains	*Exiguobacterium mexicanum* *Bacillus vallismortis*	cleanup of permafrost soils from crude oil spills	Erofeevskaya 2015
	Oleovorin	*Acinetobacter oleovorum* yeast *Candida* sp.	cleanup of soils and water from crude oil and oil products	Orlova and Stepanova 2012

Table 7.1 Contd. ...

...Table 7.1 Contd.

Biopreparation group	Brand	Microorganisms, genus/ species	Field of Application	Inventors
	MicroBak	*Pseudomonas putida* *Pseudomonas fluorescens* *Rhodococcus erythropolis* *Rhodococcus* sp.	+4 +32°C Up to 50 g/l NaCl	Filonov et al. 2007
III strains and supplements	Universal	*Pseudomonas* *Arthrobacter* yeast *Rhodotorula* *Rhodococcus* *Flavobacterium* *Kurthia*	cleanup of soils and waters from crude oil	Markarova 2004
	Biooil	yeast *Saccharomyces* sp. *Bacillus* sp. *Enterobacter* sp.	+1 + 31°C pH 4–9 up to 7.5% NaCl	Alekseev et al. 2008
	Bioprin-B	*Pseudomonas putida* *Pseudomonas fluorescens* *Micrococcus* sp. *Xanthomonas* sp.	cleanup of different ecosystems from crude oil and oil products	Saxon et al. 1998
	Biopreparation	*Arthrobacter* sp. *Rhodococcus* sp.	cleanup of soils and waters from crude oil and oil products	Karaseva et al. 2008a
	Belvitamil	*Arthrobacter* *Bacillus* *Desulfimbrio* *Pseudomonas* yeast *Candida*	cleanup of soils and water from crude oil and oil products	Onegova et al. 2003
	Biopreparation	*Bacillus cereus* *Bacillus subtilis* *Actinomyces griseus* *Actinomyces glaucus* *Pseudomonas fluorescens* *Pseudomonas mesentericus* *Pseudomonas denitrificans* *Arthrobacter globiformis*	cleanup of waters and soils from petroleum and oil products	Swarovskaya et al. 2009
	Biopreparation	*Burkholderia caryophyllii* *Pseudomonas fluorescens*	cleanup of water from petroleum hydrocarbons	Khlynovsky et al. 2012

Table 7.1 Contd. ...

...Table 7.1 Contd.

Biopreparation group	Brand	Microorganisms, genus/ species	Field of Application	Inventors
	A complex biosorbent based on bacterial, fungal and yeast strains	*Rhodococcus equi* yeast *Rhodotorula glutinis* micromycete *Trichoderma lignorum*	aquatic environment after pollution with petroleum and its products	Sharapova et al. 2011
	Microbial consortium	*Pseudoamycolata halophobica* *Kibdelosporangium aridum* *Acinetobacter oleovorum* *Rhodococcus erythropolis*	cleanup of water, soils, grounds after pollution with petroleum	Yankevich et al 2006
	Bacterial consortium	*Bacillus brevis* *Arthrobacter* sp.	removal of petroleum and oil products from water and soil	Loginov et al. 2004
	Ecosorb	*Pseudomonas* *Rhodococcus* *Mycobacterium* *Xantomonas*	sod-podzolic and sandy loam soils contaminated by petroleum	Chugunov et al. 2000
	Bioionit	*Bacillus megaterium* *Bacillus subtilis* *Pseudomonas putida* *Rhodococcus erythropolis*	cleanup of oil-polluted soils, sludge, bottom silts, wastewater sediments, reservoirs in regions with short warm period	Volkov et al. 2015 Volkov et al. 2019
	Avalon	*Serratia marcescens* *Pseudomonas fluorescens* *Acidovorax delafieldii*	cleanup of ecosystems contaminated by petroleum and oil products	Limbakh et al. 2002
	A range of biopreparations	*Bacillus vallismoris* *Exiguobacterium mexicanum* *Serratia plymuthica* *Rhodococcus* sp.	cleanup of soils after pollution with petroleum	Erofeevskaya 2014 Erofeevskaya and Glyaznetsova 2015
	IPK-N	*Rhodococcus* sp. *Rhodococcus maris* *Rhodococcus erythropolis* *Pseudomonas stutzeri* yeast *Yarrowia lipolytica*	biofilters to clean up petroleum-contaminated ecosystems	Borzenkov et al. 1994

Table 7.1 Contd. ...

...Table 7.1 Contd.

Biopreparation group	Brand	Microorganisms, genus/ species	Field of Application	Inventors
	DOP series biopreparatons	*Pseudomonas stutzeri* *Rhodococcus maris* *Rhodococcus erytropolis* yeast *Yarrovia* sp.	restoration of oil-contaminated soils and water, improvement of oil recovery, dewaxing of injection wells - low and higher temperature - NaCl content in water up to 150 g/l - pH 4.5–9.5	https://pro-ecology.ru/ru/products/biologicals/biologicals-for-remediation/item/biopreparat-dop-uni-dlya-rekultivacii-nefteshlamov https://dop-uni.ru/en Vinogradsky Insitute of Microbiology

Recently, biopreparations consisting of two or more strains are increasingly used, since the introduction of a monoculture of hydrocarbon-oxidizing microorganisms into an oil-contaminated site cannot completely solve the problem of remediation. From this point of view, the stimulation of indigenous microorganisms is in a better position, since it activates a large number of different taxonomic groups of microbiocenosis. Therefore, a promising approach may be the use of several strains that differ significantly in the spectrum of consumed substrates and metabolic characteristics.

On the other hand, it should be noted that the creation of the universal biopreparation is impossible, because firstly, crude oil consists of several hundred compounds, and oils from various fields differ in fractional and compositional composition; secondly, in the practice of bioremediation, the object has to face both mono-aspect crude oil pollution and single product (single class) contamination, with the product sharply differing in chemical properties from crude oil; thirdly, the areas of extraction, processing, and storage of oil and oil products significantly differ from each other in terms of climatic and hydrothermal conditions.

In this regard, for each type of petroleum, each region and each place of pollution, there is a need to create a specialized biopreparation, including isolation of active hydrocarbon-oxidizing microorganisms from the specific oil-contaminated object, composing a preparation, and its introduction into the initial site. Such an approach requires a fast and effective method for screening oil-degrading strains, as well as the need to study the isolated strains for subsequent compilation of a consortium.

To formulate an effective consortium, the hydrocarbon-oxidizing enzyme systems of the strains are desirable to complement each other. One of the approaches based on this principle is summarizing strains of different taxonomic groups. This approach, which is the most common and simple, is based on the fact that strains belonging to different taxa may have different hydrocarbon-oxidizing enzyme systems. Consortia often include strains of the genera *Pseudomonas*, *Rhodococcus*, *Acinetobacter*, *Arthrobacter*, *Bacillus*, and *Mycobacterium* in different ratios.

Another approach to composing microbial consortia may be a cluster analysis of dendrograms based on the studied properties of strains. Strains with similar characteristics of growth and degradation of the pollutant are located close to each other on the dendrogram, and vice versa. To create highly active consortia, the relationships between microorganisms must be considered. At the same time, it is not necessary for the consortium to consist only of active oil destructors. Perhaps the most active consortium will be one consisting of some active strains (consuming diverse fractions) and some heterotrophs that do not have a hydrocarbon-oxidizing ability, but are able to assimilate intermediate products, often toxic to hydrocarbon-oxidizing microorganisms. In addition, microbial consortia can be compiled during selection in model systems—for example, during batch cultivation in rocking flasks or multistage continuous cultivation in a flow open bioreactor bench (Kobzev 2003, Shkidchenko et al. 2007).

Currently, the following preparations were actively used—Oleovorin (*Acinetobacter oleovorans*, *Candida* sp.), Ecoil (*Mycobacterium flavescens, Pseudomonas putida, Acinetobacter* sp.), Devoroil (yeast of the genus *Candida* and bacteria *Rhodococcus* sp., *Rhodococcus morus*, *Rhodococcus erythropolis*, *Alcaligenes* sp., and *Pseudomonas stutzeri*) (Gradova et al. 2003), Destroil (*Acinetobacter* sp.), Bacispecin (*Bacillus* sp.), Universal, Lenoil (*Bacillus brevis* and *Arthrobacter* sp.) (Loginov et al. 2004), Roder (*Rhodococcus erythropolis*) (Murygina et al. 2000), Belvitamil (activated sludge from wastewater treatment plants for pulp and paper production contains hydrocarbon-oxidizing microorganisms of the genera *Arthrobacter*, *Bacillus*, *Candida*, *Desulfimbrio*, *Pseudomonas*) (Kireeva et al. 2010).

For the bioremediation of oil-contaminated chernozem (Gabbasova et al. 2001), Devoroil was used in combination with Biotrin biological supplement (protein supplement and minerals): in 110 days, 96% of the pollutant was utilized in the soil.

Gradova et al. (2003) revealed the stimulating effect of *Azotobacter chroococcum* during bioremediation in oil-contaminated soils onto hydrocarbon-oxidizing microorganisms of the Devoroil preparation. After 6 weeks in the presence of inoculant, the degree of oil biodegradation reached 69%, and 79% removal was determined after combination of the inoculant and *Azotobacter chroococcum*. Germination of wheat seeds in the treated area was 58% and 80%, respectively.

Field trials for 2 months in permafrost soil (location not indicated by the author) at an average temperature of 8–14°C using a bacterial strain *Exiguobacterium mexicanum* VKPM B-11011, isolated from oil-contaminated soil in the Amur Region, facilitated oil degradation to 87.0%, while its natural loss in permafrost soil not treated with microorganisms was 7.32% (Erofeevskaya 2016).

Using alfalfa plants (*Medicago sativa* L.) and ryegrass (*Lolium perenne* L.) and microorganisms from biopreparation Oleovorin (*Acinetobacter oleovorum,*

Candida maltosa), the possibility of creating an effective bioremediation complex for cleaning the soil (a mixture of high peats, pH 6.0–6.5) from crude oil (4% wt./wt.) was assessed. Complex consisting of *Candida maltosa* yeast strain 569 and alfalfa proved to be the most effective, and contributed to a 69% removal of oil products in the soil by the 56th day of the experiment (Orlova and Stepanova 2012).

Isolated and studied microorganisms have been used to develop a series of multifunctional biopreparations under the trademark Lenoil® (Korshunova 2019), including biopreparations Lenoil® SHP (strains *A. calcoaceticus* IB DT-5.1/1 and *O. intermedium* IB DT-5.3/2), Lenoil® Super SHP (*A. calcoaceticus* IB DT-5.1/1, *O. intermedium* IB DT-5.3/2 and *P. koreensis* IB-4), Lenoil® Grand SHP (*A. calcoaceticus* IB DT-5.1/1, *O. intermedium* IB DT-5.3/2, *P. koreensis* IB-4 and *P. ehimensis* IB739), and Lenoil® NORD SHP (*P. turukhanskensis* IB 1.1T). All these biopreparations are designed for the cleanup of oil-contaminated environments, disposal of oil-containing wastes, and soil restoration. Of all preparations, biopreparation Lenoil® NORD SHP can be used at above 0°C temperatures in Western Siberia.

In the work of Loginov et al. (Loginov et al. 2004), a comparative evaluation of the effectiveness of Lenoil and other commercial preparations for soil bioremediation was carried out. Model studies showed biopreparations based on consortia of microorganisms (Devoroil, Lenoil) to be able to adapt to high oil concentrations and effectively utilize the substrate, regardless of soil type and pollutant characteristics. Monoculture biopreparations (Bacispecin, Destroil) can be used at concentrations of the pollutant no more than 10 percent.

The Universal biopreparation (Markarova 2004) was developed on the basis of oil-oxidizing bacteria isolated from oil-contaminated soils in several regions of the Komi Republic and deposits in the Tyumen region. The first field trials of the biopreparation were carried out in the accident zone in 1994 in the Usinsky region of the Komi Republic. Oil spills covered an area of about 4 hectares, and the spills formed two water-flooded spots. The initial level of soil contamination ranged from 450 to 750 g/kg. Over two years, the decomposition of hydrocarbons in soil and water through the use of universal and additional mineral fertilizers amounted to 35–40 percent. In 2003–2004, field tests of the preparation were also carried out in the Tyumen region. Evaluation of the remediation efficiency for various objects made it possible to conclude that Universal is effective to decompose oils with different viscosity, density, composition, and can be recommended for soil and water bioremediation in the far North of Russia.

Comparative tests of eight Russian biopreparations and the UNI-REM enzyme preparation (USA) were conducted under laboratory conditions at an oil pollution level of 20–30% in the peat soil of the Samotlor field of the Khanty-Mansi Autonomous Region (Ivasishin et al. 2011) and the peat soil of the Matyushkinsky oil and gas field (Middle Priobye) (Sokolov and Khadaev 2017). With a single application, Devoroil was the most effective, twice-Destroil, being thrice applied—Rodart, Roder, and UNI-REM.

The Rhoder biopreparation (alone or in combination with the initial mechanical removal of crude oil) was described (Murygina et al. 2000) when it was used for bioremediation of petroleum-contaminated waters and soils in Moscow Region and Western Siberia during the 1994–1999 period. Rhoder proved to be effective for bioremediation of open water surfaces (100 m^2 of the Chernaya river bay, at 15–25°C, and two lakes of 5,000 m^2 each in Vyngayakhah, at 15–34°C). The degree of bioremediation was more than 99% at the initial level of pollution of 0.4–19.1 g/l. During the remediation of the swamped area (2,000 m^2) in the Urals at a temperature of 5–30°C (pollution level 10.5 g/l), after mechanical removal of oil, the cleanup efficiency was 75%; and remediation level reached 94% when Rhoder was added twice. The bioremediation efficiency for bogs (10,000 m^2) in Vyngayakhah was 65%, and only 19% for peat wetland soil (1,000 m^2) in Nizhnevartovsk at 0–27°C. This can be explained by the high initial level of oil contamination (24.3 g/l and > 750 g/g dry matter, respectively) and the absence of preliminary mechanical removal of petroleum.

The researchers (Murygina et al. 2005) also used Rhoder in field trials in the Komi Republic in the vast wetland during the 2002–2003 period, and compared the effectiveness of a bioremediation process with those after implementation of other preparations. After using Rhoder on this site of approximately 2,000 m^2 during the cold and rainy summer in 2002 (1.5 months), the level of oil pollution decreased by 20–51% of the initial level (458–738 g/kg of dry soil). In mid-September 2002, this area was covered by vegetation by 70–85 percent. After an year, the degree of cleanup was 54–79 percent. The use of Rhoder, Universal, Petrolan (unidentified microorganisms, "Pribor" Inc., Russia), and UNI-REM (enzyme-based biopreparation, BioTech Service, USA) was more effective than Bamil fertilizer (a dried and granulated sludge from agricultural wastewater treatment facilities).

Long-term studies on the isolation, study of properties of psychrotolerant microorganisms, and the possibilities of their use in bioremediation of oil-contaminated environments are carried out in the Laboratory of Plasmid Biology at G.K. Skryabin Institute of Biochemistry and Physiology of Microorganisms of the Russian Academy of Sciences, Pushchino under the guidance of Prof. A.M. Boronin, a corresponding member of RAS, and A.E. Filonov, D.Sc. in Biology. On the basis of a consortium of psychrotrophic, halotolerant hydrocarbons-degrading *Rhodococci* strains, and bioemulsifiers-producing pseudomonads, scientists developed and patented the MicroBak biopreparation for bioremediation of soils (salinity up to 5%, pH 6–8, temperature range 4–32°C, hydrocarbons content up to 15%). The strains of pseudomonads bear plasmids encoding for PAHs biodegradation catabolic routes. MicroBak field tests at low temperatures showed 50 to 90% degradation of crude oil and oil products in two months (Filonov et al. 2007, Nechaeva et al. 2009, Vetrova et al. 2013, Filonov 2016).

Thus, effective remediation of soils contaminated with petroleum and hydrocarbons can be achieved by applying biopreparations highly active in a wide range of temperatures and pH, in combination with mineral or organic fertilizers, good aeration, acidity neutralization, and the use of surfactants.

Biopreparations may come in different forms- emulsion, suspension, gel, paste, powder (Rogozina et al. 2010). To select a promising strain, it is necessary to make sure that the following criteria—high petroleum oxidizing activity, resistance to toxicants (salts of heavy metals), non-pathogenicity, and non-toxicity of the strain for humans and animals (Koronelli 1996). In addition, the cells of the strain should have high viability in order to be able to grow and utilize the oil product in a wide range of temperatures, pH, humidity, and lack of nutrients (Stabnikova et al. 1995).

The problem of maintaining the viability of microorganisms in biopreparations

A significant amount of studies focus on increasing the survival strategies of microorganisms during prolonged storage using protective media and cryoprotectants. However, since there is no sufficiently complete theoretical description of defense mechanisms, the solution to this problem remains the field of practical research when composing a biopreparation.

With long-term storage of the microbial culture, it is important to maintain its high viability and properties. With frequent re-plating, properties can change over time due to natural selection and mutations-productivity may decrease, culture heterogeneity may appear (i.e., different culture variants with different morphology and physiology) (Langeler 2005).

In practice, microbial cultures are stored in various ways. Most industrial microorganisms can be kept in active form at low positive temperatures (2–4°C) for one-two months. Another widely used preservation method is storage of cultures in semi-liquid agar under a layer of mineral oil. The oil protects the substrate from drying out and limits the access of atmospheric oxygen. Spore-forming microorganisms are very stable when stored under oil; therefore, this method is most often used for bacteria that can form spores (Langeler et al. 2005).

To save the properties of a culture without changes, it must be stored in special conditions. The long-term preservation of culture is based on cooling, freezing, or dehydration. In all these cases, cellular metabolism is limited or even stopped. Sometimes when being frozen and dehydrated, the cells come into a state of anabiosis or preanabiosis. For long-term storage of industrial microorganisms, low and ultra-low temperatures are used (Belous et al. 1987). Microorganisms are frozen at temperatures from –20°C and below. As cryoprotectants, 10–20% glycerol, 7–10% dimethyl sulfoxide, and 10–20% sucrose can be used. Rapid cooling (quick freezing) and storage are carried out either in deep cooling chambers (kelvinators) or using a refrigerant (liquid nitrogen with a temperature of –196°C, usually). The survival rate of microorganisms is maximal if they are stored in liquid nitrogen vapors, or directly in liquid nitrogen at temperatures below –30°C. Usually, after 3–5 years and 10–20 years in the cold, the survival percentage of microorganisms is 40–70% and 10–20%, respectively. Microorganisms that cannot withstand lyophilization, for example, some autotrophic bacteria, spirochetes, mycoplasmas, aquatic phycomycetes, and various viruses, are forcedly stored in liquid nitrogen (Tsutsaeva 1983).

Lyophilized cultures remain in a viable state for decades. The lyophilization method makes it possible to produce a large number of light and compact samples of each culture of microorganisms, which is very convenient. However, the percentage of viable cells of microorganisms as a result of lyophilization is often low. In the process of lyophilization, microorganisms are exposed to stressful effects of low temperatures and vacuum, with the most resistant cells selecting, which may not have the desired properties. There are a number of non-spore-forming microorganisms that usually cannot tolerate freeze-drying. Experimental data shows that in the process of lyophilization, mutations and the violation of the genetic stability of many types of microorganisms are possible. For example, after lyophilization, 99% cells *P. putida* and *Alcaligenes eutrophus* had lost the catechol-2,3-oxygenase activity encoded by a plasmid gene, and then the trait loss increased during storage (Lange and Weber 1995). Nerman and Feldblyum (1984) reported the loss of plasmids in lyophilized cultures of *B. subtilis* and *Saccharomyces cerevisiae* after 9 months of storage. Gaiek et al. (1994) found losses of more than 99% activity in lyophilized phenol-degrading mixed culture after 6 months of storage. It was also shown (Yoon 2005) that during lyophilization of cells of the strain *Pseudomonas* sp., KM12TC lost its ability to utilize phenol and resist to arsenic ions, which was associated with the elimination of plasmids carrying the responsible genes. Despite this, for a wide range of microorganisms, lyophilization is believed to mainly provide greater stability than room drying and the method of re-plating, with lyophilized cells remaining viable for many years.

Freezing is one of the most stressful effects on the cell because of ice formation inside and outside the cells. The nature of these changes depends on the sample under study, treatment with cryoprotectants, and on the cooling rate (Boldyrev 1987, Santivarangkna et al. 2008). Gram-positive bacteria proved to be more resistant to the effects of freezing and dehydration (Miyamoto-Shinohara et al. 2008). The cell wall can be ruptured because of the thin peptidoglycan layer in gram-negative microorganisms.

As for storage conditions, one of the main requirements is compliance with air-tightness and low temperatures. The oxidizing effect of oxygen accelerated by humidity adversely affects the survival of microorganisms (Miyamoto-Shinohara et al. 2000). Storage of dry material without access to light also contributes to the preservation of viability—three-hour exposure of the lyophilized *Escherichia coli* biomass to light (9×10^{-3} V/cm^2) led to a decrease in survival by two orders of magnitude (Israeli et al. 1993).

To achieve the maximal efficiency when using dry biological products, the conditions for their rehydration should be selected, since this process directly depends on how the restoration of bacterial cell viability during absorption of water will occur (Abadias et al. 2001, Costa et al. 2000).

Various cryoprotectants sucrose-dextran, sucrose-polyvinylpyrrolidone, lactose, gelatin are used during freezing and lyophilization. Typically, protective media are divided into two groups—the first one includes media in which the intensity of dehydration during lyophilization decreases due to an increase in the concentration of substances. These are penetrating (into the cell) cryoprotectants-low molecular weight substances, and this group includes monosodium glutamate, lactose, and sucrose. The second group includes media that make the rate of water evaporation constant when dried-starch, dextran, gelatin, and polyvinylpyrrolidone. The best media are those obtained by combining substances from both groups.

The protective effect of glycerol is widely known. When stored at –10°C with the addition of 15% glycerol, microorganisms of various genera, including *Pseudomonas*, remained viable for at least five months (Howard 1956). When storing lyophilized *P. fluorescens* and *Salmonella newport* cells in vacuum for five years, the use of sucrose, glutamate, and semicarbazide provided the greatest survival of microorganisms (Betty et al. 1974). A cryoprotectant (15% skim milk) can significantly increase the percentage of cells that retained the ability to utilize phenol in the presence of arsenic ions after lyophilization (Yoon 2005). Since these properties were encoded by genes located on plasmids, it can be assumed that protective media also ensured the inheritance of plasmids by bacterial cells.

It is believed that the protective mechanism of sugars is related to their ability to retain a certain amount of moisture, which is necessary for maintaining the viability of microorganisms (Leslie et al. 1995). The basis for this assumption was the results of experiments in which the survival rate of *E. coli* after long-term storage was higher when the culture was dried for a short time, and lower for long-term drying (Fry and Greaves 1951).

There is another hypothesis explaining the protective role of sugars—the loss of viability of dried microbes during storage is the result of the interaction of carbonyl groups of the drying medium with the amino groups of cellular proteins. These reactions, which occur especially intensely with reducing sugars having a free carbonyl group, are significantly accelerated with increasing temperature. Experimental data which show that the survival rates of microbes dried in sugared media sharply decreases with an increasing storage temperature, support this assumption. The neutralization of the carbonyl groups of the drying medium seems to be a reasonable method for increasing the stability of preparations (Conrad et al. 2000). Betty et al. (1974), based on the results of factor analysis, concluded that the interaction of carbonyl compounds with cellular components is the main factor affecting cell survival during storage.

Thus, when two types of cryoprotectants are combined, some of them protect bacteria from harm during freezing, and others neutralize carbonyl groups and prevent excessive dehydration during lyophilization (Santivarangkna et al. 2008).

Some other mechanisms are known to protect bacteria from exposure to low temperatures. It has been found (Kawahara 2002, Ramos et al. 2001) that some bacteria are capable of synthesizing specific proteins that differently affect the formation of ice crystals and help to minimize cell damage when the temperature decreases. Obviously, the use of such substances as components of protective environments is very promising.

Acknowledgment

The authors are thankful to Elena Demina for the help in translation.

Funding. The reported study was funded by RFBR and BRFBR, project number 20-54-00002.

References

Abadias, M., N. Teixido, J. Usall, A. Benabarre and I. Vinas. 2001. Viability, efficacy, and storage stability of freeze-dried biocontrol agent *Candida sake* using different protective and rehydration media. J Food Prot 64: 856–861.

Aislabie, J., M. McLeod and R. Fraser. 1998. Potential for biodegradation of hydrocarbons in soil from the Ross Dependency, Antarctica. Appl Microbiol Biotechnol 49: 210–214.

Alekseev, A., S. Bednarzhevsky, V. Zabelin, A. Komkova, N. Pushkarev, Yu. Rassadkin, N. Shevchenko and A. Shestopalov. 2008. A preparation for cleaning soil and water from oil and oil products. RU Patent #2337069.

Alekseev, A. 2011. Do no harm. Ind. Ecol. North. 5-6: 40–47.

Alekseev, A., E. Smorodina, L. Adamenko, E. Emelyanova, V. Zabelin, T. Ilyicheva, V.A. Reznikov and A.M. Shestopalov. 2011. Selection of the association of microorganisms-destructors of the oil fraction of solid alkanes at low positive temperatures. Modern Prob Sci Edu 6: http://www.science-education.ru/ru/article/view?id=4939.

Ananjina, L., V. Demakov, A. Nazarov and E. Plotnikova. 2010. Means for cleaning contaminated soils from oil and polycyclic aromatic hydrocarbons in conditions of increased mineralization of the environment. RU Patent #2388816.

Andreson, R., F. Khaziev, V. Deshura, F. Bagautdinov, T. Boyko and E. Novoselova. 1997. The method of recultivation of soils contaminated with oil and oil products. RU Patent #2077397.

Avchieva, P. 1998. *Candida maltosa* yeast consortium for oil pollution biodegradation. RU Patent #2114174.

Balba, T. 2003. Bioremediation of oil-contaminated sites. Case studies involving light and heavy petroleum hydrocarbons. Ph.D. Thesis. Conestoga-Rovers & Associates, Niagara Falls, NY, USA.

Betty, J., G. Coote and W. Scott. 1974. Some factors affecting the viability of dried bacteria during storage in vacuo. Appl Microbiol 4: 648–652.

Belonin, M., E. Rogozina, R. Svechina, A. Khotyanovich and N. Orlova. 1995. A biopreparation for cleaning soil and water from oil and oil products. RU Patent #2053205.

Belous, A.M., E.M. Gordienko and L.F. Rozanov. 1987. Biochemistry of membranes: freezing and cryoprotection. Vys'shaya Shkolah, Moscow.

Borzenkov, I.A., E.I. Milekhina, S.S. Belyaev and M.V. Ivanov. 1994. A consortium of microorganisms *Rhodococcus maris*, *Rhodococcus* sp., *Rhodococcus erythropolis*, *Pseudomonas stutzeri*, *Candida* sp., used to clean soil and brackish water ecosystems from oil pollution. RU Patent #2023686.

Borodavkin, P.P. 1981. Environmental protection during the construction and operation of trunk pipelines. Nedrah, Moscow.

Braddock, J.F., M.L. Ruth, J.L. Walworth and K. McCarthy. 1997. Enhancement and inhibition of microbial activity in hydrocarbon-contaminated arctic soils: implications for nutrient-amended bioremediation. Environ Sci Technol 31: 2078–2084.

Buzmakov, S.A. and I.V. Ladygin. 1993. The influence of oil fields on the flora and fauna of the Kama Urals. Proc Perm Sci Conf 1: 201–205.

Chaillan, F., C.H. Chaineau, V. Point, A. Saliot and J. Outdot. 2006. Factors inhibiting bioremediation of soil contaminated with weathered oilsand drill cuttings. Environ Pollution 144: 255–265.

Chertes, K.L., D.E. Bykov, O.V. Tupitsyna, O.A. Samarina, N.A. Uvarova, E.P. Istomina and A.M. Shterenberg. 2010. Intensive biothermal treatment of sludge waste from the oil complex. Ecol Prom Rossii 3: 36–39

Chugunov, V.A., Z.M. Ermolenko, S.K. Zhigletsova, I.I. Martovetskaya, R.I. Mironova, N.A. Zhirkova and V.P. Kholodenko. 2000. Development and testing of the Ecosorb biosorbent based on the association of oil-oxidizing bacteria for remediating oil-contaminated soils. Prikl Biokhim Mikrobiol 36: 661–665.

Chuvilin, E.M., N.S. Naletova, E.C. Miklyaeva and E.V. Kozlova. 2001. Factors affecting spreadibility and transportation of oil in regions of frozen ground. Polar Res 37: 229–238.

Conrad, P.B., D.P. Miller, P.R. Cielenski and J.J. de Pablo. 2000. Stabilization and preservation of Lactobacillus acidophilusin saccharide matrices. Cryobiology 41: 17–24.

Costa, E., J. Usall, N. Teixido, N. Garcia and I. Vinas. 2000. Effect of protective agents, rehydration media and initial cell concentration on viability of *Pantoea agglomerans* strain CPA-2 subjected to freeze-drying. J Appl Microbiol 89: 793–800.

Danilov-Daniljian, V.I. 2007. Ecological problems of the fuel and energy complex of Russia. Green World 1-2: 6–8.

Delille, D., B. Delille and E. Pelletier. 2002. Effectiveness of bioremediation of crude oil contaminated subantarctic intertidal sediment: the microbial response. Microb Ecol 44: 118–126.

Delille, D., F. Coulon and E. Pelletier. 2004. Biostimulation of natural microbial assemblages in oil-amended vegetated and desert sub-antarctic soils. Microb Ecol 47: 407–415.

Delille, D. and F. Coulon. 2007. Comparative mesocosm study of biostimulation efficiency in two different oil-amended sub-Antarctic soils. Microb Ecol 9: 1–10.

Ermolenko, Z.M., V.P. Kholodenko and V.A. Chugunov. 1997. Biological characteristics of a strain of mycobacteria isolated from oil of the Ukhta field. Mikrobiologiia 66: 650–654.

Erofeevskaya, L.A. 2014. The strain of bacteria *Exiguobacterium mexicanum*—destructor of oil and oil products. RU Patent #2523584.

Erofeevskaya, L.A. 2015. A consortium of microorganisms *Exiguobacterium mexicanum* and *Bacillus vallismortis* for cleaning permafrost soils from oil pollution. RU Patent #2565817.

Erofeevskaya, L.A. and J.S. Glyaznetsova. 2015. A biopreparation for bioremediation of oil-contaminated soils for climatic conditions of the far north. RU Patent #2565549.

Erofeevskaya, L.A. 2016. The drug for cleaning soil from oil pollution. RU Patent #2600868.

Eriksson, M., J.O. Ka and W.W. Mohn. 2001. Effects of low temperature and freeze-thaw cycles on hydrocarbon biodegradation in Arctic tundra soil. Appl Environ Microbiol 67: 5107–5112.

Escalante-Espinosa, E., M.E. Gallegos-Martinez, E. Favela-Torres and M. Gutierrez-Rojas. 2005. Improvement of the hydrocarbon phytoremediation rate by *Cyperus laxus* Lam. inoculated with a microbial consortium in a model system. Chemosphere V.59: 405–413.

Ferguson, S.H., P.D. Franzmann, I. Snape, A.T. Revil, M.G. Trefry and L.R. Zappia. 2003. Effects of temperature on mineralisation of petroleum in contaminated Antarctic terrestrial sediments. Chemosphere 52: 975–987.

Filonov, A.E., I.A. Kosheleva, A.N. Shkidchenko, I.A. Pyrchenkova, I.F. Puntus, A.B. Gafarov and A.M. Boronin. 2007. Association of bacterial strains producing bioemulsifiers for the degradation of oil and oil products in soils, fresh and sea water. RU Patent #2312891.

Filonov, A.E., A.A. Ovchinnikova, A.A. Vetrova, I.A. Nechaeva, K.P. Petrikov, E.P. Vlasova, L.I. Akhmetov, A.M. Shestopalov, V.A. Zabelin and A.M. Boronin. 2012. Oil-spill bioremediation, using a commercial biopreparation "MicroBak" and a consortium of plasmid-bearing strains "V&O" with associated plants. pp. 291–318. *In*: Romero-Zerón, L. (ed.). Introduction to Enhanced Oil Recovery (EOR) Processes and Bioremediation of Oil-Contaminated Sites. InTech, Rijeka, Croatia.

Filonov, A.E. 2016. Microbial biopreparations for cleaning the environment from oil contaminations under a moderate and cold climate. D.Sc. Thesis. Institute of Biochemistry and Physiology of Microorganisms of the Russian Academy of Sciences, Pushchino, Russia.

Foght, J.M. and D.M. McFarlane. 1999. Growth of extremophiles on petroleum. pp. 527–538. *In*: Seckbach, J. (ed.). Enigmatic microorganisms and life in extreme environments. Springer, Dordrecht, Netherlands.

Fry, R.M. and R.I. Greaves. 1951. Long term preservation of fungus culture. Natl J Hyg 49: 200.

Gabbasova, I.M., P.P. Suleymanov, F.Kh. Khaziev, T.F. Boyko, N.F. Galimzyanova, V.M. Ferdman. 2001. Recultivation of gray forest soil contaminated with oil sludge. Oil Economy 7: 8481–84.

Gaiek, R.L., C.R. Lange and A.S. Weber. 1994. The effects of freeze-drying and storage on a phenol degrading supplement. Water Environ Res 66: 698–706.

Gainutdinov, M.Z., S.M. Samosova, T.I. Artemyeva, M.Yu. Gilyazov, I.T. Hramov, I.A. Gaysin, V.I. Filchenkova and A.N. Zherebtsov. 1988. Recultivation of oil-contaminated lands of the forest-steppe zone of Tatarstan. pp. 177–197. *In*: Glazovskaya, M.A. (ed.). Restoration of Oil-contaminated Soil Ecosystems. Nauka, Moscow, Russia.

Garcia-Blanco, S., M. Motelab, A.D. Venosa, M.T. Siudan, K. Lee and D.W. King. 2001. Restoration of the oil-contaminated Saint Lawrence River shoreline: bioremediation and phytoremediation. Proc Int Oil Spill Conf pp. 303–308.

Gounot, A.M. and N.J. Russell. 1999. Physiology of cold-adapted microorganisms. pp. 33–55. *In*: Margesin, R. and F. Shinner (eds.). Cold-adapted organisms. Springer, Berlin, Heidelberg, New York.

Gradova, N.B., I.B. Gorova, R. Edaudi and R.N. Salina. 2003. The use of bacteria of the genus *Azotobacter* in bioremediation of oil-contaminated soils. Prikl Biokhim Mikrobiol 39: 318–321.

Howard, D.H. 1956. The preservation of bacteria by freezing in glycerol broth. J Bacteriol 71: 625.

https://dop-uni.ru/en.

http://home.eng.iastate.edu/~tge/ce421-521/brubaker.pdf.

https://pro-ecology.ru/ru/products/biologicals/biologicals-for-remediation/item/biopreparat-dop-uni-dlya-rekultivacii-nefteshlamov.

Ilyicheva, T.N., A.V. Mokeeva, A.M. Shestopalov, E.K. Emelyanova, A.Yu. Alekseev and V.A. Zabelin. 2014. Association of strains of oil-degrading bacteria and the method of remediation of oil-contaminated objects. RU Patent #2509150.

Ikhsanov, V.B. and N.A. Ikhsanova. 2000. The method of processing the bottom-hole zone of an oil well. RU Patent #2156353.

Israeli, E., B.T. Shaffer, J.A. Hoyt, B. Lighthart and L.M. Ganio. 1993. Survival differences among freeze-dried genetically engineered and wild-type bacteria. Appl Environ Microbiol 59: 594–598.

Ivasishin, P.L., T.A. Maryutina, E.Yu. Savonina and R.A. Talis. 2011. Effectiveness of biological preparations, humats and sorbents application for decreasing hydrocarbons residual concentrations in peat during remediation process. Environ Prot Oil Gas Complex 5: 19–23.

Jirasripongpun, K. 2002. The characterization of oil-degrading microorganisms from lubricating oil contaminated (scale) soil. Lett Appl Microbiol 35: 296–300.

Kawahara, H. 2002. The structures and functions of ice crystal-controlling proteins from bacteria. J Biosci Bioeng 94: 492–496.

Karaseva, E.V., A.A. Samkov, N.N. Volchenko, S.G. Karasev and A.A. Khudokormov. 2007. Method for microbiological remediation of oil sludge and soil contaminated with oil products (options). RU Patent #2311237.

Karaseva, E.V., A.A. Samkov, S.G. Karasev and V.Yu. Sychev. 2008a. A preparation for microbiological clean-up of oil sludge and soil contaminated with oil products. RU Patent #2317162.

Karaseva, E.V., S.G. Karasev, S.M. Samkova, V.V. Gora, N.N. Volchenko, A.A. Samkov and I.E. Golovina. 2008b. Microbiological approach to the rehabilitation of ecosystems contaminated with oil products and drilling waste during drilling operations on the offshore. Kuban Science (Nauka Kubani) 1: 14–19.

Karaseva, E.V., A.A. Samkov, A.A. Khudokormov, S.G. Karasev and N.N. Volchenko. 2009. Biopreparation for cleaning soil and water from oil and oil products. RU Patent #2365438.

Khabibullin, R.A. and M.V. Kovalenko. 1982. State of research on the assessment and elimination of the effects of soil pollution with oil, using its phytotoxicity. Land Recultivation in the USSR: Mat All-Union Sci Tech Conf Moscow V.2: 149–152.

Khlynovsky, A.M., I.V. Gordienko, N.V. Andreeva, I.V. Boykova, I.I. Novikova and M.Yu. Kozlova. 2012. Biopreparation for purifying water from hydrocarbon contaminants. Int Patent #WO 2012/082016 A 1.

Khmurchik, V.A., S.A. Ilarionov, M.Yu. Markarov and A.V. Nazarov. 2008. Oil-contaminated biogeocenoses: formation processes, scientific foundations of restoration, medical and environmental problems. *In*: Oborin, A.A. and V.N. Kataev (eds.). Perm State University, Perm, Russia.

Kireeva, N.A. 1996. Microbiological processes in oil-contaminated soils. Publishing House of Bashkirsky State University, Ufa, Russia.

Kireeva, N.A., V.V. Vodopyanov, A.S. Grigoriadi, E.I. Novoselova, G.G. Bagautdinova, A.R. Gareeva et al. 2010. The effectiveness of the use of biological products for the restoration of technologically contaminated soils. Bull Samara Sci Center Rus Acad Sci 12: 1023–1026.

Kobzev, E.N. 2003. Biodegradation of oil and oil products by microbial associations in model systems. Ph.D. Thesis. Institute of Biochemistry and Physiology of Microorganisms of the Russian Academy of Sciences, Pushchino, Russia.

Kolomytseva, M.P., I.P. Solyanikova, E.L. Golovlev and L.A. Golovleva. 2005. Heterogeneity of *Rhodococcus opacus* 1CP as a response to the stressful effects of chlorophenols. Mikrobiologiya 41: 541–546.

Koronelli, T.V. 1982. Microbiological degradation of hydrocarbons and its environmental consequences. Biol Sci 3: 5–12.

Koronelli, T.V., S.G. Dermicheva and E.V. Korotaeva. 1988. Survival of hydrocarbon-oxidizing bacteria under conditions of complete starvation. Microbiologiia 57: 298–302.

Koronelli, T.V. 1996. Principles and methods of intensifying the biological destruction of hydrocarbons in the environment. Prikl Biokhim Mikrobiol 32: 579–585.

Korshunova, T.Yu. 2019. Microbiological technologies to eliminate oil spills in various climates. D.Sc. Thesis. Ufimsky Biological Institute, Ufa, Russia.

Kovalenko, G.A., L.V. Perminova, T.V. Chuenko, I.B. Ivshina, M.S. Kuyukina, M.I. Rychkova. 2006. Carbon-containing macrostructured ceramic carriers for adsorption immobilization of enzymes and microorganisms. 5. Immobilization of yeast non-growing cells and alkanotrophic *Rhodococcus* growing cells. Biotech Rus 1: 102–113.

Kuznetsov, A.E., N.B. Gradova, S.V. Lushnikov, M. Engelhart, T. Weisser and M.V. Chebotaeva. 2010. Applied Ecobiotechnology. Textbook in 2 volumes. BINOM. Laboratory of knowledge, Moscow, Russia.

Ladygina, V.P., I.V. Trusey and Y.L. Gurevich. 2008. Bioremediation of the fuel oil contaminated subsurface. Proc 2 Int Conf Bioinf Biomed Eng. http://www.icbbe.org http://ieeexplore.ieee.org/servlet/opac?punumber=4534879.

Lange, C.R. and A.S. Weber. 1995. The loss of plasmid-encoded phenotype in *Alcaligenes eutrophus*, *Staphylococcus aureus* and *Pseudomonas putida* during freeze-drying and storage. Water Environ Res 67: 224–229.

Le Floch, S., F.X. Merlin and M. Guillerme. 1997. Bioren: recent experiment on oil polluted shoreline in temperate climate. pp. 411–417. *In*: Alleman, B.C. and A. Leeson (eds.). *In-situ* and *On-site* Bioremediation. Battelle Press, Columbus, USA.

Le Floch, S., F.X. Merlin, M. Guillerme, C. Dalmazzone and P. Le Corre. 1999. A field experiment on bioremediation. Bioren Env Tech 20: 897–907.

Lee, J.G. and E.M. Levy. 1989. Enhancement of the natural biodegradation of condensate and crude oil on beaches of Atlantic Canada. Proc 1989 Oil Spill Conf American Petroleum Institute, Washington DC: 479–486.

Langeler, J. 2005. Modern microbiology. *In*: Langeler, J., G. Drevs and G. Schlegel (eds.). Prokaryotes. 2 volumes. V. 1. Mir, Moscow, Russia.

Leslie, S.B., E. Israeli, B. Lighthart, J.H. Crow and L.M. Crowe. 1995. Trehalose and sucrose protect both membranes and proteins in intact bacteria during drying. Appl Environ Microbiol 61: 3592–3597.

Limbakh, I.Yu., G.O. Karapetyan, K.G. Karapetyan, I.I. Novikova, I.V. Boykova, I.N. Pisarev and V.A. Lednev. 2002. A biopreparation Avalon for cleaning environmental objects from oil and oil products, the method of its production. RU Patent #2181701.

Loginov, O.N., N.N. Silishchev, R.N. Churaev, T.F. Boyko, N.F. Galimzyanova, E.A. Danilova et al. 2004 A consortium of microbial strains *Bacillus brevis* and *Arthrobacter* sp., used to clean-up water and soil from crude oil and oil products. RU Patent #2232806.

Loginov, O.N., L.A. Nurtdinova, T.F. Boyko, S.P. Chetverikov and N.N. Silishchev. 2004. Efficiency of a novel preparation Lenoil in oil-polluted soil remediation. Biotech Rus 1: 77–82.

Maksimovich, N.G. and V.T. Khmurchik. 2007. A consortium of hydrocarbon-oxidizing bacteria strains *Pseudomonas aeruginosa* ND K3-1 and *Pseudomonas fluorescens* ND K3-2 as an oil products destructor and a method for cleaning oil-contaminated groundwater. RU Patent #2312719.

Margesin, R. and F. Shinner. 1999. Biological decontamination of oil spills in cold environments. J Chem Technol Biotechnol 74: 381–389.

Margesin, R. 2000. Potential of cold-adapted microorganisms for bioremediation of oil-polluted alpine soils. Int Biodeter Biodegr 46: 3–10.

Margesin, R. and F. Schinner. 2001. Bioremediation (natural attenuation and biostimulation) of diesel-oil-contaminated soil in an alpine glacier skiing area. Appl Microbiol Biotechnol 67: 3127–3133.

Margesin, R., G. Feller, C. Gerday and N.J. Russell. 2002. Cold-adapted microorganisms: adaptation strategies and biotechnological potential. pp. 871–885. *In*: Bitton, G. (ed.). The Encyclopedia of Environmental Microbiology. 2. Wiley, New York, USA.

Margesin, R., M. Hammerle and D. Tscherko. 2007. Microbial activity and community composition during bioremediation of diesel-oil-contaminated soil: effects of hydrocarbon concentration, fertilizers, and incubation time. Microb Ecol 53: 259–269.

Markarova, M. 2004. The experience of application of biological preparation "Universal" for remediation of oil contaminated lands. Bull Inst Biol Komi Sci Center Ural Bran, Rus Acad Sci 10: 21–23.

Maximovich, N.G. and V.T. Khmurchik. 2007. A consortium of hydrocarbon-oxidizing bacteria strains *Pseudomonas aeruginosa* ND K3-1 and *Pseudomonas fluorescens* ND K3-2 for oil products degradation and a method for purifying oil-contaminated groundwater. RU Patent #2312719.

Miyamoto-Shinohara, Y., T. Imaizumi, J. Sukenobe, Y. Murakami, S. Kawamura and Y. Komatsu. 2000. Survival of microbes after freeze-drying and long-term storage. Cryobiol 41: 251–255.385.

Miyamoto-Shinohara, Y., J. Sukenobe, T. Imaizumi and T. Nakahara. 2008. Survival of freeze-dried bacteria. J Gen Appl Microbiol 54: 9–24.

Mohn, W.W., C.Z. Radziminski, M.C. Fortin and K.J. Reimer. 2001. On site bioremediation of hydrocarbons-contaminated Arctic tundra soils in inoculated biopiles. Appl Microbiol Biotechnol 57: 242–247.

Morita, R.Y. 1975. Psychrophilic bacteria. Bacteriol Rev 39: 144–167.

Murygina, V., M. Arinbasarov and S. Kalyuzhnyi. 2000. Bioremediation of oil polluted aquatic systems and soils with novel preparation Rhoder. Biodeg 11: 385–389.

Murygina, V., M. Markarova and S. Kalyuzhnyi. 2005. Application preparation Rhoder for remediation of oil polluted polar marshy wetlands in Komi Republic. Env Int 31: 163–166.

Murygina, V.P., N.E. Voishvillo and S.V. Kalyuzhny. 2002 Biopreparation Roder for cleaning soils, soil, fresh and mineralized waters from oil and oil products. RU Patent #2174496.

Murzakov, B.G., A.I. Zaikina, R.A. Rogacheva and E.V. Semenova. 1996. The method of microbiological cleaning objects from oil pollution. RU Patent #2067993.

Nechaeva, I.A., A.E. Filonov, L.I. Akhmetov, I.F. Puntus and A.M. Boronin. 2009. Stimulation of microbial degradation of oil in the soil by introducing bacterial associations and mineral fertilizers in laboratory and field conditions. Biotech Rus 1: 64–70.

Nierman, W.C. and T. Feldblyum. 1984. Cryopreservation of cultures that contain plasmids. Dev Ind Microbiol 26: 423–433.

Onegova, T.S., A.A. Kalimullin, E.M. Yulbarisov, N.A. Kireeva, I.Sh. Garifullin, N.V. Zhdanova and U.N. Sadykov. 2003. A method of cleaning the soil and water from oil pollution. RU Patent # 2 198 748.

Orlova, E.V. and A.Yu. Stepanova. 2012. Estimation of possibility of creating a plant-microbial bioremediation complex from the Oleovorin preparation components. Agrochem 10: 72–78.

Pokonova, Yu.V. 2003. Oil and oil products. "Synthez", Saint-Petersburg, Russia.

Ramos, J.L., M.T. Gallegos, S. Marqués, M.I. Ramos-González, M. Espinosa-Urgel and A. Segura. 2001. Responses of gram-negative bacteria to certain environmental stressors. Curr Opin Microbiol 4: 166–171.

Rogozina, E.A., O.A. Andreeva, S.I. Zharkova, D.A. Martynova and N.A. Orlova. 2010. Comparative characteristics of native biopreparations proposed for cleanup of soils and grounds from pollution. Oil Gas Geol Theory Prac 5: 1–18.

Ron, E.Z. and E. Rosenberg. 2001. Natural roles of biosurfactants. Environ Microbiol 3: 229–236.

Santivarangkna, C., U. Kulozik and P. Foerst. 2008. Inactivation mechanisms of lactic acid starter cultures preserved by drying processes. J Appl Microbiol 105: 1–13.

Saxon, V.M., S.A. Kuznetsov, A.V. Kretov, D.P. Khromykh, I.V. Boykova, I.I. Novikova and Yu.A. Konev. 1998. A biopreparation for cleaning environmental objects from oil and oil products. RU Patent #2138451.

Sendstad, E. 1980. Accelerated biodegradation of crude oil on Antarctic shorelines. Proc 3 Arctic and marine oil spill program technical seminar, Ottawa, Canada. 1–8.

Sharapova, I.E., M.Yu. Markarova and A.V Garabadzhiu. 2011. A complex biosorbent based on bacterial and fungal strains for remediating water from oil and oil products in the presence of microalgae. RU Patent #2422587.

Shilova, I.I. 1988. Biological recultivation of oil-contaminated lands in the taiga zone. pp. 112–122. *In*: Glazovskaya, M.A. (ed.). Restoration of Oil-Spilled Soil Ecosystems. Nauka, Moscow, USSR.

Shkidchenko, A.N., A.B. Gafarov, E.S. Ivanova and T.Z. Esikova. 2007. Oil sludge biodegradation in an open flow system. Biotech Rus 6: 55–59.

Shtina, E.A. and K.A. Nekrasova. 1988. Algae of oil-contaminated soils: state of the art and objectives of the study. pp. 57–81. *In*: Glazovskaya, M.A. (ed.). Restoration of Oil-Spilled Soil Ecosystems. Nauka, Moscow, USSR.

Smolova, O.S. 2015. Biorecultivation of hydrocarbon-contaminated soils using psychrotolerant microorganisms with mycostatic activity. Ph.D. Thesis. Ufimsky Institute of Biology of the Russian Academy of Sciences, Ufa, Russia.

Sokolov, S.N. and I.R. Khadaev. 2017. Influence of biological products on the reduction of the residual concentration of oil hydrocarbons in soils. Int Res J 60: 130–136.

Soprunova, O.B., A.R. Galperina and M.A. Klyuyanova. 2011. Means for the destruction of petroleum hydrocarbons in a medium containing NaCl up to 24.0%. RU Patent #2422505.

Stabnikova, E.V., M.V. Seleznieva, O.N. Reva and V.N. Ivanov. 1995. The selection of an active hydrocarbon-degrading microorganism for cleaning oil-contaminated soils. Prikl Biokhim Microbiol 5: 534–539.

Stankevich, D.S. 2002. Use of the hydrocarbon-oxidizing bacterium *Pseudomonas* for bioremediation of oil-contaminated soils. Ph.D. Thesis. K.A. Timiryazev Agricultural Academy, Moscow, Russia.

Suni, S., A.-L. Kosunen, M. Hautala, A. Pasila and M. Romantschuk. 2004. Use of a by-product of peat excavation, cotton grass fibre, as a sorbent for oil-spills. Mar Poll Bulletin 49: 916–921.

Swarovskaya, L.I., S.I. Pisareva and L.K. Altunina. 2009. A biopreparation for cleaning soil and water from oil and oil products. RU Patent #2361686.

Teas, Ch., S. Kalligeros, F. Zanikos, S. Stournas, E. Lois and G. Anastopoulos. 2001. Investigation of the effectiveness of absorbent materials in oil spill clean up. Desalin 140: 259–264.

Tevvors, J.T and M.H.Jr. Saier. 2010. The legacy of oil spills. Water Air Soil Pollut 211: 1–3.

Thapa, B., A. Kumar and A. Ghimire. 2012. A review on bioremediation of petroleum hydrocarbon contaminants in soil. Kathmandu Univ J Sci Engin Technol 8: 164–170.

Tretiakova, M.S. 2017. Prospects for the use of endo - and rhizospheric microorganisms for the restoration of oil contaminated soils. Ph.D. Thesis. Siberian Institute of Physiology and Biochemistry of Plants of Siberian Branch of the Russian Academy of Sciences, Irkutsk, Russia.

Trusei, I.V., A.Yu. Ozerskii, V.P. Ladygina and Yu.L. Gurevich. 2009. Distribution of microorganisms in the oil-polluted ground of vadose and saturation zones. Sovrem Prob Ecol 2: 22–26.

Trusei, I. 2018. Stimulation *in situ* of indigenous psychrophilic and mesophilic microorganisms for bioremediation of soils contaminated with oil products. Ph.D. Thesis. Krasnoyarsk Scientific Center of Siberian Branch of the Russian Academy of Sciences, Krasnoyarsk, Russia.

Tsutsaeva, A.A. 1983. Cryopreservation of cell suspensions. Tsutsaeva, A.A. (ed.). Naukova Dumka, Kiev, USSR.

Van Hamme, J.D., A. Singh and O.P. Ward. 2003. Recent advances in petroleum microbiology. Microbiol Mol Biol Rev 67: 503–549.

Venosa, A.D., K. Lee, M.T. Suidan, S. Garcia-Blanco, S. Cobanli, M. Moteleb et al. 2002. Bioremediation and biorestoration of a crude oil-contaminated freshwater on the St. Lawrence River. Biorem J 6: 1–10.

Vetrova, A.A., A.A. Ivanova, A.E. Filonov, V.A. Zabelin, I.A. Nechaeva, L.Th.B. Nguet et al. 2013. Comparative efficiency of petroleum products degradation by a consortium of plasmid-containing strains of destructors and biopreparations “MicroBack”, “Biooil”. Bull Tula State Univ Nat Sci Series 2: 258–272.

Vogt, C. and H.H. Richnow. 2013. Bioremediation via *in situ* microbial degradation of organic pollutants. Adv Biochem Eng Biotechnol 142: 123–146.

Volkov, M.Yu., A.A. Ilyin and A.A. Kalilets. 2015. The preparation for biodegradation of petroleum products “Bioionite” and the method of its production. RU Patent #2571219.

Volkov, M.Yu., R.M. Abdullin, S.V. Anikin, D.A. Venkov and Z.S. Salikhov. 2019. A preparation for biodegradation of oil products and a method for its preparation. No. 2681831.

Wang, Q., S. Zhang, Y. Li and W. Klassen. 2011. Potential approaches to improving biodegradation of hydrocarbons for bioremediation of crude oil pollution. J Env Protec 2: 47–55.

Xue, J., Y. Yu, Yu. Bai, L. Wang and Y. Wu. 2015. Marine oil-degrading microorganisms and biodegradation process of petroleum hydrocarbon in marine environments: a review. Curr Microbiol 71: 220–228.

Yoon, K.P. 2005. Stabilities of artificially transconjugated plasmids for the bioremediation of cocontaminated sites. J Microbiol 43: 196–203.

Yankevich, M.I., V.V. Khadeeva, L.F. Surzhko, I.A. Afti, K.V. Kvitko, V.V. Biryukov et al. 2006. A biologically active composition to clean-up surface water, soil and grounds from oil pollution. RU Patent No. 2270808.

8

Microbial Biosurfactants Remediation of Contaminated Soils

Poulami Datta,[1] *Pankaj Tiwari*[2] and *Lalit M. Pandey*[3,*]

Introduction

Petroleum hydrocarbons are one of the primary energy sources. However, due to oil spillage, it can also be a potential environmental pollutant because of its toxicity and poor degradability. There have been numerous incidents of petroleum contamination to the ecosystem. Despite so much research, successful practical bioremediation of petroleum hydrocarbons is still a challenge (Ayotamuno et al. 2006), as petroleum is a complex mixture of hydrocarbons, namely alkanes (paraffins), cycloalkanes (napthenes), aromatic, and gaseous hydrocarbons and non-hydrocarbons. The release of these hydrocarbons contaminates the soils, which causes environmental hazards. The studies on biodegradation of soil-contaminants exhibited that there are different release rates of the compounds which are transported from the soil to aqueous phase. Due to the rate-limiting step, the feature is termed as limited bioavailability, which depends on the type and physiochemical properties of the soils and the pollutants, the microbes involved, and environmental conditions, such as temperature, presence of nutrients, and oxygen.

Often there remains a higher possibility of uptaking the harmful hydrocarbons by plants which grow in the contaminated sites. From these plants, the hydrocarbons are transferred to animal and human beings through the food chains, and imbalance the ecosystem (Alagić et al. 2015, Fismes et al. 2002). Due to the adverse effects of these chemicals on human health and environment, they have been categorized as

[1] Centre for the Environment, Indian Institute of Technology Guwahati, Guwahati, 781039, India.
[2] Department of Chemical Engineering, Indian Institute of Technology, Guwahati, 781039, India.
[3] Department of Biosciences and Bioengineering, Indian Institute of Technology, Guwahati, 781039, India.
Emails: poulami.datta@iitg.ac.in; pankaj.tiwari@iitg.ac.in
* Corresponding author: lalitpandey@iitg.ac.in

the priority environmental pollutants by the US Environmental Protection Agency (EPA 1986). Considering the harmful effects of the hydrocarbons, it became very necessary to take required measures to mitigate their presence in the environment. Bioavailability may be one of the limiting factors for the biodegradation of such compounds. Besides petroleum, the other soil pollutants may be particulates or liquids which may get absorbed or adsorbed to the soil and accumulate in the pores (Abdel-Shafy and Mansour 2016). The technologies which are usually used for soil remediation are—mechanical, evaporation, burying, dispersion, and washing (Das and Chandran 2011). However, these technologies are quite expensive and result in the incomplete decomposition/removal of the pollutants. Phyto-remediation is also used for the remediation, however the process is very slow (Cunningham et al. 1995). In this chapter, the application of biosurfactants in the remediation of soils, mechanisms of hydrocarbon uptake, and laboratory and field scale studies are discussed. A comprehensive table summarizes the recent literature towards the remediation of the contaminated soils.

Methods for the bioremediation of contaminated soils

The process of bioremediation involves the microorganisms for degradation or removing pollutants utilizing their metabolic activities (Das and Chandran 2011). Bioremediation has been considered an efficient, cost-effective, and ecologically friendly method for degrading contaminants in the polluted soils. As crude oil consists of various molecular structures, it could be assumed that mixed bacterial culture may be more effective for degrading the complex structures as compared to the pure culture. Implementing microbial consortia for the purpose of bioremediation eliminates higher cost of pure strain isolation, purification, and characterization (Kavitha et al. 2014).

Biodegradation mainly occurs through bioaugmentation and biostimulation (Simarro et al. 2013, Wu et al. 2016). Petroleum hydrocarbon contaminated sites can be reclaimed through bioremediation, which is an enhanced natural process using biosurfactant producing and oil degrading bacteria (Sharma et al. 2019b). Bioremediation technologies are highly dependent upon favorable aeration, temperature, and nutrient supplements for hydrocarbon breakdown through biological means. Bioaugmentation involves inoculation of exogenous microbes in the contaminated soils for degrading the pollutants (Taccari et al. 2012). In biostimulation technology, nutrients are supplied to the contaminated soils, which stimulate the hydrocarbon degrading ability by avoiding the metabolic limitations of the indigenous microbes already present in the soils (Wu et al. 2016). The bioremediation can be carried out *in situ* by applying microbial strains in which biosurfactant production occurs indigenously. This is quite effective in terms of technology and cost. However, the indigenous culture survival in the condition is a matter of concern. However, *Bacillus* and *Pseudomonas* strains obtained from the hydrocarbon-contaminated sites have been proven to be biosurfactant-producing and quite effective in the *in situ* conditions (Mulligan 2005).

Biosurfactant producing microbes and their screening

Surfactants are amphiphilic compounds having both hydrophobic and hydrophilic moieties. Surfactants which are synthesized chemically are called chemical surfactants, and those which are produced biologically are termed as biosurfactants. The hydrophilic moiety in biosurfactants is composed of amino acids or peptides, phosphates, carboxylic acids or alcohol, and sachharides. The hydrophobic part consists of saturated, unsaturated, or hydroxylated fatty acids. The hydrophilic component helps the surfactant to be soluble in aqueous phase and the hydrophobic part concentrates at the interfaces. The accumulation of the biosurfactant molecules at the air-water interface leads to micelles formation which reduces the surface tension and emulsifies the hydrocarbons. Potential biosurfactants reduce the surface tension from 72 mN/m to 30 ± 5 mN/m depending on their nature and concentration. The micelles assist the hydrophobic organic compounds (HOC) to increase solubility (Volkering et al. 1997).

The major types of biosurfactants include glycolipids, lipopeptides, phospholipids, fatty acids, natural lipid, and polymeric biosurfactants (Janek et al. 2010). Lipopeptides are most commonly isolated and characterized biosurfactants which exhibit excellent surface-active properties and biological activities. Biosurfactants are produced by various bacterial strains, such as *Bacillus*, *Pseudomonas*, *Arthrobacter*, *Streptomyces*, *Acinetobacter*, *Rhodococcus*, *Halomonas*, and *Enterobacter*, which can degrade or transform the harmful components of petroleum (Bezza and Chirwa 2015). Due to its origin, biosurfactants are biodegradable, non-toxic, non-hazardous, and eco-friendly compounds which can also be produced from renewable resources under *ex situ* conditions as well as *in situ* conditions (extreme environment). Several surface active properties, such as surface tension and interfacial tension reduction, emulsification index determination, and foaming stability study are significant parameters for the biosurfactant endorsement for the soil remediation process.

Characteristics of biosurfactants for the remediation of soils

It is very important to access certain criteria of the biosurfactants before utilizing them for the soil treatment process, such as toxicity nature and hydrophilic-lipophilic balance (HLB) value (Saxena et al. 2018). The toxicity of the biosurfactant and its degradation is one of the main criteria for its use for soil remediation. Biologically produced surfactants are more acceptable than the chemically synthesized surfactants due to their lower toxicity and eco-friendly nature. The CMC values of the chemical surfactants (Saha et al. 2018, 2017, 2019) are higher than the biosurfactants, so for the same action, more amount of chemical surfactant needs to be employed than biosurfactants, which impose threat to the environment. After their utilization, biosurfactants are easy to degrade or remove, whereas chemical surfactants require special techniques for their removal. HLB value decides the relative contribution of the hydrophilic moiety to the weight of the surfactant molecule, and the nature of emulsion to be formed. Surfactants with HLB value of 3 to 6 are considered to be lipophilic and form water in oil (W/O) emulsions. The surfactants with HLB value in the range of 10 to 18 are comparatively hydrophilic and form oil in water (O/W) emulsions (Witthayapanyanon et al. 2008). Surfactants which have higher HLB values (above 10) are more efficient in washing oil-contaminated soils.

Application of biosurfactant in remediation of soils

Biodegradation of the compounds present in the contaminated soil involves the interaction among the soil particles, pollutants, water, and microbes. One of the prime issues which influenced the biodegradation efficiency of complex oily compounds is its low availability of contaminants for microbial attack. Surfactants are assumed to affect the interaction and increase the bioavailability of the organic compounds for the soil remediation. The surfactants stimulate the transfer of the particles from the soil to the aqueous phase either by emulsifying them or by micellar solubilization. The bioavailability and contaminant metabolism can be enhanced by increasing substrate solubilization using the biosurfactants (Cerqueira et al. 2011), which can also enhance the solubilization of the poorly soluble compounds, such as polycyclic aromatic hydrocarbon (PAH) and assist in remediation through the following mechanisms (Pacwa-Płociniczak et al. 2011). The first one involves the increase of the substrate availability for the microorganism, while the other mechanism includes the interaction with the cell surface. The hydrophobicity of the cell surface facilitates a better interaction with the hydrocarbon. This results in decrease of surface tension as well as interfacial tension values. The degradability towards microbial attack is different for various types of hydrocarbons, which is ranked in the following order: linear alkanes > branched alkanes > small aromatics > cyclic alkanes (Ulrici 2000).

Mechanisms of hydrocarbon uptake

The bioavailability of the HOCs can be enhanced by biosurfactants through the emulsification and solubility (Fig. 8.1). Biosurfactants mainly increase the aqueous solubility of the non-aqueous phase liquid (NAPL) through reducing their surface tension/interfacial tension at the air-water and water-oil interfaces. In the presence of NAPL, biosurfactants accumulate at the liquid-liquid interface, resulting in the reduction of interfacial tension (IFT). Further, NAPL droplets are dispersed and emulsion is stabilized. This results in the enhanced solubility of hydrocarbon substrates for the microbial cells. A study revealed that the cell surface hydrophobicity (CSH) was increased by biosurfactant-producing strain than non-biosurfactant-producing strain during their growth on hexadecane. The rhamnolipid produced by *Pseudomonas aeruginosa* PG201 increased the solubility of hexadecane from 1.8 to 22.8 μg/L (Beal and Betts 2000). Solubilization occurs at a concentration above CMC, in which biosurfactants clustered together and formed micelles (Urum et al. 2003).

Ex situ washing of contaminated soil by biosurfactants

Soil washing technique is one of the promising approaches for bioremediation of oil-contaminated soil (Urum and Pekdemir 2004). The performance of soil washing depends on the characteristics of the contaminated soil, such as soil particle size distribution, organic and inorganic contents. The successful biosurfactant induced

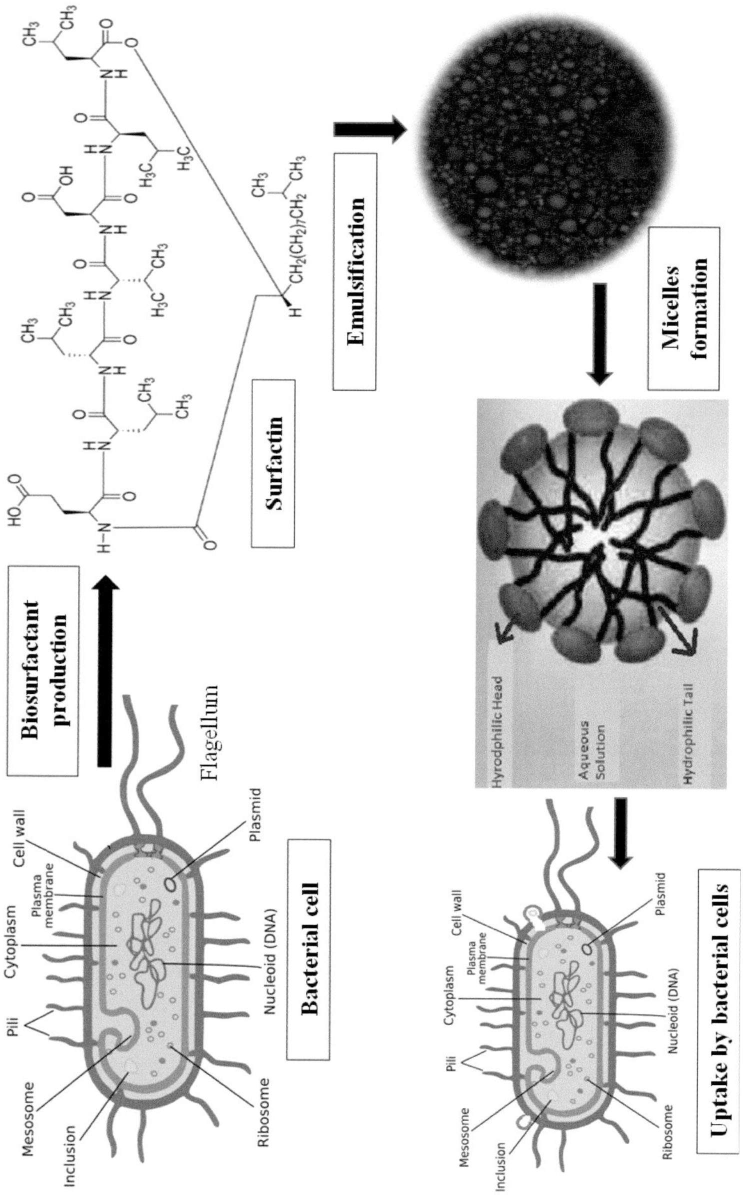

Figure 8.1. Role of biosurfactant (i.e., Surfactin) produced by *Bacillus* sp. in uptaking hydrocarbons.

soil washing is carried out in two stages: mobilization and solubilization, which occur below and above the CMC value, respectively (Mulligan et al. 2001).

A schematic of soil washing using biosurfactant is depicted in Fig. 8.2. A study of oil washing by rhamnolipid showed that it enhanced the release of poorly soluble compounds from soil. The removal efficiency depends on the biosurfactant concentration and the contact time. The rhamnolipid produced by *Pseudomonas aeruginosa* UG2 was able to remove the hydrocarbon mixture from sandy loam soil and the removal efficiency was increased from 23% to 59% after the *ex situ* washing by rhamnolipid (Scheibenbogen et al. 1994). Urum et al. explored various biosurfactants, such as lecithin, aescin, rhamnolipid, saponin, and tannin to examine the crude-oil washing from the contaminated soils in comparison with sodium dodecyl sulphate (SDS). Rhamnolipid, SDS, and saponin were able to remove more than 79% crude oil under optimum conditions, 50°C, and 10 minutes washing time (Badoga et al. 2011, Urum et al. 2003).

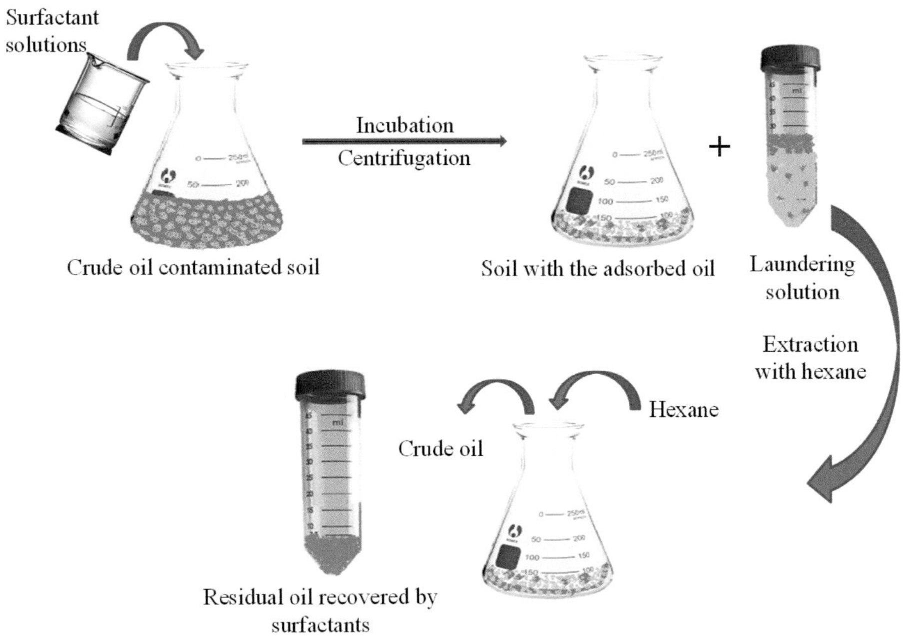

Figure 8.2. Schematic of oil washing study by microbial surfactants.

Laboratory and field scale studies

The soil contamination can occur due to different reasons, such as petroleum exploration, oil spillage, or any other type of industrial activities. Depending upon the contamination, different biosurfactants have been explored for the bioremediation of contaminated soils. Table 8.1 summarizes the studies on the effect of biosurfactants for the remediation of contaminated soils reported in the recent years. Rhamnolipid is the most predominant biosurfactant, which is produced by different strains of

Table 8.1. Effect of microbial surfactants towards remediation of the contaminated soils.

Samples	Nature of contamination	Microorganisms	Setups	Surfactant type	Treatment duration	% degradation	References
Creosote-polluted sandy loam soil from Spain	4370 mg kg^{-1} of PAHs contamination	*Mycobacterium gilvum* VM552	Flask scale (250 mL) Tenax desorption	Rhamnolipid (1 g/L)	5 months of biopiling	86.8 ± 12.7% degradation of PAH	(Posada-Baquero et al. 2019a)
Clay soil and sandy loam soil from Arkansas and California respectively	10 mg kg^{-1} Pyrene contamination	*Mycobacterium vanbaalenii* PYR-1	Flask respirometer	Rhamnolipid (1400 μg g^{-1})	50 days of bioaugmentation	58 ± 1% and 44 ± 0.3% pyrene degradation in clay and sandy loam soil respectively	(Wolf et al. 2019, Wolf and Gan 2018)
Petroleum contaminated soil from Canada	3276 mg/kg Petroleum hydrocarbon contamination	Commercial Rhamnolipid procured from Sigma-Aldrich Canada Co.	Flask scale (125 mL)	Rhamnolipid (400 mg/L)	30 minutes of soil washing	58.5% TPH reduction	(Olasanmi and Thring 2019)
PAHs, BTEX and alkanes contaminated soil of industrial site in Italy	4370 mg kg^{-1} of PAHs contamination	*Pseudomonas* and *Acinetobacter*	Tenax extraction (50 mL)	Rhamnolipid (7 mg g^{-1})	30 days of bioaugmentation	~ 98% degradation of PAHs	(Posada-Baquero et al. 2019b)
Sandy clay loam soil from gasoline refueling stations and automobile service stations in Tamil Nadu, India	10% crude oil contamination	*Shewanella* sp. BS4	Shake flask scale	2% (v/w) Rhamnolipid	135 days	75.8% hydrocarbon removal	(Joe et al. 2019)
Soil (sand, silt and clay) from Pretoria, South Africa	150 mL/kg waste oil contamination	*Pseudomonas aeruginosa*	Electrokinetic reactor (160.5 mm × 150 mm × 150 mm)	Rhamnolipid	10 days	71.4% reduction in total carbon	(Gidudu and Chirwa 2019)

Coal tar creosote contaminated soil from South Africa	3% used motor oil contamination	*Bacillus subtilis* CN2	Shake flask in 250 mL Erlenmeyer flasks	Lipopeptide (0.15% (w/v))	3 weeks	30% Naphthalene; 50% Phenanthrene; 60% Fluoranthene degradation	(Bezza and Chirwa 2015)
Artificially contaminated with hydrocarbons	10% heavy oil or diesel contamination	*Bacillus subtilis*	Flask scale (250 mL)	Cyclic lipopeptide – Surfactin ($37.5\ mg\ L^{-1}$)	24 hours	81.8% diesel and 77.8% oil removal	(Felix et al. 2019)
Kuwaiti oil-contaminated soil from Burgan	34.15 g kg^{-1} TPH contamination (n-alkanes, aromatic hydrocarbons)	*Alcaligenes, Chryseobacterium* sp.	Sequential biowashing-biopile reactor (250 mL)	Surfactin	20 days	86% TPH removal	(Kim et al. 2019)
Oil contaminated soil from Western Siberia and Kazakhstan	15% crude oil, diesel fuel, phenol; benzene and naphthalene contamination	*Rhodococcus* sp.	Flask scale	Glycolipid and trehalolipids	10 days	55–59% degradation of crude oil hydrocarbons	(Puntus et al. 2019)
Artificially diesel contaminated sand	10% diesel contamination	*Pseudomonas* sp.	Sand bioreactor	Syringafactin	28 days	70% TPH removal	(Zouari et al. 2019)
Dystrophic humicoxisol of Brazil (high-plasticity clay)	20% biodiesel (from Brazilian Petroleum Company)	*Saccharomyces cerevisiae*	5 L glass bioreactor	Extracellular biosurfactant (0.5%)	90 days	56.71% degradation	(Kreling et al. 2019)

Pseudomonas (Sharma et al. 2019a), *Acinetobacter*, and *Mycobacteria.* Recently, rhamnolipids have been used to degrade mainly the PAHs (Otenio et al. 2005), which include the harmful contaminants, such as pyrene, anthracene, phenanthrene, and benzo(a)pyrene, etc. The duration of this bioremediation treatment varied, starting from few hours to months depending on the setup chosen and the method followed. Another lipopeptide biosurfactant, surfactin (produced by *Bacillus subtilis, Alcaligenes*) (Datta et al. 2018), played an important role during the course of contaminated soil biodegradation (Montagnolli et al. 2015a, b). The bioremediation methods followed the shake flask scale for the soil washing, bioaugmentation, and *in situ* bioremediation. The experimental trials were also carried out using specialized reactor setups. Eventually combinations of two or more biosurfactants were also utilized in some of the experiments in order to achieve better and faster degradation of contaminants.

The oil removal from the oil-contaminated sites of Southern Taiwan was carried out using biosurfactants. The total petroleum hydrocarbon (TPH) removing capability from soil was analyzed using rhamnolipid and surfactin, and compared with that of the chemical surfactants (Tween 80 and Triton X 100). The biosurfactants showed better removal efficiencies 63% (rhamnolipid) and 62% (surfactin) as compared to chemical surfactants; 40% (Tween 80) and 35% (Triton X 100). This result indicates better performance of biosurfactants over synthetic surfactants in terms of mobilizing the oil pollutants from the contaminated site, and hence can be predicted as potential biostimulation agents for the bioremediation of oil-polluted soils (Lai et al. 2009). A separate biodegradation study was carried out using four different types of soils, i.e., sandy soil, fine sand soil, clay, and clay loam. Biosurfactant from marine strain *Bacillus licheniformis* MTCC 5514 was used, and the results were compared to the synthetic and commercially available surfactants (Tween 80, Triton X 100, SDS, Cetyl trimethylammonium bromide (CTAB), and Lecithin). More than 85% oil recovery could be achieved with 10% concentration of biosurfactant, whereas, SDS could recover 60% oil in case of fine sand soil. In the rest of the soil samples also a similar trend was observed, i.e., biosurfactant could remove more crude oil than the other chemical surfactants (Kavitha et al. 2014).

A laboratory scale work has been carried out to investigate the probable methods for enhancing the biodegradation rate, in which rhamnolipid was produced by *Pseudomonas aeruginosa*. The contaminated soils were supplemented by nitrogen, phosphorus, and potassium (NPK) to increase the natural fertility through biostimulation and bioaugmentation. The maximum n-alkane degradation was achieved after 56 days of treatment (Rahman et al. 2007). In a separate study, *Bacillus licheniformis* Y-1 strain from Dagang oilfield was examined to assess its potential for remediating the contaminated soil. After 30 days of remediation, TPH and PAH degradation were found to be $50.65 \pm 2.99\%$ and $40.68 \pm 3.14\%$, respectively, as measured using gas chromatography (GC). Further, the presence of biosurfactant enhanced the degradation efficiency of TPH and PAH as $54 \pm 4.39\%$ and $45.08 \pm 6.63\%$, respectively. The emulsifying property of the biosurfactant aided the petroleum hydrocarbon to be easily used up by the microorganisms, leading to bioavailability improvement (Liu et al. 2016).

The bioaugmentation and biostimulation studies have been carried out by examining the microbial community of the petroleum-contaminated soils. A microcosm study was performed to analyze the effect of bioaugmentation with *Acinetobacter* SZ-1 strain KF453955 and the biostimulation with the nutrients (nitrogen and phosphorus) on their petroleum hydrocarbon degradation potential. The study was continued for 10 weeks, in which 60% and 34% TPH degradation was achieved through biostimulation and bioaugmentation, respectively after 6 weeks of incubation. From the 7th week onwards, the degradation efficiency remained constant. Catalase activity and oil-degrader population were also determined in this study and found to be higher in biostimulation as compared to bioaugmentation (Wu et al. 2016). This study highlights the effect of nutrients (biostimulation) towards the growth of indigenous microorganisms and efficient premeditation of petroleum-contaminated soils.

The same research group performed microcosm studies for evaluating the effect of bioaugmentation and biostimulation in the oil-contaminated site of Shaanxi province, China. The degradation efficiency of various hydrocarbon contaminants was determined by infrared photometer oil content analyzer and gas chromatography-mass spectrometry (GC-MS). Contaminated soil was treated in three ways—by moisture treatment (with 20% moisture), bioaugmentation study was carried out with hydrocarbon degrading consortia containing *Pseudomonas stutzeri* GQ-4 strain KF453954, *Pseudomonas* SZ-2 strain KF453956, and *Bacillus* SQe2 strain KF453961, and biostimulation with $(NH_4)_2SO_4$ and KH_2PO_4. After 8 weeks of incubation, alkane concentration was reduced from 15400 ± 130 mg kg^{-1} to 7200 ± 280 mg kg^{-1} and 6160 ± 250 mg kg^{-1} soil, and PAH concentration was reduced from 2180 ± 20 to 1630 ± 40, and 1120 ± 40 mg kg^{-1} soil by bioaugmentation and biostimulation, respectively. Similarly, TPH degradation was 58 ± 1% and 48 ± 1% in case of bioaugmentation and biostimulation, respectively (Wu et al. 2017). This indicated bioaugmentation with microcosm to be more effective for alkane and PAH degradation than biostimulation.

Another microcosm study was conducted by isolating 39 indigenous bacterial strains (*Lysinibacillus*, *Brevibacillus*, *Bacillus*, *Paenibacillus*, *Stenotrophomonas*, *Alcaligenes*, *Delftia*, *Achromobacter*, and *Pseudomonas*) from the contaminated soils of different places of Assam, India. About 64–70% reduction of TPH was observed after 28 weeks of bioaugmentation with microcosm. Further, supplementation of nutrients to microcosm enhanced the TPH reduction to 81–84 percent. After the bioremediation, the quality of soil was monitored by earthworm mortality bioassay (Lopes et al. 2010, Tamada et al. 2012a, b) and cultivating rice (*Oryza sativa*) and mung (*Vigna radiata*) (Cruz et al. 2013, 2014). The activity of the enzymes (dehydrogenase, phosphatase, and urease) present in the soil was also examined, and showed improvement in the soil quality (Roy et al. 2014).

A long term (365 days) field study was carried to treat diesel-oil contaminated soils through bioaugmentation. The influence of bioaugmentation (*Aeromonas hydrophila, Alcaligenes xylosoxidans, Gordonia* sp., *Pseudomonas fluorescens, Pseudomonas putida, Rhodococcus equi, Stenotrophomonas maltophilia, Xanthomonas* sp.), biosurfactant application, and their combined effect (rhamnolipid aided bioaugmentation) on the petroleum contaminated site was studied. The study was performed in two states—first the microbial consortia were prepared in the laboratory conditions, whereas the next stage was continued in the fields for 1 year. TPH removal efficiency of 70% was achieved by bioaugmentation. However, rhamnolipid did not exhibit significant effect on TPH biodegradation in the long term (Szulc et al. 2014). Recently, rhamnolipid produced by *Pseudomonas aeruginosa* SR17 was employed to examine its potential for the bioremediation of oil-contaminated soils. With the application of 1.5 g/L rhamnolipid, the TPH degradation was found to be 86.1% and 80.5%, with initial concentrations of 6800 mg/L and 8500 mg/L, whereas when SDS was used, the degradation efficiencies were comparatively lesser (70.8% and 68.1%). The compositional analysis proved the presence of many PAHs, among them floranthene, benz(b)fluorene, and benz(d) anthracene were completely removed within a span of 6 months, but the remaining PAHs were degraded up to 60 to 80% (Patowary et al. 2018). These significant findings predicted the potential of biosurfactants and their producer organisms for the biodegradation of contaminated soils.

Summary

Various biosurfactants have already been examined with different types of soils and pollutants. Among them, rhamnolipids and surfactin have exhibited their potential for bioremediation of contaminated soils by degrading a large range of contaminants. Both organic and inorganic pollutants could be treated with biosurfactants by altering the bioavailability of the contaminants. Further, bioaugmentation with microcosm was found to be effective for the remediation of contaminated soils. Due to their biodegradability and lower toxicity, biosurfactants have been proven to be promising for bioremediation applications. However, further research is required to develop a suitable bioaugmentation system for the complete remediation of complex petroleum hydrocarbons.

Acknowledgment

The authors are sincerely thankful for the financial grant provided by Department of Science and Technology, Government of India (DST/INSPIRE/04/2014/002020, ECR/2016/001027, and DST/INT/UK/P-155/2017).

References

Abdel-Shafy, H.I. and M.S. Mansour. 2016. A review on polycyclic aromatic hydrocarbons: source, environmental impact, effect on human health and remediation. Egypt J Pet 25(1): 107–123.

Alagić, S.Č., B.S. Maluckov and V.B. Radojičić. 2015. How can plants manage polycyclic aromatic hydrocarbons? May these effects represent a useful tool for an effective soil remediation? A review. Clean Technol Environ Policy 17(3): 597–614.

Ayotamuno, M., R. Kogbara, S. Ogaji and S. Probert. 2006. Bioremediation of a crude-oil polluted agricultural-soil at Port Harcourt, Nigeria. Appl Energy 83(11): 1249–1257.

Badoga, S., S.K. Pattanayek, A. Kumar and L.M. Pandey. 2011. Effect of polymer–surfactant structure on its solution viscosity. Asia-Pac. J Chem Eng 6(1): 78–84.

Beal, R. and W. Betts. 2000. Role of rhamnolipid biosurfactants in the uptake and mineralization of hexadecane in *Pseudomonas aeruginosa*. J Appl Microbiol 89(1): 158–168.

Bezza, F.A. and E.M.N. Chirwa. 2015. Production and applications of lipopeptide biosurfactant for bioremediation and oil recovery by *Bacillus subtilis* CN2. Biochem Eng J 101: 168–178.

Cerqueira, V.S., E.B. Hollenbach, F. Maboni, M.H. Vainstein, F.A. Camargo, R.P. Maria do Carmo and F.M. Bento. 2011. Biodegradation potential of oily sludge by pure and mixed bacterial cultures. Bioresour Technol 102(23): 11003–11010.

Cruz, J.M., P.R.M. Lopes, R.N. Montagnolli, I.S. Tamada, N. Silva and E. Bidoia. 2013. Toxicity assessment of contaminated soil using seeds as bioindicators. J Appl Biotechnol 1(1): 1–10.

Cruz, J.M., I.S. Tamada, P.R.M. Lopes, R.N. Montagnolli and E.D. Bidoia. 2014. Biodegradation and phytotoxicity of biodiesel, diesel, and petroleum in soil. Water Air Soil Pollut 225(5): 1962.

Cunningham, S.D., W.R. Berti and J.W. Huang. 1995. Phytoremediation of contaminated soils. Trends Biotechnol 13(9): 393–397.

Das, N. and P. Chandran. 2011. Microbial degradation of petroleum hydrocarbon contaminants: an overview. Biotechnol Res Int 2011.

Datta, P., P. Tiwari and L.M. Pandey. 2018. Isolation and characterization of biosurfactant producing and oil degrading *Bacillus subtilis* MG495086 from formation water of Assam oil reservoir and its suitability for enhanced oil recovery. Bioresour Technol 270: 439–448.

EPA, U. 1986. Test methods for evaluating solid waste. SW-846, method 9081, Washington DC: Environmental Protection Agency.

Felix, A.K.N., J.J. Martins, J.G.L. Almeida, M.E.A. Giro, K.F. Cavalcante, V.M.M. Melo, O.D.L. Pessoa, M.V.P. Rocha, L.R.B. Gonçalves and R.S. de Santiago Aguiar. 2019. Purification and characterization of a biosurfactant produced by *Bacillus subtilis* in cashew apple juice and its application in the remediation of oil-contaminated soil. Colloids Surf B Biointerfaces 175: 256–263.

Fismes, J., C. Perrin-Ganier, P. Empereur-Bissonnet and J.L. Morel. 2002. Soil-to-root transfer and translocation of polycyclic aromatic hydrocarbons by vegetables grown on industrial contaminated soils. J Environ Qual 31(5): 1649–1656.

Gidudu, B. and E.M.N. Chirwa. 2019. Biosurfactant facilitated emulsification and electro-osmotic recovery of oil from petrochemical contaminated soil. Chem Eng Trans 74: 1297–1302.

Janek, T., M. Łukaszewicz, T. Rezankaa and A. Krasowska. 2010. Isolation and characterization of two new lipopeptide biosurfactants produced by *Pseudomonas fluorescens* BD5 isolated from water from the Arctic Archipelago of Svalbard. Bioresour Technol 101(15): 6118–6123.

Joe, M.M., R. Gomathi, A. Benson, D. Shalini, P. Rengasamy, A.J. Henry, J. Truu, M. Truu and T. Sa. 2019. Simultaneous application of biosurfactant and bioaugmentation with rhamnolipid-producing shewanella for enhanced bioremediation of oil-polluted soil. Appl Sci 9(18): 3773.

Kavitha, V., A.B. Mandala and A. Gnanamani. 2014. Microbial biosurfactant mediated removal and/or solubilization of crude oil contamination from soil and aqueous phase: an approach with *Bacillus licheniformis* MTCC 5514. Int Biodeter Biodegr 94: 24–30.

Kim, T., J.-K. Hong, E.H. Jho, G. Kang, D.J. Yang and S.-J Lee. 2019. Sequential biowashing-biopile processes for remediation of crude oil contaminated soil in Kuwait. J Hazard Mater 378: 120710.

Kreling, N., M. Zaparoli, A. Margarites, M. Friedrich, A. Thomé and L. Colla. 2020. Extracellular biosurfactants from yeast and soil–biodiesel interactions during bioremediation. Int J Environ Sci Technol 17: 395–408.

Lai, C.-C., Y.-C. Huang, Y.-H. Wei and J.-S Chang. 2009. Biosurfactant-enhanced removal of total petroleum hydrocarbons from contaminated soil. J Hazard Mater 167(1-3): 609–614.

Liu, B., J. Liu, M. Ju, X. Li and Q. Yu. 2016. Purification and characterization of biosurfactant produced by *Bacillus licheniformis* Y-1 and its application in remediation of petroleum contaminated soil. Mar Pollut Bull 107(1): 46–51.

Lopes, P.R.M., R.N. Montagnolli, R. de Fátima Domingues and E.D. Bidoia. 2010. Toxicity and biodegradation in sandy soil contaminated by lubricant oils. Bull Environ Contam Toxicol 84(4): 454–458.

Montagnolli, R.N., P.R.M. Lopes and E.D. Bidoia. 2015a. Assessing *Bacillus subtilis* biosurfactant effects on the biodegradation of petroleum products. Environ Monit Assess 187(1): 4116.

Montagnolli, R.N., P.R.M. Lopes and E.D. Bidoia. 2015b. Screening the toxicity and biodegradability of petroleum hydrocarbons by a rapid colorimetric method. Arch Environ Contam Toxicol 68(2): 342–353.

Mulligan, C., R. Yong and B. Gibbs. 2001. Surfactant-enhanced remediation of contaminated soil: a review. Eng. Geol. 60(1-4): 371–380.

Mulligan, C.N. 2005. Environmental applications for biosurfactants. Environ Pollut 133(2): 183–198.

Olasanmi, I.O. and R.W. Thring. 2019. Evaluating rhamnolipid-enhanced washing as a first step in remediation of petroleum-contaminated soils and drill cuttings. J Adv Res 21: 79–90.

Otenio, M.H., M.T.L.d. Silva, M.L.O. Marques, J.C. Roseiro and E.D. Bidoia. 2005. Benzene, toluene and xylene biodegradation by *Pseudomonas putida* CCMI 852. Braz J Microbiol 36(3): 258–261.

Pacwa-Płociniczak, M., G.A. Płaza, Z. Piotrowska-Seget and S.S. Cameotra. 2011. Environmental applications of biosurfactants: recent advances. Int J Mol Sci 12(1): 633–654.

Patowary, R., K. Patowary, M.C. Kalita and S. Deka. 2018. Application of biosurfactant for enhancement of bioremediation process of crude oil contaminated soil. Int Biodeter Biodegr 129: 50–60.

Posada-Baquero, R., M. Grifoll and J.-J Ortega-Calvo. 2019a. Rhamnolipid-enhanced solubilization and biodegradation of PAHs in soils after conventional bioremediation. Sci Total Environ 668: 790–796.

Posada-Baquero, R., M.L. Martín and J.-J Ortega-Calvo. 2019b. Implementing standardized desorption extraction into bioavailability-oriented bioremediation of PAH-polluted soils. Sci Total Environ 696: 134011.

Puntus, I.F., O.V. Borzova, T.V. Funtikova, N.E. Suzina, N.S. Egozarian, V.N. Polyvtseva, E.S. Shumkova, L.I. Akhmetov, L.A. Golovleva and I.P. Solyanikova. 2019. Contribution of soil bacteria isolated from different regions into crude oil and oil product degradation. J Soils Sediments 19(8): 3166–3177.

Rahman, K., T. Rahman, I. Banat, R. Lord and G. Street. 2007. Bioremediation of petroleum sludge using bacterial consortium with biosurfactant. pp. 391–408. *In*: Environmental Bioremediation Technologies, Springer.

Roy, A.S., R. Baruah, M. Borah, A.K. Singh, H.P.D. Boruah, N. Saikia, M. Deka, N. Dutta and T.C. Bora. 2014. Bioremediation potential of native hydrocarbon degrading bacterial strains in crude oil contaminated soil under microcosm study. Int Biodeter Biodegr 94: 79–89.

Saha, R., R.V. Uppaluri and P. Tiwari. 2017. Influence of emulsification, interfacial tension, wettability alteration and saponification on residual oil recovery by alkali flooding. J Ind Eng Chem 59: 286–296.

Saha, R., R.V. Uppaluri and P. Tiwari. 2018. Effects of interfacial tension, oil layer break time, emulsification and wettability alteration on oil recovery for carbonate reservoirs. Colloids Surf A Physicochem Eng Asp 559: 92–103.

Saha, R., R.V. Uppaluri and P. Tiwari. 2019. Impact of natural surfactant (Reetha), polymer (Xanthan Gum) and silica nanoparticles to enhance heavy crude oil recovery. Energy Fuels 33(5): 4225–4236.

Saxena, V., A. Hasan, S. Sharma and L.M. Pandey. 2018. Edible oil nanoemulsion: An organic nanoantibiotic as a potential biomolecule delivery vehicle. Int J Polym Mater 67(7): 410–419.

Scheibenbogen, K., R.G. Zytner, H. Lee and J.T. Trevors. 1994. Enhanced removal of selected hydrocarbons from soil by *Pseudomonas aeruginosa* UG2 biosurfactants and some chemical surfactants. J Chem Technol Biotechnol 59(1): 53–59.

Sharma, S., P. Datta, B. Kumar, P. Tiwari and L.M. Pandey. 2019a. Production of novel rhamnolipids via biodegradation of waste cooking oil using *Pseudomonas aeruginosa* MTCC7815. Biodegradation 1–12.

Sharma, S., R. Verma and L.M. Pandey. 2019b. Crude oil degradation and biosurfactant production abilities of isolated *Agrobacterium fabrum* SLAJ731. Biocatal Agric Biotechnol 21: 101322.

Simarro, R., N. González, L. Bautista and M. Molina. 2013. Assessment of the efficiency of *in situ* bioremediation techniques in a creosote polluted soil: Change in bacterial community. J Hazard Mater 262: 158–167.

Szulc, A., D. Ambrożewicz, M. Sydow, Ł. Ławniczak, A. Piotrowska-Cyplik, R. Marecik and Ł. Chrzanowski. 2014. The influence of bioaugmentation and biosurfactant addition on bioremediation efficiency of diesel-oil contaminated soil: feasibility during field studies. J Environ Manage 132: 121–128.

Taccari, M., V. Milanovic, F. Comitini, C. Casucci and M. Ciani. 2012. Effects of biostimulation and bioaugmentation on diesel removal and bacterial community. Int Biodeter Biodegr 66(1): 39–46.

Tamada, I.S., P.R.M. Lopes, R.N. Montagnolli and E.D. Bidoia. 2012a. Biodegradation and toxicological evaluation of lubricant oils. Braz Arch Biol Technol 55(6): 951–956.

Tamada, I.S., R.N. Montagnolli, P.R. Lopes and E.D. Bidoia. 2012b. Toxicological evaluation of vegetable oils and biodiesel in soil during the biodegradation process. Braz J Microbiol 43(4): 1576–1581.

Ulrici, W. 2000. Contaminated soil areas, different countries and contaminants, monitoring of contaminants. Biotechnology: Environmental Processes II 11: 5–41.

Urum, K., T. Pekdemir and M. Gopur. 2003. Optimum conditions for washing of crude oil-contaminated soil with biosurfactant solutions. Process Saf Environ Prot 81(3): 203–209.

Urum, K. and T. Pekdemir. 2004. Evaluation of biosurfactants for crude oil contaminated soil washing. Chemosphere 57(9): 1139–1150.

Volkering, F., A. Breure and W. Rulkens. 1997. Microbiological aspects of surfactant use for biological soil remediation. Biodegradation 8(6): 401–417.

Witthayapanyanon, A., J. Harwell and D. Sabatini. 2008. Hydrophilic–lipophilic deviation (HLD) method for characterizing conventional and extended surfactants. J Colloid Interf Sci 325(1): 259–266.

Wolf, D. and J. Gan. 2018. Influence of rhamnolipid biosurfactant and Brij-35 synthetic surfactant on 14C-Pyrene mineralization in soil. Environ Pollut 243: 1846–1853.

Wolf, D., Z. Cryder and J. Gan. 2019. Soil bacterial community dynamics following surfactant addition and bioaugmentation in pyrene-contaminated soils. Chemosphere 231: 93–102.

Wu, M., W.A. Dick, W. Li, X. Wang, Q. Yang, T. Wang, L. Xu, M. Zhang and L. Chen, 2016. Bioaugmentation and biostimulation of hydrocarbon degradation and the microbial community in a petroleum-contaminated soil. Int Biodeter Biodegr 107: 158–164.

Wu, M., W. Li, W.A. Dick, X. Ye, K. Chen, D. Kost and L. Chen. 2017. Bioremediation of hydrocarbon degradation in a petroleum-contaminated soil and microbial population and activity determination. Chemosphere 169: 124–130.

Zouari, O., D. Lecouturier, A. Rochex, G. Chataigne, P. Dhulster, P. Jacques and D. Ghribi. 2019. Bio-emulsifying and biodegradation activities of syringafactin producing *Pseudomonas* spp. strains isolated from oil contaminated soils. Biodegradation 30(4): 259–272.

9

Dual Benefits of Microalgae in Bioremediation

Pollutant Removal and Biomass Valorization, a Review

Maha M. Ismail

Introduction

Humans are facing a large number of health-related issues, most of which are linked to the water quality and resources, which are directly affecting human well-being. The absence of proper sanitation, lake of safe drinking water, and incomplete knowledge of the suitable wastewater (WW) disposal and treatment are among these issues (UNESCO 2009).

Besides, the exposure of water resources to pathogens and xenobiotics will indirectly affect other creatures, such as plants and aquatic and terrestrial animals. This can be a consequence of the application of this water in plant irrigation or its disposal in fresh and marine water resources, leading to bioaccumulation of the harmful chemicals by these creatures or by directly swimming into the polluted water (Schwarzenbach et al. 2006).

There are two types of aquatic chemical pollutants—macro-pollutants and micro-pollutants.

Macro-pollutants (some organic compounds that are found in the range of mg.l^{-1}). Despite the knowledge of their sources, harmful effects and availability of analytical methods to examine them, and effective technologies to remove them are still developing (Larsen et al. 2007). On the other hand, micropollutants are those

Department of Microbiology and Immunology, Faculty of Pharmacy, Cairo University, Kasr El-Aini street, Cairo Governorate, 11562, Cairo, Egypt.
Email: maha.ismail@pharma.cu.edu.eg

available in the range of ng.l^{-1} to µg.l^{-1}, and include many synthetic and natural organic and inorganic contaminants. Despite their presence at such low levels, they represent a health threat, especially when found as mixtures. Besides, their high diversity makes it extremely hard to survey their toxic impacts, which are frequently chronic (Schwarzenbach et al. 2006). Sources of these micropollutants include agricultural, industrial, municipal inputs, and oil and gas spills (Bockstaller et al. 2009). The United Nations World Water Development Report (2014) stated that in developing countries, most of municipal, agricultural, and industrial wastes is discharged into surface waters with no prior treatment. Indeed, this facilitates the contamination of usable water supply. Thus, satisfactory water and WW management is a key for human well-being and economic improvement. This represents a challenge to numerous developing countries, where serious issues regarding water supply and WW control are still clear (Wilderer and Schreff 2000).

Wastewater control involves collection, treatment, and reuse or disposal of effluent and sludge. Its control is essential to protect human beings and the environment, and save water resources, in addition to providing reuse purposes in water-scarce regions. Despite all the above-mentioned reasons, WW treatment and safe disposal is improperly implemented in many countries (Bakir 2001).

Physical, chemical, and biological techniques are used to remove contaminants from WW. In physical methods, mechanical forces are applied to remove contaminants. Chemicals methods rely on chemical reactions to change the state of the pollutant. While, in biological means, microorganisms, such as bacteria and/or microalgae, convert the organic matter into various gases and cell biomass, which is then removed in sedimentation tanks. Biological processes are usually used in conjunction with physical and chemical processes, with the main objective of reducing the organic content measured as biological oxygen demand (BOD), total organic carbon (TOC), or chemical oxygen demand (COD), and nutrient content (notably N and P) of WW (Tchobanoglous and Burton 1991).

In addition to water related issues and challenges, energy crisis is a global challenge because of the predicted depletion of the non-renewable reserves of fossil fuels in the near future (Chisti 2007). Certainly, energy crisis will negatively affect all industries and agriculture all over the world. Consequently, it is now necessary to find other renewable, eco-friendly, sustainable, and cheap energy resources. Biomasses can provide about 25% of the global energy needs, in addition to providing a resource for various industries (Briens et al. 2008).

Application of phycoremediation of wastewater

Microalgae are considered efficient solar converters; they can sequester CO_2 and remove pollutants and treat wastewater, generating beneficial biomass (Park et al. 2012). In addition, they are able to produce various useful metabolites as lipids, fatty acids, vitamins, proteins, polysaccharides, pigments, and even biogas and bio-hydrogen during cell growth (Tedesco et al. 2014). Besides, microalgae can have a promoting effect on the degrading bacterial communities indigenous to WW. Hassan et al. (2019) concluded that co-culturing *Chlorella vulgaris* in real coal coking WW could enhance the dominance of xenobiotic degrading, nitrogen fixing, bio-flocculants, and plant-growth promoting bacteria leading to enhanced treatment efficiency. Also, a negative effect on WW-borne pathogens was observed, and these findings achieved both WW detoxification and water quality enhancement. Besides, Zhou et al. (2017) and Zhang et al. (2018) reported that bacterial quorum sensing molecules in WW could trigger both microalgal self-aggregation and lipid synthesis, thus facilitating biomass harvesting and valorization.

These advantages boost the competitiveness and interests in the application of microalgae in wastewater treatment (Pittman et al. 2011). However, it is necessary to develop a suitable microalgae cultivation system with WW due to the low growth rate of microalgae species. Moreover, removal of microalgal cells from the treated wastewater is also a critical issue for affordable commercial microalgae WW treatment system.

Special cultivation systems are required for WW treatment applications using microalgae—usually open ponds (mainly raceway ponds) or closed photobioreactors (tubular, flat panel and cylindrical types) are employed. It is cheaper, simpler, and easier to operate an open pond system than a closed system. These are used for various WWs treatments (Park et al. 2011). However, there is a probability of pond contamination-diffusion of WW being treated, weather variation, temperature and illumination/darkness period fluctuation, and water evaporation, which will affect the treatment success. Closed systems offer better process controls that can avoid these disadvantages easily and guarantee the success of the process (Ugwu et al. 2008).

The design of the photobioreactor relies on several criteria, including the type of employed microalga, the ability to catch the highest illumination, good mixing, efficient temperature control system, and economic scalability (Ugwu et al. 2008).

As a result of all the aforementioned factors, employing an integrated system for WW treatment using microalgae/microalgae-bacterial consortia is an excellent option to face the issues of water and energy crisis all over the world, permitting the dual benefits of waste management and bioenergy generation (Fig. 9.1). This chapter reviews the recent studies applying phycoremediation of municipal (Table 9.1), industrial (Table 9.2), and agricultural WWs, along with the desalination concentrate (DC) (brine) as integrated systems for further valorization of the harvested biomass.

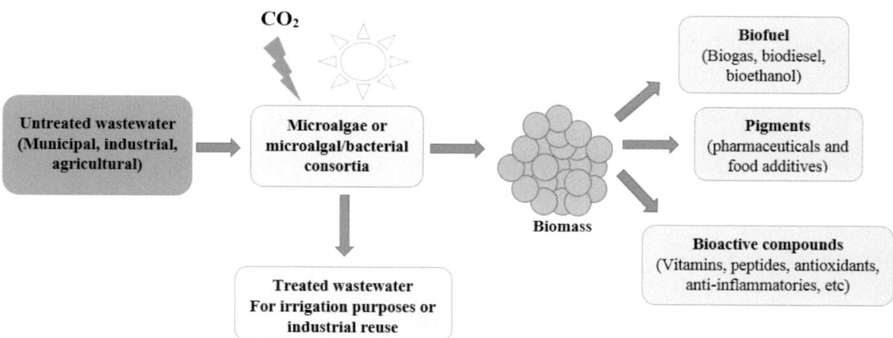

Figure 9.1. The integrated process of wastewater treatment using microalgal-based system and subsequent valorization of the harvested biomass.

Phycoremediation of urban (municipal) wastewater

Municipal WW effluent contains human and other organic and inorganic-waste, urea, proteins, nutrients, in addition to trace elements and heavy metals, such as lead, zinc, and copper (Shen et al. 2015). It is considered a rich source of nitrogen and phosphorus. These compounds offer an attractive medium for microalgal cultivation and biomass generation (Hena et al. 2015). Many studies have reported variable microalgal species for efficient municipal WW treatment, coupled with analyzing the biomass contents of bioactive compounds for further applications, such as biofuels and pharmaceuticals industries (Kothari et al. 2013, Malla et al. 2015).

In a phycoremediation study performed by Caporgno et al. (2015), municipal WW was used for the cultivation of two freshwater microalgae species (*Chlorella kessleri* and *Chlorella vulgaris*) and one marine species (*Nannochloropsis oculata*) using a flat-panel photobioreactor (PBR) operated in a batch mode. The freshwater species could remove total nitrogen (TN) and total phosphorus (TP) with high efficiency without prior acclimation, in spite of the higher N/P ratio in this WW than the ratio reported to be required for growth of freshwater microalgae (Wang et al. 2010). The authors noticed that when both *Chlorella* spp. were cultivated in WW, they underwent a progressive N-starvation state, leading to the accumulation of lipids, which is beneficial for biofuel production purposes. After that, both *Chlorella* spp. were anaerobically digested for the production of methane. Both gave higher yields than that reported in the literature using other microalgal types. Regarding *N. oculata*, it could not remove TN efficiently, and this was attributed to the high N/P ratio of 100, which is higher than the optimal ratio of 16 required for marine microalgae growth, and that this had an effect on cell division, growth rate, and consequently, N-removal. It was concluded also that the aim of WW utilization determines the choice of the microalgal type-whether freshwater or marine. In case the reduction of nutrient level is required, then freshwater species are preferable, while in case biofuel or valuable products are required, then marine microalgae is preferred, but with optimization of cultivation conditions, such as salinity (Caporgno et al. 2015).

Arora et al. (2016) reported that cultivation of *Chlamydomonas debaryana* new isolate for integrated domestic WW remediation and biodiesel production yielded a promising quality biodiesel for application in motor engines. Álvarez-Díaz et al. (2017) examined the growth of seven microalgae species in synthetic medium and in real secondarily treated urban WW. All the microalgae grew well in the synthetic media, while only three showed better growth in the WW and yielded higher values of biomass and lipid productivities and lipid content. Generally, the microalgae did not consume the entire available nitrogen, although they consumed all phosphorus completely, and so they failed to reach N-starvation state in order to accumulate lipids, unlike what was observed in the synthetic medium. Supplementation of WW with nutrients allows for better microalgal growth, while reducing its lipid content significantly—an observation reported by Eladel et al. (2019). These authors reported enhanced lipid accumulation, lipid productivity, nitrogen and phosphorus removal, and higher saturated fatty acids in the microalgal biomass cultivated in the tested

Table 9.1. Recent studies applying microalgae-based systems for municipal wastewater treatment with subsequent evaluation and valorization of the generated biomass.

Microalgae	Cultivation system	Nutrients removal	Biomass/Bioactive compound	Reference
Scenedesmus sp.	Batch mode; indoor reactor	7% N and 0.7% P removal, > 88% Fe, Zn and Cd	BP[a] 130 mg.l^{-1}.d^{-1} FAME[b] yield of 12%	McGinn et al. (2012)
	Continuous mode; indoor chemostat vessels	90% N and P removal, 100% Fe, Zn and Cd	BP up to 267 mg.l^{-1}.d^{-1} FAME yield of 5%	
Consortium of 17 microalgal genera	Outdoor reactor	85% TOC, 90% TN[*], 70% TP[**]	BP 122 mg.l^{-1}.d^{-1} Lipids 28%	Mahapatra et al. (2014)
Chlorella kessleri	Indoor flat-panel airlift PBR	> 96% TN, 99% TP	BC[c] 2.34 g.l^{-1} within 9 d Saponifiable lipid 7.4% Protein 36.7%, Carbohydrate 44.6% Biogas 346 ml	Caporgno et al. (2015)
Chlorella vulgaris		95% TN, > 98% TP	BC 2.63 g.l^{-1} within 8 d Saponifiable lipid 11.3% Protein 35.2% Carbohydrate 36.2% Biogas 415 ml	
Nannochloropsis oculata		47% TN, > 96% TP	BC 1.05 g.l^{-1} within 14 d	
Chlamydomonas debaryana IITRIND3	Indoors one liter shake flasks	86% COD, 71% TOC, 72% TN and 80% TP	BP 188 mg.l^{-1}.d^{-1} Total pigments 12 μg.ml^{-1} Protein 30% Carbohydrate ~ 28% Lipid 40%	Arora et al. (2016)

Chlorella vulgaris	Indoors two liter flasks, batch cultivation	93.7 to 95.7% TN, complete TP removal	BP 107 $mg.l^{-1}.d^{-1}$ Lipid content 17.2% LP[d] 18.4 $mg.l^{-1}.d^{-1}$	Álvarez-Díaz et al. (2017)
Chlorella kessleri			BP 71 $mg.l^{-1}.d^{-1}$ Lipid content 19% LP 13.3 $mg.l^{-1}.d^{-1}$	
Chlorella sorokiniana			BP 76 $mg.l^{-1}.d^{-1}$ Lipid content 25.6% LP 19.5 $mg.l^{-1}.d^{-1}$	
Scenedesmus obliquus			BP 81 $mg.l^{-1}.d^{-1}$ Lipid content 36.7% LP 29.8 $mg.l^{-1}.d^{-1}$	
Chlorella vulgaris	Cylindrical glass PBR with CO_2-augmented air	76.3% COD, 94.2% NH_3 and 94.8% P	-	Chaudhary et al. (2018)
Scenedesmus obliquus		76% COD, 92.6% NH3 and 93.1% P		
Chlorella sorokiniana	Indoors five liter transparent carboys	74% NO_3-N, 83% NH_3-N and 78% TP removal	BC 0.74 $g.l^{-1}$ LP 16.2 $mg.l^{-1}.d^{-1}$ Proteins 26.5% Carbohydrate 21.4% Lipids 22.1%	Eladel et al. (2019)
Nannochloropsis oceanica	1500 liter PBR	Complete N, P removal	Proteins 36.6% Carbohydrate 23.5% Lipids 17.9% Energy 4.9 $kcal.g^{-1}$	Silkina et al. (2019)

*Total nitrogen, ** Total phosphorous, [a] Biomass productivity, [b] Fatty acid methyl esters, [c] Biomass concentration, [d] Lipid productivity.

municipal WW without extra nutrient supplementation than when supplemented with Bold's Basal medium elements. Another study examined the effect of CO_2-augmented air supplementation on the nutrient removal from municipal WW by each of two microalgal spp. *Chlorella vulgaris* and *Scenedesmus obliquus*. The authors observed enhanced nutrient removal when 5% CO_2-augmented air was supplemented, rather than normal air (Chaudhary et al. 2018).

Phycoremediation of industrial wastewater

There are various types of industrial WW depending on the type of industry- mostly all of them contain heavy metals, toxic compounds, and less nitrogen and phosphorus than municipal WW. Because of their complexity and toxicity, they require special process conditions and choice of the right type of microalga in order to achieve adequate treatment (Chinnasamy et al. 2010). Among the industrial WW that have been studied for phycoremediation are textile, pharmaceutical, petroleum, paper, distillery, and some food industries WW (Delrue et al. 2016). Among the most polluting and potable water consuming industries all over the world is the textile industry, which is responsible for releasing effluents consisting of recalcitrant dyes, metals, and other toxic chemicals. Besides, COD, BOD, and pH values of this WW are high (Yaseen and Scholz 2016), In addition, composition of the textile effluents are variable according to the applied process, used equipment, mill, fabrics, and chemicals (Brik et al. 2006). This necessitates adequate treatment before discharge, and subsequent reuse of the treated effluent for the same industry (Yaseen and Scholz 2018).

In a study performed by Wu et al. (2017), the effect of various experimental conditions for optimization of *Chlorella* sp. culture in the textile WW was examined. These authors found that the highest yield percentage (20%) of fatty acid methyl esters (FAME) was obtained at culture conditions of adding extra 4 $mg.l^{-1}$ K_2HPO_4 and 1 $g.l^{-1}$ urea at pH 10 with CO_2 sparging. In contrast, Kumar et al. (2018b) examined the remediation of textile WW using a microalgal consortium. Without extra-added nutrients, they performed a five-cycle fed-batch operation. They noticed that the duration of each cycle is further reduced until it reached 10 days, and this resulted in a high biomass productivity of 480 $mg.l^{-1}.d^{-1}$. This could indicate the capability of the microalgae to adapt to this toxic WW as a sole nutrients source. These results indicate that WW supplementation with extra nutrients sometimes may have a potential role in optimization of the remediation and microalgal lipid accumulation. Lin et al. (2017) examined the microalgal remediation of textile desizing of anerobically digested WW effluent pretreated by granular activated charcoal. Based on the results of this study, these authors predicted that the resulting biomass and biogas would achieve 1400 $kg.d^{-1}$ and 2×10^7 $kJ.d^{-1}$, respectively, provided the treatment be performed at effluent flowrate of 1000 $m^3.d^{-1}$. Behl et al. (2019) developed a novel technology for the remediation of DR-31 azo dye. They could fabricate an effective, up to three times recyclable and eco-friendly bionanocomposite (composed of graphene oxide/*Desmodesmus* sp.), which facilitated the transfer of the biologically generated electrons from the microalga directly to DR-31 azo dye to reduce it into amines that are readily utilized by the microalga as a N-source. This led to enhanced lipid content of 11% than 9% when applying the microalga alone.

Pharmaceutical WW is another type of industrial WWs. Pharmaceutical compounds are biologically active substances intended to improve the quality of human health. However, recent studies suggest those pharmaceuticals and their residues to be resistant compounds persisting in the environment and affecting other non-target species in aquatic and terrestrial ecosystems (Vieno et al. 2007). Since the level of pharmaceuticals is not regulated by law and they are considered as a threat to the environment, attention has been paid to the environmental fate of these compounds in order to improve their efficient removal from wastewater treatment plants (WWTPs) and prevent their continuous introduction to the aqueous environment (Collado et al. 2012). Some studies examined the potential removal of pharmaceuticals from WW using microalgae. It was reported that abiotic photodegradation is the main mechanism of environmental removal of pharmaceuticals, especially antibiotics. However, application of microalgae-based removal systems offers subsequent biomass utilization for additional benefit (Diaz 2016). López-Serna et al. (2019) examined the removal efficiency of 19 veterinary medicines from piggery WW using two PBRs—one operated using microalgal-bacterial consortium and the other operated using purple photosynthetic bacteria. They noticed that the removal of the drugs was more efficient by the first PBR. Ismail et al. (2016) observed the ability of microalgal-bacterial consortium to degrade ketoprofen with increase in the measured microalgal chlorophyll content, while the microalga (*Chlorella* sp.) alone could not degrade ketoprofen. Other authors have also reported that genus *Chlorella* could only tolerate, not degrade organic pollutants (Rakaiby et al. 2012). The removal of the mixture of analgesics (paracetamol, ketoprofen, and aspirin) from a synthetic WW using the same microalgal-bacterial consortium at fed-batch mode was examined by Ismail et al. (2017). The whole consortium showed a better removal capacity for pharmaceuticals and their metabolites (*p*-aminophenol and salicylic acid), which could not be removed by the bacteria alone. Besides, the harvested biomass was rich in various bioactive compounds for a wide range of useful applications following wastewater treatment.

Diaz (2016) investigated the removal capacity of estrogenic compounds by two microalgae spp. with or without the addition of anaerobic digester centrate (ADC) obtained from an urban WWTP. The author found that addition of ADC to BG-11 medium could enhance the biomass production of *Raphidocelis subcapitata* from 150 to 370 $mg.l^{-1}$ within 4 days. This was explained via the ability of this microalga to grow mixotrophically, utilizing light and glucose in BG-11 medium. In the same study, *Chlamydomonas reinhardtii* showed the best removal capacity of nine antibiotics and one antidepressant—five antibiotics were removed by 63–93%, the other four were removed by up to 41%. Meawhile, the antidepressant drug was more persistent and removed by only 13% (Diaz 2016).

Table 9.2. Recent studies applying microalgae-based systems for industrial wastewater treatment with subsequent evaluation and valorization of the generated biomass.

Microalgae	WW	Cultivation system	Nutrient removal	Pollutant removal	Biomass/Bioactive compound	Reference
Chlorella vulgaris	Textile	Erlenmeyer flasks	53.7% COD	72% color removal	BC 0.77g.l^{-1} (at 5% TW[b], 10 g l^{-1} $NaHCO_3$),	El-Kassas and Mohamed (2014)
Desmodesmus sp.	Sugarcane vinasse WW	Stirred batch reactor	52% TN, 36% COD	-	BP 101 mg.l^{-1}.h^{-1} BC 4000 mg.l^{-1}	De Mattos and Bastos (2015)
Scenedesmus sp.	Meat processing 1[ry] effluent	Bubble column PBR	44% COD, 100% TN, 47% TP	-	BP 4200 mg.m^2.d^{-1} LP 10 mg.l^{-1}.d^{-1} Lipid 7% CH_4 905 ml	Assemany et al. (2016)
Raphidocelis subcapitata	Synthetic WW spiked with 17β-estradiol (E2) and 17α-ethinyl estradiol (EE2) in Erlenmeyer flasks	BG-11 medium	-	88–100% E2, 60% EE2 removal	-	Diaz (2016)
		BG-11 + ADC[i]	-	100% E2, 95% EE2, 100% BPA[n] removal	-	
Chlamydomonas reinhardtii		P49 medium	-	100% E2 and EE2 removal	-	
		P49 medium + ADC	-	100% E2, 76% EE2, 100% BPA removal	-	
Chlorella minutissimma + *Citrobacter* sp. + *P. aeruginosa* + *Bacillus subtilis*	Petroleum based effluent from petroleum industry		-	92% degradation of total PAH[m] except acenaphthyle-ne, anthracene and benzo (a) anthracene	-	Omojevwe and Ezekiel (2016)

Microalgal-bacterial consortia: *Aphanocapsa* sp. + *Citrobacter* sp. + *P. aeruginosa,* + *Bacillus subtilis*		Erlenmeyer flasks	-	47% degradation of total PAH	-	
Microalgal-bacterial consortia: *Chlorella minutissima* + *Aphanocapsa* sp. + *P. aeruginosa* + *Bacillus subtilis* + *Citrobacter* sp.		Erlenmeyer flasks	-	68% degradation of total PAH	-	
Chlorella vulgaris NIOCCV	Tannery (50% diluted)	Erlenmeyer flasks	100% TN and TP, 88.3% COD, 27.3% BOD	-	Chl a[c] 2.2 mg. l^{-1} 400 × 10^3 cells.ml^{-1}	Das et al. (2017)
Chlorella sp. G23	Textile	Glass tube	~ 50% NH_4^+-N, 90% COD	50% color removal	FAME[d] 20% (at 4 mg.l^{-1} K_2HPO_4, 1 g.l^{-1} urea, pH 10 and CO_2 sparging)	Wu et al. (2017)
Scenedesmus sp.	GAC[e] treated starch-containing textile WW	Lab-scale reactors	COD 89.5%, Carbohydrates 97.4% Organic acids 94.7%	92.4% color removal	BP 20 mg.l^{-1}.d^{-1} Total H, CH_4 and C_2H_5OH energy productivity 16.9 kJ.l^{-1}.d^{-1}	Lin et al. (2017)

Table 9.2 Contd. ...

...Table 9.2 Contd.

Microalgae	WW	Cultivation system	Nutrient removal	Pollutant removal	Biomass/Bioactive compound	Reference
Chlorella sp. and 4 Gram negative bacteria spp.	Synthetic pharmaceutical WW	Stirred tank PBR	~ 100% COD	96–99% of all drugs/ metabolites removal	Carbohydrates 16% Lipids 7% Blend of SFAs, MUFAs and PUFAs Protein 50% Essential (65.5%), non-essential (29%) and conditional amino acids (5.5%) Total phenolics 5.8 $\mu g.mg^{-1}$ biomass Total antioxidants ~ 60 µmol. mg^{-1} biomass	Ismail et al. (2017)
Microalgal consortium dominated by *Chlorella* and *Scenedesmus* spp.	Textile	Indoor PBR	93% TN, 100% TP, 78% COD	-	BC[b] 1.9 $g.l^{-1}$ in terms of total solids and 1.1 $g.l^{-1}$ in terms of volatilized solids, Protein 2.5 $g.l^{-1}$, Carbohydrate 2.4$g.l^{-1}$	Huy et al. (2018)
Microalgal consortium dominated by *Chlorella* and *Scenedesmus* spp.	Textile	Indoor fed-batch PBR	99.6% TN[*], 100% TP[**], 71% COD at end of 3rd cycle	72% color removal at end of 5th cycle	BP[a] 480 $mg.l^{-1}.d^{-1}$ by 5th cycle	Kumar et al. (2018b)
Graphene oxide/ *Desmodesmus* sp. bionanocomposite	Textile	-	-	90% DR-31 dye removal	Lipid 11% SFAs[f] 53% MUFAs[g] 35% PUFAs[h] 12%	Behl et al. (2019)

Chlorella sp.	Textile WW effluent and flue gas	Batch PBR	84% TOC 92%, COD, 96%, PO_4, 100% NH_4, and 99% NO_3	-	BP 209 $mg.l^{-1}.d^{-1}$	Yadav et al. (2019)
Chlorococcum sp.			73% TOC 85.2 COD, 75% PO_4, 100% NH_4, and 86% NO_3	-	BP 105 $mg.l^{-1}.d^{-1}$	
Platymonas subcordiformis	Surfactant industry WW	Erlenmeyer flasks	-	91% nonylphenol removal	-	Wang et al. (2019)
Tribonema minus	Tofu industry WW	Column PBR	~ 93% TN, 73% TP and 85% COD	-	BC 777 $mg.l^{-1}$ Carbohydrate 30% Protein 15.6% Neutral lipid 30% Palmitoleic acid 16% Chrysolaminarin 7% BP 450 $mg.l^{-1}.d^{-1}$ LP[o] 133 $mg.l^{-1}.d^{-1}$ Palmitoleic acid and Chrysolaminarin productivity 67, 30 $mg.l^{-1}.d^{-1}$	Wang et al. (2019)
Microalgal-bacterial consortium	Vet drugs in Piggery WW	Open PBR	Average of 76% TOC, 68% TN, 86% TP	Removal by 93, 95% (Oxycycline, doxycycline) < 70% (danofloxacin, sulfadimidine), > 70% (other drugs)	-	López-Serna et al. (2019)

* Total nitrogen, ** Total phosphorous, [a]Biomass productivity, [b]Biomass concentration, [c]Chlorophyll a content, [d]Fatty acid methyl esters, [e]Granular activated charcoal, [f]Saturated fatty acids, [g]Monounsaturated fatty acids, [h]Polyunsaturated fatty acids, [i]Anaerobic digester centrate, [j]Primary effluent wastewater, [k]Ciprofloxacin, [l]High rate algal pond, [n]Bisphenol A, [m]Polyaromatic hydrocarbon, [o]Lipid productivity.

Phycoremediation of agricultural wastewater

Industrial and agricultural WW are highly polluted and poorly treated using conventional treatment methods (Delrue et al. 2016). Agricultural WW consists of animal manure, plant parts, livestock and poultry wastes, and other agricultural activities. Microalgae can be successfully applied in agricultural WW remediation. For example, a *Scenedesmus obliquus* strain could remove up to 70% COD of a previously anaerobically digested cattle wastewater under batch operation conditions with protein rich biomass productivity of 213–358 $mg.l^{-1}.d^{-1}$ (De Mendonça et al. 2018). Another *Scenedesmus obliquus* strain was examined for its ability to remediate poultry WW effluents. There was more than 97% removal of nitrogen and phosphorus, with production of up to 29 and 36% saponifiable matter and sugars, respectively (Oliveira et al. 2018).

Swine WW is a type of agricultural WW characterized by a high value of COD and nitrogen. If discharged without proper treatment, it causes natural water contamination (Wang et al. 2012). Swine manure was used for cultivation of a mixed microalgae dominated by *Scenedesmus* and *Chlorella* spp. This consortium could remove up to 50% TN, and TP with subsequent biohydrogen production of 116 $ml.g^{-1}$ biomass after a fermentation period of about 36 hours (Kumar et al. 2018a). In a batch culture of *Chlorella* sp. in undiluted piggery WW, biomass productivity of 681 $mg.l^{-1}.d^{-1}$ was observed. At 25% dilution, the highest values of lipid content and productivity were obtained (29% and 155 $mg.l^{-1}.d^{-1}$, respectively). In semi-continuous culture at 25 to 75% dilution, Kuo et al. (2015) reported *Chlorella* sp. biomass and lipid productivity values higher than 852 and 128 $mg.l^{-1}.d^{-1}$, respectively. In a study performed by López-Pacheco et al. (2019), different combination ratios of Nejayote and swine wastewater were used as a cost-effective culture medium for *Arthrospira maxima* and *Chlorella vulgaris* cultivation. Interestingly, in case of *A. maxima*, 92% TN, 75% TP, and 96% COD were removed. *C. vulgaris* could remove 91% TN, 85% TP, and 96% COD.

Another type of agricultural WW is the aquaculture one. It is characterized by high content of COD, BOD, and nutrients such as nitrogen and phosphorus responsible for eutrophication. Besides, it is an important growing industry that provides humans with about 50% of fish for consumption, leading to production of large WW volumes. That is why it is necessary to be treated to avoid such negative environmental impacts and to be effectively reused (FAO 2012). In a study performed by Andreotti et al. (2017), three microalgae species—*Tetraselmis suecica*, *Isochrysis galbana*, and *Dunaliella tertiolecta* were investigated for their bioremediation capacity of grey mullet *Mugil cephalus* wastewater under batch conditions. *T. suecica* could remove more than 90% nitrogen and phosphorus, in addition to producing the highest biomass yield than the other two microalgae types. Khatoon et al. (2016) compared the growth of *Chaetoceros calcitrans, Nannochloris maculata*, and *Tetraselmis chuii* in Conway culture medium and in a shrimp pond aquaculture WW. The authors did not find a significant difference between biomass obtained from both cultures, indicating the suitability of aquaculture WW for successful microalgal cultivation. However, aquaculture WW was superior regarding lipid productivity of

N. maculate, protein, and lipid content of *N. maculata* and *T. chuii*, as well. These results suggest the superiority of aquaculture WW as a cost-effective medium for microalgae growth, with no need for extra nutrients supplementation. Besides, there was subsequent gain of enhanced biomass and bioactive compounds production for various applications. On the other hand, Ansari et al. (2017) and Guldhe et al. (2017) found that the use of Nile tilapia aquaculture WW supplemented with $NaNO_3$ has led to more enhanced microalgal biomass, lipid, and protein productivities than without supplementation. This can be explained in terms of nutrient deficiency of this type of aquaculture WW.

Desalination concentrate as an alternative for microalgal cultivation

Desalination concentrate (DC or brine) is a highly concentrated water containing minerals and elements resulting from desalination technologies. The characteristics of brine depend on the raw water quality and the treatment plant recovery (Zarzo 2018).

Discharge of the brine in surface water, sea, and lakes is not a suitable option, since it causes change in the salinity levels of aquatic systems, and causes a negative environmental impact (Williams 2001). Many studies reported the application of hyper-salinity water concentrate as an alternative medium for microalgal growth and analysis of their content of lipid and protein. Matos et al. (2017) reported the genera able to adapt to this type of WW as *Dunaliella, Scenedesmus, Arthrospira, Nannochloropsis*, and *Chlorella* spp. In addition, Sánchez et al. (2015) proposed an integrated approach to use the reject brine originated from inland desalination plant in Brazil. They reported the capability of growing *Arthrospira* sp. in the brine as a substitute or support to fish farming, owing to its ability to provide valuable nutrients for fish and for humans, as well.

Some studies reported brine to be nutrient deficient, and thus requiring supplementation with nutrients and growth promoters to stimulate microalgal growth. Zarzo et al. (2014) reported brine (from brackish water reverse osmosis (RO) process) to be a more favorable growth media for *Nannochloropsis salina* than artificial freshwater after supplementation with phosphorus, vitamins, and molibdene metal. They also found that *Oocystis* sp. could remove 45% of NO_3 in an outdoor pilot plant, and was also the best one regarding biomass production. Matos et al. (2015b) reported that use of inland DC mixed with F/2 medium (ratio 3:1) for cultivation of *Nannochloropsis gaditana* led to best biomass concentration and lipid productivity. When the concentration of DC was increased, it reduced the biomass content of polyunsaturated fatty acids (PUFA) and increased the saturated fatty acids (SFA) level, which is suitable for biodiesel production. Another study (Matos et al. 2015a) reported *Chlorella vulgaris* to be able to grow in brackish water RO DC mixed with optimized Bold's basal medium at variable concentrations. However, a combination of 25% DC + 75% optimized BBM showed the best biomass production and lipid content, while higher DC concentrations inhibited microalgal growth, biomass production, and lipid content. In a study performed by ElBarmelgy

(2019), *Nannochloropsis* sp. was adapted to grow under salt stress conditions in both sea salt-based F/2 medium at salt concentrations up to 80 $g.l^{-1}$ and in DC-based optimized medium of 70 $g.l^{-1}$ salinity. The microalga grew well under all salt stress conditions, except in DC, when it was employed as a sole source of nutrients. Silkina et al. (2019) proposed a sustainable idea of diluting various types of WW with brine from a desalination plant. They diluted each of agricultural, municipal, and aquaculture WW with F/2 medium to raise the WW salinity in order to be suitable for cultivation of the marine *Nannochloropsis oceanica* instead of using deionized or tap water. They reported complete removal of nitrogen and phosphorus from all types of WWs. The biochemical composition of the biomass generated from the three F/2-diluted WWs were comparable to what was obtained from F/2 medium. Thus, these authors achieved more sustainable and economic waste-based systems for the cultivation of microalga with adequate remediation and biomass valorization. It can be concluded that microalage are considered a promising solution for water and energy related issues when they are successfully employed in the treatment of various WW types. Considering some factors is critical for the success of the process. According to the WW type, one can predict the success probability of the microalgae-based treatment. Nitrogen and phosphorus-rich WW is easier to be effectively treated by this system, owing to its ability to provide the microalgae with its growth requirements. Certain WW types can be deficient in nutrients required for microalgal growth and biomass production, thus, supplementation with nutrients such as nitrogen, phosphorus and carbon-sources is necessary. Besides, dilution of the WW might be important to reduce the level of toxic chemicals, thus protecting the microalgae from being harmed during cultivation. For economic purposes, it is recommended to perform dilution by another WW (e.g., brine for marine microalgae cultivation or municipal WW, which is rich in nutrients for microalgal growth). The choice of the right microalgal type can significantly affect the sustainability of the employed system. For example, oleaginous marine microalgae are more suitable for WW with high salinity, since salt stress enhances lipid productivity and diversity of the fatty acids, which allows for better introduction of the microalgal biomass in the suitable application. Besides, the purpose of the cultivation system determines whether an axenic monoalgal culture or a mixed microalgal consortia will be applied. If the purpose is to produce certain bioactive compounds known to belong to certain microalga sp., then a monoalgal culture is employed, taking into consideration that this will need controlled conditions to avoid its contamination or loss. In case a synergistic effect is intended, then a co-culture consortium is the suitable choice—by virtue of being able to perform in different mechanisms and behaviors for the waste treatment, and also showing improved yields of biomass production and variable bioactive compounds than an axenic culture.

Acknowledgment

I am thankful to Dr Abdel-gawad Hashem, Professor of Microbiology and Immunology, Faculty of Pharmacy, British University in Egypt, for reviewing the chapter.

References

Álvarez-Díaz, P.D., J. Ruiz, Z. Arbib, J. Barragán, M.C. Garrido-Pérez and J.A. Perales. 2017. Freshwater microalgae selection for simultaneous wastewater nutrient removal and lipid production. Algal Res 24: 477–485.

Andreotti, V., A. Chindris, G. Brundu, D. Vallainc, M. Francavilla and J. García. 2017. Bioremediation of aquaculture wastewater from *Mugil cephalus* (Linnaeus, 1758) with different microalgae species. Chem Ecol 33: 750–761.

Ansari, F.A., P. Singh, A. Guldhe and F. Bux. 2017. Microalgal cultivation using aquaculture wastewater: integrated biomass generation and nutrient remediation. Algal Res 21: 169–177.

Arora, N., A. Patel, K. Sartaj, P.A. Pruthi and V. Pruthi. 2016. Bioremediation of domestic and industrial wastewaters integrated with enhanced biodiesel production using novel oleaginous microalgae. Environ Sci Pollut Res 23: 20997–21007.

Assemany, P.P., M.L. Calijuri, M.D. Tango and E.A. Couto. 2016. Energy potential of algal biomass cultivated in a photobioreactor using effluent from a meat processing plant. Algal Res 17: 53–60.

Bakir, H.A. 2001. Sustainable wastewater management for small communities in the Middle East and North Africa. J Environ Manag 61: 319–328.

Behl, K., M. Joshi, M. Sharma, S. Tandon, A.K. Chaurasia, A. Bhatnagar and S. Nigam. 2019. Performance evaluation of isolated electrogenic microalga coupled with graphene oxide for decolorization of textile dye wastewater and subsequent lipid production. Chem Eng 375: 121950.

Bockstaller, C., L. Guichard, O. Keichinger, P. Girardin, M.B. Galan and G. Gaillard. 2009. Comparison of methods to assess the sustainability of agricultural systems. A review. Agron Sustain Dev 29: 223–235.

Briens, C., J. Piskorz and F. Berruti. 2008. Biomass valorization for fuel and chemicals production—a review. Int J Chem React Eng 6: 1–49.

Brik, M., P. Schoeberl, B. Chamam, R. Braun and W. Fuchs. 2006. Advanced treatment of textile wastewater towards reuse using a membrane bioreactor. Proc Biochem 41: 1751–1757.

Caporgno, M.P., A. Taleb, M. Olkiewicz, J. Font, J. Pruvost, J. Legrand and C. Bengoa. 2015. Microalgae cultivation in urban wastewater: Nutrient removal and biomass production for biodiesel and methane. Algal Res 10: 232–239.

Chaudhary, R., Y.M. Tong and A.K. Dikshit. 2018. CO_2-assisted removal of nutrients from municipal wastewater by microalgae *Chlorella vulgaris* and *Scenedesmus obliquus*. Int J Environ Sci Technol 15: 2183–2192.

Chinnasamy, S., A. Bhatnagar, R.W. Hunt and K.C. Das. 2010. Microalgae cultivation in a wastewater dominated by carpet mill effluents for biofuel applications. Bioresour Technol 101: 3097–3105.

Chisti, Y. 2007. Biodiesel from microalgae. Biotechnol Adv 25: 294–306.

Collado, N., G. Buttiglieri, L. Ferrando, S. Rodriguez-Mozaz, D. Barceló, J. Comas and I. Rodriguez-Roda. 2012. Removal of ibuprofen and its transformation products: experimental and simulation studies. Sci Total Environ 433: 296–301.

Das, C., K. Naseera, A. Ram, R.M. Meena and N. Ramaiah. 2017. Bioremediation of tannery wastewater by a salt-tolerant strain of *Chlorella vulgaris*. J Appl Phycol 29: 235–243.

De Mattos, L.F.A. and R.G. Bastos. 2015. COD and nitrogen removal from sugarcane vinasse by heterotrophic green algae *Desmodesmus* sp. Desalin. Water Treat 57: 9465–9473.

De Mendonça, H.V., J.P.H.B. Ometto, M.H. Otenio, I.P.R. Marques and A.J.D. Dos Reis. 2018. Microalgae-mediated bioremediation and valorization of cattle wastewater previously digested in a hybrid anaerobic reactor using a photobioreactor: Comparison between batch and continuous operation. Sci Total Environ 633: 1–11.

Delrue, F., P.D. Álvarez-Díaz, S. Fon-Sing, G. Fleury and J.F. Sassi. 2016. The environmental biorefinery: using microalgae to remediate wastewater, a win-win paradigm. Energies 9: 132.

Diaz, A.H. 2016. Degradation of pharmaceutical compounds by microalgae: photobioreactor wastewater treatment, biomass harvesting and methanization. Ph.D. Thesis, Universitat Autònoma de Barcelona, Barcelona, Spain.

Eladel, H., A.E.-F. Abomohra, M. Battah, S. Mohmmed, A. Radwan and H. Abdelrahim. 2019. Evaluation of *Chlorella sorokiniana* isolated from local municipal wastewater for dual application in nutrient removal and biodiesel production. Bioproc Biosyst Eng 42: 425–433.

ElBarmelgy, A.H. 2019. Sustainable utilization of desalination concentrate. MSc. Thesis, American University in Cairo, Cairo, Egypt.

El-Kassas, H.Y. and L.A. Mohamed. 2014. Bioremediation of the textile waste effluent by *Chlorella vulgaris*. Egypt J Aquat Res 40: 301–308.

FAO. 2012. The State of World Fisheries and Aquaculture, Food and Agriculture Organization of the United Nations, FAO, Rome, Italy.

Guldhe, A., F.A. Ansari, P. Singh and F. Bux. 2017. Heterotrophic cultivation of microalgae using aquaculture wastewater: A biorefinery concept for biomass production and nutrient remediation. Ecol Eng 99: 47–53.

Hassan, M., T. Essam, A. Mira and S. Megahed. 2019. Biomonitoring detoxification efficiency of an algal-bacterial microcosm system for treatment of coking wastewater: harmonization between *Chlorella vulgaris* microalgae and wastewater microbiome. Sci Total Environ 677: 120–130.

Hena, S., S. Fatimah and S. Tabassum. 2015. Cultivation of algae consortium in a dairy farm wastewater for biodiesel production. Water Resour Ind 10: 1–14.

Huy, M., G. Kumar, H.-W. Kim and S.-H. Kim. 2018. Photoautotrophic cultivation of mixed microalgae consortia using various organic waste streams towards remediation and resource recovery. Bioresour Technol 247: 576–581.

Ismail, M.M., T.M. Essam, Y.M. Ragab and F.E. Mourad. 2016. Biodegradation of ketoprofen using a microalgal–bacterial consortium. Biotechnol Lett 38: 1493–1502.

Ismail, M.M., T.M. Essam, Y.M. Ragab, A.B.E. El-Sayed and F.E. Mourad. 2017. Remediation of a mixture of analgesics in a stirred-tank photobioreactor using microalgal-bacterial consortium coupled with attempt to valorise the harvested biomass. Bioresour Technol 232: 364–371.

Khatoon, H., S. Banerjee, M.S. Syahiran, N.B. Mat Noordin, A.M.A. Bolong and A. Endut. 2016. Re-use of aquaculture wastewater in cultivating microalgae as live feed for aquaculture organisms. Desalin Water Treat 57: 29295–29302.

Kothari, R., R. Prasad, V. Kumar and D.P. Singh. 2013. Production of biodiesel from microalgae *Chlamydomonas polypyrenoideum* grown on dairy industry wastewater. Bioresour Technol 144: 499–503.

Kumar, G., D.D. Nguyen, P. Sivagurunathan, T. Kobayashi, K. Xu and S.W. Chang. 2018a. Cultivation of microalgal biomass using swine manure for biohydrogen production: Impact of dilution ratio and pretreatment. Bioresour Technol 260: 16–22.

Kumar, G., M. Huy, P. Bakonyi, K. Bélafi-Bakó and S.-H. Kim. 2018b. Evaluation of gradual adaptation of mixed microalgae consortia cultivation using textile wastewater via fed batch operation. Biotechnol Rep 20: e00289.

Kuo, C.-M., T.-Y. Chen, T.-H. Lin, C.-Y. Kao, J.-T. Lai, J.-S. Chang and C.-S. Lin. 2015. Cultivation of *Chlorella* sp. GD using piggery wastewater for biomass and lipid production, Bioresour Technol 194: 326–333.

Larsen, T.A., M. Maurer, K.M. Udert and J. Lienert. 2007. Nutrient cycles and resource management: implications for the choice of wastewater treatment technology. Water Sci Technol 56: 229–237.

Lin, C.-Y., M.-L.T. Nguyen and C.-H. Lay. 2017. Starch-containing textile wastewater treatment for biogas and microalgae biomass production. J Clean Prod 168: 331–337.

López-Pacheco, I.Y., D. Carrillo-Nieves, C. Salinas-Salazar, A. Silva-Núñez, A. Arévalo-Gallegos, D. Barceló, S. Afewerki, H.M.N. Iqbal and R. Parra-Saldívar. 2019. Combination of nejayote and swine wastewater as a medium for *Arthrospira maxima* and *Chlorella vulgaris* production and wastewater treatment. Sci Total Environ 676: 356–367.

López-Serna, R., D. García, S. Bolado, J.J. Jiménez, F.Y. Lai, O. Golovko, P. Gago-Ferrero, L. Ahrens, K. Wiberg and R. Muñoz. 2019. Photobioreactors based on microalgae-bacteria and purple phototrophic bacteria consortia: A promising technology to reduce the load of veterinary drugs from piggery wastewater. Sci Total Environ 692: 259–266.

Mahapatra, D.M., H.N. Chanakya and T.V. Ramachandra. 2014. Bioremediation and lipid synthesis through mixotrophic algal consortia in municipal wastewater. Bioresour Technol 168: 142–150.

Malla, F.A., S.A. Khan, S.G.K. Rashmi, N. Gupta and G. Abraham. 2015. Phycoremediation potential of *Chlorella minutissima* on primary and tertiary treated wastewater for nutrient removal and biodiesel production. Ecol Eng 75: 343–349.

Matos, Â.P., W.B. Ferreira, R.C. Torres, L.R.I. Morioka, M.H.M. Canella, J. Rotta, T. da Silva, E.H.S. Moecke and E.S. Sant'Anna. 2015a. Optimization of biomass production of *Chlorella vulgaris* grown in desalination concentrate. J Appl Phycol 27: 1473–1483.

Matos, Â.P., R. Feller, E.H.S. Moecke and E.S. ant'Anna. 2015b. Biomass, lipid productivities and fatty acids composition of marine *Nannochloropsis gaditana* cultured in desalination concentrate. Bioresour Technol 197: 48–55.

Matos, Â.P., E.H.S. Moecke and E.S. Sant'Anna. 2017. The use of desalination concentrate as a potential substrate for microalgae cultivation in Brazil. Algal Res 24: 505–508.

McGinn, P.J., K.E. Dickinson, K.C. Park, C.G. Whitney, S.P. MacQuarrie, F.J. Black, J.-C. Frigon, S.R. Guiot and S.J.B. O'Leary. 2012. Assessment of the bioenergy and bioremediation potentials of the microalga *Scenedesmus* sp. AMDD cultivated in municipal wastewater effluent in batch and continuous mode. Algal Res 1: 155–165.

Oliveira, A.C., A. Barata, A.P. Batista and L. Gouveia. 2018. *Scenedesmus obliquus* in poultry wastewater bioremediation. Environ Technol 40: 3735–3744.

Omojevwe, E.G. and F.O. Ezekiel. 2016. Microalgal-bacterial consortium in polyaromatic hydrocarbon degradation of petroleum-based effluent. J Bioremediat Biodegrad 7: 359.

Park, J.B.K., R.J. Craggs and A.N. Shilton. 2011. Wastewater treatment high rate algal ponds for biofuel production. Bioresour Technol 102: 35–42.

Park, J., J. Seo and E.E. Kwon. 2012. Microalgae production using wastewater: effect of light-emitting diode wavelength on microalgal growth. Environ Eng Sci 29: 995–1001.

Pittman, J.K., A.P. Dean and O. Osundeko. 2011. The potential of sustainable algal biofuel production using wastewater resources. Bioresour Technol 102: 17–25.

Rakaiby, M.E.L., T. Essam and A. Hashem. 2012. Isolation and characterization of relevant algal and bacterial strains from Egyptian environment for potential use in photosynthetically aerated wastewater treatment. J Bioremed Biodegrad S8: 001.

Sánchez, A.S., I.B.R. Nogueira and R.A. Kalid. 2015. Uses of the reject brine from inland desalination for fish farming, *Spirulina* cultivation, and irrigation of forage shrub and crops. Desalination 364: 96–107.

Schwarzenbach, R.P., B.I. Escher, K. Fenner, T.B. Hofstetter, C.A. Johnson, U. Von-Gunten and B. Wehrli. 2006. The challenge of micropollutants in aquatic systems. Science 313: 1072–1077.

Shen, Q.-H., J.-W. Jiang, L.-P. Chen, L.H. Cheng, X.-H. Xu and H.-L. Chen. 2015. Effect of carbon source on biomass growth and nutrients removal of *Scenedesmus obliquus* for wastewater advanced treatment and lipid production. Bioresour Technol 190: 257–263.

Silkina, A., N.E. Ginnever, F. Fernandes and C. Fuentes-Grünewald. 2019. Large-scale waste bio-remediation using microalgae cultivation as a platform. Energies 12: 2772.

Tchobanoglous, G. and F.L. Burton. 1991. Wastewater Engineering: Treatment Disposal Reuse. 3rd ed. McGraw-Hill, New York.

Tedesco, S., T. Marrero-Barroso and A.G. Olabi. 2014. Optimization of mechanical pre-treatment of Laminariaceae spp. biomass-derived biogas. Renew Energ 62: 527–34.

Ugwu, C.U., H. Aoyagi and H. Uchiyama. 2008. Photobioreactors for mass cultivation of algae. Bioresour Technol 99: 4021–4028.

UNESCO. 2009. The United Nations World Water Development Report3: Water in a Changing World. Paris/New York: UNESCO/Berghahn Books.

United Nation World Water Development Report. United Nations Educational, Scientific and Cultural Organization. 2014. ISBN 978-92-3-104259-1.

Vieno, N., T. Tuhkanen and L. Kronberg. 2007. Elimination of pharmaceuticals in sewage treatment plants in Finland. Water Res 41: 1001–1012.

Wang, L., M. Min, Y. Li, P. Chen, Y. Chen, Y. Liu, Y. Wang and R. Ruan. 2010. Cultivation of green algae *Chlorella* sp. in different wastewaters from municipal wastewater treatment plant. Appl Biochem Biotechnol 162: 1174–1186.

Wang, H., H. Xiong, Z. Hui and X. Zeng. 2012. Mixotrophic cultivation of *Chlorella pyrenoidosa* with diluted primary piggery wastewater to produce lipids. Bioresour Technol 104: 215–220.

Wang, L., H. Xiao, N. He, D. Sun and S. Duan. 2019. biosorption and biodegradation of the environmental hormone nonylphenol by four marine microalgae. Sci Rep 9.

Wilderer, P.A. and D. Schreff. 2000. Decentralized and centralized wastewater management: a challenge for technology developers. Water Sci Technol 41: 1–8.
Williams, W.D. 2001. Anthropogenic salinisation of inland waters. Dev Hydrobiol 162: 329–337.
Wu, J.-Y., C.-H. Lay, C.-C. Chen and S.-Y. Wu. 2017. Lipid accumulating microalgae cultivation in textile wastewater: Environmental parameters optimization. J Taiwan Inst Chem E 79: 1–6.
Yadav, G., S.K. Dash and R. Sen. 2019. A biorefinery for valorization of industrial waste-water and flue gas by microalgae for waste mitigation, carbon-dioxide sequestration and algal biomass production. Sci Total Environ 688: 129–135.
Yaseen, D.A. and M. Scholz. 2016. Shallow pond systems planted with *Lemna minor* treating azo dyes. Ecol Eng 94: 295–305.
Yaseen, D.A. and M. Scholz. 2018. Textile dye wastewater characteristics and constituents of synthetic effluents: a critical review. Int J Environ Sci Technol 16: 1193–1226.
Zarzo, D., E. Campos, D. Prats, P. Hernandez and J.A. Garcia. 2014. Microalgae production for nutrient removal in desalination brines. IDA J Desal Water Reuse 6: 61–68.
Zarzo, D. 2018. Beneficial uses and valorization of reverse osmosis brines. pp. 365–397. *In*: Gude, V.G. (ed.). Emerging Technologies for Sustainable Desalination Handbook. Butterworth-Heinemann, Oxford, England, United Kingdom.
Zhang, C., Q. Li, L. Fu, D. Zhou and J.C. Crittenden. 2018. Quorum sensing molecules in activated sludge could trigger microalgae lipid synthesis. Bioresour Technol 263: 576–582.
Zhou, D., C. Zhang, L. Fu, L. Xu, X. Cui, Q. Li and J.C. Crittenden. 2017. Responses of the microalga chlorophyta sp. to bacterial quorum sensing molecules (n-acylhomoserine lactones): aromatic protein-induced self-aggregation. Environ Sci Technol 51: 3490–3498.

10

Bioremediation and Biodegradation of Crude Oil Polluted Soil

Modupe Elizabeth Ojewumi

Introduction

Bioremediation can be defined as the biological clean-up process in which the degrading property of microorganisms, such as fungi and bacteria is employed for the degradation or decrease of harmful substances that pollutes either the soil or water [environment]. It can also be defined as a process used to treat contaminated media, including water, soil, and subsurface material, by altering environmental conditions to stimulate growth of microorganisms and degrade the target pollutants. For instance, hydrocarbons at various degrees, metals, pesticides, and mechanical solvents (Korda et al. 1997, Ojewumi et al. 2018a, 2019a). Of all the technologies and methods researched into in recent years for oil clean up, bioremediation is still the best approach to remediating polluted soil or water by hydrocarbons due to its ability to prevent and inhibit the accumulation of contaminants. This technique is very effective in cleaning up petroleum hydrocarbon pollution, and it is also very cheap (El-Nawawy et al. 1992). It is a modern technique whereby the natural degrading ability of microorganisms is used to reduce the concentration and/or toxicity of a wide range of chemical substances that are released into the environment. The remediation process in question works by stimulating the microbes that occur naturally in the environment to degrade organic wastes found in soil and groundwater. The microorganisms that degrade crude oil use the hydrocarbon breakdown as their source of chemical energy. Some microorganisms can naturally degrade petroleum hydrocarbons by utilizing the carbon within them to survive. The hydrocarbons that

Chemical Engineering Department, Covenant University, P.M.B. 1023, Ota, Ogun State, Nigeria.
Email: modupe.ojewumi@covenantuniversity.edu.ng

exist in crude oil serve as substrates for the microorganisms. Normally without any external enhancement or intervention, as soon as an oil spill occurs, there is a rise in the population of microbes that degrade hydrocarbons within the ecosystem.

Since the 19th century, petroleum has been utilized for a very long time for power generation and lubrication. The innovation of the motor engine and its quick appropriation in all vehicle structures broadened the occupation of this natural resource. This caused the expansion of the petroleum business. One of the major concerns of the oil industry today is how to improve the recovery of a large percentage of oil remaining unrecovered in the old and new depleted producing fields (Ojewumi et al. 2017, 2018b, c). These pollution problems can be minimized, yet not completely eliminated, and hence, bring on several issues for the environment (Pala et al. 2006). The famous bioremediation is an amazing spill clean-up technique, in which the normal degrading ability of microorganisms is harnessed for the degradation and decrease of the harmful substances that pollute the environment. Some substances in this category include petroleum subsidiaries, aliphatic and sweet-smelling hydrocarbons, mechanical solvents, pesticides, and metals (Korda et al. 1997).

Petroleum (i.e., rock oil) is a source of energy which is extensively used for lighting, heating, and in internal combustion engines, when and after being explored, this crude oil flows through piping and vessels under pressure from the reservoir or from mechanically assisted pumps (Eneh 2011). The discharge of this liquid petroleum hydrocarbon into marine (offshore) or land (onshore) is injurious to living biota in the ecosystem of the spill, even as it exhibits potential risks of normal soil processes interference, fire hazards, and further contaminations of air and water (Mahjoubi et al. 2018, Ojewumi et al. 2018d, Brakstad et al. 2017, Maddela et al. 2016, Akhundova and Atakishiyeva 2015, Ekperusi and Aigbodion 2015). Oil spills by any accidental leakages, material ruptures, or improper handling from production installations, such as the well-heads leading to raw crude oil spillage, or from flow lines towards storage and refinement, after the removal of impurities, translates to treated oil spillage (Liu et al. 2017, Obayori et al. 2012). Hydrocarbons, the major constituents from petroleum, persist in the environment as recalcitrant contaminants, whose removal and treatment are highly problematic and constitute challenges to stakeholders and researchers globally (Liu et al. 2017, Hassanshahian et al. 2012, Tang et al. 2011). Soil amendment technologies, known as remediation, include physical separation, chemical treatments, photo-degradation, and biologically-mediated remediation (bioremediation), but among these, the use of bioremediation techniques is attracting preference (Zhang et al. 2012, Mohsenzadeh and Rad 2015, Ibrahim et al. 2016). This is due to the combined advantages of relatively lower cost, higher effectiveness, and resultant lower adverse impacts on the ambient environment, ensuing from the use of bioremediation instead of the other methods that could rather lead to further environmental-toxicity (Liu et al. 2017, Ibrahim et al. 2016). The technique of bioremediation exhibits these potencies from the abilities of employing metabolic activities of animals, plants, and microorganisms for removal or conversion from toxic to non-toxic compounds, i.e., detoxification, of the recalcitrant hydrocarbon pollutants from oil spills (Abioye 2011, Tang et al. 2011, Mohsenzadeh and Rad 2015).

The always increasing global energy demand makes it imperative that the world is still highly dependent on petroleum products for meeting energy needs in many ramifications of livelihood, a condition that necessitates continuous extraction/production of petroleum from its location deep down within the earth (Abioye 2011, Minai-Tehrani et al. 2015, Liu et al. 2017). The situation ensuing from this includes crude petroleum oil spill that could be through uncontained excessive pressure from production installations/platforms, e.g., raw crude oil from well-heads, blowouts, etc., or from transportation or improper handling, e.g., of treated crude oil in flow lines or storage tanks (Liu et al. 2017, Ite and Semple 2012). The resulting oil spill that could be into marine (offshore) or soil (onshore) environments are very toxic and hazardous to the environmental ecosystem, and could adversely affect the well-being of living organs, air, water, and soil processes, as well as the potential of fire hazards (Akhundova and Atakishiyeva 2015, Brakstad et al. 2017, Hassanshahian et al. 2012). Onshore spill of crude oil affects healthy living in the society, agricultural productivity, groundwater/sources for potable water, and living biota in flowing streams/rivers, among others (Philp and Atlas 2005, Maddela et al. 2016, Van der Perk 2013). Avoiding or mitigating these adverse effects from crude oil spillage situation necessitates the need for amending the soil via the procedure known as remediation.

Among known methods for remediating crude oil polluted soil, including physical separation, chemical degradation, photo-degradation, and bioremediation, the method of bioremediation is attracting preference due to its comparative effectiveness, relatively low cost, and eco-friendliness, compared to the other techniques (Ibrahim et al. 2016) . Unlike bioremediation, other methods that could be used for oil polluted soil remediation have also been recognized with the potential of leaving daughter compounds, i.e., secondary residuals, after the parent/primary crude oil pollutant has been removed, which can even exhibit higher toxicity levels than the parent crude oil pollutant (Ibrahim et al. 2016). In contrast, bioremediation technique usage detoxifies contaminants in crude oil and effectively removes pollutants by destroying them in the stead of transferring them to another medium (Ibrahim et al. 2016, Perelo 2010, Okoh and Trejo-Hernandez 2006).

Studies have employed plants species for bioremediation, in a process known as phytoremediation (Minoui et al. 2015), but the use of microorganisms as biologically-mediated remediation of crude oil polluted soil is still linked to the effectiveness of phytoremediation systems. This is due to the fact that they have been used in reported works for effective repair of crude oil polluted soil (Liu et al. 2017, Zhang et al. 2012). However, there is paucity of reported work employing *Pseudomonas aeruginosa* for the bioremediation of Escravos Light crude oil blend obtainable in Nigeria.

Lucas and MacGregor (2006) reported that pollution caused by crude oil spill happens as a result of land runoff, accidents that occur involving pipelines and oil carrying vessels, oil exploration and production engineering operations, oil shipping activities, and improper effluent discharge into the environment. These spills have negative and disastrous economic, environmental, and social effects on the society. As a result, oil spill accidents have created some media interest and have brought many together in a battle regarding the response of the government to oil spill incidents and the actions that are being taken to curb the menace (Broekema 2016). The type of oil that spills, the quantity or severity of the oil spill, the environment in which the spill occurs, and the prevailing weather conditions are factors that must be considered before choosing the most effective technique to clean up the spill (Choi and Cloud 1992). Raw crude oil is basically crude coming directly from the depths of the earth through the wellhead. It may or may not have some amount of water and dissolved natural gas within it. Treated crude oil in this case is not necessarily crude oil that has gone through the refining process. Treated crude refers to crude oil that has been passed through an oil production platform. These platforms are located in central locations of major oil producing fields. Raw crude from the different wellheads is channeled to the platform. Here, the crude is stripped off dissolved natural gas, excess water which forms emulsion with the crude, and condensates. Raw crude also comes with a high level of sand, and this is also removed.

Before the knowledge of bioremediation came into the limelight, there had been the existence of some crude oil spill clean-up techniques, some of which are in use till this day. According to Larson (2010) these remediation techniques are:

1. The physical remediation methods
2. The chemical remediation methods
3. The thermal remediation methods

A lot of microorganisms naturally have the enzymatic or catalytic ability to degrade or simplify petroleum hydrocarbons. Some of the microbes are specific in their action. Alkanes may be degraded by certain types of microorganisms. Other microorganisms may degrade only aromatic compounds. It has been experimentally proven that alkane compounds that range from C10 to C26 are the most easily biodegradable. Very toxic aromatic compounds, such as benzene, toluene are also easily biodegraded by microorganisms. More complex structures are very resistant to biodegradation. There are only a few microbes that can degrade these compounds, and the biodegradation rate is much lower than in pure alkanes (Dang and Lovell 2016). A mixed microbial population may guarantee a higher level of biodegradation. The speed and efficiency of the clean-up of a soil contaminated with petroleum and petroleum products largely depends on the presence of hydrocarbon-degrading microorganisms in the soil. The factors that are necessary for populating microbial growth are temperature, oxygen, pH, content of nitrogen and phosphorus.

Bioremediation kinetics seeks to investigate how different experimental conditions can influence the rate of the microbial degradation of the crude oil. With the study of microbial degradation kinetics, it is possible to know and understand the kinetics of soil bioremediation and also to determine the quantity of crude oil left at a particular time (Agarry and Jimoda 2013, Agarry et al. 2013).

Bioremediation/biodegradation organisms

Crude oil constitutes another material that pollutes the environment. This can be remediated using microorganisms (Ojewumi et al. 2018a, 2019b, c) such as: bacteria and fungi (*Aspergillus niger, Pseudomonas aeruginosa, Saccharomyces cerevisiae, Acetobacter*, etc.), *A. niger, P. aeruginosa* have been extensively used in research.

A. *A. niger* as a crude oil degrading microorganism is a haploid fungi that is filamentous in nature. It is an important microorganism in the field of biotechnology. *A. niger* has found wide relevance in bio-transformations and management, such as the conversion of waste paper to fermentable sugar, orange peel to bioethanol, etc. (Ojewumi et al. 2018a, b). The fungi is popularly found in deteriorating or decaying vegetation (Schuster et al. 2002). Fungi perform a key function in degrading hydrocarbons by releasing some capable enzymes which act on them. This is probably as a result of their aggressive growth, extensive hyphal growth in soil, and greater biomass production. Fungi has a very high tendency for biodegradation technology (Hammel 1995).

B. *P. aeruginosa* as hydrocarbon degrading microorganism is a rod-shaped, gram-negative bacterium with an unbelievable nutritional versatility. *P. aeruginosa*, as well as many other species of *Pseudomonas*, degrade hydrocarbons and are also capable of breaking down toluene. This bacteria degrades toluene, which is the simplest form of methylbenzene through the oxidation of methyl group to alcohol, aldehyde, and acid, which is then converted to catechol. Due to this characteristic, *P. aeruginosa* can be used in environmental pollution control (Johnson and Olsen 1997, Pala et al. 2006). It is naturally found in some environments, such as soil, sewage, plants, water, humans, animals, and hospitals (Pala et al. 2006). *P. aeruginosa* is a bacteria of the order *Pseudomonadales*, family *Pseudomononadaceae*, and genus *Pseudomonas*. *P. aeruginosa* and other bacteria can be used for various experimental activities, such as bioremediation and bioconversion of various waste materials (Ojewumi et al. 2018a, 2019b, c).

Acknowledgments

Authors would like to appreciate the contributions of Ejemen Valentina Anenih, a Process Engineer at National Engineering and Technical Company, Lagos, and the proofreading of Dr O.O. Awolu of the Department of Food Science and Technology, Federal University of Technology, Akure.

References

Abioye, O.P. 2011. Biological remediation of hydrocarbon and heavy metals contaminated soil. Soil contamination. IntechOpen.

Agarry, S., M. Aremu and O. Aworanti. 2013. Kinetic modelling and half-life study on enhanced soil bioremediation of bonny light crude oil amended with crop and animal-derived organic wastes. J Pet Environ Biotechnol 4: 137.

Agarry, S.E. and L.A. Jimoda. 2013. Application of carbon-nitrogen supplementation from plant and animal sources in *in-situ* soil bioremediation of diesel oil: experimental analysis and kinetic modelling. J Environ Ear Sc 3: 51–62.

Akhundova, E. and Y. Atakishiyeva. 2015. Interaction between plants and biosurfactant producing microorganisms in petroleum contaminated Absheron soils. Phytoremediation for Green Energy. Springer.

Brakstad, O.G., S. Lofthus, D. Ribicic and R. Netzer. 2017. Biodegradation of petroleum oil in cold marine environments. Psychrophiles: From Biodiversity to Biotechnology. Springer.

Broekema, W. 2016. Crisis-induced learning and issue politicization in the EU: The braer, sea empress, erika, and prestige oil spill disasters. Pub Admin 94: 381–398.

Choi, H.M. and R.M. Cloud. 1992. Natural sorbents in oil spill cleanup. Environ Sci Technol 26: 772–776.

Dang, H. and C.R. Lovell. 2016. Microbial surface colonization and biofilm development in marine environments. Microbiol Mol Biol Rev 80: 91–138.

Ekperusi, O. and F. Aigbodion. 2015. Bioremediation of petroleum hydrocarbons from crude oil-contaminated soil with the earthworm: Hyperiodrilus africanus. 3 Biotech 5: 957–965.

El-Nawawy, A., I. El-Bagouri, M. Abdal and M. Khalafawi. 1992. Biodegradation of oily sludge in Kuwait soil. Wor J Microbiol Biotech 8: 618–620.

Eneh, O.C. 2011. A review on petroleum: Source, uses, processing, products, and the environment. J Appl Sc 11: 2084–2091.

Hammel, K.E. 1995. Mechanisms for polycyclic aromatic hydrocarbon degradation by ligninolytic fungi. Environ Hea Pers 103: 41–43.

Hassanshahian, M., G. Emtiazi and S. Cappello. 2012. Isolation and characterization of crude-oil-degrading bacteria from the Persian Gulf and the Caspian Sea. Mar Poll Bull 64: 7–12.

Ibrahim, M., R. Shuaibu, S. Abdulsalam and S. Giwa. 2016. Remediation of Escravous crude oil contaminated soil using activated carbon from coconut shell. J Biorem Biodeg 7.

Ite, A.E. and K.T. Semple. 2012. Biodegradation of petroleum hydrocarbons in contaminated soils. Microbial Biotechnology: Ener Envrion 250–278.

Johnson, G.R. and R.H. Olsen. 1997. Multiple pathways for toluene degradation in Burkholderia sp. strain JS150. Appl Environ Microbiol 63: 4047–4052.

Korda, A., P. Santas, A. Tenente and R. Santas. 1997. Petroleum hydrocarbon bioremediation: sampling and analytical techniques, *in situ* treatments and commercial microorganisms currently used. Appl Micro Biotech 48: 677–686.

Larson, H. 2010. Responding to oil spill disasters: The regulations that govern their response. Retrieved on 26th February.

Liu, Y., C. Li, L. Huang, Y. He, T. Zhao, B. Han and X. Jia. 2017. Combination of a crude oil-degrading bacterial consortium under the guidance of strain tolerance and a pilot-scale degradation test. Chin J Chem Engr 25: 1838–1846.

Lucas, Z. and C. Macgregor. 2006. Characterization and source of oil contamination on the beaches and seabird corpses, Sable Island, Nova Scotia, 1996–2005. Mar Poll Bull 52: 778–789.

Maddela, N.R., R. Burgos, V. Kadiyala, A.R. Carrion and M. Bangeppagari. 2016. Removal of petroleum hydrocarbons from crude oil in solid and slurry phase by mixed soil microorganisms isolated from Ecuadorian oil fields. Inter Biodet Biodeg 108: 85–90.

Mahjoubi, M., S. Cappello, Y. Souissi, A. Jaouani and A. Cherif. 2018. Microbial bioremediation of petroleum hydrocarbon–contaminated marine environments. Rec Ins Pet Sc Engr 325. DOI: 10.5772/intechopen.72207.

Minai-Tehrani, D., S. Minoui and M.H. Shahriari. 2015. Reciprocal effects of oil-contaminated soil and festuca (Tall fescue). Phyto for Gr Ener. Springer. 141–148.

Minoui, S., D. Minai-Tehrani and M.H. Shahriari. 2015. Phytoremediation of crude oil-contaminated soil by Medicago sativa (Alfalfa) and the effect of oil on its growth. Phyto for Gr Ener. Springer. 123–129.

Mohsenzadeh, F. and A.C. Rad. 2015. Bioremediation of petroleum polluted soils using Amaranthus retroflexus L. and its rhizospheral funji. Phyto for Gr Ener. Springer. 131–139.

Obayori, O.S., L.B. Salam and I.M. Omotoyo. 2012. Degradation of weathered crude oil (Escravos Light) by bacterial strains from hydrocarbons-polluted site. Afri J Microbiol Res 6: 5426–5432.

Ojewumi, M.E., M.E. Emetere, D.E. Babatunde and J.O. Okeniyi. 2017. *In situ* bioremediation of crude petroleum oil polluted soil using mathematical experimentation. Inter J Chem Engr. Volume 2017, Article ID 5184760, 11 pages. https://doi.org/10.1155/2017/5184760.

Ojewumi, M.E. and V.A. Ejemen. 2018a. The *ex-situ* bioremediation kinetics of raw and treated crude oil polluted soil using *Aspergillus niger* and *Pseudomonas aeruginosa*. MDPI 2: 1–17.

Ojewumi, M.E., E.V. Anenih, O.S. Taiwo, B.T. Adekeye, O.O. Awolu and E.O. Ojewumi. 2018b. A bioremediation study of raw and treated crude petroleum oil polluted soil with *Aspergillus niger* and *Pseudomonas aeruginosa*. J Ecol Engr 19: 226–235.

Ojewumi, M.E., B.I. Obielue, M.E. Emetere, O.O. Awolu and E.O. Ojewumi. 2018c. Alkaline pre-treatment and enzymatic hydrolysis of waste papers to fermentable sugar. J Ecol Engr 19: 211–217.

Ojewumi, M.E., J.O. Okeniyi, J.O. Ikotun, E.T. Okeniyi, V.A. Ejemen and A.P.I. Popoola. 2018d. Bioremediation: data on Pseudomonas aeruginosa effects on the bioremediation of crude oil polluted soil. DIB 19: 101–113.

Ojewumi, M.E., J.O. Okeniyi, E.T. Okeniyi, J.O. Ikotun, V.A. Ejemen and E.T. Akinlabi. 2018e. Bioremediation: data on biologically-mediated remediation of crude oil (Escravos light) polluted soil using *Aspergillus niger*. Chem Dat Coll 17-18(2018): 196–204.

Ojewumi, M.E., V.E. Anenih, E.E. Alagbe and E.A. Oyeniyi. 2019a. Kinetics study of biologically remediated crude oil polluted soil using a bacteria and fungi. J Phys: Conference Series, 1299 (2019). IOP Publishing, 012001. 1299. doi:10.1088/1742-6596/1299/1/012001.

Ojewumi, M.E., M.E. Emetere, C.E. Amaefule, B.M. Durodola and O.D. Adeniyi. 2019b. Bioconversion of Orange Peel waste by *Escherichia coli* and *Saccharomyces cerevisiae* to ethanol. Inter J Pharm Sc Res 10(3): 1246–1252.

Ojewumi, M.E., P.C. Ogele, D.T. Oyekunle, J.A. Omoleye, S.O. Taiwo and Y. Obafemi. 2019c. Co-digestion of cow dung with organic kitchen waste to produce biogas using *Pseudomonas aeruginosa*. J Phys: Conference Series 1299(2019). IOP Publishing, 012011. doi:10.1088/1742-6596/1299/1/012011.

Ojewumi, M.E., O.E. Kolawole, D. Oyekunle, O.S. Taiwo and A. Adeyemi. 2019d. Bioconversion of waste foolscap and newspaper to fermentable sugar. J Ecol Engr 20: 35–41.

Okoh, A. and M. Trejo-Hernandez. 2006. Remediation of petroleum hydrocarbon polluted systems: exploiting the bioremediation strategies. Afri J Biotech 5(25): 2520–2525.

Pala, D.M., De D.D. Carvalho, J.C. Pinto and G.L. Sant'anna Jr. 2006. A suitable model to describe bioremediation of a petroleum-contaminated soil. Inter Biodet Biodeg 58: 254–260.

Perelo, L.W. 2010. *In situ* and bioremediation of organic pollutants in aquatic sediments. J Haz Mat 177: 81–89.

Philp, J.C. and R.M. Atlas. 2005. Bioremediation of contaminated soils and aquifers. pp. 139–236. Bioremediation. *In*: Atlas, R. and J. Philip (eds.). Bioremediation. ASM Press, Washington, DC. doi: 10.1128/9781555817596.ch5.

Schuster, E., N. Dunn-Coleman, J. Frisvad and P. Van Dijck. 2002. On the safety of *Aspergillus niger*—a review. Appl Microbiol Biotech 59: 426–435.

Tang, J., M. Wang, F. Wang, Q. Sun and Q. Zhou. 2011. Eco-toxicity of petroleum hydrocarbon contaminated soil. J Environ Sci 23: 845–851.

Van Der Perk, M. 2013. Soil and Water Contamination. CRC Press.

Zhang, X., D. Xu, C. Zhu, T. Lundaa and K.E. Scherr. 2012. Isolation and identification of biosurfactant producing and crude oil degrading *Pseudomonas aeruginosa* strains. Chem Engr J 209: 138–146.

11

Microbial Recycling of 'Sustainable' Bioplastics

A Rational Approach?

*Mansi Rastogi** and *Sheetal Barapatre*

"Sustainable bioplastic", an introduction

Owing to its widespread applications, plastic is rendered as an indispensable part of modern life. It is a petroleum-based product, and its excessive production calls for sustainable alternative solutions. Petroleum-based plastics are made from about 4% of the oil that the world uses every year, and have severe impacts on environment and human health, which include carbon dioxide emissions and accumulation in the environment due to non-biodegradability (Pathak et al. 2014, Jain and Tiwari 2015). About 34 million tons of plastic waste is produced annually across the world, 93% of which finds its way into oceans and landfills (Pathak et al. 2014), estimated to increase two-fold in the next 20 years. Despite all attempts to develop reusing and recycling practices, developing countries still rely on regular landfilling. The estimated amount of virgin plastics produced till date is accounted to be around 8.3 billion metric tons, out of which 6.3 billion metric tons has become waste. About 12% of this waste was incinerated, 9% recycled, whilst the remaining 79% still rests in the landfills (Geyer et al. 2017).

Plastic disposal has become a global concern, owing to the release of carbon dioxide during the production of plastics. Around 400 million tons of CO_2 is produced every year during the process of plastic manufacturing, and 4% of these produced throughout the globe are eventually found lying in seas and oceans, hence posing a

Department of Environment Sciences, Maharshi Dayanand University, Delhi Bypass, Rohtak, Haryana 124001, India.
Email: barapatesheetal26@gmail.com
* Corresponding author: rastogimansi86@gmail.com

great risk for the marine life and environment present in these oceans (Strategy 2018). These plastics, when ingested by the marine organisms, threaten their life, and also result in bioaccumulation in food chain (Jain and Tiwari 2015, Pathak et al. 2014). Hence, alternative ways to dispose of these plastics, for lowering the environmental and health impacts are deemed as the need of the hour.

Different types of plastics with varying compositions might have different recycling rates, which also defines the extent of the problem. Recycling rates vary greatly depending on the country, the management policies, and the nature of the plastic item. Among them, Polyethylene Terephthalate (PET) has the greatest recovery rate among packaging plastics (70% in Europe). Only 30% of PET bottles were recycled in the year 2015 in the US, while 34% of High Density Poly Ethylene (HDPE), 18% of polypropylene, 4.1% of low-density polyethylene (LDPE), and 3.3% of polyvinyl chloride (PVC) were recycled (ACC and APR 2016). These recycling rates in the US are much lower than countries like Germany, where over 92% of PET bottles were recycled, and 26% of the recycled plastic has been used for new PET bottles (GVM 2016).

In this chapter, we have discussed an alternative strategy based on renewable sources, which are commercially and environmentally acceptable, and could be effectively rendered to replace petroleum-based plastics. There is a need for switching to the use of biodegradable plastics or bioplastics that are sustainable by-products with a good potential for successful recycling, biodegradation (with the aid of microbial enzymes), and incineration. Despite being eco-friendly, bioplastics have certain drawbacks, such as high production cost and inferior mechanical properties. While agro-waste can be used to manage the high cost of production, it is also suggested that bioplastics such as Poly Lactic Acid (PLA) and Poly hydroxyalkonates possess optimum properties, such as high tensile strength and modulus (Tabasi and Ajji 2015). Furthermore, Polyhydroxybutyrate (PHB) bioplastics have gained the attention of the scientific community owing to their low CO_2 emission.

Biopolymers/plastic recycling

Bioplastics is a wider term that is used for materials whose origin can be attributed to biomass, which are biodegradable, and may be a combination of both these attributes (Van den Oever et al. 2017). Bioplastics are processed from plant biomass, including starch, cellulose, polyesters produced by living organisms, and synthetic green polymers. These biopolymers are organic compounds produced due to metabolic reactions, and could be fossil-based (can be degraded), bio-based substances that cannot be degraded (Fig. 11.1), or found in combination with fossil-based plastics. Although bioplastics possess the characteristics of conventional plastics, they are more eco-friendly compared to petroleum formulated plastics and can be degraded by bacteria, fungi, algae, and protozoa within a short period of time (Okan et al. 2019). The by-products formed as a result of decomposition of bioplastics are carbon dioxide, water, methane, and residual biomass, which do not harm the environment.

At a stage where the global economies are generating enormous magnitudes of waste, we need to contemplate and restructure our strategies, enabling sustainable

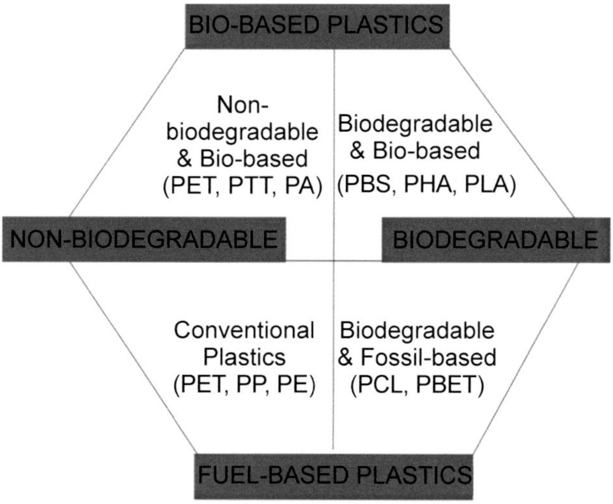

Figure 11.1. Types of plastics.

product development for a better future. We need to look up to the concept of circular economy and strike a balance between resources and waste products by recuperating waste as resources. We need to maximize our prospects for material recovery as well as deliberation for substitutes for plastics. Bioplastics can be classified as non-biodegradable and biodegradable plastics. Both these categories hold importance from the perspective of sustainability. Similarly, bio-based plastics can be formulated to create either completely biodegradable plastics that produce CO_2 as their end product, or their non biodegradable counterparts that trap carbon and can be used for making sewer pipes, roofing material, and even road surfaces. Likewise, cyanobacteria and higher plants also enhance photosynthetic processes to produce feed stocks which can further be used to create an array of biodegradable plastics and infrastructure that aids to capture carbon (Balaji et al. 2013). At present, majority of the bioplastics are created from agricultural feed stocks, but they do not comply with the governmental policies, as they consume arable land and fresh water. It has been found that microalgae can be used to produce bioplastics and overcome all these issues by increasing overall photosynthetic competence, resolving desertification, using non-arable lands and enhancing the conversion of CO_2. They also ease eutrophication by consuming wastewater and recovering nutrients, such as nitrogen and phosphorus. Microalgae-based bioplastics can be premediated to degrade in natural and industrial conditions. Such bioplastics can be employed to generate products that have small shelf-life, and can be degraded completely to carbon dioxide.

There are four types of degradable bioplastics:

- Biodegradable bioplastics: can be degraded completely under natural conditions by microbes and do not produce any toxic by-products. They are the most preferred ones due to their potential for consumption by microbes.

- Compostable bioplastics: are degraded biologically by composting and do not produce any toxic products. Bioplastics can be termed compostable bioplastics depending upon their degradability, rate of disintegration, and eco-toxicity of their by-products.
- Photodegradable bioplastics have a photo perceptive group attached to the polymer. UV radiations can be used to split the polymer structure, making it susceptible to microbial degradation.
- The Working Group for Safer Chemicals and Sustainable Materials have characterized bio-based bioplastics as "plastics in which 100% carbon can be obtained from renewable agricultural and forestry resources, for instance, corn starch, soybean protein, and cellulose" (Thakur et al. 2018).

Bioplastics with a number of merits possess a foremost advantage that they are manufactured from renewable resources and not fossils, and their production requires less energy as compared to their fossil-based counterparts. They have a smaller carbon footprint, which is due to their potential for carbon sequestration that remains sequestered even if the plastic is recycled again and again (Fig. 11.2). Even the non-biodegradable bioplastic hoards carbon dioxide perpetually. Bioplastics emit lesser greenhouse gases, reported to produce only 0.49 kg CO_2, which is emitted from the production of 1 kg of resin. This value is quite less than 2 ~ 3 kg CO_2 produced by petroplastics, thereby reducing 80% of the global warming potential (Yu and Chen 2018).

On the contrary, few disadvantages are also associated with the production and consumption of bioplastics. They incur high cost-nearly double the cost of

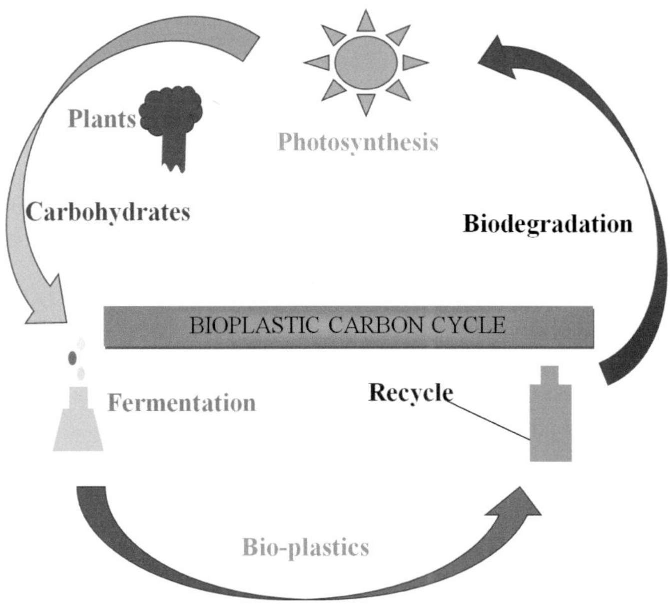

Figure 11.2. Bioplastic carbon cycle.

manufacturing the conventional plastics. However, it is expected that owing to the large scale production of the bioplastics, a reduction in their cost is acclaimed. As for the management of waste, we need to develop strategies for efficient segregation of bioplastics from conventional plastics, which might otherwise corrupt the recycling process. In order to ensure that production of bioplastics does not compromise with agricultural produce, we need to ensure that bioplastics are manufactured from agro-residues and food by-products.

Plastic pollution

Plastics impose a grave threat to the environment, human health, and wildlife. Essentially, animals are threatened when they ingest plastics or get entrapped in them. The United Nations Convention on Biological Diversity states that around 600 species of marine organisms are affected due to ingestion of plastics in seas and oceans, ranging from microbes to whales (Auta et al. 2017). Just like air pollutants, plastic waste do not respect national or state boundaries, and they move via water bodies, and finally deposit in the ocean floors and sea beds. Jambeck et al. 2015 have reported that nearly 4.4–12.7 million MT plastic is accumulated in the world's oceans annually. Additionally, microplastics are minute particles with diameters less than 5 mm, and are most commonly found in the benthic zones, and account for around 13.2% of the total plastic waste. Microplastics have raised health concerns due to their ability to enter the food chain. Microplastics damage the reproduction and development processes, and also cause modifications in gene expression and tissues, which proves to be fatal, thereby, changing the size of population and modifying community structure. They find their way in human diet through fish and crustaceans, which are found to ingest these microplastics. In humans, they are responsible for causing cancer, inflammation, and infections due to bioaccumulation.

Due to escalating demand for plastics and their slow degradation, the need for sustainable plastics has risen. The fact that renewable resources can serve to provide marketable plastics has reduced the dependence on petrochemicals and also associated after effects, such as CO_2 emissions and bioaccumulation of microplastics. Earlier, the major concern was to find substitutes for fossil-based plastics, where bioplastics were considered as the apt solution. However, most of the bioplastics which were manufactured using crop residues were used by blending them with fossil-based plastics, and were introduced in the commercial markets as "plastics with enhanced biodegradability". Eventually it was noticed that they did not serve the very target of sustainability, as they were only partially degradable, leaving microplastics behind.

Waste management systems, lack of infrastructure, and availability of resources for the same are different in different parts of the world. Thus, they impact the potential of different countries to combat plastic pollution, causing accumulation of plastics in landfills or hazardous chemical emissions due to open burning. The severed connection between production of plastics and management of waste has rendered governments of various countries ineffective in controlling plastic pollution. It is imperative for international communities to come forward and join hands to cap these harmful emissions by formulating effective policies considering the various life stages of a plastic lifecycle and connecting production and waste management practices, thereby closing the loop to encourage a circular economy.

Role of microbiota in the biodegradation process

Bioremediation strongly relies on microorganisms found in the plastic contaminated sites. These microorganisms secrete enzymes, which readily degrade the organic contaminants and utilize these contaminants as food for their growth. Certain studies carried out to use non-native and genetically modified microorganisms to degrade contaminants at given sites reported that organic contaminants act as a carbon source for the microbes (Bandopadhyay et al. 2018, Radhika and Murugesan 2012).

The factors vital for biodegradation of polymers include pH, moisture, temperature, and oxygen content (Massardier-Nageotte et al. 2006, Kale et al. 2007). The biodegradation of plastics depends on its chemical structure, crystallinity, and the intricacy of polymeric chain. If the polymeric chain is small and less intricate with amorphous groups attached to it, it can be degraded effectively, amenable to degradation by microbes. The affinity of enzymes towards specific functional groups can be used to escalate their disintegration. Polymeric biodegradation includes biodeterioration, biofragmentation, and bioassimilation, wherein biodeterioration, the physicochemical and mechanical properties of the polymer are altered with the help of microorganisms (Luyt 2017). These polymers are then broken down due to the action of microbes to give oligomers and finally monomers. This is termed as biofragmentation, and is followed by assimilation (Fig. 11.3), where the carbon present in the plastic is now converted to CO_2, water, and biomass by microorganisms (Lucas et al. 2008).

As far as degradation of plastic is concerned, oxidative degradation is the key method which can decrease the molecular weight of the material. Such polymers are converted to monomers, dimmers, and oligomers due to the secretion of intracellular and extracellular enzymes by the microorganisms. The number of enzymes secreted varies depending upon the microbial species and strains. Also, enzymes are specific to specific substrates (Underkofler et al. 1958). The by-products thus generated are used as energy sources. Enzymes that degrade plastics utilize the substrate as a source of carbon and energy which assist in biodegradation.

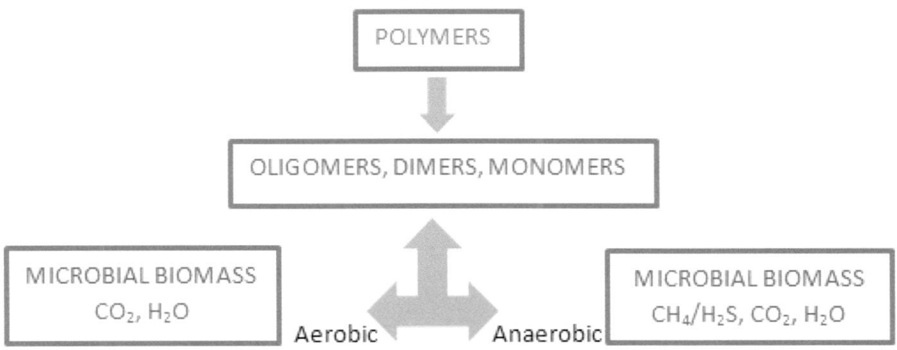

Figure 11.3. Breakdown of polymers during biodegradation.

Over 90 kinds of microbes have been found reliable for achieving degradation of bioplastics. These include photosynthetic bacteria, aerobes, anaerobes, and archaebacteria. They can be isolated from soil and other compost materials (Kumaravel et al. 2010, Accinelli et al. 2012). Microbial consortiums are expected to produce improved biodegradation of bioplastics due to production of wide range of enzymes and by utilizing the intermediate products for the degradation process. Besides environmental factors, such as moisture content, aeration, optimum pH, and temperature, the composition and structure of the polymer have a great role to play in the biodegradation of bioplastics. The composition of the biopolymer can be customized to improve its degradation by amending it with materials possessing highly soluble sugar content. The biocomposite thus produced might enhance its tensile strength, but it may also interfere with the degradation of the polymer during certain phases. To overcome this issue and achieve a more stable product, biocomposite blend must be optimized.

Formation of clear zones and estimating the zone diameters formed in petri plates where bioplastic serves as a sole source of carbon for bacterial or fungal colony indicates bioplastic degradation is possible using isolated microbes (Emadian et al. 2017, Trivedi et al. 2016, Brodhagen et al. 2015). Further observations using scanning electron microscope (SEM) confirm the alterations in the polymeric structure as achieved by microbial amendment to the plastic biodegradation process (Shen et al. 2015). Besides this, Fourier Transform Infrared (FTIR) spectroscopy can also be used to identify variation in bond intensity caused by microbial degradation (Phukon et al. 2012). *Cupriavidus necator* and *Pseudomonas chlororaphis* have been found to accelerate decomposition of bioplastics manufactured from Polylactid acid (PLA).

Factors affecting the rate of degradation

Many factors contribute individually and collectively to affect the rate of polymer degradation during its recycling or treatment (Fig. 11.4). Among them, environmental factors, such as temperature, low pH, moisture, etc. and chemical compositions of polymers are the major ones affecting the degradation of polymers (Laoutid et al. 2009). Furthermore, factors such as molecular weight, size of the polymer crystallinity, and co-polymer composition can also be accounted (Nair and Laurencin 2007). Describing the environment factors, water, gaseous exposure (mainly oxygen), and moisture gain can essentially induce effectual biodegradation of polymers, fulfilling the necessitated microbial growth and reproduction conditions (Idumah et al. 2019). As the moisture and humidity levels up, it accelerates the microbial activity, speeding the polymers degradation mechanism (Ho et al. 2009). Hence, the polymer degradation will be more rapid in the moisture laden conditions compared to dry conditions. A low pH within the environment directly affects the polymer biodegradation process. The acidic nature is thought to affect the microbial growth and make changes to reaction rates during hydrolysis of polymers (Masutani and Kimura 2014).

Temperature critically substantiates the degradation of polymers, which is likely less possible without the required temperature (Caulfield et al. 2002). An optimum

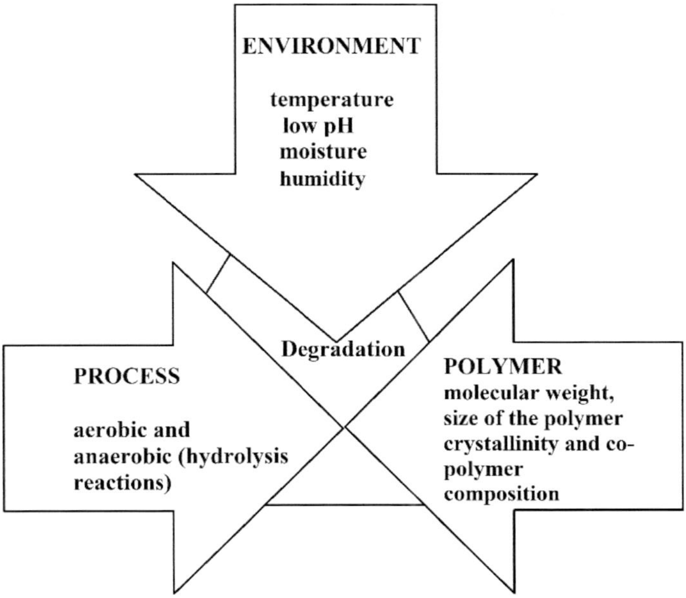

Figure 11.4. Factors contributing to affect polymer degradation during its recycling.

temperature supports microbial growth, but a higher one relegates the microbial action or might even stop it (Pathak 2017). Usually, the microorganism continues to grow with temperature until it attains a range where their growth impedes. A noticeable increase was observed for the hydrolysis rate in an environmental chamber, with temperature being increased to the glass transition temperature (Thakur et al. 2018), essentially for the polylactic acid film, where the different temperatures were 28, 40, and 55°C at 50 and 100% relative humidity (Ho et al. 1999).

Apart from the above stated factors, aerobic and anaerobic environment also influence the polymer degradation. In anaerobic polymer degradation, the hydrolysis reactions are hastened, owing to the flexibility of the polymer chain, resulting in fast tracked biodegradability (Thakur et al. 2018). Additionally, hybrid or co-polymer configuration, referring to a distant co-monomer penetrating into the polymer structure, decreases the crystallinity of the polymer, subsiding the biodegradability rate. For example, addition of polycaprolactone into lignin sharply increased the biodegradation rate (Chiellini and Corti 2003).

In case of molecular weight, a larger polymer will possess lesser flexibility and a greater polymer's glass transition temperature (Adhikari et al. 2016, Tabasi and Ajji 2015). Besides, the polymers with higher molecular weights will have lesser water solubility with reduced microbial efficacy. At last, a control on the size and shape of a polymer is required, where a larger sized particle slows down the biodegradability. A larger polymer persuades a higher degradability rate with a more accessible surface area for the microbes (Mihai et al. 2014, Li et al. 2016, Trivedi et al. 2016).

Biodegradation pathways

Considering the biodegradability of a polymer, it depends on the intricacy, crystallinity, and length of the polymer chain (Emadian et al. 2017). In this regard, less is more for polymer biodegradation, such that shorter molecular chains have clean structures, exhibiting low crystallinity. The biodegradation pathways for polymers can be classified into biotic and abiotic degradation (Lucas et al. 2008). The biotic biodegradation is facilitated by microbes (Sen and Raut 2015), while abiotic polymer degradation can ensue by ultraviolet radiation, high temperature, heat, and physical abrasion (da Costa et al. 2016). However, abiotic degradation and biotic/biodegradation often co-exist due to the unsuitable sterile conditions present on Earth. The mechanism of abiotic degradation offers a prerequisite for biotic degradation tailoring the polymer structure, enhancing the accessibility of microbes. Generally, degradation of a polymer is a very slow process co-played by the environmental factors and microbial actions (Wierckx et al. 2018).

The process of biodegradation mainly involves four mechanisms—photodegradation, thermo-oxidative degradation, hydrolytic degradation, and microbial degradation (Fig. 11.5). Photodegradation is a natural mechanism, where UV light received from the sun provides the activation energy to initiate the polymer degradation process (Azwa et al. 2013). This is followed by thermo-oxidative degradation as oxygen atoms combine with the polymer, making the plastic brittle, making way for hydrolysis. The high-molecular weight polymer primarily undergoes oxidation or hydrolysis through enzymes to generate functional groups, improving the hydrophobicity. The fractured pieces of the polymer chains possess a sufficiently low molecular weight, further metabolized by active microbes. The microbes either

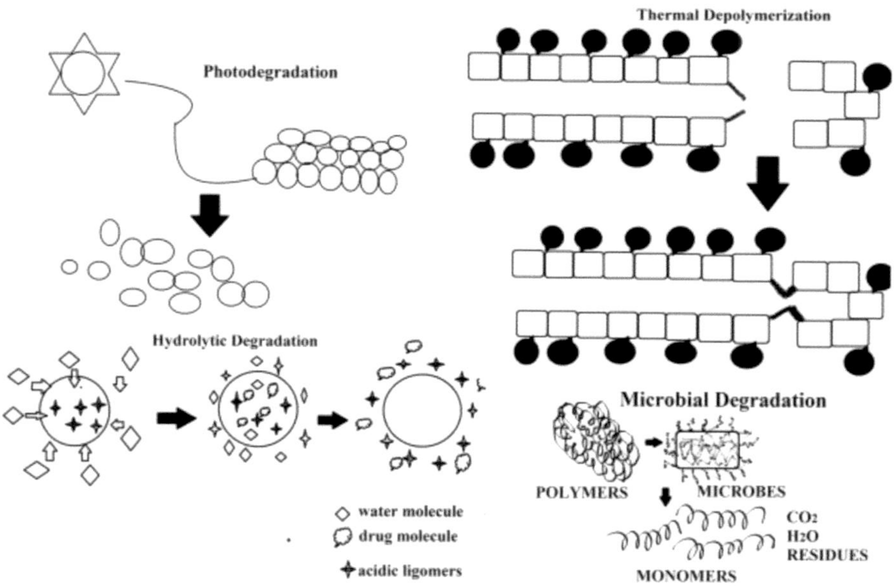

Figure 11.5. Polymer biodegradation mechanisms.

consume the carbon present in the polymer chains, or incorporate it into biomolecules with the evolution of carbon dioxide.

Additionally, certain physical parameters (morphology, surface area, crystallinity, etc.) co-affect this process by controlling the rate of degradation properties (Huang 1995). For specific polymers, to make them biodegradable, modifications are applied to the mechanical properties, crystalline structure, or molecular weight to reduce the resistance towards degradation (Elsawy et al. 2017).

Degradation technologies

Mechanical recycling

This seems a favorable option in case of recycling, essentially for lower volumes of a specific waste instead of mixed consumer wastes. This process doesn't involve any alterations to the polymer and only thermoplastic polymers can be utilized, as they could be re-melted or reprocessed to obtain end products (Rudnik 2019). To specify the case, it entails recycling of large 'point source' generated waste, such as PLA cups. For this process, physical method involves cutting, shredding, or washing of the plastic wastes to form granulates or pellets of desired size and quality (for manufacturing), further melted by method of extrusion to generate new products.

An excellent product, "biocomposites" can be generated by blending the reprocessed material with virgin material. During biocomposite preparation, a précise dispersion is targeted, utilizing the developed synergisms between blended materials to increase the interfacial area for a better polymer-filler interaction (Deepalekshmi et al. 2013). However, this method can deteriorate the product quality owing to heterogeneity of the solid waste and chain-scission reactions (water and traces acidic impurities), producing an inferior recycled resin. For this, intensive drying of the polymer is recommended, through chain extender compounds or method of vacuum degassing (Meng et al. 2012).

Four main methodologies can be followed to derive the desired biocomposites (Fig. 11.6). Firstly, (i) solution method, where the polymers are dissolved in adequate solvents with blender additives accompanied by evaporation or precipitation of the solvent. Secondly, (ii) melt-mixing, involves a direct mixing and melting of polymers with the additives. Third is, (iii) *in situ* polymerization, where the fillers or additives are dispersed in a monomer solution and polymerized in the presence of blended particles; and lastly, (iv) template synthesis, includes synthesis of blended polymers from a precursor solution with polymers as templates (Fawaz and Mittal 2015).

Organic recycling (composting or anaerobic digestion): composting

In view of the environment, composting is a preferred method for disposing biodegradable polymers, as most of them lend themselves to being accepted well to composting systems. Compostable plastic is defined as a plastic that undergoes biological degradation during composting to generate consistent end products in form of carbon dioxide, water, biomass, and inorganic compounds, with no toxic residues a plastic that undergoes (ASTM D6400-04, 2004). While a compostable polymer is always biodegradable, a biodegradable polymer cannot always be composted (Kale et al. 2007). Thus, biopolymer products that are certified compostable, can

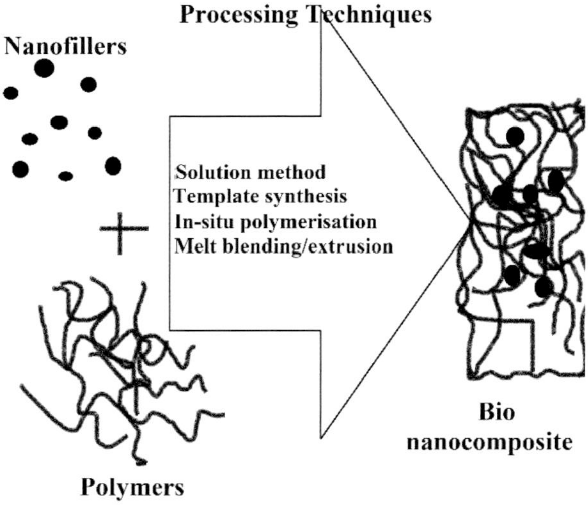

Figure 11.6. Methodologies to derive biocomposites.

only be successfully directed to the composting waste stream. Generally, bio-based and petroleum-based bioplastics (e.g., PLA, PHA, PBS, starch-based, and PCL) are susceptible to biodegradation through composting, at optimized environmental conditions (such as pH, moisture content, and temperature). These biopolymers are proficiently degraded via composting, in the presence of microbes to reduce the plastic burden on the environment. Perhaps, environmentally degradable/compostable polymers could profoundly contribute to the existing landfill capacity by accelerating the polymer breakdown process, releasing additional materials for biodegradation.

Compost is often used as a microbial enrichment to degrade different biopolymers (Hayes et al. 2012). For most operations, sludge and solid waste combined together provides effectual operation. In addition, sewage and activated sewage sludge can be excellent enhancers for biodegradation of polymers, due to preponderant microbes and higher nitrogen and phosphorus content. Primarily, hydrolysis in combination with aerobic and anaerobic microbial actions is accountable for a supreme mechanism of polymer degradation (Mohan 2011). Aerobes and anaerobes, such as bacteria, fungi, or actinomycetes, consume the chain structure of a compostable polymer through enzymes. The enzymes "digest" these polymers, producing end-products, such as water, CO_2, and biomass (Fig. 11.7). The efficacy of process or quality of product is demonstrated satisfactorily, when less than 10% of polymer's original dry weight is left after sieving. That is, about 90% of the test compost passes well through a 2.0 mm sieve (Korner et al. 2005).

Aerobic composting

In aerobic composting, oxidation of the test material occurs in the presence of oxygen, releasing CO_2 and H_2O (gas) (Awasthi et al. 2014). Sufficient aeration is required to optimize the prevalent moisture conditions, ensued by forced or

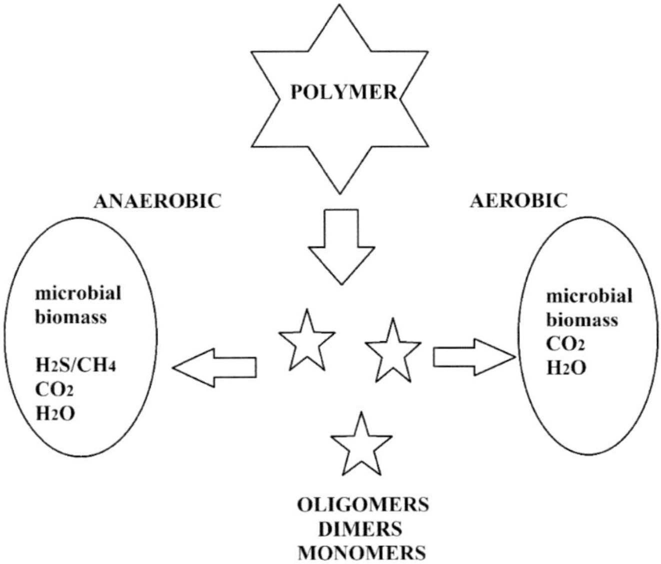

Figure 11.7. Organic recycling (composting or anaerobic digestion): composting.

natural aeration within the compost mix (Onwosi et al. 2017, Tatàno et al. 2015). Moreover, continuous monitoring of the composting parameters (pH, temperature, C/N moisture content, volatile solids) is needed during the course of composting (Getahun et al. 2012, Rastogi et al. 2019). For bioplastics, biodegradation rate is tested using different standardized equipment and protocols, such as testing aerobic biodegradability of bioplastic in soil through ASTM D 5988-12, ASTM WK 29802, ISO 17566, etc.

Anaerobic composting

Anaerobic digestion, generally applied to the organic fraction of municipal solid waste both alone and mixed with sludge from wastewater treatment plants, involves a complex ecosystem of anaerobic bacteria and methanogenic archaea (Ren et al. 2018). Microbes convert various types of biomass and organic waste into biogas (60–70% methane, 30–40% carbon dioxide, traces of hydrogen and hydrogen sulphide), leaving a nutrient-rich digestate for land application (Sheets et al. 2015). Anaerobic digestion may be carried out either in a single-phase or two-phase system. In the two-phase system, hydrolysis and acidogenesis react in the first reactor, and the utilization of those acids during methanogenesis takes place in the second reactor (Kondusamy and Kalamdhad 2014). However, it is reported that majorly single-phase systems are used in anaerobic digestion plants for organic waste. Also, the simulations at laboratory scale for assessing degradation of bioplastics generally use a single-phase system. The main parameters monitored are temperature, pH, ammonia, volatile fatty acids (Yirong et al. 2017), and production and composition of biogas (Novais et al. 2018). Anaerobic digestion, using thermophilic microbes to produce methane and compost, is also gaining support as an alternative to landfills. Methane production may be faster, more efficient, and more predictable in this system, and a useful end-product. Anaerobic microbes in the presence of water in the landfill will consume natural products and produce methane, CO_2, and humus.

As discussed below, temperature seems to be the discriminatory variable in the biodegradation of bioplastics both under aerobic and anaerobic conditions. Regarding the environmental conditions influencing the biodegradation process, the relevance of the temperature in both aerobic composting and anaerobic conditions has already been discussed. In particular, the initial thermophilic phase seems to play a fundamental role in making the process start by hydrolyzing the complex molecules into more readily biodegradable oligomers, assimilable by mesophilic microorganisms (Emadian et al. 2017). Research carried out under anaerobic conditions for biodegradation of bioplastics is less prioritized than that in aerobic conditions, and only at laboratory scale.

Enzymatic depolymerization (enzyme catalyzed and non-enzymatic hydrolysis)

Depending on the type of chemical bonds present in the polymer, plastics such as PET, PU, and PE can be modified or even completely degraded by enzymes (Wei and Zimmermann 2017a). Their enzymatic degradability greatly depends on the type of molecular bonds present in the polymer. Plastics containing hydrolyzable bonds in their backbones, e.g., ester or urethane bonds, are depolymerized by polyester hydrolases, lipases, and proteases (Fig. 11.8). Aromatic moieties in the backbones of, e.g., PET and PU, result in a higher resistance to biodegradation compared to their analogs containing aliphatic building blocks (Wei and Zimmermann 2017 b). Synthetic polymers, such as PE containing only carbon-carbon bonds in their backbones, are recalcitrant to biological attack (Wei and Zimmermann 2017a), and, as explained before, their degradation in the environment has mainly been observed as the result of a combination of abiotic and biotic effects (Lucas et al. 2008). Among the biotic factors, several oxidoreductases have been shown to be involved in the biodegradation of PE (Restrepo-Flórez et al. 2014).

Enzymatic depolymerization is a rather new recycling approach that has several advantages, including low energy consumption, mild reaction conditions, and the possibility for stereo-specific biopolymer degradation, and enzymatic repolymerization of the resultant monomers (Ignatyev et al. 2014). Depolymerization of PLA and other polymers and reutilization of monomers for new plastic synthesis are attractive recycling options. Physical, thermal, and chemical depolymerization have been considered and sometimes implemented by industry.

Enzymes which can be either intracellular or extracellular, are responsible for enzymatic degradation of bioplastics. Depolymerases which can be obtained from bioplastic-degrading microorganisms were investigated as enzymes, and play a significant role in bioplastics biodegradation (Chua et al. 2013). Many studies have been conducted on depolymerase purification from bioplastic-degrading microorganisms. Intercellular depolymerase from Rhodospirillum rubrum were investigated as PHB-degrading enzymes. The depolymerase enzyme responsible for PCL degradation was isolated from *Streptomyces thermoviolaceus* subsp. Thermoviolaceus 76T-2 (Chua et al. 2013). Other enzymes, such as lipase from Alcaligenes faecalis, estrase

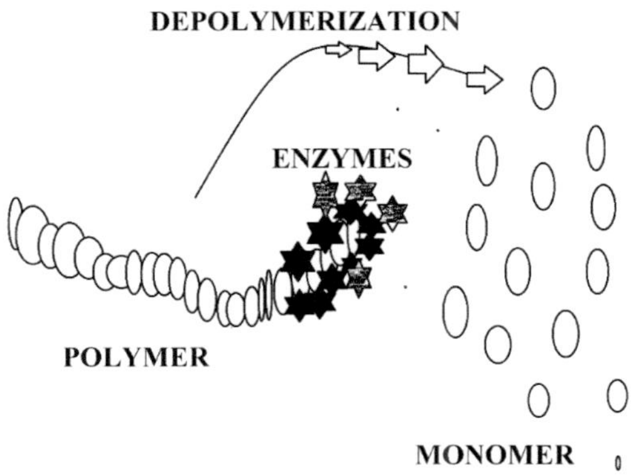

Figure 11.8. Enzymatic depolymerization mechanism in a polymer.

from *Comomonas acidivorans*, and serine from *Pestalotiopsis* microspora were also produced, which were involved in bioplastic biodegradation (Trivedi et al. 2016).

Life cycle assessments

The environmental impact associated with a product ('biobased plastics') can be measured by a tool "LCA", covering all the stages from extraction to waste disposal (Hottle et al. 2013). However, using a renewable resource does not guarantee sustainability of the material, and is suggestive of a plastic being environment friendly over its entire life cycle. To study the recyclability potential of polymers such as bioplastics and its derivatives, assessing the durability or the service life (thermal and hydrothermal ageing) of that polymer to emulate the mechanical recycling by multiple extrusions is suggested (Badia et al. 2017). All the bioplastics affect the environment differently during the course of their life cycles, making the study of cradle-to-grave assessment essential. Therefore, LCA includes all life cycle phases, such as production, use, methodology, and end-of-life scenario (Fig. 11.9).

Certain explicit questions are—When to use (biodegradable) bio-based plastics, depending on gauged CO_2 reduction and GHG emissions? - Recyclability? – Role of bioplastics in minimizing plastic soup risks? – The extent of influence on end-of-life disposal options, and how? – Policy implementation by consumers for proper disposal?

Sometimes in LCA studies, we elaborately assess the amount of biomass absorbed CO_2 used for the synthesis of a bio-based plastic compared to when it is incinerated or biodegraded. This is called the cradle-to-grave analysis where, the CO_2 uptake and CO_2 emission are accounted for in the agricultural phase during the end-of-life phase (Pawelzik et al. 2013). Essentially, on a product level, the bio-carbon uptake is taken into account, where the product serves as a carbon sink (temporarily). In the order of priority, the method of mechanical recycling majorly influences the GHG balance, creating a lower demand for the raw materials (Zhuo and Levendis 2014). The process of Incineration or digestion being environmentally safe contributes well to energy production, with energy recovery. Another process, 'composting' offers a neutral approach to biodegradable bio-based plastics for CO_2

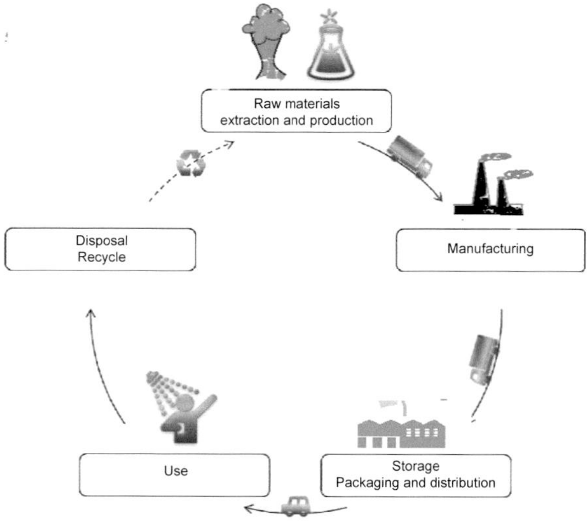

Figure 11.9. Life cycle assessment of a bioplastic from cradle to grave.

emissions with no compost generation. However, composting for bio-based plastics can be practiced only when it has added value and has co-benefits, contributing indirectly to reduction in greenhouse gas emissions. These include increased amount of the food waste collected to be composted with reduced fossil plastics at the end.

For government, industry, and NGOs, the sustainability criteria as prerequisite to support bioplastics includes:

1) Meeting the sustainability criteria, by obtaining certification for a set of sustainability and quality criteria.
2) Criteria check for sustainable production of biomass by Energy Agreements and the Green Deal Green Certificates
    a) A minimum CO_2-eq reduction percentage bio-based content
    b) A ban on direct land-use change
    c) Follow the mandatory rules for sustainable agricultural practices

Conclusion

Due to good quality, commercial and environmental suitability of the bio-based plastics derived from renewable resources, they have great acceptability in our day-to-day life. They possess good recycling capabilities, as well as triggered biodegradability, but its natural degradation is still a great threat. To amplify the bioplastic degradation process, incorporation of different microorganisms is found to play an important role during various stages. Microbial degradation of plastics favored by environmental conditions is considered as the cheapest, efficient, and eco-friendly acceptable method. A better insight towards the biodegradation of bioplastics can be obtained by studying the synergism between the involved microbiota. A high-molecular weight and hydrophobic surface makes the microbial degradation of these materials a difficult process. The microbe-aided (new explored strains inoculation) degradation, screened from soil, marine environment, or plastic waste dump site could lead to superior performance and form stable biofilms. These microbes release specific extracellular enzymes that amplify the plastic degradation rate by aerobic and anaerobic mechanisms. Thus, rationale for microbiota-mediated recycling approach appears as a positive and achievable agenda. However, a closer research to detect and characterize the efficient plastic degradation at molecular level is required.

Regarding sustainability potential of the bioplastics, a check on the production costs and inclusion of externality costs, including recycling, environmental degradation, and health-related costs is required. If adequately enacted (considering bioplastics), this could significantly increase the transition speed towards a better renewable circular bioeconomy.

Acknowledgment

I would like to express my special thanks to the Editor (Renato Montagnolli) and publisher (Taylor & Francis/CRC Press), who gave me the golden opportunity to do this wonderful chapter for the book entitled—Biodegradation, Pollutants, and Bioremediation.

References

ACC, APR. 2016. The 2015 US national postconsumer plastics bottle recycling rate report American Chemical Council & Assoc. Plastic Recyclers. https://plastics.americanchemistry.com/2015-United-States-National-Postconsumer-Plastic-Bottle-Recycling-Report.pdf.

Accinelli, C., M.L. Saccà, M. Mencarelli and A. Vicari. 2012. Deterioration of bioplastic carrier bags in the environment and assessment of a new recycling alternative. Chemosphere 89(2): 136–143.

Adhikari, D., M. Mukai, K. Kubota, T. Kai, N. Kaneko, K.S. Araki and M. Kubo. 2016. Degradation of bioplastics in soil and their degradation effects on environmental microorganisms. J Agric Chem Environ 5(01): 23.

Auta, H.S., C.U. Emenike and S.H. Fauziah. 2017. Distribution and importance of microplastics in the marine environment: a review of the sources, fate, effects, and potential solutions. Environ Int 102: 165–176.

Awasthi, M.K., A.K. Pandey, J. Khan, P.S. Bundela, J.W. Wong and A. Selvam. 2014. Evaluation of thermophilic fungal consortium for organic municipal solid waste composting. Bioresour Technol 168: 214–221.

Azwa, Z.N., B.F. Yousif, A.C. Manalo and W. Karunasena. 2013. A review on the degradability of polymeric composites based on natural fibres. Mater Des 47: 424–442.

Badia, J.D., O. Gil-Castell and A. Ribes-Greus. 2017. Long-term properties and end-of-life of polymers from renewable resources. Polym Degrad 137: 35–57.

Balaji, S., K. Gopi and B. Muthuvelan. 2013. A review on production of poly β hydroxybutyrates from cyanobacteria for the production of bio plastics. Algal Res 2(3): 278–285.

Bandopadhyay, S., L. Martin-Closas, A.M. Pelacho and J.M. DeBruyn. 2018. Biodegradable plastic mulch films: Impacts on soil microbial communities and ecosystem functions. Front Microbiol 9: 819.

Brodhagen, M., M. Peyron, C. Miles and D.A. Inglis. 2015. Biodegradable plastic agricultural mulches and key features of microbial degradation. Appl Microbiol Biot 99(3): 1039–1056.

Caulfield, M.J., G.G. Qiao and D.H. Solomon. 2002. Some aspects of the properties and degradation of polyacrylamides. Chem Rev 102(9): 3067–3084.

Chiellini, E. and R. Solaro. (eds.). 2012. Biodegradable Polymers and Plastics. Springer Science & Business Media.

Chua, C.M., W. Leggat, A. Moya and A.H. Baird. 2013. Near-future reductions in pH will have no consistent ecological effects on the early life-history stages of reef corals. Mar Ecol Prog Ser 486: 143–151.

da Costa, J.P., P.S. Santos, A.C. Duarte and T. Rocha-Santos. 2016. (Nano) plastics in the environment–sources, fates and effects. Sci Total Environ 566: 15–26.

Deepalekshmi, P., P.M. Visakh, A.P. Mathew, A.K. Chandra and S. Thomas. 2013. Advances in elastomers: their composites and nanocomposites: state of art, new challenges and opportunities. pp. 1–9. *In*: Advances in Elastomers II. Springer, Berlin, Heidelberg.

Elsawy, M.A., K.H. Kim, J.W. Park and A. Deep. 2017. Hydrolytic degradation of polylactic acid (PLA) and its composites. Renew Sust Energ Rev 79: 1346–1352.

Emadian, S.M., T.T. Onay and B. Demirel. 2017. Biodegradation of bioplastics in natural environments. Waste Manage 59: 526–536.

Fawaz, J. and V. Mittal. 2015. Synthesis of polymer nanocomposites: review of various techniques. Synthesis Techniques for Polymer Nanocomposites 992–1057.

Getahun, T., A. Nigusie, T. Entele, T. Van Gerven and B. Van der Bruggen. 2012. Effect of turning frequencies on composting biodegradable municipal solid waste quality. Resour Conserv Recycl 65: 79–84.

Geyer, R., J.R. Jambeck and K.L. Law. 2017. Production, use, and fate of all plastics ever made. Sci Adv 3(7): e1700782.

GVM. 2016. Aufkommen und Verwertung von PET-Getränkeflaschen in Deutschland 2015. Gesellschaft für Verpackungsmarktforschung mbH, Mainz. http://www.kunststoffverpackungen.de/show.php?ID=5961&PHPSESSID=apceu6k6r1irm4q7qff60ofp50.

Hayes, D.G., S. Dharmalingam, L.C. Wadsworth, K.K. Leonas, C. Miles and D.A. Inglis. 2012. Biodegradable agricultural mulches derived from biopolymers. pp. 201–223. *In*: Degradable Polymers and Materials: Principles and Practice (2nd Edition). American Chemical Society.

Ho, K.L.G., A.L. Pometto and P.N. Hinz. 1999. Effects of temperature and relative humidity on polylactic acid plastic degradation. J Environ Polym Degrad 7(2): 83–92.

Ho, K.S., Y.K. Han, Y.T. Tuan, Y.J. Huang, Y.Z. Wang, T.H. Ho and Lin, S.C. 2009. Formation and degradation mechanism of a novel nanofibrous polyaniline. Synth Met 159(12): 1202–1209.

Hottle, T.A., M.M. Bilec and A.E. Landis. 2013. Sustainability assessments of bio-based polymers. Polym Degrad 98(9): 1898–1907.

Huang, S.J. 1995. Polymer waste management–biodegradation, incineration, and recycling. J Macromol Sci A 32(4): 593–597.

Idumah, C.I., A. Hassan and D.E. Ihuoma. 2019. Recently emerging trends in polymer nanocomposites packaging materials. Polym-Plast Technol 58(10): 1054–1109.

Ignatyev, I. A., W. Thielemans and B. Vander Beke. 2014. Recycling of polymers: a review. Chem Sus Chem 7(6): 1579–1593.

Jain, R. and A. Tiwari. 2015. Biosynthesis of planet friendly bioplastics using renewable carbon source. J Environ Health Sci Eng 13(1): 11.

Jambeck, J.R., R. Geyer, C. Wilcox, T.R. Siegler, M. Perryman, A. Andrady, Ramani, Narayan and K.L. Law. 2015. Plastic waste inputs from land into the ocean. Sci 347(6223): 768–771.

Kale, G., T. Kijchavengkul, R. Auras, M. Rubino, S.E. Selke and S.P. Singh. 2007. Compostability of bioplastic packaging materials: an overview. Macromol Biosci 7(3): 255–277.

Kerr, R.S. 1994. Handbook of Bioremediation. Lewis Publishers, NY.

Kondusamy, D. and A.S. Kalamdhad. 2014. Pre-treatment and anaerobic digestion of food waste for high rate methane production—A review. J Environ Chem Eng 2(3): 1821–1830.

Korner, I., K. Redemann and R. Stegmann. 2005. Behavior of biodegradable plastics incomposting facilities. Waste Manage 25: 409–415.

Kumaravel, S., R. Hema and R. Lakshmi. 2010. Production of polyhydroxybutyrate (bioplastic) and its biodegradation by *Pseudomonas lemoignei* and *Aspergillus niger*. E-j of Chem, 7.

Laoutid, F., L. Bonnaud, M. Alexandre, J.M. Lopez-Cuesta and P. Dubois. 2009. New prospects in flame retardant polymer materials: from fundamentals to nanocomposites. Mater Sci Eng R Rep 63(3): 100–125.

LI, W.C., H.F. Tse and L. FOK. 2016. Plastic waste in the marine environment: A review of sources, occurrence and effects. Sci Total Environ 566: 333–349.

Lucas, N., C. Bienaime, C. Belloy, M. Queneudec, F. Silvestre and J.E. Nava-Saucedo. 2008. Polymer biodegradation: Mechanisms and estimation techniques—A review. Chemosphere 73(4): 429–442.

Luyt, A.S. 2017. Editorial corner—a personal view are biodegradable polymers the solution to the world's environmental problems? Express Polym Lett 11(10): 764–764.

Massardier-Nageotte, V., C. Pestre, T. Cruard-Pradet and R. Bayard. 2006. Aerobic and anaerobic biodegradability of polymer films and physico-chemical characterization. Polym Degrad 91(3): 620–627.

Masutani, K. and Y. Kimura. 2014. PLA synthesis. From the monomer to the polymer.

Meng, Q., M.C. Heuzey and P.J. Carreau. 2012. Control of thermal degradation of polylactide/clay nanocomposites during melt processing by chain extension reaction. Polym Degrad 97(10): 2010–2020.

Mihai, M. and N.A. Alemdar Legros. 2014. Formulation-properties versatility of wood fiber biocomposites based on polylactide and polylactide/thermoplastic starch blends. Polym Eng Sci 54: 1325–1340.

Mohan, K. 2011. Microbial deterioration and degradation of polymeric materials. J Biochem Technol 2(4): 210–215.

Nair, L.S. and C.T. Laurencin. 2007. Biodegradable polymers as biomaterials. Prog Polym Sci 32(8-9): 762–798.

Novais, R.M., T. Gameiro, J. Carvalheiras, M.P. Seabra, L.A. Tarelho, J.A. Labrincha and I. Capela. 2018. High pH buffer capacity biomass fly ash-based geopolymer spheres to boost methane yield in anaerobic digestion. J Clean Prod 178: 258–267.

Okan, M., H.M. Aydin and M. Barsbay. 2019. Current approaches to waste polymer utilization and minimization: A review. J Chem Technol Biot 94(1): 8–21.

Onwosi, C.O., V.C. Igbokwe, J.N. Odimba, I.E. Eke, M.O. Nwankwoala, I.N. Iroh and L.I. Ezeogu. 2017. Composting technology in waste stabilization: on the methods, challenges and future prospects. J Environ Manage 190: 140–157.

Pathak, V. M. 2017. Review on the current status of polymer degradation: a microbial approach. Bioresour Bioprocess 4(1): 15.

Pawelzik, P., M. Carus, J. Hotchkiss, R. Narayan, S. Selke, M. Wellisch and M.K. Patel. 2013. Critical aspects in the life cycle assessment (LCA) of bio-based materials—Reviewing methodologies and deriving recommendations. Resour Conserv Recycl 73: 211–228.

Phukon, P., J.P. Saikia and B.K. Konwar. 2012. Bio-plastic (P-3HB-co-3HV) from *Bacillus circulans* (MTCC 8167) and its biodegradation. Colloids Surf. B Biointerfaces 92: 30–34.

Radhika, D. and A.G. Murugesan. 2012. Bioproduction, statistical optimization and characterization of microbial plastic (poly 3-hydroxy butyrate) employing various hydrolysates of water hyacinth (Eichhornia crassipes) as sole carbon source. Bioresour Technol 121: 83–92.

Rastogi, M., M. Nandal and L. Nain. 2019. Additive effect of cow dung slurry and cellulolytic bacterial inoculation on humic fractions during composting of municipal solid waste. Int J Recycl Org Waste Agric 1–8.

Ren, Y., M. Yu, C. Wu, Q. Wang, M. Gao, Q. Huang and Y. Liu. 2018. A comprehensive review on food waste anaerobic digestion: Research updates and tendencies. Bioresour Technol 247: 1069–1076.

Restrepo-Flórez, J.M., A. Bassi and M.R. Thompson. 2014. Microbial degradation and deterioration of polyethylene—A review. Int Biodeterior 88: 83–90.

Rudnik, E. 2019. Compostable Polymer Materials. Newnes.

Sen, S.K. and S. Raut. 2015. Microbial degradation of low density polyethylene (LDPE): A review. J Environ 3(1): 462–473.

Sheets, J.P., X. Ge and Y. Li. 2015. Effect of limited air exposure and comparative performance between thermophilic and mesophilic solid-state anaerobic digestion of switchgrass. Bioresour Technol 180: 296–303.

Shen, H., J. Guo, H. Wang, N. Zhao and J. Xu. 2015. Bioinspired modification of h-BN for high thermal conductive composite films with aligned structure. ACS Appl Mater 7(10): 5701–5708.

Standard, A.S.T.M. 2004. D6400-04. Standard Specification for Compostable Plastics.

Strategy, P. 2018. A European strategy for plastics in a circular economy. Communication from the Commission to the European Parliament, the Council, the European Economic and Social Committee and the Committee of the Regions. Brussels.

Swati Pathak, C.L.R. Sneha and Blessy Baby Mathew. 2014. Bioplastics: Its timeline based scenario & challenges. J of Polym & Biopolym Phy Chem 2(4): 84–90. doi: 10.12691/jpbpc-2-4-5.

Tabasi, R.Y. and A. Ajji. 2015. Selective degradation of biodegradable blends in simulated laboratory composting. Polym Degrad Stab 120: 435–442.

Tatàno, F., G. Pagliaro, P.D. Giovanni, E. Floriani and F. Mangani. 2015. Biowaste home composting: Experimental process monitoring and quality control. Waste Manage 38: 72–85.

Thakur, S., J. Chaudhary, B. Sharma, A. Verma, S. Tamulevicius and V.K. Thakur. 2018. Sustainability of bioplastics: Opportunities and challenges. Curr Opin Green Sustain Chem 13: 68–75.

Trivedi, P., A. Hasan, S. Akhtar, M.H. Siddiqui, U. Sayeed and M.K.A. Khan. 2016. Role of microbes in degradation of synthetic plastics and manufacture of bioplastics. J Chem Pharm Res 8: 211–216.

Underkofler, L.A., R.R. Barton and S.S. Rennert. 1958. Production of microbial enzymes and their applications. Appl Microbiol 6(3): 212.

Van den Oever, M., K. Molenveld, M. van der Zee and H. Bos. 2017. Bio-based and biodegradable plastics: facts and figures: focus on food packaging in the Netherlands (No. 1722), Wageningen Food & Biobased Research.

Wei, R. and W. Zimmermann. 2017a. Microbial enzymes for the recycling of recalcitrant petroleum-based plastics: how far are we? Microb Biotechnol 10(6): 1308–1322.

Wei, R. and W. Zimmermann. 2017b. Biocatalysis as a green route for recycling the recalcitrant plastic polyethylene terephthalate. Microb Biotechnol 10(6): 1302–1307.

Wierckx, N., T. Narancic, C. Eberlein, R. Wei, O. Drzyzga, A. Magnin, H. Ballerstedt, S.T. Kenny, E. Pollet, L. Avérous and K.E. O'Connor. 2018. Plastic biodegradation: Challenges and opportunities. Consequences of Microbial Interactions with Hydrocarbons, Oils, and Lipids: Bioremediat Biodegrad, pp. 1–29.

Yirong, C., W. Zhang, S. Heaven and C.J. Banks. 2017. Influence of ammonia in the anaerobic digestion of food waste. J Environ 5(5): 5131–5142.

Yu, J. and L.X.L. Chen. 2008. The greenhouse gas emissions and fossil energy requirement of bioplastics from cradle to gate of a biomass refinery. Environ Sci Technol 42: 6961–6.

Zhuo, C. and Y.A. Levendis. 2014. Upcycling waste plastics into carbon nanomaterials: A review. J Appl Polym Sci 131(4).

12

Hydrogels and Nanocomposite Hydrogels for Removal of Dyes and Heavy Metal Ions from Wastewaters

Mohammad Sirousazar,[1,*] *Ehsan Roufegari-Nejhad*[2] and *Elham Jalilnejad*[1]

Introduction

Environmental pollution is one of the critical issues faced by modern civilization because it causes harmful and long-lasting damage to the earth and ecological systems. Rapid industrialization and huge demands of the resources leads to discharge of large volumes of waste product into the water bodies (Hosseinzadeh and Tabatabai Asl 2019, Ahmad et al. 2020). Various hazardous chemicals, including both the organic and inorganic wastes, are discharged into the environment via several industrial processes, e.g., metal plating, fertilizer industry, mining operations, dying in textile industries, automobile fuel, battery manufacturers, paper and pulp industries, and ammunition industries (Kasgoz and Durmus 2008, Crini et al. 2019, Samaddar et al. 2019, Shalla et al. 2019).

Different types of toxic pollutants, such as heavy metals (HMs) and cationic and anionic dyes, exist in wastewater due to industrial and agricultural activities. HMs are generally considered to be the metal elements with high atomic weight whose density exceeds 5 g/cm^3. A large number of elements fall into this category,

[1] Faculty of Chemical Engineering, Urmia University of Technology, Band Highway, 57155-419, Urmia, Iran.

[2] Faculty of Chemical Engineering, Sahand University of Technology, New Sahand Town, 51335/1996, Tabriz, Iran.

Emails: e_roufegarinejhad98@sut.ac.ir; e.jalilnejad@uut.ac.ir

* Corresponding author: m.sirousazar@uut.ac.ir

but arsenic, cadmium, chromium, copper, nickel, zinc, lead, and mercury are those relevant in the environmental context (Barakat 2011). The pollution of HM ions from industries, including paper, plastics, textile, and printing, has already turn into a global problem that endangers human health and disturbs the ecological equilibrium (Qi et al. 2019). HMs can coexist with other ions and easily form complexes with the complexing agents, such as ethylene diamine tetra acetic acid, which exacerbated the toxicity and environmental risk of HMs (Wang et al. 2019).

Nowadays, the total annual production of dyes is above $7*10^5$ tons, with almost 10,000 types of dyes. The problem is that more than 15% of these dyes are lost in industrial effluents during manufacturing and processing operations, which are very dangerous (Anfar et al. 2019). Dyes pollution of different water sources is one of the main concerns owing to the toxic effect on human beings (Salleh et al. 2011, Zeng et al. 2017, Nakhjiri et al. 2019). Dyes may affect the human and animal bodies and cause harmful effects, such as memory loss, mental confusion, and rashes (Hassan et al. 2019, Kumar et al. 2019, Mallakpour and Tabesh 2019). Synthetic dyes represent a relatively large group of organic chemicals that have complex aromatic molecular structures that are stable and resistant to biodegradation. With the growing use of a wide variety of dyes, the pollution of dye-contaminated wastewaters is becoming a major environmental problem (Kasgoz and Durmus 2008, Li et al. 2008).

For the treatment of wastewaters and removal of dyes and HMs, many materials have been utilized, in which the polymer-based hydrogels and nanocomposite hydrogels (NCHs) are more effective and favorable. Hydrogels and NCHs show excellent results as multifunctional materials in adsorption of dyes and HMs from water solutions and wastewaters (Hassan et al. 2019, Kumar et al. 2019, Mallakpour and Tabesh 2019). Various techniques have been employed for the treatment of dye or HM contaminated water, such as electrocoagulation, electrofloatation, electrodeposition, chemical precipitation, ion exchange, and adsorption (Baghbadorani et al. 2019, Kodoth and Badalamoole 2019, Kong et al. 2019). In the following section, the basic concepts of the hydrogels and NCHs are briefly introduced.

Hydrogels and NCHs

Hydrogels

Polymeric hydrogels are the typical fascinating and versatile soft materials with wide potential applications in drug delivery and tissue engineering, cell modulating substances, as well as the adsorption of dyes and HMs (Dong et al. 2018, Bao et al. 2019, Huang et al. 2019, Roufegari-Nejhad et al. 2019, Shojaeiarani et al. 2019). The hydrogel materials have crosslinked three-dimensional network structures with high water-retention and sorption capacities. Upon appropriate modification, the functional groups in their network structures, such as carboxyl (-COOH), hydroxyl (-OH), amine ($-NH_2$), and thiol (-SH) can provide the necessary chelating sites for the adsorption of dyes and HM ions. They are heterogeneous mixtures of two phases. The dispersed phase is water and the solid phase is a three-dimensional solid network (Nascimento et al. 2018). The "network" indicates that crosslinks hinder the dissolution of the hydrophilic polymer chains into the aqueous phase. Hydrogels are smart materials that may respond to environmental stimuli (temperature, pH, ionic strength, electric field, the presence of enzyme, etc.) and swell or shrink accordingly (Asadi et al. 2018, Mahinroosta et al. 2018, Nascimento et al. 2018, Bao et al. 2019, Pakdel and Peighambardoust 2018, Roufegari-Nejhad et al. 2019, Samaddar et al. 2019, Shalla et al. 2019). The hydrophilic nature of the hydrogel component governs the water sorption, and the swelling response of the hydrogel in the swollen stage is attributed to the large free space between crosslinked network (Shojaeiarani et al. 2019). When the network of a hydrogel is held together by molecular entanglements or secondary forces, such as ionic, H-bonding, or hydrophobic forces, it is called "reversible" or "physical" hydrogel. The hydrogel is called "permanent" or "chemical" hydrogel when it is crosslinked together covalently (Asadi et al. 2018).

The hydrogels are synthesized by various methods, such as freeze-drying, porogenation, microemulsion formation, and phase separation. They are prepared according to their application in the form of bead, film, ring, and hollow fiber (Pakdel and Peighambardoust 2018, Crini et al. 2019). There are various ways to classify hydrogels. Figure 12.1 shows the most paramount parameters for the classification of hydrogels (Sirousazar et al. 2014, Chen et al. 2018, Mahinroosta et al. 2018). In terms of ionic charges present in a polymer network, hydrogels are classified into

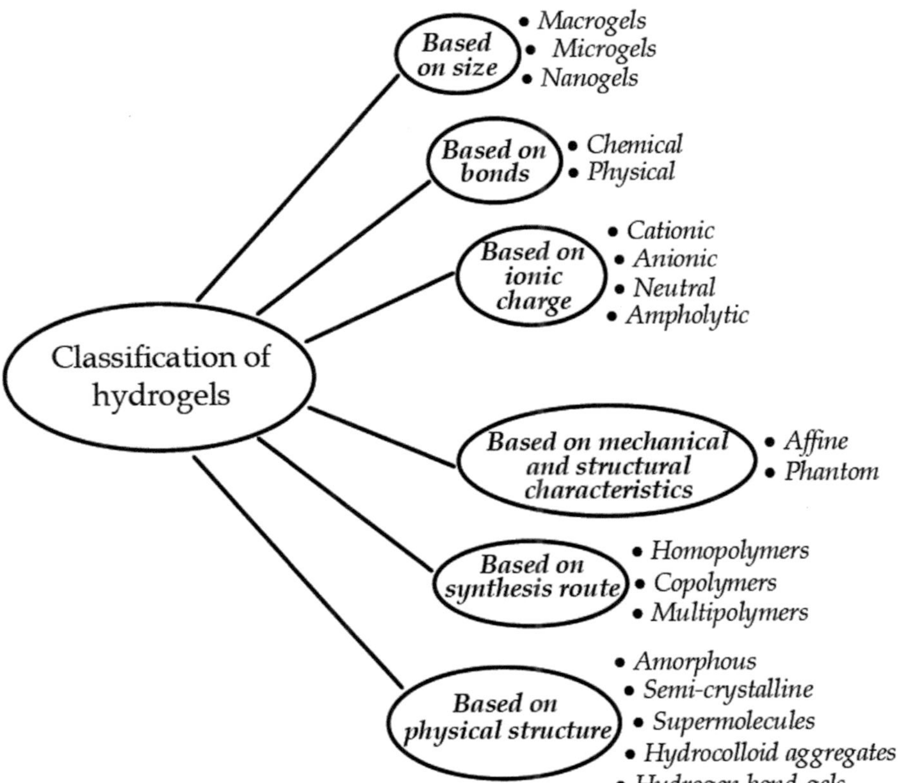

Figure 12.1. Classification of hydrogels based on different parameters.

anionic, cationic, neutral, and ampholytic hydrogels. As Fig. 12.2 illustrates, anionic hydrogels typically contain negative ions attached to the polymer network, whereas cationic hydrogels include fixed positive ions. However, neutral hydrogels must have equal numbers of positive and negative ions. Generally, it is assumed that a neutral hydrogel has a fairly uniform concentration distribution of fixed ions throughout the polymer matrix (Mahinroosta et al. 2018).

Hydrogels have attracted special attention as they have superior mechanical properties and tolerance of external vigorous stirring, which is important for the reuse of adsorbents (Shalla et al. 2019). Furthermore, they have attracted considerable attention for their potential applications in agriculture, biomedical, cosmetics, and food industry (Fig. 12.3) (De France et al. 2018, Bao et al. 2019, Roufegari-Nejhad et al. 2019, Shalla et al. 2019), but our focus will be application of hydrogel for wastewater treatment. Hydrogels with hydrophilic polymer chains are effective adsorbent of dyes and HMs because they can adsorb and trap ionic metals and dyes (Firdaus et al. 2019, Samaddar et al. 2019). Nowadays, numerous investigations of HM ions removal using different types of hydrogels, such as polyvinyl alcohol,

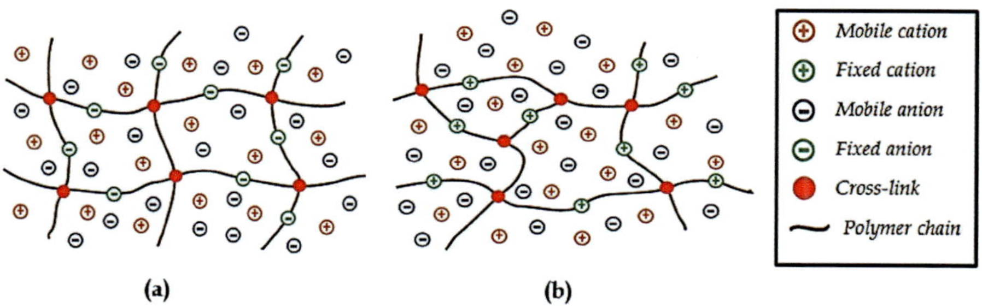

Figure 12.2. Schematic of the structure of (a) anionic and (b) cationic hydrogels.

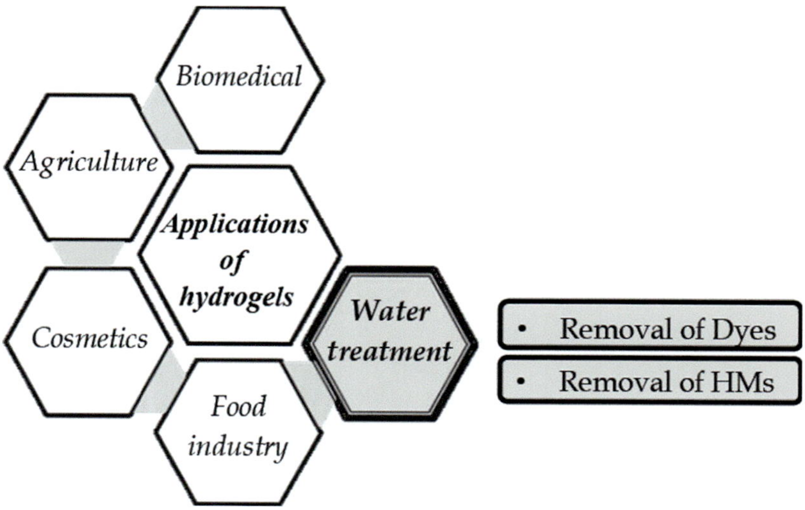

Figure 12.3. Applications of hydrogels.

cellulose, carboxymethyl cellulose, starch, and polyacrylamide have been conducted (Firdaus et al. 2019).

NCHs

The combination of nanotechnology with other fields of science has attracted increasing attention during the past decades. There have been numerous approaches to incorporate nanoscale methods with conventional methods toward manufacturing improved materials, such as polymer-based nanocomposites. In general, polymer nanocomposites refer to a kind of composite material, which has a multi-phase structure, and one of the phases possesses a particle size of less than 100 nm (Sirousazar et al. 2007, Bao et al. 2019). NCHs are one example of such a combination of nanotechnology and polymer hydrogels (Sirousazar 2013, Shaabani et al. 2016a, Rafieian et al. 2019).

As mentioned in the previous section, hydrogels have promising applications in several fields because of their tunable physical, chemical, and biological properties, high biocompatibility, and versatility in fabrication. In spite of these significant features, hydrogels may possess some shortcomings, e.g., poor mechanical strength, low strain, and low thermal stability, which restrict the applications of hydrogels (Sirousazar et al. 2012, Jahani-Javanmardi et al. 2016). As a result of these limitations, trials have been made by researchers for redesigning new hydrogels, named as NCHs, with improved chemical, physical, and mechanical properties.

The NCHs consist of immiscible stiff or soft components and complex nanoscale structures (Chen et al. 2018). A wide range of nanoparticles (NPs), including ceramic NPs (e.g., silica and titanium oxide), metal or metal-oxide NPs (e.g., gold, silver, and iron oxide), carbon-based nanomaterials (e.g., carbon nanotubes and graphene), polymeric NPs (e.g., micelles, dendrimers, and nanogels), and clays (e.g., montmorillonite, laponite, and kaolinite) were successfully introduced into NCH networks by *in situ* polymerization, *in situ* growth of the NPs, or physical mixing (Shaabani et al. 2016b, Sirousazar et al. 2016, Chen et al. 2018).

Comparing with neat polymeric hydrogels, the NPs play a significant role in the enhancement of the structural stability of NCHs due to the multiple interactions between NPs and polymers, such as hydrogen bonds, van der Waals interactions, and electrostatic forces (Chen et al. 2018). The NPs in the three-dimensional polymer network of NCHs not only serve as crosslinkers to reinforce hydrogels, but also endow the hydrogels with their characteristic functionalities.

Methods for removing pollutants from wastewaters

For many years, researchers have used different methods for removing dyes and HMs from different wastewaters. These methods can be classified as the electrochemical treatments, including the electrocoagulation, electrofloatation, and electrodeposition techniques, and the physicochemical processes, including the chemical precipitation, ion exchange, and adsorption (Fig. 12.4) (Azimi et al. 2017, Zeng et al. 2017, Zare et al. 2018).

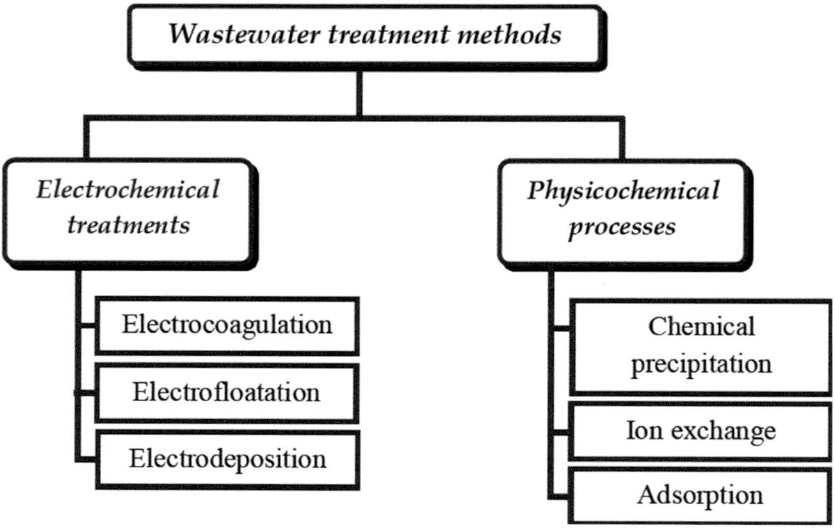

Figure 12.4. Tree diagram for wastewater treatment methods.

Electrochemical treatments

The three major technologies of electrochemical therapy are electrocoagulation, electroplating, and electrodeposition. Electrochemical treatments of wastewaters have not received great attention so far because of the need for large capital investments and expensive electricity supply. The electrochemical technologies are in a situation that they are not only comparable with other technologies in terms of costs, but they also are more efficient and more compact. In some cases, the electrochemical technologies may be an essential step that cannot be neglected in treating wastewaters containing refractory pollutants (Azimi et al. 2017).

Physicochemical processes

Chemical precipitation

Chemical precipitation is a simple, easily automated treatment method. This treatment method is widely used in removing dyes and HMs from wastewaters. Chemical precipitation needs a lot of chemicals to reduce pollutants to an acceptable limit for discharge, but it sometimes fails to reach this point. These chemicals will be a large source of further pollution (Azimi et al. 2017, Bolisetty et al. 2019). In precipitation, contaminants that are either dissolved or suspended in the solution are forced to settle out of solution as a solid precipitate, which can then be filtered, centrifuged, or otherwise separated from the liquid portion. In this process, pH adjustment of HMs and reaction with chemical reagents (for example, lime, hydroxides, and sulfides) result in the formation of insoluble particles, which are then removed by simple sedimentation. The major advantage of this process is its inexpensiveness and simplicity of operation. The major drawback, however, is that the HM concentration does not reach the range acceptable for discharge, mandating additional post-treatments (Bolisetty et al. 2019).

Ion exchange

Ion exchange treatment is based on a reversible interchange of ions between the solid and liquid phases. The whole procedure begins with ion-exchange reactions, then the HM ions will be physically adsorbed, and a complex is formed between the counterion and the functional group. At the end, hydration occurs at the surface of the solution or pores of the adsorbent. Different factors, such as pH, temperature, the initial concentration of the adsorbent and sorbate, and contact time affect the ion-exchange operation (Kansara et al. 2016, Azimi et al. 2017). Ion exchange is a well-established method commonly applied to drinking water treatment for hardness removal, but it is also increasingly being studied for the removal of HM ions (Bolisetty et al. 2019). Ion exchange resin comprises a mineral and carbon-based network structure with attached functional groups, while synthetic organic resins are commonly used for ion exchange. Resins that exchange positive ions are called cationic resins, and those that exchange negative ions are anionic resins. These resins are synthesized by the polymerization of carbon-based material into a porous three-dimensional structure. Cationic resins containing acidic functional groups, such as sulphonic groups, are exchanged for hydrogen or sodium, whereas anions are exchanged for hydroxyl ions, and they have basic functional groups, such as amines. This treatment method is highly selective for the detoxification of certain HMs ions, such as Cr^{6+}, Pb^{2+}, Zn^{2+}, Cu^{2+}, and Cd^{2+} (Kansara et al. 2016, Bolisetty et al. 2019).

Adsorption

Adsorption is another important physicochemical method for the removal of dyes and HMs from wastewaters. It is a removal process on the basis of mass transfer between the liquid phase and the solid phase, called adsorbent. Adsorption is recognized as a simple and easy operation process with high efficiency (Renge et al. 2012). It has become a well-known, effective, and economical method to remove HMs and dyes due to the high removal efficiency, the possibility of regenerating adsorbents, and the flexibility in design and operation. In adsorption, no secondary pollution is produced and adsorbents can be regenerated by a suitable desorption process. For the sorption of pollutants onto an adsorbent, three key stages are involved: (i) the penetration of the pollutant from the bulk solution to the adsorbent surface; (ii) adsorption of the pollutant on the adsorbent surface, and (iii) penetration in the adsorbent structure.

As the specific area is a primary factor in adsorption, nanoparticles with a high surface/volume ratio are ideal candidates for this process, and technologies producing new types of adsorbents with high chemical activity and specific surface area are rapidly expanding. Adsorbents primarily used in wastewater treatment are categorized as carbon-, metal-, and zeolite-based nanosorbents and also hydrogels and NCHs (Azimi et al. 2017, Crini et al. 2018, Bolisetty et al. 2019). Biosorbents which are prepared from natural materials, such as biomass and cellular products and also natural hydrogels, have shown high efficiency in the removal of dyes and HMs from wastewater through a low-cost operation (Crini et al. 2018). In the following section, some recent researches performed on the application of hydrogel and NCHs as novel adsorbents for removal of dyes and HMs from aqueous solutions and wastewaters are summarized.

Hydrogel and NCHs adsorbents in removal of dyes and HMs ions

The three-dimensional network, porous structure, and swelling ability of hydrogels and NCHs make them excellent candidates as functional adsorbents to be used in the adsorption process. In recent years, NCHs have been successfully utilized as effective adsorbents for the removal of dyes and HMs from wastewater. Yan et al. studied the adsorption of methylene blue (MB) cationic dye onto polyaniline (PANI) hydrogel fabricated using phytic acid as both dopant and crosslinking agent. They found that with the increase of pH, the adsorption capacity of PANI hydrogel for MB had a significant increasing trend, while it had low adsorption capacities for MB when the pH was lower than 3. It was attributed to the fact that the adsorbent had a more positive potential on the surface when the pH value was lower. The maximum adsorption of 71.2 mg/g was achieved at a pH value of 6.5. Based on the adsorption behavior and the chemical properties of the adsorbent and adsorbate, they proposed an adsorption mechanism for MB adsorption on the PANI hydrogel, as shown in Fig. 12.5. It seems that the electrostatic forces between the MB dye and the PANI emeraldine base lead to the MB adsorption on the PANI hydrogel (Yan et al. 2015).

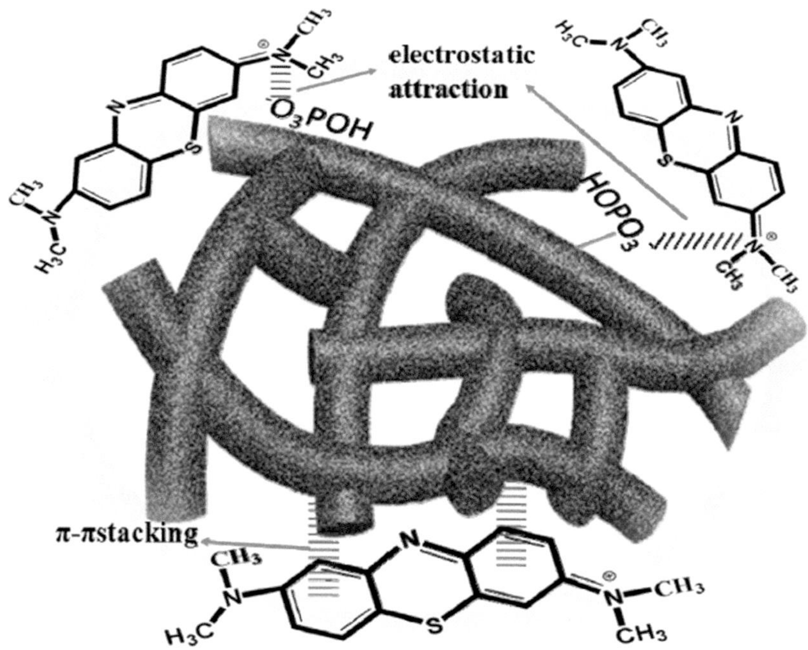

Figure 12.5. Adsorption mechanism for the removal of MB dye by PANI hydrogel (Yan et al. 2015).

In another work, Sharma et al. synthesized a starch/poly(alginic acid-cl-acrylamide) nano-hydrogel (ST/PL(AA-cl-AAm) NHG) by co-polymerization method, and utilized it as an effective adsorbent for the removal of coomassie brilliant (CB) blue R-250 dye from the aqueous solution. The effect of initial CB concentration was carried out by varying the dye concentration from 20 to 120 mg/L. It is clear from Fig. 12.6a that the percentage of dye uptake decreased progressively with increasing initial dye concentration up to 120 mg/L. The higher percentage of dye removal at low concentration could be related to the low ratio of the initial number of dye molecules on the surface site of the hydrogel. However, at relatively higher concentrations, the removal percentage was decreased due to the availability of lesser adsorption sites (Sharma et al. 2017). Figure 12.6b indicates that an increase in adsorbent loading increased the uptake percentage of CB, which was due to the availability of greater active sites, as well as the optimal surface area containing copious pores with available volume supplying more functional groups boosting the adsorption of CB molecules. However, after 0.1 g/50 mL of adsorbent loading, no significant percentage uptake was observed. Actually, adsorbent particles grouped together to form the cluster, so there was no significant increase in specific surface area of hydrogel (Sharma et al. 2017). Although the efficiency increased with adsorbent amount, but the amount of CB adsorbed per unit mass (q_e) diminished, as shown in Fig. 12.6b. It was ascribed to the unsaturation of active surface sites

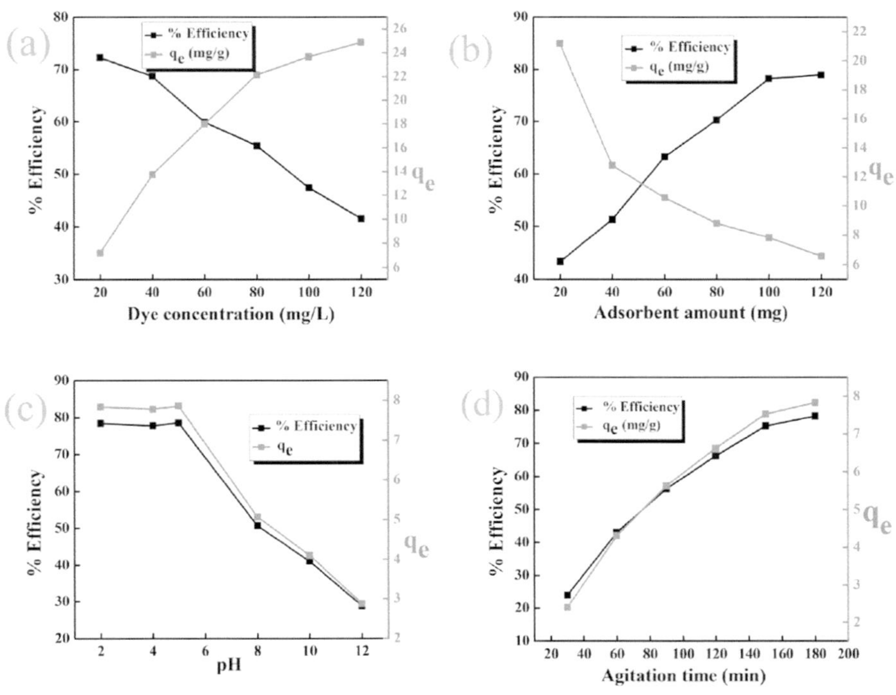

Figure 12.6. The effect of (a) initial dye concentration, (b) adsorbent amount, (c) pH, and (d) agitation time on the adsorption of CB dye onto ST/PL(AA-*cl*-AAm) NHG (Sharma et al. 2017).

on hydrogel adsorbent due to the increase in the ratio of adsorption sites to the dye molecules (Sharma et al. 2017). The effect of pH on the CB removal by the hydrogel adsorbent was illustrated in Fig. 12.6c. The maximum removal of anionic CB dye was observed during the pH range of 2.0–5.0, and then efficiency percentage declined sharply from 78.56 to 28.78 percent. At lower pH (2.0–5.0), the fraction of functional groups accommodated on the surface of hydrogel gets protonated and acquires a positive charge. Thus, the electrostatic interaction occurred between the hydrogel and anionic CB dye. However, at higher pH, the concentration of hydrogen ion became lower, and competition of H^+ ion with the anionic dye molecules becomes negligible (Sharma et al. 2017). The effect of the agitation time for the removal of dye by adsorbent was investigated in the range of 30–180 min at room temperature (Fig. 12.6d). The adsorption percentage was increased with increasing the contact time and reached a plateau after 150 minutes. At the starting, more vacant surface sites were available for the adsorption of CB and with the passage of time, the remaining vacant sites were difficult to be occupied due to repulsive forces between the solute molecules onto adsorbent and liquid phase (Sharma et al. 2017).

Zare et al. developed novel poly(aniline-co-m-phenylenediamine) (PAMpDA) and poly(aniline-co-3-aminobenzoic acid) (PA3ABA) copolymer nanoparticles synthesized through copolymerization approach for the removal of Cd(II), Co(II), and Pb(II) from aqueous solutions. They investigated the effect of pH, adsorption dosage, contact time, and initial concentration of HM ions on the adsorption process. The maximum adsorption percentages for Cd(II), Co(II), and Pb(II) were determined to be 85%, 90%, and 80% at pH 6, adsorption dosage of 50 mg, contact time of 70 min, and initial concentration of 50 ppm. A mechanism based on the chelating effect between NH_2/COO^- groups and Pb(II) ions was also suggested. They expressed that amine and carboxyl groups in the main chain structure of copolymer were complexed to Pb(II) ions (Zare et al. 2016).

Ghorbani and Eisazadeh fabricated the polyaniline/Rice husk ash (PANI/RHA) nanocomposites having the diameter of ca. 50–100 nm via cast method for the removal of chemical oxygen demand (COD), dyes, anions, Cd(II), and Cu(II) from cotton textile wastewater. The transmission electron microscopy (TEM) image of the prepared PANI/RHA nanocomposite adsorbent is shown in Fig. 12.7. They proposed an adsorption mechanism for the complexation between anions or HM ions and the nitrogen atoms of the –N=C– groups through sharing their lone pair of electrons (Fig. 12.8). They also employed the PANI/RHA nanocomposite adsorbents for Hg(II) and Zn(II) removal from aqueous solution, and studied the influence of various experimental parameters such as, pH, adsorbent dosage, contact time, and rotating speed. The removal percentages of 95% for Hg(II) and 85% for Zn(II) were reported at optimum conditions, including pH 3–9, adsorbent dosage of 10 g/L, equilibrium contact time of 20 min, and rotating speed of 400 (Ghorbani and Eisazadeh 2013).

More recently, Roufegari-Nejhad et al. prepared poly(vinyl alcohol)/montmorillonite (PVA/MMT) NCH adsorbents using a cyclic freezing-thawing process for adsorption of MB dye from aqueous solutions. Based on the dye adsorption experiments, it was found that the MB removal from aqueous solution could be promoted by increasing the MMT loading level and the temperature of the

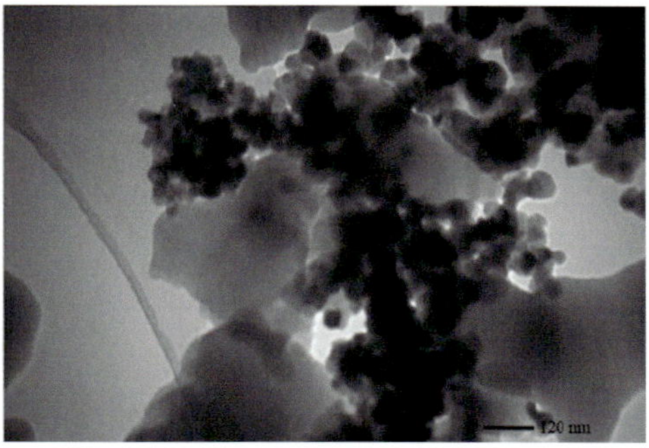

Figure 12.7. TEM image of PANI/RHA nanocomposite adsorbent (Ghorbani and Eisazadeh 2013).

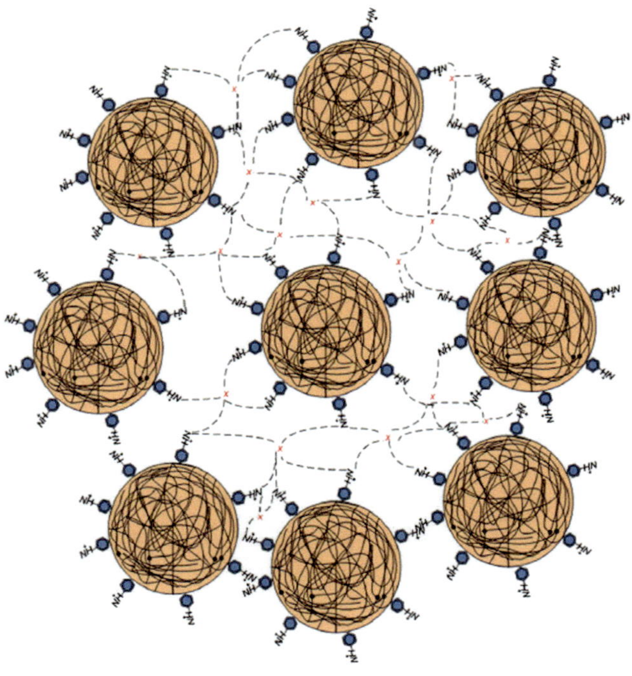

Figure 12.8. Possible complexations and ion exchanges between anions or HM ions and –NH– groups on the PANI/RHA nanocomposite chains (Ghorbani and Eisazadeh 2013).

adsorption process, as shown in Figs. 12.9a, b. According to the obtained results, the maximum adsorption was achieved in a weakly alkaline environment at a pH of 9 (Fig. 12.9c). The results showed that by increasing the initial concentration of MB in adsorption solution, the adsorption process was restricted and the removal percentage of MB decreased (Fig. 12.9d). Finally, it was concluded that the prepared PVA/MMT adsorbents were suitable adsorbents for adsorption of MB from aqueous solutions and wastewaters, and they can be considered useful adsorbents in practical applications (Roufegari-Nejhad et al. 2019).

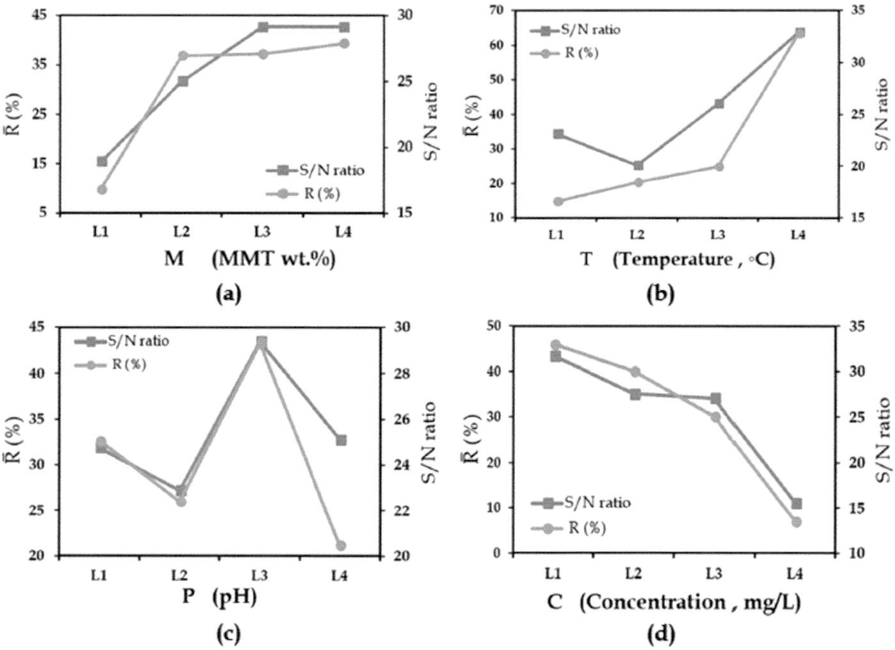

Figure 12.9. The effect of (a) MMT loading level in adsorbent, (b) temperature, (c) pH, and (d) initial MB concentration in solution on the percentage removal of MB using PVA/MMT NCH adsorbents (Roufegari-Nejhad et al. 2019).

Conclusions

The chemical contamination of water by dyes and HMs is a serious environmental problem owing to their potential human toxicity. Therefore, the development of technologies that can remove toxic pollutants from wastewaters is required. Adsorption on hydrogels and NCHs is a cheaper and popular method for the removal of pollutants from the wastewater. In this chapter, the recent advances in the field of hydrogel and NCHs for removal of dyes and HM ions from wastewater were reviewed, along with a brief summary of the basic topics, including the fundamental concepts of hydrogels, NCHs, and effective methods to treat wastewater. Hydrogels and NCHs with hydrophilic polymer chains are effective adsorbents for the removal of dyes and HMs from wastewater because they can easily adsorb and trap ionic metals and dyes. However, most common hydrogels and NCHs are on the basis of the synthetic polymers that have low stability and are non-environment friendly materials. Therefore, the preparation of novel hydrogel and NCHs on the basis of biocompatible materials with improved properties seems to be the upcoming challenge for the researchers in the field of production of hydrogel and NCHs adsorbents for industrial wastewater purification.

References

Ahmad, N., S. Sultana, M.Z. Khan and S. Sabir. 2020. Chitosan based nanocomposites as efficient adsorbents for water treatment. pp. 69–83. *In*: Oves, M., M. Ansari, M. Zain Khan, M. Shahadat and M.I. Ismail (eds.). Modern Age Waste Water Problems. Springer, Cham. Switzerland.

Anfar, Z., H. Ait Ahsaine, M. Zbair, A. Amedlous, A. Ait El Fakir, A. Jada and N.E. Alem. 2019. Recent trends on numerical investigations of response surface methodology for pollutants adsorption onto activated carbon materials: A review. Critical Reviews in Environ Sci Technol (in press).

Asadi, N., E. Alizadeh, R. Salehi, B. Khalandi, S. Davaran and A. Akbarzadeh. 2018. Nanocomposite hydrogels for cartilage tissue engineering: a review. Artif Cell Nanomed Biotechnol 46: 465–471.

Azimi, A., A. Azari, M. Rezakazemi and M. Ansarpour. 2017. Removal of heavy metals from industrial wastewaters: a review. Chem Bio Eng Reviews 4: 37–59.

Baghbadorani, N.B., T. Behzad, N. Etesami and P. Heidarian. 2019. Removal of Cu^{2+} ions by cellulose nanofibers-assisted starch-g-poly (acrylic acid) superadsorbent hydrogels. Composites Part B: Eng 176: 107084.

Bao, Z., C. Xian, Q. Yuan, G. Liu and J. Wu. 2019. Natural polymer-based hydrogels with enhanced mechanical performances: preparation, structure, and property. Adv Healthc Mater 8: 1900670.

Barakat, M.A. 2011. New trends in removing heavy metals from industrial wastewater. Arab J Chem 4: 361–377.

Bolisetty, S., M. Peydayesh and R. Mezzenga. 2019. Sustainable technologies for water purification from heavy metals: review and analysis. Chem Soc Rev 48: 463–487.

Chen, T., K. Hou, Q. Ren, G. Chen, P. Wei and M. Zhu. 2018. Nanoparticle–polymer synergies in nanocomposite hydrogels: from design to application. Macromol Rapid Commun 39: 1800337.

Crini, G., E. Lichtfouse, L.D. Wilson and N. Morin-Crini. 2018. Adsorption-oriented processes using conventional and non-conventional adsorbents for wastewater treatment. pp. 23–71. *In*: Crini, G. and E. Lichtfouse (eds.). Green Adsorbents for Pollutant Removal. Environmental Chemistry for a Sustainable World, vol 18. Springer, Cham. Switzerland.

Crini, G., G. Torri, E. Lichtfouse, G.Z. Kyzas, L.D. Wilson and N. Morin-Crini. 2019. Cross-linked chitosan-based hydrogels for dye removal. J Sustain Agr 36: 381–425.

De France, K.J., F. Xu and T. Hoare. 2018. Structured macroporous hydrogels: Progress, challenges, and opportunities. Adv Healthcare Mater 7: 1700927.

Dong, C., J. Lu, B. Qiu, B. Shen, M. Xing and J. Zhang. 2018. Developing stretchable and graphene-oxide-based hydrogel for the removal of organic pollutants and metal ions. Applied Catalysis B Environ 222: 146–156.

Firdaus, F., M.S.F. Idris and S.F.M. Yusoff. 2019. Adsorption of nickel ion in aqueous using rubber-based hydrogel. J Polym Environ 27: 1770–1780.

Ghorbani, M. and H. Eisazadeh. 2013. Removal of COD, color, anions and heavy metals from cotton textile wastewater by using polyaniline and polypyrrole nanocomposites coated on rice husk ash. Compos Part B: Eng 45: 1–7.

Hassan, H., A. Salama, A.K. El-ziaty and M. El-Sakhawy. 2019. New chitosan/silica/zinc oxide nanocomposite as adsorbent for dye removal. Int J Biol Macromol 131: 520–526.

Hosseinzadeh, H. and M.M. Tabatabai Asl. 2019. Enhanced removal of Cr (VI) from aqueous solutions using poly (Pyrrole)-g-Poly (Acrylic Acid-co-Acrylamide)/Fe_3O_4 magnetic hydrogel nanocomposite adsorbent. Chem Biochem Eng Q 33: 19–33.

Huang, S., L. Wu, T. Li, D. Xu, X. Lin and C. Wu. 2019. Facile preparation of biomass lignin-based hydroxyethyl cellulose super-absorbent hydrogel for dye pollutant removal. Int J Biol Macromol 137: 939–947.

Jahani-Javanmardi, A., M. Sirousazar, Y. Shaabani and F. Kheiri. 2016. Egg white/poly(vinyl alcohol)/MMT nanocomposite hydrogels for wound dressing. J Biomater Sci Polym Ed 27: 1262–1276.

Kansara, N., L. Bhati, M. Narang and R. Vaishnavi. 2016. Wastewater treatment by ion exchange method: a review of past and recent researches. Environ Sci Indian J 12: 143–150.

Kasgoz, H. and A. Durmus. 2008. Dye removal by a novel hydrogel-clay nanocomposite with enhanced swelling properties. Polym Adv Technol 19: 838–845.

Kodoth, A.K. and V. Badalamoole. 2019. Silver nanoparticle-embedded pectin-based hydrogel for adsorptive removal of dyes and metal ions. Polymer Bulletin (in press).

Kong, W., M. Chang, C. Zhang, X. Liu, B. He and J. Ren. 2019. Preparation of Xylan-g-/P (AA-co-AM)/GO nanocomposite hydrogel and its adsorption for heavy metal ions. Polymers 11: 621–635.

Kumar, A., G. Sharma, Thakur and D. Pathania. 2019. Sol–gel synthesis of polyacrylamide-stannic arsenate nanocomposite ion exchanger: binary separations and enhanced photo-catalytic activity. SN Applied Sci 1: 862.

Li, P., N.H. Kim, S.B. Heo and J.H. Lee. 2008. Novel PAAm/Laponite clay nanocomposite hydrogels with improved cationic dye adsorption behavior. Composites: Part B 39: 756–763.

Mahinroosta, M., Z.J. Farsangi, A. Allahverdi and Z. Shakoori. 2018. Hydrogels as intelligent materials: a brief review of synthesis, properties and applications. Mater Today Chem 8: 42–55.

Mallakpour, S. and F. Tabesh. 2019. Tragacanth gum based hydrogel nanocomposites for the adsorption of methylene blue: Comparison of linear and non-linear forms of different adsorption isotherm and kinetics models. Int J Biol Macromol 133: 754–766.

Nakhjiri, M.T., G.B. Marandi and M. Kurdtabar. 2019. Adsorption of methylene blue, brilliant green and rhodamine B from aqueous solution using collagen-gp (AA-co-NVP)/Fe_3O_4@ SiO_2 nanocomposite hydrogel. J Polym Environ 27: 581–599.

Nascimento, D.M., Y.L. Nunes, M.C. Figueirêdo, H.M. de Azeredo, F.A. Aouada, J.P. Feitosa, M.F. Rosa and A. Dufresne. 2018. Nanocellulose nanocomposite hydrogels: technological and environmental issues. Green Chem 20: 2428–2448.

Pakdel, P.M. and S.J. Peighambardoust. 2018. Review on recent progress in chitosan-based hydrogels for wastewater treatment application. Carbo Polym 201: 264–279.

Qi, X., R. Liu, M. Chen, Z. Li, T. Qin, Y. Qian, S. Zhao, M. Liu, Q. Zeng and J. Shen. 2019. Removal of copper ions from water using polysaccharide-constructed hydrogels. Carbohydr Polym 209: 101–110.

Rafieian, S., H. Mirzadeh, H. Mahdavi and M.E. Masoumi. 2019. A review on nanocomposite hydrogels and their biomedical applications. Sci Eng Compos Mater 26: 154–174.

Renge, V.C., S.V. Khedkar and S.V. Pande. 2012. Removal of heavy metals from waste waters using low cost adsorbents: a review. Sci Revs Chem Commun 2: 580–584.

Roufegari-Nejhad, E., M. Sirousazar, V. Abbasi-Chiyaneh and F. Kheiri. 2019. Removal of methylene blue from aqueous solutions using poly (vinyl alcohol)/montmorillonite nanocomposite hydrogels: Taguchi optimization. J Polym Environ 27: 2239–2249.

Salleh, M.A.M., D.K. Mahmoud, W.A. Karim and A. Idris. 2011. Cationic and anionic dye adsorption by agricultural solid wastes: a comprehensive review. Desalination 280: 1–13.

Samaddar, P., S. Kumar and K.H. Kim. 2019. Polymer hydrogels and their applications toward sorptive removal of potential aqueous pollutants. Polym Reviews 59: 418–464.

Shaabani, Y., M. Sirousazar and F. Kheiri. 2016a. Crosslinked swellable clay/egg white bionanocomposites. Appl Clay Sci 126: 287–296.

Shaabani, Y., M. Sirousazar and F. Kheiri. 2016b. Synthetic–natural bionanocomposite hydrogels on the basis of polyvinyl alcohol and egg white. J Macromol Sci B Phys 55: 849–865.

Shalla, A.H., Z. Yaseen, M.A. Bhat, T.A. Rangreez and M. Maswal. 2019. Recent review for removal of metal ions by hydrogels. Sep Sci Technol 54: 89–100.

Sharma, G., M. Naushad, A. Kumar, S. Rana, S. Sharma, A. Bhatnagar, F.J. Stadler, A.A. Ghfar and M.R. Khan. 2017. Efficient removal of coomassie brilliant blue R-250 dye using starch/poly (alginic acid-cl-acrylamide) Nano hydrogel. Process Safe Environ Protec 109: 301–310.

Shojaeiarani, J., D. Bajwa and A. Shirzadifar. 2019. A review on cellulose nanocrystals as promising biocompounds for the synthesis of nanocomposite hydrogels. Carbo Polym 216: 247–259.

Sirousazar, M., M. Yari, B.F. Achachlouei, J. Arsalani and Y. Mansoori. 2007. Polypropylene/montmorillonite nanocomposites for food packaging. E-polymers 7: 305–313.

Sirousazar, M., M. Kokabi, Z.M. Hassan and A.R. Bahramian. 2012. Nanoporous nanocomposite hydrogels composed of polyvinyl alcohol and Na-montmorillonite. J Macromol Sci B Phys 51: 1583–1595.

Sirousazar, M. 2013. Mechanism of gentamicin sulphate release in nanocomposite hydrogel drug delivery systems. J Drug Deliv Sci Technol 23: 619–621.

Sirousazar, M., M. Forough, K. Farhadi, Y. Shaabani and R. Molaei. 2014. Hydrogels: properties, preparation, characterization and biomedical applications in tissue engineering, drug delivery and wound care. pp. 295–357. *In*: Tiwari, A. (ed.). Advanced Healthcare Materials. John Wiley & Sons Inc., Hoboken, NJ, USA.

Sirousazar, M., A. Jahani-Javanmardi, F. Kheiri and Z.M. Hassan. 2016. *In vitro* and *in vivo* assays on egg white/polyvinyl alcohol/clay nanocomposite hydrogel wound dressings. J Biomater Sci Polym Ed 27: 1569–1583.

Wang, L., Y. Wang, F. Ma, V. Tankpa, S. Bai, X. Guo and X. Wang. 2019. Mechanisms and reutilization of modified biochar used for removal of heavy metals from wastewater: a review. Sci Total Environ 668: 1298–1309.

Yan, B., Z. Chen, L. Cai, Z. Chen, J. Fu and Q. Xu. 2015. Fabrication of polyaniline hydrogel: synthesis, characterization and adsorption of methylene blue. Appl Surf Sci 356: 39–47.

Zare, E.N., M. Mansour Lakouraj and A. Ramezani. 2016. Efficient sorption of Pb (ii) from an aqueous solution using a poly (aniline-co-3-aminobenzoic acid)-based magnetic core–shell nanocomposite. New J Chem 40: 2521–2529.

Zare, E.N., A. Motahari and M. Sillanpaa. 2018. Nano adsorbents based on conducting polymer nanocomposites with main focus on polyaniline and its derivatives for removal of heavy metal ions/dyes: a review. Environ Res 162: 173–195.

Zeng, G., Z. Ye, Y. He, X. Yang, J. Ma, H. Shi and Z. Feng. 2017. Application of dopamine-modified halloysite nanotubes/PVDF blend membranes for direct dyes removal from wastewater. Chem Eng J 323: 572–583.

13

Review on Period of Biodegradability for Natural Fibers Embedded Polylactic Acid Biocomposites

*Arun Y. Patil, Nagaraj R. Banapurmath** and *Sunal S.*

Introduction

Today plastic has become a part of our life, and plays a significant role in meeting various human necessities. In the course of its extensive usage, people are exposed to BISPHENOL A (BPA), leading to heart disease, cancer, diabetes, and a few other common ill effects, such as hair loss, hypertension, and metabolic problems. Tolerable Daily Intake (TDI) norms set for various countries are viz. Europe: 5, US FDA: 5, Australia: 50, China: 50, Korea: 50 μg/kg bw/day, respectively, whereas for India, the standards are not yet stated. Recent studies showed nearly 92% of Americans aged 6 or older test positive for BPA. On the other hand, plastic is a highly rigid and tough material, and burying it deep within the earth's crust will still affect the environment. Plastic takes up to 1,000 years to degrade in the landfill. Even CO_2 emission is a major concern during the manufacturing of plastic, particularly as it releases hazardous substances (chemical pollutants) during its course of processing. Nowadays, the focus is on adopting 'green manufacturing' to provide a safer environment for future generations. Hence, biodegradable and biocompatible materials play a vital role as alternatives to conventional plastic. Today, natural fibers are considered extensively for building bi-polymer based biocomposites with a view that degradation of natural fiber was a pivotal parameter to decide the life of biocomposites. From the literature survey, it was noted that very few papers have

Centre for Material Science, School of Mechanical Engineering B.V.B. College of Engineering and Technology, KLE Technological University, Vidya Nagar, Hubli, Karnataka 580031, India.

* Corresponding author: nrbanapurmath@gmail.com

emphasized on the aspect of degradation time of natural fibers, and hence the focus of this study was on the degradation time of plant-based natural fibers. The aim was to do an extensive literature survey on Polylactic Acid (PLA) as the holding matrix and various fibers treated and untreated as fillers used in recent times, affecting mechanical, physical, and chemical behavior of the developed biocomposites, and on the method of fabrication of biocomposites.

PLA—Next generation bioplastic

Bioplastics are alternatively termed as 'organic plastics', and are a type of plastic whose basic ingredients are made out of bio materials, viz. vegetable oil or recyclable substances. The plant-based ingredients are listed as follows:

a) Potatoes
b) Sugar cane
c) Wheat and
d) Corn (A Market Insight Report 2013)

Potatoes are presently under consideration, and seem to have good potential.

The base material for PLA is lactic acid, and the consumption of lactic acid around the world is illustrated in Fig. 13.1 for the year 2015. From Fig. 13.1, it can be observed that the major consumer of lactic acid was North America. Food and beverage industries were the highest consumers of lactic acid and followed by PLA production. By 2020, it is expected that PLA production will become the single largest outcome product, followed by food and beverages using lactic acid.

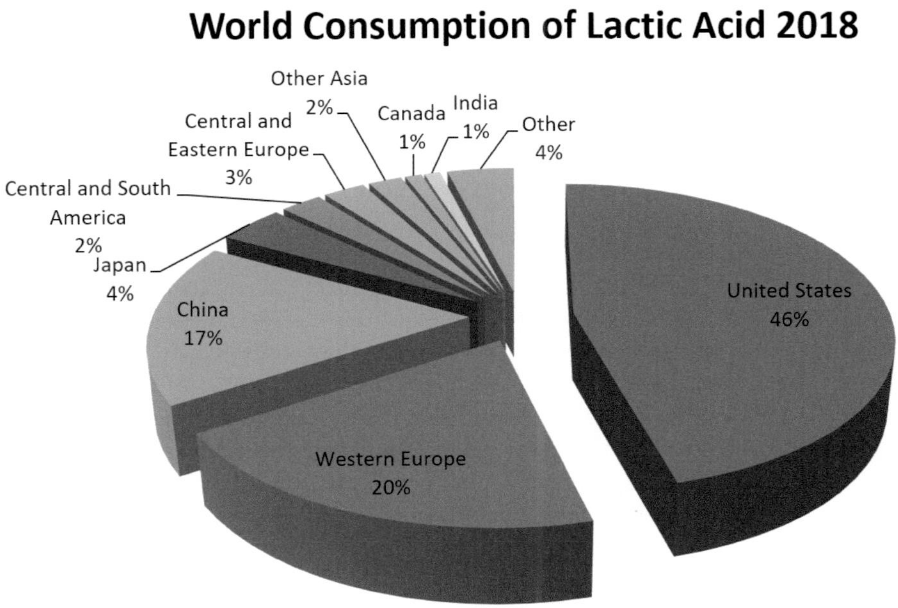

Figure 13.1. World consumption of lactic acid in 2018 [IHS Report 2018].

From 2015 to 2020, expected lactic acid consumption would grow at the rate of 6% per year.

PLA with natural fiber

In recent years, PLA was extensively used as a novel biopolymer, in many diversified applications, and there is a lot of confusion in the minds of researchers to choose the appropriate natural fibers as fillers for the required applications. The mechanical properties of natural fibers available from the works of earlier researchers are presented in Table 13.1 (Sathishkumar et al. 2012).

From Table 13.2, the inference was given on natural fibers with treated/untreated condition, varied volume/weight percentage, and method of fabrication for PLA-based polymer matrix. The detailed discussion was envisaged based on parameters mentioned below:

- Degradation time of natural fiber and polymer
- Weight or volume fraction
- Surface treatment of natural fiber
- Method of fabrication for fiber reinforced matrix

Degradation time of natural fiber and polymer

Degradation period of biopolymer was elaborated with PLA and various other natural fibers, as in the Table 13.3. Recently, natural fiber selection was of concern for researchers, as pristine fibers were subjected to natural degradation, early failure, debonding with substrate and so on. Today, the prime factor with natural fiber was to use any type of chemical to degrade the matrix, or a pristine polymer matrix itself will be able to degrade on its own. To enlighten on such factors, Fig. 13.2 illustrates the broad category of natural fibers based on their origin.

Natural fibers were prone to early degradation in comparison to synthetic fiber because of its ingredient constituents, viz, cellulose, hemi cellulose, lignin, pectin, wax, moisture, and specific gravity percentage. Degradation occurs by means of three mechanisms. Firstly, there is chemical degradation mediated by water, enzymes, microorganisms. Secondly, there is physical degradation, which occurs by two means—bulk erosion and surface erosion. Polyesters, PLA, PLGA, and polylactones are few examples for bulk erosion. In case of surface erosion, degradation is much faster than bulk erosion, for example polyanhydrides, polyorthoesters, etc. Finally, morphological degradation happens based on swelling, deformation, bubbling, and disappearance. Patil et al. 2018b concluded that degradation is affected by lignin and hemi cellulose content for natural fibers, such as sweet lime and lemon fillers embedded in polymer substrate. However, prediction of degradation with the help of chemical composition was carried out by Sen and Reddy (2011) for jute and hemp. It was clear from the study of water hyacinth and reed fibers, that hemi cellulose were the most affected parameters for thermal degradation when compared to other constituents. More the content of hemi cellulose, higher the scarcity of water absorption, leading to early degradation (Methacanon et al. 2010). Manfredi et al. 2006 revealed that among flax, jute, and sisal fibers, low lignin content at higher

Table 13.1. Mechanical properties of natural fibers.

Fiber name	Density (Kg/m³)	Diameter (μm)	Tensile strength (MPa)	Tensile modulus (GPa)	% Elongation	Reference
Nettle	1510–1610	-	560–1600	24.5–87	2.1–2.5	Huang 2005
Cotton	1500–1600	-	287–600	5.5–12.6	3.0–10.0	Bajracharya et al. 2017
Ramie	1500–1560	-	220–940	44–128	2.0–3.0	Huang 2005
Flax	1500–1540	-	345–1500	27.6–80	1.2–3.2	Rytlewski et al. 2018
Hemp	1470–1480	-	550–900	70	1.6	Lopez et al. 2017
Jute	1440–1460	-	393–800	10.0–30.0	1.5–1.8	Katogi et al. 2017
Sisal	1450–1500	50–300	227–400	9.0–20.0	2.0–14.0	Badia et al. 2017
Pineapple	1440	20–80	413–1627	34.5–82.5	0.8–1.0	Jia et al. 2014
Kenaf	1400	81	250	4.3	-	Kurniawan et al. 2015
Banana	1350	80–250	529–759	8.2	1–3.5	Vigneswaran et al. 2015
Coir	1150	100–460	108–252	42159	15–40	Monteiro et al. 2008
Root	1150	100–650	157	6.2	3	Mallillin et al. 2008
Palymyrah	1090	70–1300	180–215	7.4–604	7.0–15.0	Arumuga Prabu et al. 2016
Date	990	-	309	11.32	2.73	Mrabet et al. 2017
Bamboo	600–910	-	503	35.91	1.4	Qian et al. 2017
Talipot	890	200–700	143–294	9.3–13.3	2.7–5	Pothan 2009
Snake grass	887	45–250	278.82	9.71	2.87	Sathishkumar et al. 2012
Elephant grass	817	70–400	185	7.4	2.5	Rao et al. 2007
Petiole bark	690	250–650	185	15	2.1	Sain et al. 2014
Rachilla	650	200–400	61	2.8	8.1	Quattrocchi et al. 2017
Rachis	605	350–408	73	2.5	13.5	Lingham-Soliar et al. 2017
Coconut	352	-	119.8	18	5.5	Chanap et al. 2012
Red banana	-	-	482–567	-	30.6	Kiruthika et al. 2009

Table 13.1 Contd. ...

...Table 13.1 Contd.

Fiber name	Density (Kg/m^3)	Diameter (μm)	Tensile strength (MPa)	Tensile modulus (GPa)	% Elongation	Reference
Nendran	-	-	407–505	-	28.3	Kiruthika et al. 2009
Rasthaly	-	-	304–388	-	27.8	Preethi et al. 2012
Egg Shell	210 to 220	-	-	0.3	-	Patiño et al. 2002
Orange Peel	1087	-	35–45	-	-	Patil et al. 2018a
Lemon Peel	1098	-	45–55	-	-	Patil et al. 2018a
Onion Peel	847	-	20–25	0.659	-	Patil et al. 2018a
Potato Peel	1075	-	18–21	0.52	-	Patil et al. 2018a
Carrot Peel	751	-	17–21	0.515	-	Patil et al. 2018a

temperature conditions resulted in early degradation of jute fibers in comparison to the rest of the fibers.

Clarity on natural fibers was observed with the help of subsequent Figs. 13.3 to 13.9 enlighten on their degradation time.

Weight or volume fraction

The most important part of biocomposite was to identify the percentage of inclusion onto the substrate, and several researchers have worked on this critical area, as highlighted below.

Flax

Currently, flax fiber is in demand, due to its quick degradation, with wet fiber having 20% higher tensile strength than dry fiber. The weight percent variation ranges from 1% to 65% of PLA substrate. The enhancement in degree of crystallinity was observed, with 20% of weight fraction yielded into average cell size reduction, resulting in cell density increase (Pilla et al. 2008). For 10% weight fraction embodiment into PLA, they illustrate non-linear behavior, inhibiting micro-damage in the biocomposite material developed (Varna et al. 2012). An extreme case of 65% of weight fraction of PLA used resulted in comparative analysis with carbon/epoxy laminate, showing better absorbed energy and normalised residual strength (Lopez et al. 2017). In a latest work 20 wt% inclusion enhanced the tensile strength of biocomposite by 20% (Rytlewski et al. 2018).

Hemp

Hemp is another such material which has captured the market throughout the world, where the weight percent variation range from 1% to 48% with PLA substrate. It

Table 13.2. Comparative study of PLA with various natural fibers and other latest research works.

Sl. No	Substrate	Filler	Percentage	Fiber Treatment	Method of fabrication	Result	References
1	PLA	Bamboo fiber	60 wt% fiber	Untreated	Hot pressing	Thermal conductivity is smaller in comparison to GFRP and CFRP.	Takagi et al. 2007
2	PLA	Flax	0 to 20 wt%	Silane Treated	Injection molding method	The average cell size decreased while the cell density increased with the fiber content. The degree of crystallinity increased with the fiber content.	Pilla et al. 2008
3	PLA	Soft wood saw dust	60 to 70%	Heat-treated	Extrusion	Cyclic treatment of the studied composites resulted in 20 60% loss of flexural strength, depending on type of composite.	Butylina et al. 2010
4	PLA	Kenaf	NA	Untreated	Injection molding method	Interfacial shear strength of PLA–KF is 5.41 ± 2.23 MPa. The value is comparable as those measured for hemp/PP and flax/Mater-Bi.	Anuar et al. 2011
5	PLA	Wood	60%	Not mentioned	Hot pressing	Absorption of moisture led to softening of the wood fibers and PLA matrix.	Yaowen Liu et al. 2018
6	PLA	Pine wood floor	15 to 35%	NA	Extrusion blending and Injection molding	The impact strength and elongation at the break of the PLA-based WPCs toughened with SBS were higher than their non-toughened counterparts, in contrast to the decrease in the tensile strength.	Qiang et al. 2012
7	PLA	polysaccharide	10 to 20 vol%	Untreated	Electro spinning	Variation in the average fiber diameters alone did not always affect sheet wettability.	Sunthornvarabhas et al. 2012
8	PLA	Sisal	6 wt%	Caustic soda	Compression Molding	Caustic soda treated fibers, uneven spherulitic growth was observed at crystallization temperatures at and above 125°C.	Prajer et al. 2012

Table 13.2 Contd. ...

...Table 13.2 Contd.

Sl. No	Substrate	Filler	Percentage	Fiber Treatment	Method of fabrication	Result	References
9	PLA	Flax	10 wt%	Untreated	Injection molding method	The observed non-linearity was attributed to micro damage, visco-elastic and visco-plastic response, suggesting Schapery's type of model for visco-elasticity and Zapas' model for visco-plasticity. After loading at stress levels below the maximum possible, the elastic modulus is not affected, and therefore, damage does not need to be included in the material model.	Varna et al. 2012
10	PLA	Tea poly phenol	10 to 50 vol%	Untreated	Electro-spinning	Positive rate of PLA/TP composite nanofiber films was greater than that of pure PLA nano fibrous films, increasing from 1.45 and 0.78% to 9.26 and 6.47% against *S. aureus* and *E. coli*, respectively.	Fei et al.2013
11	PLA	Hydroxyapatite	15 wt%	Untreated	Injection molding method	Mechanical properties, such as flexural strength and the modulus of bimodal-HA/PLA composite with both 5 and 1 μm of representative size, were higher than those of mono modal-HA/PLA composites with 5 or 1 μm of representative size.	Takayama et al. 2013
12	PLA	Wood Plastic Composite	NA	NA	Extrusion blending and Injection molding	The impact toughness and interface adhesion of the PLA-based WPCs toughened with LLDPE were significantly improved, compared with their unmodified counterparts.	Qiang et al. 2013

13	PLA	$CaCO_3$	5%	Not mentioned	Compression Molding	PLA impact resistance of the nanocomposites was also improved by 1.6 times with the addition of 20 phr Poly Ethylene Glycol (PEG) plasticiser and by 1.4 times with the addition of 20 phr TbC plasticiser.	Varna et al. 2012
14	PLA	Egg Shell Powder	1 to 5%	Untreated	Composite film casting	The tensile strength and modulus of the composite films were found to be higher than those of PLA and increased with ESP content up to 4 wt% and then decreased. The thermal stability of the composite films increased with ESP content.	Ashok et al. 2013
15	PLA	Dura fiber and Tenera fiber palm press fibers	10%	Not mentioned	Injection molding method	The flexural modulus of the composites increased by 98%, while the impact strength increased by 52 percent.	Akos et al. 2013
16	PLA	TiO_2 Nano particle	0.5, 2, 5, 10%	propanoic acid	Solution blending method	Increased degree of crystallinity from 7.4% to 34 percent.	Buzarovska et al. 2013
17	PLA	Flax	10%	Dicumyl peroxide (DCP)	Injection molding method	Dicumyl peroxide (DCP) is commonly applied as a cross-linking agent varied with 0.5 and 2.5 wt%.	Rytlewski et al. 2014
18	PLA	Hemp	10%	Dicumyl peroxide (DCP)	Injection molding method	Dicumyl peroxide (DCP) is commonly applied as a cross-linking agent varied with 0.5 and 2.5 wt%.	Rytlewski et al. 2014
19	PLA	PLA	10-40%	Not mentioned	Film stacking and Hot pressing	Initial modulus is 2–6 times higher than that measured for PLA and PBS films.	Jia et al. 2014
20	PLA	PBS	10-40%	Not mentioned	Film stacking and Hot pressing	Initial modulus is 2–6 times higher than that measured for PLA and PBS films.	Jia et al. 2014

Table 13.2 Contd. ...

...Table 13.2 Contd.

Sl. No	Substrate	Filler	Percentage	Fiber Treatment	Method of fabrication	Result	References
21	PLA	Silica	7%	Untreated	Hot pressing	Relatively better dispersion was found for PLA/PEGME-silica nanocomposites, the unmodified silica at high dosages tended to have more aggregates formation and induced heterogeneity within the PLA matrix, stemming from the incompatibility.	Imre et al. 2014
22	PLA	Oil palm empty fruit bunch	Up to 40%	Ultrasound and polydimethylsiloxane	Melting Cast Method	The analysis revealed that polydimethylsiloxane treatment composites show reduced wettability with increased crystallinity.	Akindoyo et al. 2014
23	PLA	Cotton-comber pulp	5.50%	Chloroform	Electro-spinning	The strength parameters increased by 17% on average, and included, for example, 7% increase in tearing resistance, and even 35% increase in air permeability.	Modzelewska et al. 2014
24	PLA	Polystyrene	0 to 100%	Dimethyl sulfoxide	Compression Molding	Properties change accordingly, the blend containing the smallest dispersed particles has the largest tensile strength, while PLA/PS blends with the coarsest structure have the smallest. The latter blends are also very brittle.	Imre et al. 2014
25	PLA	Polycarbonate	0 to 100%	Dimethyl sulfoxide	Compression Molding		Imre et al. 2014
26	PLA	Poly Methyl Metha Acrylate (PMMA)	0 to 100%	Dimethyl sulfoxide	Compression Molding		Imre et al. 2014

27	PLA	Glass Fiber	10 to 30 wt%	Not mentioned	Injection molding method	The mechanical performance and heat resistance of PLA-PC alloy were improved significantly by adding glass fiber. The tensile strength, flexural strength, flexural modulus, notched izod impact strength, and HDT were 124.7 MPa, 173.2 MPa, 7.3 GPa, 8.1 kJ = m^2, and 143 C, respectively, at the loading of 30 phr GF.	Lin et al. 2014
28	PLA	Aluminium hypophosphite	5 to 30 wt%	Not mentioned	Injection molding method	The mechanical performance and heat resistance of PLA-PC alloy were improved significantly by adding glass fiber. The tensile strength, flexural strength, flexural modulus, notched izod impact strength, and HDT were 124.7 MPa, 173.2 MPa, 7.3 GPa, 8.1 kJ = m^2, and 143 C, respectively, at the loading of 30 phr GF.	Lin et al. 2014
29	PLA	Hemp	48%	Untreated	Compression Molding & micro-braiding	Weight-averaged molecular weight affected the mechanical properties of monolithic PLA and also the transverse properties of PLA.	Kobayashi et al. 2014
30	PLA	Basalt	25.00%	Silane Treated	Hot pressing	Silane treated and the atmospheric plasma polymerized composites have 26 and 22% higher strength, respectively, as compared with the untreated one.	Kurniawan et al. 2015
31	PLA	polyethylenimine	NA	Untreated	Electro-spinning	To incorporate PEI with PLA as ultra-fine fibers to diminish the acidic inflammation caused by biodegradation of PLA. The fibrous composite membrane of PVA/PEI-PLA could provide better biocompatibility and would be used as drug-delivery carriers or tissue-engineering scaffolds.	Dong et al. 2015

Table 13.2 Contd. ...

...Table 13.2 Contd.

Sl. No	Substrate	Filler	Percentage	Fiber Treatment	Method of fabrication	Result	References
32	PLA	Kenaf	25 wt%	Alkali and Untreated	Dry blending, twin screw extrusion and compression molding	Alkali treatment improved interfacial adhesion between PLA and kenaf.	Kurniawan et al. 2015
33	PLA	Ammonium Polyphosphate	10 to 20%	Alkali and Untreated	Dry blending, twin screw extrusion and compression molding	SEM analysis indicates good dispersion of APP in PLA matrix, however, interfacial adhesion between PLA and matrix decreased with increasing APP content.	Kurniawan et al. 2015
34	PLA	Sisal	NA	Alkali and Untreated	Hot pressing	Influence of fiber content and fabrication methods, which can significantly affect the mechanical properties of sisal fiber-reinforced polymer composites.	Ibrahim et al. 2015
35	PLA	Halloysite	2 wt%	Untreated	Compression Molding	The impact strength and elongation at break of PLA/HNC nanocomposites were increased significantly by the addition of SEBS-g-MAH. The melting behavior, degree of crystallinity, and cold crystallization temperature of PLA/HNC composites were affected by the SEBS-g-MAH content. The thermal stability of PLA nanocomposites was increased in the presence of SEBS-g-MAH.	Tham et al.2015
36	PLA	Jute Spun Yarn	41 to 43 wt%	Alkali treatment	Hot pressing	Fatigue strength at 10^6 cycles was 55% of the ultimate strength, which is an almost identical percentage to that of GFRP.	Katogi et al. 2017

37	PLA	Pennisetum purpureum or Napier grass or elephant grass or Uganda grass	10 to 30 wt%	Scaffold	Particulate leaching method	The compression strength of the scaffolds was found to increase from 1.94 to 9.32 MPa, while the compressive modulus increased from 1.73 to 5.25 MPa as the fillers' content increased from 0 wt% to 30 wt%.	Revati et al. 2017
38	PLA	Sisal	10 to 30 wt%	MAH	Compression Molding	The addition of fiber was more relevant in terms of crystallinity degree- which increased according to the amount of sisal, due to the augment of nucleation sites, than in terms of molar mass– which was quite similar between biocomposites, and just remarkably different in comparison to neat PLA.	Badia et al. 2017
39	PLA	Bamboo cellulose Nano whiskers	1 to 16 wt%	Trimethoxysilane Silane treated	Solution casting	Elongation at break increased to 250.8% after the 4 wt% A-189 treatment compared to 12.35% with untreated composites. The C/O ratio of BCNW decreased after coupling agent treatment.	Qian et al. 2017
40	PLA	Basalt	4.25 wt%	Untreated	Injection molding method	The composites with pristine BFs present evidently increased strength and modulus relative to neat PLA. After surface treatment, BF shows more evident reinforcing effect because the interfacial energy between BF and PLA decreases sharply from 27.40 to 8.58 mJ m^{-2}.	Ying et al. 2017
41	PLA	Wood fiber maleic anhydride	30 wt% wood fiber 5 % Maleic anhydride	Untreated	Injection molding method	Melt flow index was reduced by 10–16% for compatibilized composites, implying the cross-linking of the polymer. Compatibilized composites of PLA, Bioflex, and PHBV exhibited improved thermal and strength properties, and reduced water absorption.	Yatigala et al. 2017

Table 13.2 Contd. ...

...Table 13.2 Contd.

Sl. No	Substrate	Filler	Percentage	Fiber Treatment	Method of fabrication	Result	References
42	PLA	biobased silica/ carbon hybrid nanoparticles	0.5 and 1.0 wt%	Untreated	3D Printing technique	TGA and Tensile tests revealed significant enhancement in thermal stability, maximum strain, and strain to failure properties due to the integration of 0.5 and 1.0 wt% of silica/ carbon nanoparticles (SCNPs).	Biswas et al. 2017
43	PLA	Burma Padauk Sawdust	30 wt.%	Untreated	Injection molding method	Maximum water absorption around 4.5% was observed when sawdust was added at 30 wt%. Under water immersion and sunlight exposure, the tensile and flexural strengths tended to reduce slightly, due to the plasticization effect.	Petchwattana et al. 2017
44	PLA	Petioles of red rhubarb	NA	NA	NA	The impact resistance of bio-inspired composites exceeded that of composites with identical fiber-fraction, but random fiber-distribution by more than a factor of two, while their tensile and flexural properties did not differ significantly, suggesting promising new routes for the design of tough, bio-compatible composites.	Graupner et al. 2017
45	PLA	Thermoplastic polyurethane/ cellulose-nanofiber	2.5 wt.%	Untreated	Compression Molding	Adding the CNFs to the PLA-TPU blends had a considerable effect on the force recovery of the cylindrical shell SMPs in a way that the presence of 4% CNF within the polymeric matrix increased the force recovery up to 55 percent.	Barmouz et al. 2017

46	PLA	Cellulose nano fiber	0.1 wt% CNFs	Compatibalizer Glycidyl methacrylate (GMA)	NA	PLA/PLA-g-GMA/CNFs biocomposites exhibited a slight decrease in the very early stages of crystallization, then become time-independent. The fiber content of 0.1 wt% was found to give the highest crystallization rate of both PLA/CNFs and PLA/PLA-g-GMA/CNFs biocomposites.	Nguyen et al. 2017
47	PLA	Kenaf Fiber/ Multiwalled Carbon Nanotube	1.0 wt%	NaOH and Acetic acid silane/acetone	Compression Molding	Annealing procedure may be a route to tune the microstructure and physical properties of a polymer, which might open a promising door to produce the natural fiber/ PLA green composites without using any environmentally harmful reagents.	Chen et al. 2017
48	PLA	chitosan, carboxymethyl cellulose and silver nano particle	0, 1%, 2.5%, 5%, 10%	Acetic acid	Freeze drying method	Incorporation of nanocomposite during scaffolds preparation helped in achieving the desirable 80–90% porosity, with pore diameter ranging between 150 and 500 μm, and mechanical strength was also significantly improved, matching with the mechanical strength of cancellous bone.	Hasan et al. 2017
49	PLA	Flax	65.00%	Untreated	Compression Molding	The results were compared with carbon/ epoxy laminates, showing some important advantages in terms of absorbed energy and normalized residual strength.	Lopez et al. 2017
50	PLA	Cotton gin waste and flax fibers	10 to 30 wt%	Maleic anhydride polypropylene (MAPP)	Compression Molding	Flexural modulus increased by 42% with the addition of 30 wt% of CGW as compared to neat PLA. Furthermore, the improvement of modulus of elasticity of composites containing 30 wt% of CGW is comparable to hybrid composites reinforced with different percentage of flax fibers.	Bajracharya et al. 2017

Table 13.2 Contd. ...

...Table 13.2 Contd.

Sl. No	Substrate	Filler	Percentage	Fiber Treatment	Method of fabrication	Result	References
51	PLA	Hemp Hurd	10 wt% and 20 wt%	Grafted glycidyl methacrylate	Injection molding	Mechanical analysis, which showed increases in the glycidyl methacrylate-grafted 20% (w/w) hemp hurd/poly(lactic acid) biocomposite, retaining 94% of the neat polymer strength, with increases in crystallinity, and showing a range of thermomechanical properties desirable for rigid biocomposite applications.	Khan et al. 2017
52	PLA	Nettle (Girardinia diversifolia)	Till 50 wt%	Alkali treatment	Compression Molding	The tensile, bending, and impact properties of the biocomposites initially increased with the increase of nettle fiber content and decreased afterwards. The maxima of these properties were obtained at equal weight proportion of nettle and poly(lactic acid) fibers. The thermogravimetric analysis inferred that the biocomposites were thermally enough stable and their thermal stability increased with the increase of nettle fiber content.	Kumar et al. 2017
53	PLA	cellulosic fibers pulp and paper solid waste	2 wt%	Enzymatic treatment	Not mentioned	Mechanical properties of biocomposites with 2 % (w/w) of treated cellulosic fibers (Young's Modulus 887.83 MPa with tensile strain at breakpoint of 7.22%, tensile stress at yield 41.35 MPa) was enhanced in comparison to the recycled PLA (Young's Modulus 644.47 ± 30.086 MPa, with tensile strain at breakpoint of 6.01 ± 0.83%, tensile stress at yield of 29.49 ± 3.64 MPa).	Laadila et al. 2017

54	PLA	Rosin Films	1, 3, 5, 10, and 20 wt%	Chloroform	Solution-casting method	The PLA-RS-10 film exhibits a maximum improvement in elongation-at-break (%) by ~ 67%. The presence of RS in PLA matrix reduces the passage of oxygen through the biocomposite films. It is found that PLA-RS-10 biocomposite film shows reasonable reduction in OTR by ~ 41%. PLA-RS-20 film shows remarkable reduction in transmission of UV-B, UV-A, and visible light by ~ 98%, 92%, and 53%, respectively.	Narayanan et al. 2017
55	PLA	cellulosic fibers	5 wt%, 10 wt% and 15 wt%	Not mentioned	Injection Molding	Compared to classical fibers (such as carbon and glass fibers), natural fibers can suffer from twisting phenomena during the extrusion and molding stages.	Gigante et al. 2017
56	PLA	Aspen wood fibers	0.5,1,2,5 wt %	Hydrothermally treated	microinjection molding machine	Compared to neat PLA, the Young's modulus, 15 elongation to break, and toughness of PLA/2%HLCNCs were improved by 14, 77, and 30%, 16, respectively.	Wei et al. 2017
57	PLA	Ag nano particles	1% wt. of CNC (PLA/1CNC), 0.5% wt	Not mentioned	Solvent casting method	Grafting of silver nanoparticles on cellulose nanocrystals positively affected the thermal degradation and cold crystallization processes. Results of antibacterial tests showed that PLA nanocomposite films displayed a good antibacterial activity.	Yalcinkaya et al. 2017
58	PLA	Mg alloy	96%Mg, 3%Al and 1%Zn all in wt%	Hydrofluoric acid	Injection Molding	PLA-clad Mg rod with intermediate coating showed much higher tensile and bending strength and better corrosion resistance in SBF solution in comparison with PLA-clad Mg rod without coating.	Butt et al. 2017

Table 13.2 Contd. ...

...Table 13.2 Contd.

Sl. No	Substrate	Filler	Percentage	Fiber Treatment	Method of fabrication	Result	References
59	PLA	polyamide (PA) micro fibrillar composites (MFCs)	3, 9, and 18 wt%	Not mentioned	Compression molding	The oxygen transmission rates (OTRs) and the water vapor transmission rates (WVTRs) of the MFCs with 3 wt% and 9 wt% of PA microfibrils were significantly (up to 38% for OTR and 28% for WVTR) lower than the pure PLA.	Kakroodi et al. 2017
60	PLA	chitosan, Rosin modified cellulose nanofiber carboxymethyl cellulose and	0 wt%, 2 wt%, 5 wt%, 8 wt%, and 10 wt%	sonication	Not mentioned	The increase in mechanical properties was observed for samples with R-CNF reinforcement. The optimum mechanical properties of the composite film were achieved with a percolation threshold of 8% of R-CNF loading.	Niu et al. 2017
61	PLA	Biomass wastes	Not mentioned	Not mentioned	Not mentioned	A study has demonstrated the viability of using waste fibers as filler for a PLA matrix and achieve materials that would minimize the environmental load associated with fossil fuels.	Spiridon et al. 2017
62	PLA	Biobased PC glass fiber	32 wt% PLA 3251D, 62 wt% PC 6 wt% Elvaloy PTW EBA-GMA	Not mentioned	Injection molding	Glass fiber composites based on branched PC/PLA matrix demonstrate superior mechanical and heat resistance properties over linear PC/PLA matrix composites. A competitive weight to strength ratio, excellent heat resistance (as compared to commercial PC/ABS composites).	Yuryev et al. 2017
63	PLA	bleached kraft soft wood (BKSW)	Water/ diglyme (35/65 wt%)	Untreated	kinetic mixing and Injection molding	Fiber addition promoted the crystallization of PLA by favoring the heterogeneous nucleation and accelerating the crystallization kinetic, without adversely altering the thermal stability.	Espinach et al. 2017

64	PLA	Flax	20 wt%	Untreated	Extrusion and injection molding	Composites were examined using mass loss analysis, DSC, and photoelectron spectroscopy (XPS) in order to determine structural changes induced by biodegradation. Application of TAIC and EB at dose of 40 kGy led to the increase of tensile strength of about 20 percent.	Rytlewski et al. 2018
65	PLA	Gelatin-forsterite	NA	Acetic acid	FDM technique	Hierarchical scaffolds, fabricated using FDM and ES, revealed 52% increase in elastic modulus rather than the pure one owing to the presence of forsterite nano powder incorporated within gelatin matrix.	Naghieh et al. 2018
66	PLA	graphene nanoplatelets	0.125 wt%, 0.25 wt%, 0.5 wt% and 1.0 wt%	Acetone (Ac)and Chloroforms (TCM)	Electro-spinning	Tensile tests demonstrated that the reinforcing effect of GnP when added to the PLA matrix was more than three times higher in the aligned systems if compared with the respective randomly oriented mats.	Scaffaro et al. 2018
67	PLA	hydroxyapatite	10 wt%	Ethanol and ethyl acetate	Injection molding	Fourier transforms infrared showed that surface of the HA was effectively modified and the characteristic chemical structure of the HA was not disrupted.	Akindoyo et al. 2018
68	PLA	Cotton fabric	10%, 20%, and 30%	2% NaOCl	Hot pressing	Differential pore sizes affect both the number of drugs loaded and released, as well as mechanical tensile strength and water absorption properties. Fractional weight loss due to degradation seemed to be consistent with the amount of PLA in the composites.	Macha et al. 2018

Table 13.2 Contd. ...

...Table 13.2 Contd.

Sl. No	Substrate	Filler	Percentage	Fiber Treatment	Method of fabrication	Result	References
69	PLA	Nanostructured amorphous magnesium phosphate	20 wt%	Not mentioned	Not mentioned	The PLA film showed an average coating thickness of 500 nm. With the well dispersed AMP nanoparticles in PLA matrix, the thickness of nAMP/PLA composite film slightly increased to 700 nm. Both PLA and nAMP/PLA coated AZ31 samples underwent significantly low degradation in SBF, according to the results of electrochemical measurements by potentiodynamic and EIS, and immersion test.	Ren et al. 2018
70	PLA	Leaves of Posidonia oceanica (PO)	10 wt% or 20 wt%	Untreated	Compression Molding	The results highlighted that the use of PO leaves as filler influenced all the investigated properties. In particular, by increasing the PO content, both tensile and flexural moduli increased.	Scaffaro et al. 2018
71	PLA	Thermoplastic starch (TPS) native cassava starch	NA	Coating with beeswax (BW)	Flat extrusion, calendering and thermopressing	It was also possible to coat these trays with beeswax emulsion for reducing their water vapor permeability (WVP). These trays can be used to package fresh fruits and vegetables because of their suitable mechanical properties and WVP.	Reis et al. 2018
72	PLA	Clay (Na+-montmorillonite)	5 wt%	Untreated	Compression Molding	Presence of montmorillonite (MMT) in the polymer film can reduce the lipid oxidation of processed meat products, extending their shelf life, and thus, suggesting that the new film is a potential good alternative to conventional bioplastics.	Vilarinho et al. 2018

73	PLA	Wheat Straw	10–40 wt%	Treated	Injection-molding machine	PLA/wheat straw composites exhibited faster crystallization rate of less than 1 min and higher crystallinity, which reduced the molding time. Composites exhibited major enhancement in flexural modulus and exceed the known value reported for PLA/wheat straw composite with PLA-g- MA by melt mixing.	Yang et al. 2018
74	PLA	Galactoglucomanan (GGM), wood cellulose	20 wt%, up to 25%	Not mentioned	Hot melt extrusion (HME) and 3D printing with the FDM™ technique	The application of hemicelluloses as bioplastic is still undermined. We have demonstrated a new route to use such side-stream and biorenewable wood biopolymer as feedstock materials for FDM 3D printing. On one hand, incorporation of wood hemicelluloses may reduce the application of PLA; on the other hand, hemicelluloses with versatile active sites may be used as carriers or molecular anchors to introduce desired functionality and features.	Xu et al. 2018
75	PLA	Saw dust particle	1.5% casein 131 sodium salt, 1.5% yeast extract and 1.0% glucose	Not mentioned	Not mentioned	Suggested high potential application of 63 PLA/SP + Bac7293 as a good antimicrobial packaging for pangasius fish fillets.	Woraprayote et al. 2018

Table 13.3. Various methods of treatment currently in use (Natural fiber report 2016).

Sl. No	Type of treatment	Description	Chemical structure
1	Alkaline Treatment	Disruption of hydrogen bonding in the network structure	NA
		Surface roughness enhanced for composite material	
		Removes certain amount of lignin, wax, and oils covering external surface of the fiber cell wall	
		Depolymerises cellulose	
		Exposes the short length crystallites	
		Changes the orientation of the highly packed crystalline cellulose order	
		Forms amorphous region	$CH_2CHSi(OC_2H_5)_3 \xrightarrow{H_2O} CH_2CHSi(OH)_3 + 3C_2H_5OH$
2	Silane Treatment	Lets natural fibers adhere to a polymer matrix	$CH_2CHSi(OH)_3 + Fiber - OH$
		Stabilization of composite matrix	
		Silanol reacts with hydroxyl group of fiber, forming a stable covalent bond	
3	Acetylation Treatment	Esterification method	$Fiber - OH + CH_3 - C(=O) - O - C(=O) - CH_3$
		Cellulosic fibers of composite material are plasticized	
		Chemical modification with acetic anhydride substitutes the polymer hydroxyl groups of wall cell with acetyl group	
4	Benzoylation Treatment	Modifying the properties of these polymers so that they become hydrophobic	$Fiber - OH + NaOH \rightarrow Fiber - O^-Na^+ + H_2O$
		Decreased hydrophilic nature of the treated fiber	
		Increasing the strength of composite	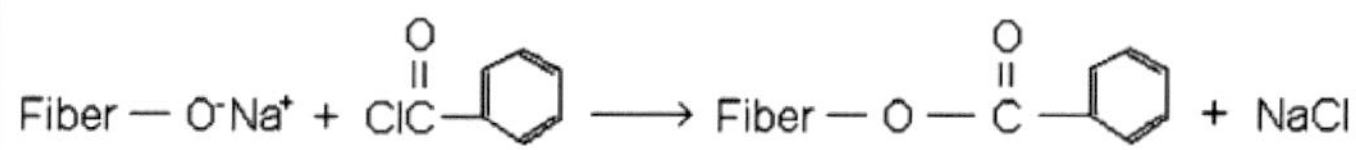
		Decreasing its water absorption	
		Improving its thermal stability	

5	Acrylation Treatment	Initiated by free radicals of the cellulose molecule	$Fiber - OH + CH_2 = CHCN \rightarrow Fiber - OCH_2CH_2CN$
		Cellulose is treated with high energy radiation to generate radicals with chain scission	
6	Coupling Agents	Maleated coupling agents are widely used to strengthen composite containing fillers and fiber reinforcements	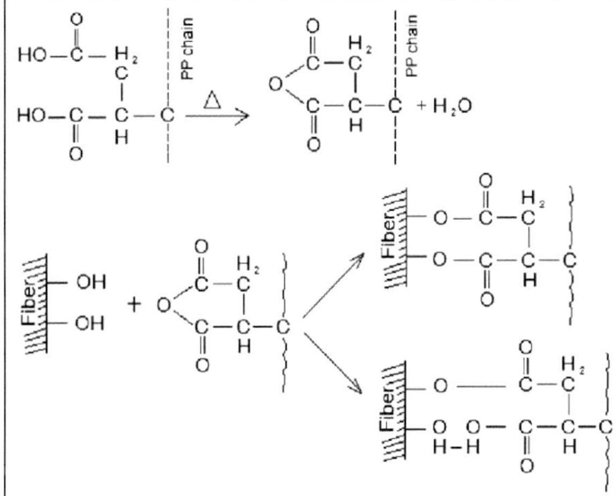
7	Iso cyanate Treatment	Highly susceptible to reaction with hydroxyl groups of cellulose and lignin in fibers	$R-N=C=O + HO-Fiber \longrightarrow R-N(H)-C(=O)-O-Fiber$
		Act like a coupling agent used in fiber reinforced composites	
8	Permanganate Treatment	Leads to the formation of cellulose radical through MnO_3 ion formation	$Cellulose-H + KMnO_4 \longrightarrow Cellulose-H-O-Mn(=O)_2-OK^+$
		Highly reactive Mn^{3+} ions are responsible for initiating graft copolymerization	

Table 13.3 Contd. ...

...Table 13.3 Contd.

Sl. No	Type of treatment	Description	Chemical structure
9	Sodium Chloride Treatment	Bleaching of natural fiber with NaCl, which cleans the fiber thoroughly	NA
		Composite materials are made rough, which is responsible for better fiber- matrix adhesion resulting into interlocking of rough fiber surface and matrix polymer chain	
10	Peroxide Treatment	Fibers are coated with Benzoyl peroxide or Dicumyl peroxide chemicals in acetone for about 30 min after alkali treatment	$RO - OR \rightarrow 2RO^{\cdot}$ $RO^{\cdot} + PE - H \rightarrow ROH + PE$ $RO^{\cdot} + Cellulose - H \rightarrow ROH + Cellulose$ $PE + Cellulose^{\cdot} \rightarrow PE - Cellulose$

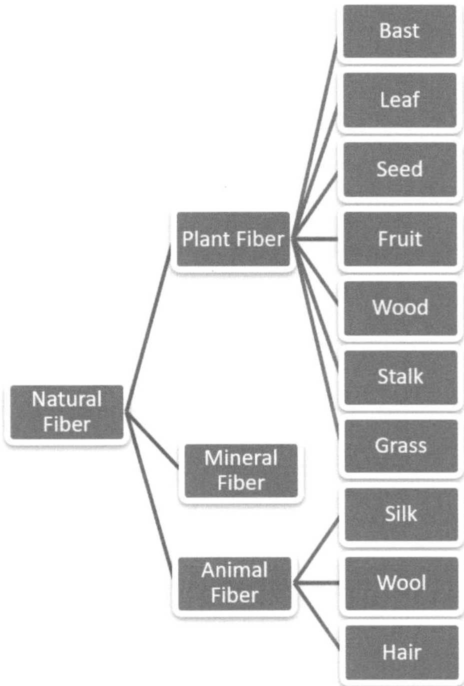

Figure 13.2. Classification of natural fibers.

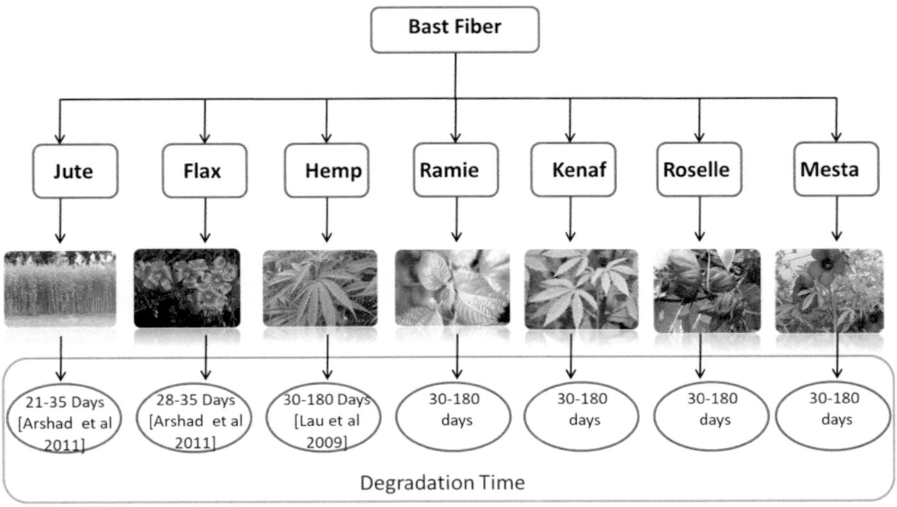

Figure 13.3. Classification of Bast fiber along with degradation time.

has unique qualities, such as becoming water vapor after burning, and it is strong and durable enough to compete with subordinate fibers. Hemp fiber replenishes the soil rather than depleting it, and can be used as efficient bio-fuel. Ten percent weight fraction of PLA resulted in better cross-linking, with agents such as Dicumyl Peroxide

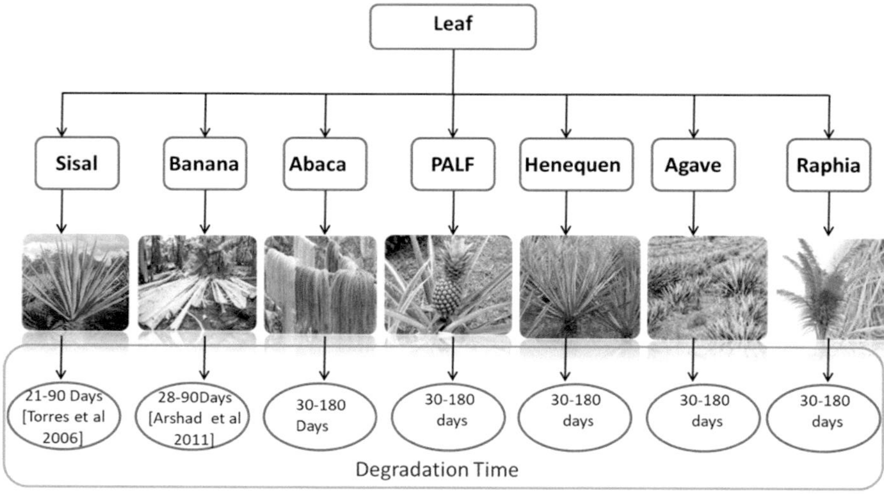

Figure 13.4. Classification of leaf-based natural fiber along with degradation time.

Figure 13.5. Classification of seed-based natural fiber along with degradation time.

(DCP) yielding better mechanical properties (Rytlewski et al. 2014). Monolithic PLA with transverse direction of embedment of fiber at 48% of weight fraction resulted in vague outcome. However the understand with research work gave a new insight in the comparative study (Kobayashi et al. 2014). Meanwhile, a recent study revealed that 10 to 20 wt% fraction helped the matrix to achieve 94% of polymer base material strength, with increase in crystallinity and better thermo-mechanical properties as well (Khan et al. 2017).

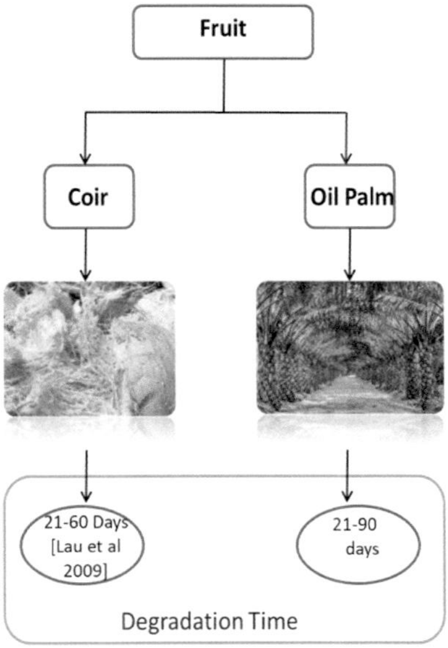

Figure 13.6. Classification of fruit-based NF.

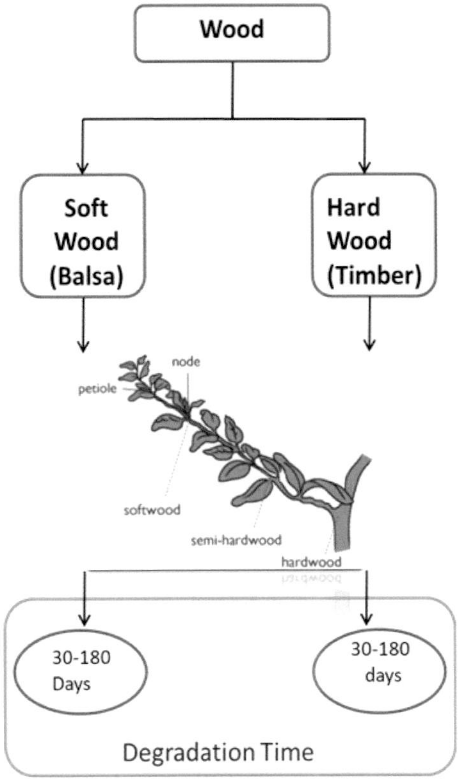

Figure 13.7. Classification of wood-based NF.

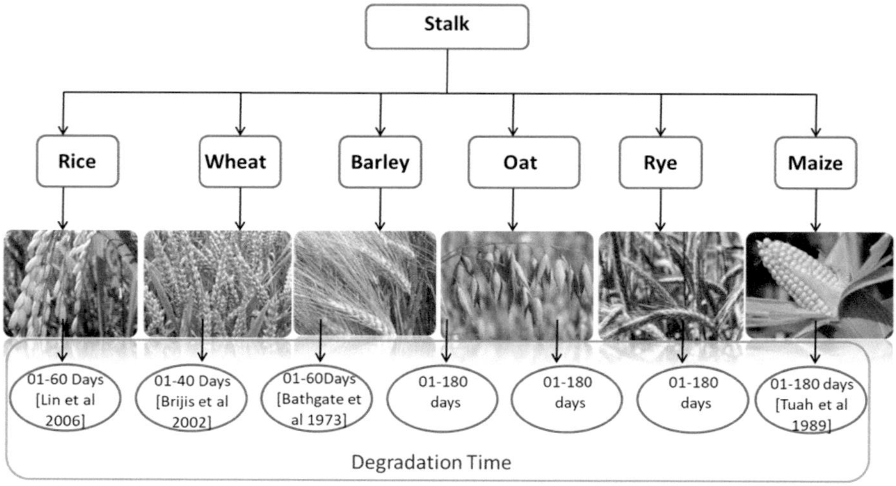

Figure 13.8. Classification of stalk-based natural fiber along with degradation time.

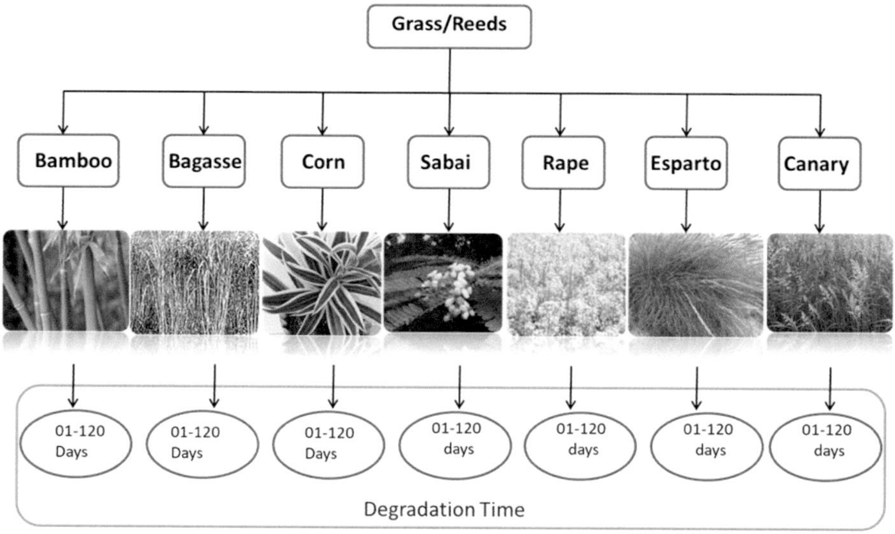

Figure 13.9. Classification of grass-based natural fiber along with degradation time.

Kenaf

Kenaf fiber production was banked in China (44%) and India (39%), making them the largest producers of natural fibers in the world. Obviously, its usage is also higher in Asian countries. The advantage of Kenaf fiber was that it has carbon sequester potential and high heat combustion property when compared to its counterpart. The percentage of weight fraction was varied from 1% to 25% in PLA matrix. The results for 1% inclusion of wt percent showed promising way to reproduce green composites without any harmful environmental reagents (Chen et al. 2017). Interfacial adhesions got improved with 25% weight fraction by inclusion between PLA and Kenaf (Kurniawan et al. 2015).

Jute

Jute fiber known as 'Golden fiber' was considered as an economically viable and abundantly available natural resource throughout Asian countries, and its highlighting properties were high tensile strength, good resistance to microorganisms, low flexibility, and poor elasticity. Many researchers have worked on the variation of weight fraction or volume fraction. In one of the cases, weight fraction was varied from 41% to 43% of PLA, which resulted in fatigue strength of 10^6 cycles with 55% of ultimate strength enhancement, which matches with GFRP (Katogi et al. 2017).

Sisal

Sisal fibers were currently the best choice in house panels, and the strength was similar to fiber glass. The sisal fiber with weight fraction or volume fraction percentage was embedded with PLA as substrate. The weight percentage of sisal varying from 6% to 30% of PLA was adopted. At 6 wt% fraction caustic soda treated fibers exhibited uneven spherulitic growth at temperature 125°C (Prajer et al. 2012). Fiber content of 10% to 30% resulted in increasing degree of crystallinity and molar mass of matrix variation (Badia et al. 2017).

Bamboo

Bamboo was considered to be an ancient fiber. Its use was first found 5,000 years back in India. Fiber has unique properties, viz antibacterial, UV transmittance is 0.06%, best fiber for moisture absorption and permeability. The percentage weight fraction considered was 1% to 60% with PLA. Bamboo nano cellulose whiskers with 1 to 16 wt% fraction showed 250% increase in elongation before break (Qian et al. 2017). In case of 60 wt% fraction, the recorded thermal conductivity was lower than GFRP and CFRP (Takagi et al. 2007).

Surface treatment of natural fiber

Natural fibers in pristine condition were not susceptible to be used as reinforcements or fillers in the holding matrix, and in some scenarios, they need a treatment to strengthen the bonding between fiber and matrix. Various types of treatments were discussed in Table 13.3, depending upon the fiber and PLA as holding matrix. Lack of adhesion in between fibers and polymer matrix resulted in void spaces around the fibers in a matrix, leading to additional water consumption (Hamid et al. 2012). Potassium hydroxide (KOH) or sodium hydroxide (NaOH) were the surface treatment alternatives, with which hydrogen bonding of cellulose diminishes and yields to elimination of open hydroxyl groups. Otherwise, they tend to adhere with water molecules (Dittenber and GangaRao 2012). Silane treatment is best suited to equilibrate holding matrix with natural fiber against water interfering. Silicon gets settled in the cell lumina and at the edge fibers are restricted for water penetration due to hydrophobic behavior of the treatment. This treatment is far more superior than surface coating, as the latter doesn't reduce the amount of moisture absorption; instead, it reduces the absorption rate (Azwa et al. 2013).

Fabrication method for fiber reinforced matrix

Fibers were embedded into the substrate in either particulate form, wired/filament form, or film stacked form. Fabrication of biocomposite was made through various methods, and they were bifurcated based on type of filler, environmental condition, application, thickness, and many other properties. The current fabrication processes used in industry are as discussed below.

Compression molding

Recently, compression molding methods are suited for kenaf fiber at a temperature of 160°C, with pressure of 10 MPa at a time period of 10 minutes (Ochi et al. 2018). Cotton, lyocell, and hemp are subjected to temperature range of 180°C with pressure of pressing of 4.2 MPa for a time period of 20 minutes (Graupner et al. 2009). Jute fiber embedded in PLA has to reach a temperature range of 185–195°C at a pressure of 1.3 MPa for a period of 8 min (Memon et., 2013). On similar note, flax fiber is kept at a temperature of 180–200°C at a pressure of 5 MPa for a time span of 5 to 15 minutes (Alimuzzaman et al. 2013).

Foam injection molding

Polymer was heated to 10°C above its melting temperature to form a viscous liquid, then the particles were mixed into the polymer substrate and fed to the Injection Molding machine. Uniform distribution of a drug can be realized based on the polymer and how many contour shapes can possibly be created. Thermal stability and cost of the setup were concerns with this type of polymer making composites. Cellulosic fiber was considered with PLA at a temperature of 170°C with barrel pressure of 18 MPa (Ding et al. 2015) for fabrication of biocomposite using foam injection molding. This type of equipment calls for a number of advantages, such as material consumption (Lee et al. 2008), improved dimensional stability (Yoon et al. 2009), a shorter cycle time (Lee et al. 2008, Pierick and Jacobsen 2001), and lower injection pressures and clamp forces (Pierick et al. 2001). However, equipments are compatible with better thermal and acoustic insulation properties (Jahani et al. 2015, Ameli et al. 2014), and improved mechanical properties (Sun et al. 2015, Wong et al. 2008).

Solvent casting

In this method, particles were dissolved in polymer with addition of an organic solvent, and the entire solution was transferred to the mold and kept under chilled condition. After it is allowed to evaporate, then formation of particle matrix occurs. The method was considered as the simplest among all, and suitable for heat sensitive blends. The drawbacks include non-uniform mixing of particulates and composite subjected to many voids. In particular, solvent casting is preferred for unfilled PLA film, for easy insertion of fillers, and can be further set for hot pressing (Olivieri et al. 2016).

Film stacking and hot pressing

In the film stacking method, films were stacked layer by layer, considering the thickness and orientation as their prime parameters. The thin films orientations were considered from 0 to 90° as per the requirement from angle of inclination. Once the number of layers were identified and assigned to the machine, the process was further subjected to hot pressing, in which computer controlled precise pneumatic or hydraulic pressing was undertaken. The identification of the temperature inside the setup was observed based on the inside chamber with the help of thermo couple or thermo sensors installed. The hot press method is suitable for sheet-like structures, in particular for meso and nano-sized fibers for polymer composite materials (Chen et al. 2017).

Electrospinning

In electrospinning, high electrical potential was applied to a syringe as a droplet at the end of the tip. At a distance of 7 cm away from the electrospun nanofibers were collected. All electrospinning processes run at an ambient temperature (You et al. 2006). PLLA/PDLA (1:1) and pure PLLA were considered for electrospinning by dissolving the cast films in chloroform, as its evaporation rate was matched suitably.

Green manufacturing

Green manufacturing has been the buzz word in the industry, particularly in metal or non-metal production, and eventually yields lesser CO_2 emission, and consumption of energy is required to be of nominal or low capacity. Figure 13.10 illustrates the energy consumption for various types of polymers compared to PLA.

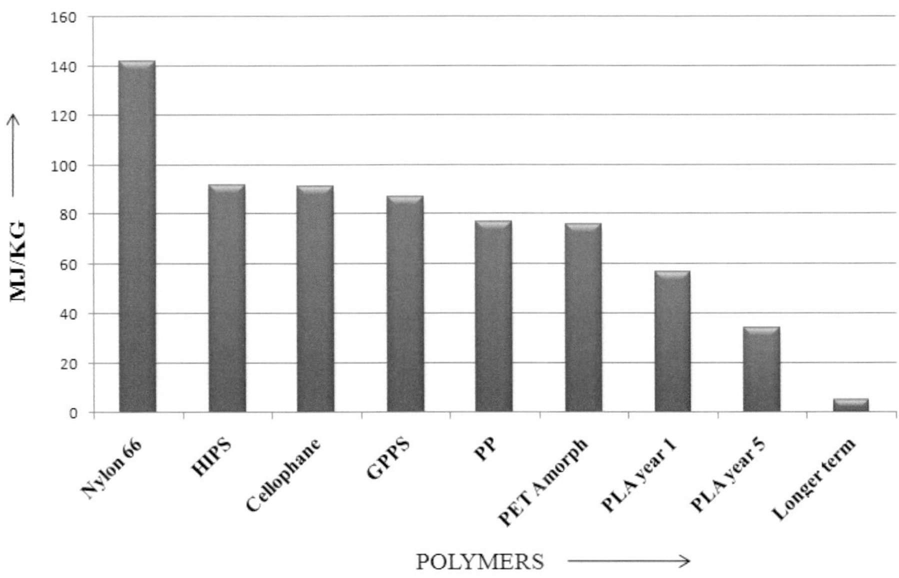

Figure 13.10. Energy for production of polymer.

Applications of PLA with natural fibers

By 2016 to 2030, the annual growth of bio-chemicals will rise up to 22%, biomaterial by 25%, and bio-fuels by 36 percent. Recent trends have shown that there is a lot of space for natural fibers to enter the industry in untouched areas. The details of applications of natural fibers with PLA as bio-polymer composites are highlighted in Table 13.4. Researchers are working with novel natural fiber materials with biodegradable, bio-compatible aspects as their main objectives.

Table 13.4. Applications of natural fiber with PLA as bio-polymer.

Sl. No	Sector	Application components	Filler
1	Automotive Industry	Door panel	Kenaf/Hemp wood fiber
2		Spare wheel pan	Abaca
3		Other interior trim	Kenaf/Flax
4		Spare tire cover	Flax
5		Seatbacks	Flax
6		Rare parcel shelves	Kenaf/Flax
7	Construction Industry	Decking	Wood flour fiber
8		Railing System	
9		Window frame	
10		Fencing	Flax
11		Panels	Rice husk, Bagasse
12	Electronic Industry	Mobile case	Kenaf/Flax
13		Laptop case	Flax
14	Food Serviceware	Cups	Jute, Banana
15		Cans	Jute, Hemp, Banana
16	Package Industry	Electronic gadgets	Coir, Hemp, Sisal, Banana
17		Consumer Products	Coir, Hemp, Sisal
18		Kitchen utensils	Coir, Hemp, Sisal, Banana
19	Textile Industry	Men/Women trousers	Cotton, Wool
20		Napkins, Towels	Cotton, Wool

Challenges and futuristic applications

The usage of natural fiber by 2030 will be 22% in biochemical, 25% in biomaterial, and 36% in bio-fuel segments, leading to prospective arena for researchers. On a similar note, biodegradable/bio-compatible polymers were the current trends when it comes to biocomposite arena. PLA was the first choice when it comes to bio-polymer selection, in particular for biodegradable and biocompatible aspects. Homogeneous blend mix of PLA with fillers or additives or in the form of reinforcement were the current challenges, as PLA in its pristine form behaves like brittle material. To restrict the early degradation of PLA with any kind of natural fiber is another challenge to work on. The embodiment of new and novel fibers has exhibited good specific properties as resulting into variability in the other properties. The major concern for researchers in biodegradable and biocomposite polymers is non-linear behavior of matrix, average to minimal impact strength, and poor long-term performance (durability).

In recent days, usage of PLA in automotive, packaging, household applications, agriculture, and textiles is quite common. Even the medical arena is also coming out with applications, such as stents for blood vanes with minimum inflammation and infection to envisage early healing. Even the orthopaedic surgeons are preferring these sort of materials as cap for knee joints (surgical pins, rods, screws, and plates), eliminating the need to reopen the plates once healed properly by operation. Other medical areas, such as sutures, prosthesis and drug delivery systems, microcapsules, microspheres, and thin coatings are a few areas to list that are benefitting from this. The futuristic applications can be in the arena of general purpose construction applications- in dental arena as either denture base material or dental implantation, competing with current materials, such as polymethylmethaacrylate (PMMA) and ceramic material, more into bio-medical applications.

Conclusion and remarks

The extensive literature survey on recent trends in utilization of natural fibers as alternative fillers reinforced in PLA matrix was critically revealed in a categorical way, focusing on various operational parameters that affect biodegradable materials and their biocompatible composites. From the exhaustive review of the literature work carried out, the following conclusions were drawn.

- From the sustainability aspect, choosing the right kind of natural fiber based on its degradation time and considering appropriate surface treatment processes, the problems such as poor long-term performance and non-linear behavior of biocomposites can be addressed.
- Manufacturing process and production of biodegradable or biocomposites were of high concern and should lead to less CO_2 emission with cleaner production In this regard, PLA is considered to be the best choice, and various natural fiber behaviors were discussed. During the course of review, authors found that fibers, such as solid house hold wastes (fruit peels and vegetable peels), hemp, jute, sisal, bamboo, and wood fibers were right blends with PLA to yield minimum CO_2 emission and achieve green manufacturing concept.
- The study also revealed the surface treatment process and weight and volume fraction utilization of filler in holding matrix to attain high strength, stiffness, and better thermal properties. With the literature it was clear that silane and alkaline treatments were the best methods to enhance the properties of natural fiber and percentages from 10 to 30 of filler in holding matrix resulted in convincing outcomes.

- PLA with natural fiber have reached many diversified areas. Till date, there are few arenas in which its impact has not yet been realised, whether it be generic construction and dental application, in particular, raw material used as construction ingredients and implantation material.

References

Abdessalem Mrabet, Guillermo Rodríguez-Gutiérrez, Rocío Rodríguez-Arcos, Rafael Guillén-Bejarano, Ali Ferchichi, Marianne Sindic and Ana Jiménez-Araujo, 2016. Quality characteristics and antioxidant properties of muffins enriched with date fruit (*Phoenix Dactylifera* L.) Fiber Concentrates 39(4): 237–244.

Abilash Noorul Islam and Kumaracoil M. Sivapragash. 2013. Environmental benefits of eco-friendly natural fiber reinforced polymeric composite materials. International Journal of Application or Innovation in Engineering & Management (IJAIEM), Volume 2, Issue 1, January 2013.

Ahmed Khan, Belas. 1999. Non-linear behaviour of unidirectional filament wound COTFRP, CFRP, and GFRP composites. pp. 537–43. *In*: Proceedings of World Engineering Congress, WEC 99, Mechanical and Manufacturing Engineering, KualaLumpur.

Ahmed Khan, Belas, Haining Na, Venkata Chevali, Philip Warner, Jin Zhu and Hao Wangates. 2017. Glycidyl methacrylate-compatibilizedpoly(lacticacid)/hemp hurd biocomposites: processing, crystallization, and thermo-mechanical response. J Mater Sci Technol 34(2): 387–397. http://dx.doi.org/10.1016/j.jmst.2017.03.004.

Aida C. Mallillin, Trinidad P. Trinidad, Ruby Raterta, Kevin Dagbay and Anacleta S. Loyola. 2008. Dietary fiber and fermentability characteristics of root crops and legumes. Br J Nutr 100(3): 485–8.

Akos, N.I., MatUzir Wahit, Rahmah Mohamed and Abdirahman Ali Yussuf. 2013. Comparative studies of mechanical properties of poly(ε-caprolactone) and poly(lactic acid) blends reinforced with natural fibers. Compos Interfaces 20(7): 459–467. DOI:10.1080/15685543.2013.807158.

Alimuzzaman, S., R.H. Gong and M. Akonda. 2013. Nonwoven polylactic acid and flax biocomposites. Polym Composites 34(10): 1611–1619.

Almgren, K.M. and E.K. Gamstedt. 2012. Characterization of interfacial stress transfer ability by dynamic mechanical analysis of cellulose fiber based composite materials. Compos Interfaces 17(9): 845–861. DOI: 10.1163/092764410X539235.

Ameli, A., D. Jahani, M. Nofar, P.U. Jung and C.B. Park. 2014. Development of high void fraction polylactide composite foams using injection molding: Mechanical and thermal insulation properties. Compos Sci Technol 90: 88–95.

Anuar, H., A. Zuraida, B. Morlin and J.G. Kovács. 2011. Micromechanical property investigations of poly(lactic acid)–kenaf fiber biocomposites. J Nat Fibers 8(1): 14–26. DOI:10.1080/15440478.2011.550765.

Arumuga Prabu, V., V. Manikandan, R. Venkatesh, P. Vignesh, S. Vignesh, K. Siva Sankar, P. Sripathy and E. Subburaj. 2015. Influence of redmud filler on palmyra fruit and palmyra fiber waste reinforced polyester composite: hardness, tensile and impact studies. Mater Phys Mech 24(2015): 41–49.

Ashok, B., S. Naresh, K. Obi Reddy, K. Madhukar, J. Cai, L. Zhang and A. Varada Rajulu. 2013. Tensile and thermal properties of poly(lactic acid)/eggshell powder composite films. Int J Polym Anal Ch 19(3): 245–255. DOI: 10.1080/1023666X.2014.879633.

Athijayamania, A., M. Thiruchitrambalam, U. Natarajana Pazhanivel. 2009. Effect of moisture absorption on the mechanical properties of randomly oriented natural fibers/polyester hybrid composite. Mater Sci Eng 517: 344–53.

Avella, M., Jan J. De Vlieger, Maria Emanuela Errico, Sabine Fischer, Paolo Vacca and Maria Grazia Volpe. 2005. Biodegradable starch/clay nanocomposite films for food packaging applications. Food Chem 93: 467–74.

Azwa, Z.N., B.F. Yousif, A.C. Manalo and W. Karunasena. 2013. A review on the degradability of polymeric composites based on natural fibres. Adv Mater Res-Switz 47: 424–442.

Badia, J.D., T. Emma Strömberg and E.K. Kittikorn, Monica. 2017. Relevant factors for the eco-design of polylactide/sisal biocomposites to control biodegradation in soil in an end-of-life scenario. Polym Degrad Stabil. doi: 10.1016/j.polymdegradstab.2017.06.004.

Barmouz Mohsen and Amir Hossein Behravesh. 2017. Shape memory behaviors in cylindrical shell PLA/TPU-cellulose nanofiber bio-nanocomposites: Analytical and experimental assessment. Composites: Part A 101: 160–172.

Belmares Hector, Arnoldo Barrera and Margarita Monjaras. 1983. New composite materials from natural hard fibres. Part 2: Fatigue studies and a novel fatigue degradation model. Ind Eng Chem Prod Res Dev 22: 643–52.

Bharath, K.N. 2008. Experimental investigation on biodegradable and swelling properties of natural fibers reinforced urea formaldehyde composites. IJAEA 1(4): 13–16.

Brian P. Mooney. 2009. The second green revolution? Production of plant-based biodegradable plastics. Biochem J 418(2): 219–232. DOI: 10.1042/BJ20081769" Prog Polym Sci 38: 1629–52.

Kristof Brijs Filip, Delvaux Viky Gilis and Jan A. Delcour. 2012. Solubilisation and degradation of wheat gluten proteins by barley matt proteolytic enzymes. J I Brewing 108(3): 348–354. https://doi.org/10.1002/j.2050-0416.2002.tb00560.x.

Butylina, S., O. Martikka and T. Kärki. 2010. Comparison of water absorption and mechanical properties of wood–plastic composites made from polypropylene and polylactic acid. Wood Mater Sci. Eng. 5(3-4): 220–228. DOI:10.1080/17480272.2010.532233.

Buzarovska, A. 2013. PLA nanocomposites with functionalized TiO_2 nanoparticles. Polym-Plast Technol 52(3): 280–286. DOI:10.1080/03602559.2012.751411.

Cazaurang, M.N., P.J. Herrera-Franco P.I. Gonzalez-Chi and M. Aguilar-Vega. 1991. Physical and mechanical properties of henequen fibers. J Appl Polym Sci 43: 749–56.

Chen, Po-Yuan, Hong-Yuan Lian, Yeng-Fong Shih, Su-Mei Chen-Wei and Ru-Jong Jeng. 2017. Preparation, characterization and crystallization kinetics of kenaf fiber/multiwalled carbon nanotube/polylactic acid (PLA) green composites. Mater Chem Phys 196: 249–255. doi: 10.1016/j.matchemphys.2017.05.006.

Conn, R.E., J.J. Kolstad, J.F. Borzelleca, D.S. Dixler, L.J. Filer Jr, B.N. Ladu Jr and M.W. Pariza. 1995. Safety assessment of polylactide (PLA) for use as a food-contact polymer. Food Chem Toxicol 33: 273–83.

Dittenber, D.B. and H.V.S. GangaRao. 2012. Critical review of recent publications on use of natural composites in infrastructure. Compos A Appl Sci Manuf 43: 1419–29.

Dong, C., X. Yuan, M. He and K. Yao. 2015. Preparation of PVA/PEI ultra-fine fibers and their composite membrane with PLA by electrospinning. J Biomat Sci, Polymer Edition, 17(6): 631–643. DOI:10.1163/156856206777346287.

Espinach, F.X., S. Boufi, M. Delgado-Aguilar, F. Julián, P. Mutjé and J.A. Méndez. 2017. Composites from poly(lactic acid) and bleached chemical fibres: Thermal properties. Composites Part B. doi: 10.1016/j.compositesb.2017.09.055.

Fuad, M.Y.A., S. Rahmad and M.R.N. Azlan. 1998. Filler content determination of bio-based thermoplastics composites by thermogravimetric analysis. pp. 268–75. *In*: Proceedings of the Fourth International Conference on Advances in Materials and Processing Technologies, 98, Kuala Lumpur.

Gigante, V., L. Aliotta, V.T. Phuong, M.B. Coltelli, P. Cinelli and A. Lazzeri. 2017. Effects of waviness on fiber-length distribution and interfacial shear strength of natural fibers reinforced composites. Compos Sci Technol 152: 129–138.

Graupner, N., A.S. Herrmann and J. Müssig. 2009. Natural and man-made cellulose fibre-reinforced poly (lactic acid) (PLA) composites: An overview about mechanical characteristics and application areas. Compos Part A-Appl S 40(6): 810–821.

Graupner, N., D. Labonte and J. Müssig. 2017. Rhubarb petioles inspire biodegradable cellulose fibre-reinforced PLA composites with increased impact strength. Compos Part A-Appl S 98: 218–226.

Hamid, M.R.Y., M.H. Ab Ghani and S. Ahmad. 2012. Effect of antioxidants and fire retardants as mineral fillers on the physical and mechanical properties of high loading hybrid biocomposites reinforced with rice husks and sawdust. Ind Crops Prod 40: 96–102.

Hasan, A., G. Waibhaw, V. Saxena and L.M. Pandey. 2018. Nano-biocomposite scaffolds of chitosan, carboxymethyl cellulose and silver nanoparticle modified cellulose nano-whiskers for bone tissue engineering applications. Int J Biol Macromol 11: 923–934.

Hassan, E., Y. Wei, H. Jiao and Y.M. Huo. 2012. Plant fibers reinforced poly (lactic acid) (PLA) as a green composites: review. International Journal of Engineering Science and Technology (IJEST).

Huang, G. 2005. Nettle (*Urtica cannabina* L) fibre, properties and spinning practice. The Journal of the Textile Institute 96: 1, 11–15, DOI: 10.1533/joti.2004.0023.

Imre, B., K. Renner and B. Pukanszky. 2014. Interactions, structure and properties in poly(lactic acid)/thermoplastic polymer blends. eXPRESS Polymer Letters 8(1): 2–14. Available online at www.expresspolymlett.comDOI: 10.3144/expresspolymlett.2014.2.

Jahani, D., A. Ameli, M. Saniei, W. Ding, C.B. Park and H.E. Naguib. 2015. Characterization of the structure, acoustic property, thermal conductivity, and mechanical property of highly expanded open-cell polycarbonate foams. Macromol Mater Eng 300(1): 48–56.

James A. Schrader, Kenneth G. McCabe and David Grewell. 2015. Bioplastics and biocomposites for sustainable horticultural containers: Performance and biodegradation in home compost. Acta Hortic 1170: 1101–1108. DOI: 10.17660/ActaHortic.2017.1170.142.

Jia Horng Lin, Chien Lin Huang, Chih Kuang Chen, Jo Mei Liao and Ching Wen Lou. 2014. Biodegradable fibre reinforced composites composed of polylactic acid and polybutylene succinate. Plast Rubber Compos 43(3): 82–88. DOI: 10.1179/1743289813Y.0000000070.

John O. Akindoyo, Mohammad, Dalour Hossen Beg, Suriati Binti Ghazali, Muhammad Remanul Islam and Abdullah Al Mamun. 2015. Preparation and characterization of poly(lactic acid)-based composites reinforced with poly dimethyl siloxane/ultrasound-treated oil palm empty fruit bunch. Polym-Plast Technol 54(13): 1321–1333. DOI: 10.1080/03602559.2015.1010219.

Kakroodi, A.D., Y. Kazemi, M. Nofar and Chul B. Park. 2017. Tailoring poly(lactic acid) for packaging applications via the production of fully bio-based *in situ* microfibrillar composite films. Chem Eng J 308: 772–782.

Katogi, H., Y. Shimamura, K. Tohgo, T. Fujii and K. Takemura. 2017. Effect of matrix ductility on fatigue strength of unidirectional jutespun yarns impregnated with biodegradable plastics. Adv Compos Mater 27(3): 235–247. DOI:10.1080/09243046.2017.1381896.

Khalid, A.A., B. Sahari and Y.A. Khalid. 1998. Environmental effects on the progressive crushing of cotton and glass fibre/epoxy composite cones. pp. 680–89. *In*: Proceedings of the Fourth International Conference on Advances in Materials and Processing Technologies, 98, Kuala Lumpur.

Khubaib, A. and M. Mujahid. 2011. Biodegradation of Textile Materials, Degree of Master in Textile Technology, The Swedish School of Textiles.

Kiruthika, A.V. and K. Veluraja. 2009. Experimental studies on the physico-chemical properties of banana fibre from various varieties. Fibers Polym 10: 193–199. https://doi.org/10.1007/s12221-009-0193-7.

Kumar, N., D. Das and B. Neckář. 2017. Effect of fiber orientation on tensile behavior of biocomposites prepared from nettle and poly(lactic acid) fibers: Modeling & experiment. Compos Part B 138: 113–121.

Kurniawan, D., Byung Sun Kim, Ho Yong Lee and Joong Yeon Lim. 2015. Towards improving mechanical properties of basalt fiber/polylactic acid composites by fiber surface treatments. Compos Interface 22(7): 553–562. DOI:10.1080/09276440.2015.1054743.

Laadila, M.A., K. Hegde, T. Rouissi, S.K. Brar, R. Galvez, L. Sorelli, R.B. Cheikh, M. Paiva and K. Abokitse. 2017. Green synthesis of novel biocomposites from treated cellulosic fibers and recycled bio-plastic polylactic acid. J Clean Prod, 164: 575–586. doi: 10.1016/j.jclepro.2017.06.235.

Laly A. Pothan. 2009. Natural Fiber Reinforced Polymer Composites: From Macro to Nano scale, Éd. des Archives Contemporaines.

Lau, A.K., W.W. Cheuka and K.V. Lo. 2009. Degradation of greenhouse twines derived from natural fibers and biodegradable polymer during composting. J Environ Manage 668–671.

Layth Mohammed, M.N.M. Ansari, Grace Pua, Mohammad Jawaid and M. Saiful Islam. 2015. A review on natural fiber reinforced polymer composite and its applications. Natural Fiber Reinforced Polym Composite, 2015, doi.org/10.1155/2015/243947.

Lee, J.W.S., J. Wang, J.D. Yoon and C.B. Park. 2008. Strategies to achieve a uniform cell structure with a high void fraction in advanced structural foam molding. Ind Eng Chem Res 47(23): 9457–64.

Lingham-Soliar, T. 2017. Microstructural tissue-engineering in the rachis and barbs of bird feathers. Sci Rep 7: 45162. https://doi.org/10.1038/srep45162.

López, A.R., J. Artero-Guerrero, J. Pernas-Sánchez and C. Santiuste. 2017. Compression after impact of flax/PLA biodegradable composites. Polym Test 59: 127–135.

Luo, S. and A.N. Netravali. 1999. Mechanical and thermal properties of environmentally friendly green composites made from pineapple leaf fibres and poly(hydroxybutyrate-co-valerate) resin. Polym Compos 20(3): 367–78.

Macha, I.J., M.M. Muna and J.L. Magere. 2018. *In vitro* study and characterization of cotton fabric PLA composite as a slow antibiotic delivery device for biomedical applications. J Drug Deliv Sci Tec 43: 172–177.

Manfredi, L.B., E.S. Rodríguez, M. Wladyka-Przybylak and A. Vázquez. 2006. Thermal degradation and fire resistance of unsaturated polyester, modified acrylic resins and their composites with natural fibres. Polym Degrad Stabil 91: 255–61.

Manik C. Biswas, Shaik Jeelani and Vijaya Rangar. 2017. Influence of bio-based silica/carbon hybrid nanoparticles on thermal and mechanical properties of biodegradable polymer films. Composites Communications 4: 43–53.

Memon, A. and A. Nakai. 2013. Fabrication and mechanical properties of jute spun yarn/PLA unidirection composite by compression molding. Energy Proced 34: 830–838.

Methacanon, P., U. Weerawatsophon, N. Sumransin, C. Prahsarn and D.T. Bergado. 2010. Properties and potential application of the selected natural fibers as limited life geotextiles. Carbohydr Polym 82: 1090–6.

Modzelewska, I., E. Patelski and E. Tyrolczyk. 2014. Certain properties of cotton paper with the addition of nanofibers in the form of PLA/PHB composite. J Nat Fibers 12(1): 52–60. DOI: 10.1080/15440478.2014.892463.

Monteiro, S.N., L.A.H. Terrones and J.R.M. D'Almeida. 2008. Mechanical performance of coir fiber/polyester composites. Polym Test 27(5): 591–595.

Muhammad Shoaib Butt, Jing Bai, Xiaofeng Wanc, Chenglin Chu, Feng Xue, Hongyan Ding and Guanghong Zhou. 2017. Mg alloy rod reinforced biodegradable poly-lactic acid composite for load bearing bone replacement. Surf Coat Tech 309: 471–479. http://dx.doi.org/10.1016/j.surfcoat.2016.12.005.

Naghieh, S., E. Foroozmehr, M. Badrossamay and M. Kharaziha. 2018. Combinational processing of 3D printing and electro-spinning of hierarchical poly(lactic acid)/gelatin-forsterite scaffolds as a biocomposite: Mechanical and biological assessment. Adv Mater Res-Switz 133(2017): 128–135.

Narayanan, M., S. Loganathan, R.B. Valapa, S. Thomas and T.O. Varghese. 2017. UV protective poly(lactic acid)/rosin films for sustainable packaging. International Journal of Biological Macromolecules. http://dx.doi.org/10.1016/j.ijbiomac.2017.01.152.

Navin Chand, Mainul Islam Fahim, Prabhat Sharma and M.N. Bapat. 2012. Influence of foaming agent on wear and mechanical properties of surface modified rice husk filled polyvinylchloride Laboratory of Microbial Technology, Dept. of Bio science and biotechnology, Kyushu university, Japan.

Nawadon Petchwattana and Sirijutaratana Covavisaruch. 2014. Mechanical and morphological properties of wood plastic biocomposites prepared from toughened poly(lactic acid) and rubber wood Sawdust (Hevea brasiliensis) J Bionic Eng 11(4): 630–637.

Nguyen, T.C., C. Ruksakulpiwat, S. Rugmai, S. Soontaranon and Y. Ruksakulpiwat. 2017. Crystallization behavior studied by synchrotron small-angle X-ray scattering of poly (lactic acid)/cellulose nanofibers composites. Compos Sci Technol 143: 106–115.

Niaounakis, M. 2013. Biopolymers: Reuse, Recycling and Disposal. Elsevier, Amsterdam.

Niu, X., Y. Liua, Y. Songa, J. Hanb and H. Pana. 2017. Rosin modified cellulose nanofiber as a reinforcing and co-antimicrobial agents in polylactic acid/chitosan composite film for food packaging. Carbohyd Polym. https://doi.org/10.1016/j.carbpol.2017.11.079.

Ochi, S. 2008. Mechanical properties of kenaf fibers and kenaf/PLA composites. Mechanics of Materials 40(4): 446–452.

Olivieri, R., L. Di Maio, P. Scarfato and L. Incarnato. 2016. Preparation and characterization of biodegradable PLA/organosilylated clay nanocomposites. AIP Conference Proceedings 1736: 020102. https://doi.org/10.1063/1.4949677.

Patil, A.Y., N. Umbrajkar Hrishikesh, G.D. Basavaraj, Gireesha R. Chalageri and Krishnaraja G. Kodancha. 2018. Influence of bio-degradable natural fiber embedded in polymer matrix. Materials Today: Proceedings 5: 7532–7540. Elsevier.

Patiño, S. and J. Grace. 2002. The cooling of convolvulaceous flowers in a tropical environment. Plant, Cell and Environ 25(1): 41–51.

Pierick, D. and K. Jacobsen. 2001. Injection molding innovation: The microcellular foam process. Plast Eng 57(5): 46–51.

Patil, A.Y., N. Umbrajkar Hrishikesh, G.D. Basavaraj, Gireesha R. Chalageri and Krishnaraja G. Kodancha. 2008. Microcellular and solid polylactide–flax fiber composites. Compos Interface 16(7-9): 869–890. DOI:10.1163/092764409X12477467990283.

Pilla, S., A. Kramschuster, J. Lee, G.K. Auer, S. Gong and L.S. Turng. 2009. Microcellular and solid polylactide-flax fiber composites. Compos Interfaces 16(7-9): 869–90.

Prajer, M. and Martin P. Ansell. 2012. Observation of transcrystalline growth of PLA crystals on sisal fibre bundles and the effect of crystal structure on interfacial shear strength. Compos Interface 19(1): 39–50. DOI: 10.1080/09276440.2012.688398.

Preethi, P. and G. Balakrishnamurthy. 2013. Physical and Chemical Properties of Banana Fiber Extracted from Commercial Banana Cultivars Grown in Tamilnadu State, Published 2013, Mathematics.

Qian, J.D., E. Strömberg, T. Kittikorn and M. Ek. 2017. Relevant factors for the eco-design of polylactide/sisal biocomposites to control biodegradation in soil in an end-of-life scenario. Polym Degrad Stabil. doi: 10.1016/j.polymdegradstab.2017.06.004.

Qiang, T., D. Yu, H. Gao and Y. Wang. 2013. Polylactide-based wood plastic composites toughened with SBS. Polym-Plast Technol 51(2): 193–198. DOI: 10.1080/03602559.2011.618518.

Rahul Chanap. 2012. Study of Mechanical and Flexural Properties of Coconut Shell Ash Reinforced Epoxy Composites, Thesis.

Rao, K.M.M., A.V.R. Prasad, M.N.V.R. Babu, K.M. Rao and A. Gupta. 2007. Tensile properties of elephant grass fiber reinforced polyester composites. J Mater Sci 42: 3266–3272. https://doi.org/10.1007/s10853-006-0657-8.

Reis, M.O., Juliana Bonametti Olivato, Ana Paula Bilck, Juliano Zanela, Maria Victoria Eiras Grossmann and Fabio Yamashita. 2018. Biodegradable trays of thermoplastic starch/poly (lactic acid) coated with beeswax. Ind Crop Prod 112: 481–487.

Ren, Y., E. Babaie and Sarit B. Bhaduri. 2018. Nanostructured amorphous magnesium phosphate/poly (lactic acid)composite coating for enhanced corrosion resistance and bioactivity of biodegradable AZ31 magnesium alloy. Prog Org Coat 118: 1–8.

Report on BISPHENOL-A (BPA) IN Teethers—an Indian perspective, Dec 2016.

Revathi, S., G. Ravipravin, C. Rajesh, S. Mohanapriya and N.S. Raghavee. 2017. Analysis of radio over fiber link based on optical carrier suppression 2017. International conference on Microelectronic Devices, Circuits.

Rhim, J.-W., H.-M. Park and C.-S. Ha. 2013. Bio-nanocomposites for food packaging applications. Prog Polym Sci 38: 1629–52.

RITMIR030: Polylactic Acid—A Market Insight Report, July 2013.

Rohan M. Bajracharya, Dilpreet S. Bajwa and Sreekala G. Bajwa. 2017. Mechanical properties of polylactic acid composites reinforced with cotton gin waste and flax fibers. Procedia Eng 200: 370–376.

Rytlewski Piotr, Krzysztof Moraczewski, Rafał Malinowski and Marian Żenkiewicz. 2014. Assessment of dicumyl peroxide ability to improve adhesion between polylactide and flax or hemp fibres. Compos Interfaces 21(8): 671–683, DOI: 10.1080/15685543.2014.927262.

Sain-Zygmunt Hejnowicz and Wilhelm Barthlott. 2005. Structural and mechanical peculiarities of the petioles of giant leaves of Amorphophallus (Araceae). Am J Bot 92(3): 391–403. doi: 10.3732/ajb.92.3.391.

Sathishkumar, T.P., P. Navaneethakrishnan and S. Shankar. 2012. Tensile and flexural properties of snake grass natural fiber reinforced isophthallic polyester composites. Compos Sci Technol 72: 1183–1190.

Sathishkumar, T.P., N. Kumar and D. Das. 2017. Fibrous biocomposites from nettle (Girardiniadiversifolia) and poly(lactic acid) fibers for automotive dashboard panel application. Compos Part B. doi:10.1016/j.compositesb.2017.07.059.

Satoshi Kobayashi and Keita Takada. 2015. Transverse properties of hemp/PLA composite fabricated with micro-braiding technique. Advan Comp Mater 24(6): 509–518, DOI: 10.1080/09243046.2014.935076.

Satyanarayana, K.G., S.G.K. Pillai, B.C. Pai and K. Sukumaran. 1990. Natural fiber–polymer composite. Cem Compos 12: 117–36.

Scaffaro, R. and F. Lopresti. 2018. Properties-morphology relationships in electrospun mats based on polylactic acid and graphene nanoplatelets. Composites: Part A. doi: https://doi.org/10.1016/j.compositesa.2018.02.026.

Sen, T. and H. Jagannatha Reddy. 2011. Application of sisal, bamboo, coir and jute natural composites in structural upgradation. Int J Innov Manage Technol 2: 186–91.

Spiridon, I., Raluca Nicoleta, Darie-Nita and Adrian Bele. 2017. New opportunities to valorize biomass wastes into green materials. II. Behaviour to accelerated weathering. J Clean Prod 172(2018): 2567–2575.

Sreenivasan, V.S., D. Ravindran, V. Manikandan and R. Narayanasamy. 2011. Mechanical properties of randomly oriented short sansevieria cylindrica fibre/polyester composites. Mater Des 32: 2444–55.

Sun, X., H. Kharbas, J. Peng and L.-S. Turng. 2015. A novel method of producing lightweight microcellular injection molded parts with improved ductility and toughness. Polymer 56: 102–10.

Takagi, H., S. Kako, K. Kusano and A. Ousaka. 2007. Thermal conductivity of PLA-bamboo fiber composites. Adv Compos Mater 16(4): 377–384. DOI: 10.1163/156855107782325186.

Takayama, T., K. Uchiumi, H. Ito, T. Kawai and M. Todo. 2013. Particle size distribution effects on physical properties of injection molded HA/PLA composites. Adv Compos Mater 22(5): 327–337. DOI:10.1080/09243046.2013.820123.

Tham, W.L., Z.A. Mohd Ishak and W.S. Chow. 2015. Mechanical and thermal properties enhancement of poly(lactic acid)/halloysite nanocomposites by maleic-anhydride functionalized rubber. J Macromol Sci B 53(3): 371–382. DOI:10.1080/00222348.2013.839314.

Torres, F.G., O.H. Arroyo, C. Grande and E. Esparza. 2006. Bio- and photo-degradation of natural fiber reinforced starch-based bio-composites. Int J Polym Mater 55: 1115–1132.

Tuah, A.K. and E.R. Orskov. The degradation of untreated and treated maize cobs and cocoa pod husks in the rumen, FAO Corporate Document Repository.

Umberto Quattrocchi. 2017. CRC World Dictionary of Palms, CRC Press, 03-Aug-2017, Science 2753 pp.

Varna, J., L. Rozite, R. Joffe and A. Pupurs. 2012. Non-linear behaviour of PLA based flax composites. Plastics, Rubber and Composites 41(2): 49–60. DOI:10.1179/1743289811Y.0000000007.

Vert, M., Yoshiharu Doi, Karl-Heinz Hellwich, Michael Hess, Philip Hodge, Przemyslaw Kubisa, Marguerite Rinaudo and François Schué. 2012. Terminology for biorelated polymers and applications (IUPAC Recommendations 2012). Pure Appl Chem 84: 377–410.

Vigneswaran, C., V. Pavithra, V. Gayathri and K. Mythili. 2015. Banana Fiber: Scope and Value Added Product Development 9(2).

Vilarinho, F., Mariana Andrade, Giovanna G. Buonocore, Mariamelia Stanzione, M. Fátima Vaz and Ana Sanches Silva. 2018. Monitoring lipid oxidation in a processed meat product packaged with nanocomposite poly(lactic acid) film. Eur Polym J 98: 362–367.

Wei Liqing, Umesh P. Agarwala, Laurent Matuanab, Ronald C. Saboa and Nicole M. Stark. 2018. Performance of high lignin content cellulose nanocrystals inpoly(lactic acid). Polymer 135: 305–313.

Wittek, T. and T. Tanimoto. 2008. Mechanical properties and fire retardancy of bidirectional reinforced composite based on biodegradable starch resin and basalt fibres. Express Polym Lett 2(11): 810–822. Available online at www.expresspolymlett.com doi: 10.3144/expresspolymlett.2008.94.

Wong, S., J.W.S. Lee, H.E. Naguib and C.B. Park. 2008. Effect of processing parameters on the mechanical properties of injection molded thermoplastic polyolefin (TPO) cellular foams. Macromol Mater Eng 293(7): 605–13.

Woraprayote, Weerapong, Laphaslada Pumpuang, Amonlaya Tosukhowong, Takeshi Zendo, Kenji Sonomoto, Soottawat Benjakul and Wonnop Visessanguan. 2018. Antimicrobial biodegradable food packaging impregnated with Bacteriocin 7293 for control of pathogenic bacteria in pangasius fish fillets. LWT - Food Sci Technol-Leb (2017), doi: 10.1016/j.lwt.2017.10.026.

Xu, Wenyang, Pranovich Andrey, Uppstu Peter, Wang Xiaoju, Kronlund Dennis, Hemming Jarl, Öblom Heidi, Moritz Niko, Preis Maren, Sandler Niklas, Willför Stefan and Xu Chunlin. 2018. Novel bio renewable composite of wood polysaccharide and polylactic acid for three dimensional printing. Carbohyd Polym 187: 51–58.

Yalcinkaya, E.E., D. Puglia, E. Fortunati, F. Bertoglio, G. Brunie, L. Visai and J.M. Kenny. 2017. Cellulose nanocrystals as templates for cetyltrimethylammonium-bromide mediated synthesis of Ag nanoparticles and their novel use in PLA films. Carbohyd Polym 157: 1557–1567.

Yang Shuangqiao and Shibing Bai Qi Wang. 2018. Sustainable packaging biocomposites from polylactic acid and wheatstraw: Enhanced physical performance by solid state shear milling process. Compos Sci Technol 158: 34–42.

Yanna Fei, Hongbo Wang, Weidong Gao, Yuqin Wan, Jiajia Fu and Ruihua Yang. 2014. Antimicrobial activity and mechanism of PLA/TP composite nanofibrous films. J Textile Institute 105(2): 196–202, Received 15 Mar 2013, Accepted 09 Aug 2013, Published online: 01 Oct 2013.

Yaowen Liu, Xue Liang, Shuyao Wang, Wen Qin and Qing Zhang. 2018. Electrospun antimicrobial polylactic acid/tea polyphenol nanofibers for food-packaging applications. Polymers (Basel) 10(5): 561.

Yatigala, N.S., Dilpreet S. Bajwa and Sreekala G. Bajwa. 2017. Compatibilization improves physico-mechanical properties of biodegradable bio-based polymer composites. Compos Part A-Appl S 315–325.

Ying Zeren, Defeng Wua, Ming Zhanga and Yaxin Qiu. 2017. Polylactide/basalt fiber composites with tailorable mechanical properties: Effect of surface treatment of fibers and annealing composite structures. Clin Podiatr Med Surg 176: 1020–1027. http://dx.doi.org/10.1016/j.cpm.2017.08.009.

Yoon, J.D., T. Kuboki, P.U. Jung, J. Wang and C.B. Park. 2009. Injection molding of wood-fiber/plastic composite foams. Compos Interfaces 16(7-9): 797–811.

Yuryev Yury, Amar K. Mohanty and Manjusri Misra. 2017. Novel biocomposites from biobased PC/PLA blend matrix system for durable applications. Compos Part B 130: 158–166.

Zhu, J., X. Li, C. Huang, L. Chen and L. Li. 2014. Structural changes and triacetin migration of starch acetate film contacting with distilled water as food simulant. Carbohydr Polym 104: 1–7.

14

New Approaches on Phytoremediation of Soil Cultivated with Sugarcane with Herbicide Residues and Fertigation

Luziane Cristina Ferreira and *Paulo Renato Matos Lopes**

Introduction

Sugarcane belongs to the genus *Saccharum* and the family Poaceae. This vegetal presents an adequate growth, with conditions of high luminosity, temperature, and soil humidity (Silva and Silva 2012). Among the biotic factors that can negatively interfere with its productivity, presence of weeds is highlighted as the competition for all available resources (Kuva et al. 2003, Squassoni 2012).

Hence, herbicides are widely used to control these weeds, and one of the most applied molecule in the crops is tebuthiuron. This compound has a pre-emergent action that inhibits the plant photosystem II (Brasil 2018). Despite the importance of pesticides in production activities, it's important to note that their excessive use can result in problems, such as residual effect (Mancuso et al. 2011), intoxication to non-target organisms (Silva et al. 2014), leaching, and water contamination (Souza et al. 2001).

Another organic compound widely used in Brazilian sugarcane fields is vinasse, whose fertilizing action is exploited in fertigation. It is the main residue from the ethanol distillation process. Vinasse use in fertigation can cause different changes in soil, such as increased bioavailability of some ions, higher cation exchange capacity, increased water retention, improvement of physical structure, and higher microbial

College of Technology and Agricultural Sciences – São Paulo State University (UNESP), Rodovia Comandante João Ribeiro de Barros, km 651 – Dracena, SP, Brazil.
Email: Lluzicferreira@gmail.com
* Corresponding author: prm.lopes@unesp.br

activity (Rezende 1979, Nunes et al. 1981, Glória and Orlando Filho 1983, Silva and Ribeiro 1997, Silva et al. 2007).

However, sugarcane vinasse application can also result in bad impacts on the environment due to its high biological oxygen demand (BOD), acid pH, and high corrosivity (Andrade 2007, Santos et al. 2009, España-Gamboa et al. 2011, Lima et al. 2016). Furthermore, its prolonged use may result in saturation of cations in soil, and in leaching and subsequent contamination of water (Morillo and Villaverde 2017).

Thus, bioremediation is presented as a sustainable alternative to remediate agricultural soil with organic compounds. This process can include different techniques—biodegradation (microorganisms), vermicomposting (earthworms), and phytoremediation (plants). Bioremediation to eliminate organic pollutants is characterized as an economical and environmentally friendly technique (Morillo and Villaverde 2017), and offers the possibility to recover contaminated areas by reducing the concentration and toxic effects of polluting agents (Fasanella and Cardoso 2016).

As herbicide tebuthiuron presents a widespread use in sugarcane crops, its high persistence in soil and high toxic potential may cause serious environmental impacts (Ibama 2009, Tonieto 2014). Therefore, it is necessary to evaluate new alternatives of treatment as phytoremediation in order to establish efficient methods to reduce these negative effects on the environment. Also, the association of this pesticide with vinasse application in sugarcane crops could present a synergy which can maximize its harmful effects on the soil.

Sugarcane

Brazil is the largest producer of sugarcane in the world, and contributes to 25% of total world production (Sindhu et al. 2016). As a result, Brazil also stands out as the world's largest producer of sugar and ethanol, with about 660 million tons of sugarcane produced per year (Mazutti et al. 2006, Mapa 2016).

Thus, the country has a great advantage for its edaphoclimatic potential, which greatly favors agricultural productions, such as sugarcane. However, there are some factors that can cause losses for this production (Nascimento 2016).

One of these concerns is the extensive use of pesticides. According to Victoria Filho and Christoffoleti (2004), weed control must be done properly as these plants cause difficulty in crop harvesting and serious losses in productivity. Moreover, Brazilian Agricultural Economics Institute (IEA 2008) stated that sugarcane is in the second place in herbicide use, and the crop that uses the most is soybean in Brazil.

Herbicides

Weed community has caused serious productivity losses when it is not adequately controlled in several crops. Despite presenting a highly efficient photosynthetic pathway (C4), sugarcane is affected by the presence of these weeds, which compete for all available resources (Victoria Filho and Christoffoleti 2004, Sandaniel et al. 2008).

Therefore, a technological package for the proper management of weeds is necessary to obtain high productivity, such as herbicides use (Kuva et al. 2008, Oliveira and Brighenti 2011). This chemical method is the most used in conventional agriculture because it has easy access and low cost compared to other control techniques (Kuva et al. 2008).

Among the herbicides most used in sugarcane field, tebuthiuron is highlighted due to the high amount applied in this crop. Its molecule (1-[5-tert-butyl-1,3,4-thiadiazol-2-yl]-1,3-dimethylurea) is included in the chemical group of substituted urea, and it is a selective herbicide with a systemic action (Fig. 14.1).

Tebuthiuron presents a toxicological classification from moderate to extreme (Rodrigues and Almeida 2011), and it is considered dangerous for the environment–classified as Class III according to Ibama (2009). This molecule also has high solubility (Franco-Bernardes et al. 2014, Tonieto 2014) and very high half-life in the soil, whose persistence can find residues for two years or more after its application (Rodrigues and Almeida 2011).

This herbicide promotes the interruption of electron transport in plants. Its action mode is by inhibiting photosynthesis because the molecule binds to the binding site of plastoquinone QB and protein D1 of photosystem II. In this process, there is also an electron transport block, which results in the paralysis of CO_2 fixation and ATP and $NADHP_2$ productions that are fundamental elements for plant growth (Oliveira Jr. et al. 2011, Silva et al. 2014).

Furthermore, sugarcane crops receive the application of other organic compounds that can also result in environmental impacts. Therefore, the use of different substances could interfere in the environmental behavior of these molecules. Among the compounds most used in sugarcane plantations is vinasse, represented by the fertigation technique.

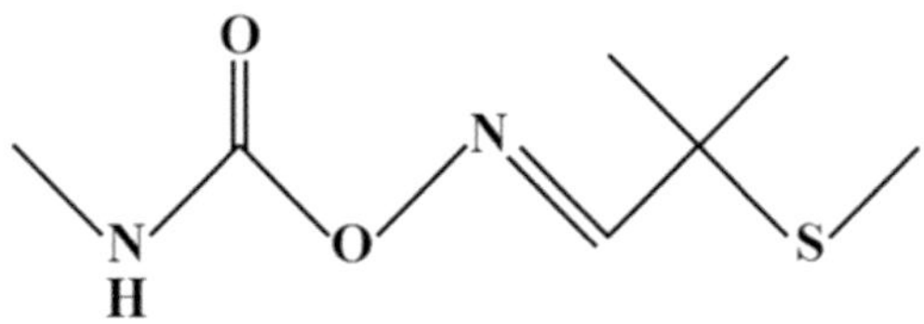

Figure 14.1. Structural formula of the herbicide tebuthiuron (Ferreira 2019).

Fertigation

Fertigation is based on the use of a by-product of sugar and alcohol powerplants as fertilizer in sugarcane fields (Silva 2011). This cost-effective strategy promotes good results to sugarcane, as vinasse also has abundance of organic matter (Hoaran et al. 2018).

In view of its chemical composition, vinasse presents high content of potassium, nitrogen, phosphorus, sulfates, chlorides, and others elements (Andrade 2007). Therefore, a sustainable alternative in the sugarcane ethanol production chain is using this residue to replace applications of mineral fertilizer (Santos et al. 2009).

Vinasse represents the main residue from ethanol distillation, whose generation is between 10 to 14 liters for each liter of biofuel produced (Assad 2017). This fact highlights its polluting potential, and consequently, the importance of its proper disposal in the environment.

Souza et al. (2015) observed that fertigation with vinasse promoted gains in soil fertility, favored stalks production, and increased the final product quality. Nevertheless, authors recognized that this practice can cause serious problems to the environment depending on the volume applied and the applied management.

Thus, the vinasse reuse in fertigation has limitations regarding the dosage in soil. If the limit is exceeded, it will not represent a sustainable strategy due to its negative effects on the environment and on the culture itself (Lima et al. 2016).

Phytoremediation

Recently, numerous anthropogenic actions are getting attention of experts with their impact on ecosystems. Regarding agricultural practices, pesticides are responsible for this worldwide concern, because of its indiscriminate use and inadequate management of applications in crops. Therefore, these poorly conducted practices can result in contamination of soil and surface and underground waters (Oliva Junior 2012).

However, many researchers are searching for efficient and ecologically viable solutions for the treatment of these areas. Therefore, metabolism of living organisms has several advantages for remediation of soils with high concentration of agricultural pesticide (Pires et al. 2003, 2005a, 2005b, Carmo et al. 2008, Pires et al. 2008, Santos et al. 2010, Belo et al. 2011, Madalão et al. 2012, 2013, Monquero et al. 2013, Araújo and Orlanda 2014, Azubuike et al. 2016, Melo et al. 2017, Morillo and Villaverde 2017, Villaverde et al. 2017, Wang et al. 2018).

Bioremediation offers different possibilities for destroying or making contaminants harmless substances by physiology of microorganisms, plants, or earthworms in microbial biodegradation, phytoremediation, and vermicomposting, respectively (Morillo and Villaverde 2017).

Araújo and Orlanda (2014) used bioremediation in Brazilian savanna soil contaminated by 2,4-dichlorophenoxyacetic (2,4-D) using bacterial strains. Results revealed a high percentage of inactivation of this herbicide by microorganisms. Also, the cultivation of some plant species in the impacted soil can stimulate microbial growth and induce the degradation of polluting molecules by microbiota capable of (co-)metabolizing them (Melo et al. 2017).

Herbicide phytoremediation has been widely studied in Brazil (Pires et al. 2006, Carmo et al. 2008, Belo et al. 2011, Madalão et al. 2012, Monquero et al. 2013). In this process, plants are used to reduce the concentration and toxicity of contaminants in the environment (Ali et al. 2013). Researches related to this technique seek to understand the plant-contaminant interaction (Vasconcellos et al. 2012), whose plant physiology is capable of promoting environmental decontamination (Souza et al. 2011).

Phytoremediation depends on contaminant characteristics and plant species, and can be by different mechanisms—extraction, stabilization, volatilization, accumulation, degradation, or stimulation of associated indigenous microbiota (phytostimulation). Plants are capable of supporting large microbial populations in the rhizosphere by producing exudates in roots (Turpault et al. 2007, Azabuike et al. 2016, Santos et al. 2017). Therefore, it represents an important remediation method for areas impacted with highly persistent herbicides (Santos et al. 2010).

Therefore, studies have been performed in order to reduce this residual effect of herbicides by using plant species that accelerate the decrease of their levels in soil. Results also showed agricultural benefits by using species of agricultural interest, such as forage plants, and by applying them to green manures (Pires et al. 2003, 2005a, b, 2008, Melo et al. 2017).

A recent research by Ferreira (2019) used the potential of plant species in decreasing tebuthiuron concentration in the soil with vinasse application (Fig. 14.2). The author tested four plant species in phytoremediation (*Canavalia ensiformis, Cajanus cajan, Pennisetum glaucum*, and *Mucuna pruriens*), and evaluated the process efficiency using *Crotalaria juncea* as bioindicator. Results showed that *M. pruriens* was the most suitable for use in sugarcane fields reforming as green manure based on plant development. According to the development of *C. juncea*, it is observed that the vessels that were previously occupied by *M. pruriens* allowed a better development and production of biomass for the bioindicator species.

Furthermore, several studies have observed a reduction in pesticides concentration in soil using phytoremediation (Pires et al. 2003, 2005a, b, 2008, Carmo et al. 2008, Madalão et al. 2013, Melo et al. 2017). In spite of these good results, the toxicity of soil samples before and after treatment was not analyzed. Thus, it is essential to evaluate the ecotoxicological potential in the environment to demonstrate the success in bioremediation process (Banks and Schultz 2005). A concern about bioremediation of organic compound without ecotoxicology analysis owing to the degradation of these molecules can generate intermediate chemical species that are more toxic than the original compound (Rocha et al. 2018).

Figure 14.2. Phytoremediation research using four plant species in soil with tebuthiuron and vinasse (Ferreira 2019).

Ecotoxicity

Ecotoxicology is the area of toxicology that assesses the toxic effects of natural or synthetic compounds in ecosystems. The action of these molecules can result in impacts on fauna, flora, and/or microbiota of contaminated area (Forbes and Forbes 1994).

Adverse effects caused by chemical agents on living beings are evaluated through ecotoxicity tests, which evaluate these damages on test organisms or non-target species of the substance under analysis (Hagner et al. 2010, Cardoso and Alves 2013).

Bioremediation studies have been using seeds as test-organisms in ecotoxicological evaluation to assess the efficiency of treatments in environments contaminated by different compounds—polycyclic aromatic hydrocarbons – HPAs (Carvalho 2010), automotive lubricant oils (Lopes et al. 2010), pharmaceuticals (Rede 2011), vegetable oils and biodiesel (Tamada et al. 2012), petroleum and derivatives (Cruz et al. 2014), fire foams (Montagnolli et al. 2017), among others.

Regarding the impact of organic compounds on terrestrial environments, the effects of agricultural pesticides must be determined in order to know their environmental behavior and ecotoxicity for bioindicators in acute and chronic stress. These molecules widely used around the world constantly cause contamination problems because of their high persistence and toxicity. Thus, bioassays with different living beings are able to assess the impacts these products have on non-target organisms, and also predict the efficiency of a treatment method adopted (Grisolia 2005).

Compounds' permanence in soil can be quantitatively determined by chromatographic techniques. On the other hand, ecotoxicity bioassays detect the impact of pesticides and their residues on biota with several methodologies and test-organisms. Experimental protocols of ecotoxicological evaluation generally present good results, simplicity, and low cost. However, results are semi-quantitative and specific to molecule and test organism used (Beyer et al. 1988).

In relation to this attention to ecotoxicity bioassays, it was shown that researches on soil treatment with pesticides using microorganisms (Villaverde et al. 2017, Wang et al. 2018) and plants (Pires et al. 2003, 2005a, b, 2008, Carmo et al. 2008, Madalão et al. 2013, Melo et al. 2017) generally do not assess the soil toxicity before and after the biological process. Therefore, they are not able to certify the bioremediation efficiency due to the absence of comparative ecotoxicological data.

Ferreira (2019) proposed to evaluate the phytoremediation potential of species of agronomic interest for associations between tebuthiuron and vinasse. Ecotoxicological effects of different treatments were determined at the initial (zero) and final (50 days) of remediation using lettuce seeds (*Lactuca sativa*) as test organisms (Fig. 14.3).

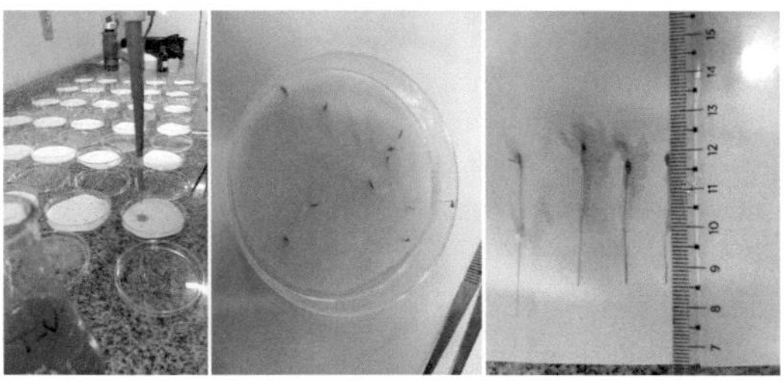

Figure 14.3. Ecotoxicological evaluation with lettuce seeds as test-organisms of soil with tebuthiuron and vinasse using phytoremediation (Ferreira 2019).

Firstly, results revealed two bean species (*Canavalia ensiformis* and *Cajanus cajan*) did not resist the presence of the herbicide even though they are not the target species of the herbicide. Regarding the ecotoxicity tests, the presence of tebuthiuron in soil revealed a toxic potential for *L. sativa* seeds, but the application of vinasse to the herbicide favored the reduction of this ecotoxicity in the initial time.

Conclusion

The presence of persistent compounds in the environment has induced several political, social, economic, and mainly environmental problems. Thus, the indiscriminate use of pesticides and the incorrect management of crops cause serious contamination in soil and water. Therefore, the balance between agricultural productivity and environment sustainability must be considered one of the priorities of modern society.

Brazilian prominence in sugarcane production consequently promotes the greater use of herbicides for weeds control. Sugarcane crop is also an important production system in reusing agroindustrial wastes. An example is fertigation with vinasse for its fertilizer potential. However, the association of these compounds with pesticides in soil can cause environmental effects not yet known to science.

The use of microorganisms and plants in bioremediation is a way of treating these areas. This technique represents a sustainable and low cost alternative for the degradation of persistent and highly toxic compounds in agricultural soils. However, the evaluation of bioremediation efficiency must be complete when there is an ecotoxicological study in soil samples before and after treatment. The quantification of polluting compound is not conclusive to certify that the adopted method was efficient.

Therefore, conducting research on new forms of treatment based on phytoremediation in agricultural soils is fundamental for the development of a productive, conscious, and sustainable society.

References

Ali, H., E. Khan and M.A. Sajad. 2013. Phytoremediation of heavy metals—concepts and applications. Chemosphere 91: 869–881.

Andrade, J.M.F. 2007. Environmental impacts of sugarcane agroindustry: subsidies for management (Impactos ambientais da agroindústria da cana-de-açúcar: subsídios para a gestão). Specialization Monograph, University of São Paulo, Piracicaba, SP, Brazil.

Araújo, L.C.A. and J.F.F. Orlanda. 2014. Biodegradation of 2,4-D herbicide using bacteria selected from the soil of the Cerrado of Maranhão (Biodegradação do herbicida 2,4-D utilizando bactérias selecionadas do solo do cerrado maranhense). Rev Ecotox Meio Amb 24: 21–32.

Assad, L. 2017. Use of waste from the sugar and alcohol sector challenges companies and researchers (Aproveitamento de resíduos do setor sucroalcooleiro desafia empresas e pesquisadores). Cienc Cultura 69: 13–16.

Azubuike, C.C., C.B. Chikere and G.C. Okpokwasili. 2016. Bioremediation techniques classification based on site of application: principles, advantages, limitations and prospects. World J Microb Biotech 32: 180.

Banks, M.K. and K.E Schultz. 2005. Comparison of plants for germination toxicity tests in petroleum-contaminated soils. Water Air Soil Poll 167: 211–219.

Belo, F.A., A.T.C.P. Coelho, L.R. Ferreira, A.A. Silva and J.B. Santos. 2011. Potential of plant species in the remediation of soil contaminated with sulfentrazone (Potencial de espécies vegetais na remediação de solo contaminado com sulfentrazone). Planta Daninha 29: 821–828.

Beyer, E.M., M.J. Duffy, J.V. Hay and D.D. Schlueter. 1988. Sulfunylureia. pp. 117–189. *In*: Kearney, P.C. and D.D. Kaufman (eds.). Herbicides: Chemistry, Degradation, and Mode of Action. New York: M. Dekker.

Brasil. Ministério da Agricultura, Pecuária e Abastecimento. Agrofit, 2018. Brasília, DF.

Cardoso, E.J.B.N. and P.R. Alves. 2013. Soil ecotoxicology. pp. 27–50. *In*: Ghousia, B. (ed.). Ecotoxicology. InTechOpen, London, UK.

Carmo, M.L., S.D.O. Procópio, F.R. Pires, A. Cargnelutti Filho, A.L.D. Barroso, G.P. Silva and L. Pacheco. 2008. Selection of plants for phytoremediation of soils contaminated with picloram (Seleção de plantas para fitorremediação de solos contaminados com picloram). Planta Daninha 26: 301–313.

Carvalho, M.V.F. 2010. Chemical and toxicological evaluation of PAA-contaminated soil submitted to biodegradation by the basidiomycete fungus *Pycnoporus sanguineus* (Avaliação química e toxicológica de solo contaminado por HPAs submetido à biodegradação pelo fungo basidiomiceto *Pycnoporus sanguineus*). Masters dissertation, Feira de Santana State University, Feira de Santana, BA, Brazil.

Cruz, J.M., I.S. Tamada, P.R.M. Lopes, R.N. Montagnolli and E.D. Bidoia. 2014. Biodegradation and phytotoxicity of biodiesel, diesel, and petroleum in soil. Water Air Soil Poll 225: 1962.

Espana-Gamboa, E., J. Mijangos-Cortes, L. Barahona-Perez, J. Dominguez-Maldonado, G. Hernández-Zarate and L. Alzate-Gaviria. 2011. Alzate-Gaviria Vinasses: characterization and treatments. Waste Manage Res 29: 1235–1250.

Fasanella, C.C. and E.J.B.N. Cardoso. 2016. Bioremediation (Biorremediação). pp. 197–210. *In*: Andreote F.D. and Cardoso E.J.B.N. (eds.). Soil microbiology (Microbiologia do solo). ESALQ, Piracicaba, Brazil.

Ferreira, L.C. 2019. Soil phytoremediation with application of tebuthiuron and vinasse by species of agronomic interest (Fitorremediação de solo com aplicação de tebuthiuron e vinhaça por espécies de interesse agronômico). Masters dissertation, São Paulo State University, Ilha Solteira, SP, Brazil.

Forbes, V.E. and T.L. Forbes. 1994. Ecotoxicology in Theory and Practice. Chapman and Hall, London, UK.

Franco-Bernardes, M.F., L.R. Maschio, M.T.V. Azeredo Oliveira and E.A. Almeida. 2014. Biochemical and genotoxic effects of a commercial formulation of the herbicide tebutiurom in *Oreochromis niloticus* of different sizes. Ecotox Environ Contam 9: 59–67.

Glória, N.A. and J. Orlando Filho. 1983. Application of vinasse as a fertilizer (Aplicação de vinhaça como fertilizante). Boletim técnico PLANALSUCAR 5: 1–38).

Grisolia, C.K. 2005. Pesticides: mutation, cancer and reproduction (Agrotóxicos: mutação, câncer e reprodução). Editora Universidade de Brasília, Brasília, DF, Brazil.

Hagner, M., O.P. Penttinen, T. Pasanen, K. Tiilikkala and H. Setälä. 2010. Acute toxicity of birch tar oil on aquatic organisms. Agri Food Sci 19: 24–32.

Hoaran, J., Y. Caro, I. Grondin and T. Petit. 2018. Sugarcane vinasse processing: toward a status shift from waste to valuable resource. A review. J Water Process Eng 24: 11–25.

Ibama - Instituto Brasileiro do Meio Ambiente e dos Recursos Naturais Renováveis. Manual para requerimento de avaliação ambiental: agrotóxicos e afins. Brasília-DF: IBAMA, p.180, 2009. Available in: <http://ibama.gov.br/sophia/cnia/livros/ManualparaRequerimentodeAvaliacaoAmbiental.pdf>. Access in: April 20, 2018.

IEA – Instituto De Economia Agrícola. Defensivos Agrícolas: Cana-de-açúcar, 2018. Disponível em: <http://ciagri.iea.sp.gov.br/nia1/cadeia/cadeiaCana.aspx>. Acesso em: May 08, 2018.

Kuva, M.A., R. Gravena, R.A. Pitelli, P.J. Christoffoleti and P.L.C.A. Alves. 2003. Weed interference periods in sugarcane culture: III—brachiaria grass (*Brachiaria decumbens*) and coloniz grass (*Panicum maximum*) (Períodos de interferência das plantas daninhas na cultura da cana-de-açúcar: III—capim-braquiária (*Brachiaria decumbens*) e capim-colonião (*Panicum maximum*)). Planta Daninha 21: 37–44.

Kuva, M.A., A.S. Ferraudo, R.A. Pitelli, P.L.C.A. Alves and T.P. Salgado. 2008. Infestation patterns of weed communities in raw sugarcane agroecosystem (Padrões de infestação de comunidades de plantas daninhas no agroecossistema de cana-crua). Planta Daninha 26: 549–557.

Lima, F.A., A.C Santos Jr, L.C. Martins, B. Sarrouh and R.C.Z. Lofrano. 2016. Review of toxicity and environmental impacts related to effluent vinasse from the sugar and alcohol industry (Revisão sobre a toxicidade e impactos ambientais relacionados à vinhaça efluente da indústria sucroalcooleira). Cadernos UniFOA 11: 27–34.

Lopes, P.R.M., R.N. Montagnolli, R.F. Domingues and E.D. Bidoia. 2010. Toxicity and biodegradation in sandy soil contaminated by lubricant oils. B Environ Contam Tox 84: 454–458.

Madalão, J.C., F.R. Pires, K. Chagas, A. Cargnelutti Filho and S.O. Procópio. 2012. Use of legumes in phytoremediation of soil contaminated with sulfentrazone (Uso de leguminosas na fitorremediação de solo contaminado com sulfentrazone). Pesqui Agropecu Trop 42: 390–396.

Madalão, J.C., F.R. Pires, A. Cargnelutti Filho, A.F. Nascimento, K. Chagas, R.S. Araújo, S.O. Procópio and R. Bonomo. 2013. Susceptibility of plant species with phytoremediation potential of the herbicide sulfentrazone (Susceptibilidade de espécies de plantas com potencial de fitorremediação do herbicida sulfentrazone). Rev Ceres 60: 111–121.

Mancuso, M.A.C., E. Negrisoli and L. Perim. 2011. Residual effect of herbicides on the soil ("carryover") (Efeito residual de herbicidas no solo ("carryover")). Rev Bras Herb 10: 151–164.

Mazutti, M., J.P. Bender, H. Treichel and M. Di Luccio. 2006. Optimization of inulinase production by solid-state fermentation using sugarcane bagasse as substrate. Enzyme Microb Tech 39: 56–59.

Melo, C.A.D., W.M.D. Souza, F.P.D. Carvalho, A.M. Massenssini, A.A.D. Silva, L.R. Ferreira and M.D. Costa. 2017. Microbial activity of soil with sulfentrazone associated with phytoremediator species and inoculation with a bacterial consortium. Bragantia 75: 300–310.

Ministério Da Agricultura, Pecuária E. Abastecimento (MAPA). Plano Nacional de Agroenergia 2015–2016. 110p Secretaria de Produção e Agroenergia. 2ed. revisada. Brasilia: Embrapa Informação tecnológica, 2016.

Monquero, P.A., M.C. Côrrea, L.N. Barbosa, A. Gutierrez, I. Orzari and A.C.S. Hirata. 2013. Selection of green manure species for phytoremediation of diclosulam (Seleção de espécies de adubos verdes visando à fitorremediação de diclosulam). Planta Daninha 31: 127–135.

Montagnolli, R.N., P.R.M. Lopes, J.M. Cruz, E.M.T. Claro, G.M. Quiterio and E.D. Bidoia. 2017. The effects of fluoride based fire-fighting foams on soil microbiota activity and plant growth during natural attenuation of perfluorinated compounds. Environl Toxicol Phar 50: 119–127.

Morillo, E. and J. Villaverde. 2017. Advanced technologies for the remediation of pesticide-contaminated soils. Sci Total Environ 586: 576–597.

Nascimento, A. 2016. Efficacy of herbicides applied in pre-planting incorporated in the culture of sugarcane (Eficácia de herbicidas aplicados em pré-plantio incorporado na cultura da cana-de-açúcar). Masters dissertation, São Paulo State University, Ilha Solteira, SP, Brazil.

Nunes, M.R., A.C.X. Velloso and J.R. Leal. 1981. Effect of vinasse on exchangeable cations and other soil chemical elements (Efeito da vinhaça nos cátions trocáveis e outros elementos químicos do solo). Pesqui Agropecu Bras 16: 171–176.

Oliva Júnior, E.F. 2012. Environmental impacts of anthropic action at the source of Piauí River (Os impactos ambientais decorrentes da ação antrópica na nascente do rio Piauí). Rev Eletrônica Faculdade José Augusto Vieira 5: 1–17.

Oliveira, M.F. and A.M. Brighenti. 2008. Comportamento dos herbicidas no ambiente. pp. 549–557. *In:* Oliveira Jr, R.S., J. Constantin and M.H. Inoue. Biologia e manejo de plantas daninhas. Edição 2011. Daninha. Curitiba: Ominipax.

Oliveira Jr., R.S. 2011. Mecanismos de ação de herbicidas. pp. 141–192. *In*: Oliveira Jr. R.S., J. Constantin and M.H. Inoue (eds.). Biologia e manejo de plantas daninhas. Ominipax, Curitiba, PR, Brazil.

Pires, F.R., C.M. Souza, A.A. Silva, M.E.L.R. Queiroz, S.O. Procópio, J.B. Santos and P.R. Cecon. 2003. Selection of plants with potential for phytoremediation of tebuthiuron (Seleção de plantas com potencial para fitorremediação de tebuthiuron). Planta Daninha 21: 451–458.

Pires, F.R., C.M. Souza, P.R. Cecon, J.B. Santos, M.R. Totola, S.O. Procópio, A.A. Silva and C.S.W. Silva. 2005a. Inferences on rhizospheric activity of species with potential for phytoremediation of the herbicide tebuthiuron (Inferências sobre atividade rizosférica de espécies com potencial para fitorremediação do herbicida tebuthiuron). Rev Bras Cienc Solo 29: 627–634.

Pires, F.R., C.M. Souza, A.A. Silva, P.R. Cecon, S.O. Procópio, J.B. Santos and L.R. Ferrreira. 2005b. Phytoremediation of tebuthiuron-contaminated soils using species grown for green manure

(Fitorremediação de solos contaminados com tebuthiuron utilizando-se espécies cultivadas para adubação verde). Planta Daninha 23: 711–717.

Pires, F.R., S.O. Procópio, C.M. Souza, J.B. Santos and G.P. Silva. 2006. Adubos verdes na fitorremediação de solos contaminados com o herbicida tebuthiuron. Caatinga 19(1): 92–97.

Pires, F.R., S.D. Oliveira Procópio, J. Barbosa Dos Santos, C.M.D., Souza and R.R. Dias. 2008. Phytoremediation evaluation of tebuthiuron using Crotalaria juncea as an indicator plant (Avaliação da fitorremediação de tebuthiuron utilizando Crotalaria juncea como planta indicadora). Rev Cienc Agron 39: 245–250.

Rede, D.S.G.M. 2011. Ecotoxicological evaluation of soils contaminated by ibuprofen (Avaliação ecotoxicológica de solos contaminados por ibuprofeno). Ph.D. Thesis, Polytechnic Institute of Porto, Porto, Portugal.

Rezende, J.O. 1979. Consequences of vinasse application on some physical properties of an alluvial soil: a case study (Consequências da aplicação de vinhaça sobre algumas propriedades físicas de um solo aluvial: estudo de um caso). Ph.D. Thesis, University of São Paulo, Piracicaba, SP, Brazil.

Rocha, R.S., A.A.G.F. Beati, R.B. Valim, J.R. Steter, R. Bertazzoli and M.R.V. Lanza. 2018. Evaluation of degradation by-products of the herbicide ametrine obtained via advanced oxidative processes (Avaliação dos subprodutos de degradação do herbicida ametrina obtidos via processos oxidativos avançados). Rev Bras Eng Biossist 12: 52–67.

Rodrigues, M.N. and F.S. Almeida. 2011. Tebuthiuron and hexazinone (Tebutiuron e hexazinona). pp. 102–697. *In*: Rodrigues, M.N. and F.S. Almeida (eds.). Herbicide Guide (Guia de herbicidas). IAPAR, Londrina, Brazil..

Sandaniel, C.R., L.B. Fernandez and A.L.L. Barroso. 2008. Sugarcane weed control with herbicides applied in pre-emergence (Controle de plantas daninhas em cana soca com herbicidas aplicados em pré-emergência). Núcleos 5: 1–10.

Santos, A.R., M.L. Sales and M.L. Campolino. 2017. Seeds of *Lactuca sativa* (lettuce) as a bioindicator of water toxicity in urban streams JK and Interlagos, southeastern region of Sete Lagoas, MG (Sementes de Lactuca sativa (alface) como bioindicador da toxicidade da água dos córregos urbanos JK e Interlagos, região sudeste de Sete Lagoas, MG). Rev Bras Cienc Vida 5: 1–14.

Santos, E.A., C.M. Dutra, L.R. Ferreira, M. Rodrigues Dos Reis, A. Cabral França and J. Barbosa Dos Santos. 2010. Rhizospheric activity of herbicide-treated soil during *Stizolobium aterrimum* remediation process (Atividade rizosférica de solo tratado com herbicida durante processo de remediação por *Stizolobium aterrimum*). Pesq Agropecu Trop 40: 1–7.

Santos, T.M.C., M.A.L. Santos, C.G. Santos, V.R. Santos and D.S. Pacheco. 2009. Effect of fertigation with vinasse on soil microorganisms (Efeito da fertirrigação com vinhaça nos microrganismos do solo). Rev Caatinga 22: 155–160.

Silva, M.A.S., N.P. Griebeler and L.C. Borges. 2007. Use of vinasse and impacts on soil and groundwater properties (Uso de vinhaça e impactos nas propriedades do solo e lençol freático). Rev Bras Eng Agricol Amb 11: 108–114.

Silva, A.M.P. 2011. Fertigation with the use of vinasse in the cultivation of sugarcane and its effect on the soil (Fertirrigação com o uso da vinhaça na cultura da cana-de-açúcar e seu efeito no solo). Masters dissertation, Anhanguera University - UNIDERP, Campo Grande, MS, Brazil.

Silva, G.S., C.A.D. Melo, C.M.T. Fialho, L.D. Tuffi Santos, M.D. Costa and A.A. Silva. 2014. Impact of sulfentrazone, isoxaflutole and oxyfluorfen on the microorganisms of two forest soils. Bragantia 73: 292–299.

Silva, J.P.N. and M.R.N. Silva. 2012. Notions of sugar cane culture (Noções da cultura da cana-de-açúcar). Instituto Federal de Educação, Ciência e Tecnologia de Goiás, Inhumas, Brazil.

Sindhu, R., E. Gnansounou, P. Binod and A. Pandey. 2016. Bioconversion of sugarcane crop residue for value added products—An overview. Renewable Energy 98: 203–215.

Souza, J.K.C., F.O. Mesquita, J. Dantas Neto, M.M.A. Souza, C.H.A. Farias, H.C. Mendes, R.M.A. Nunes and J.K.C. Souza. 2015. Fertigation with vinasse in sugarcane production (Fertirrigação com vinhaça na produção de cana-de-açúcar). Agropecu Cient Semi-Árido 11: 07–12.

Souza, L.A.D., S.A.L.D. Andrade, S.C.R.D. Souza and M.A. Schiavinato. 2011. Tolerance and phytoremediation potential of Stizolobium aterrimum associated with the arbuscular mycorrhizal fungus Glomus etunicatum in soil contaminated by lead (Tolerância e potencial fitorremediador

de *Stizolobium aterrimum* associada ao fungo micorrízico arbuscular *Glomus etunicatum* em solo contaminado por chumbo). Rev Bras Cienc Solo 35: 1441–1451.
Souza, M.D., R.C. Boeira, M.A.F. Gomes, V.L. Ferracini and A.H.N. Maia. 2001. Adsorption and leaching of tebuthiuron in three types of soil (Adsorção e lixiviação de tebuthiuron em três tipos de solo). Rev Bras Cienc Solo, Viçosa 25: 1053–1061.
Squassoni, V.L. 2012. Monitoring of the weed community in sugarcane and the efficiency of chemical control using multivariate data analysis techniques (Monitoramento da comunidade de plantas daninhas na cana-de-açúcar e da eficiência de controle químico por meio de técnicas de análise multivariada de dados). Masters dissertation, São Paulo State University, Jaboticabal, SP, Brazil.
Tamada, I.S., P.R.M. Lopes, R.N. Montagnolli and E.D. Bidoia. 2012. Toxicological evaluation of vegetable oils and biodiesel in soil during the biodegradation process. Braz J Microbiol 43: 1576–1581.
Tonieto, T.A.P. 2014. Dynamics of tebutiuron and hexazine herbicides in the raw cane system (Dinâmica dos herbicidas tebutiurom e hexazina no sistema de cana crua). Masters dissertation, São Paulo State University, Piracicaba, SP, Brazil.
Turpault, M.P., G.R. Gobran and P. Bonnaud. 2007. Temporal variations of rhizosphere and bulk soil chemistry in a Douglas fir stand. Geoderma 137: 490–496.
Vasconcellos, M.C., D. Pagliuso and V.S. Sotomaior. 2012. Phytoremediation: A proposal for soil decontamination (Fitorremediação: Uma proposta de descontaminação do solo). Estud Biologia 34: 261–267.
Victoria Filho, R. and P.J. Christoffoleti. 2004. Weed management and sugarcane productivity (Manejo de plantas daninhas e produtividade da cana). Visão Agrícola 3: 32–37.
Villaverde, J., M. Rubio-Bellido, F. Merchan and E. Morillo. 2017. Bioremediation of diuron contaminated soils by a novel degrading microbial consortium. J Environ Manage 188: 379–386.
Wang, X., X. Hou, S. Liang, Z. Lu, Z. Hou, X. Zhao and H. Zhang. 2018. Biodegradation of fungicide tebuconazole by *Serratia marcescens* strain B1 and its application in bioremediation of contaminated soil. Int Biodeter Biodegr 127: 185–191.

15

Bioactivity and Degradability Study of the Bone Scaffold Developed from *Labeo Rohita* Fish Scale Derived Hydroxyapatite

Payel Deb,[1,*] *Emon Barua,*[1] *Sumit Das Lala*[2] and *Ashish B. Deoghare*[1]

Introduction

Over the last few decades, there has been a growing demand and interest towards the development of bone substitute materials. The necessity is due to the injury in bone tissues because of accidents or inherited diseases (Bhat and Kumar 2012). The therapies available for the treatment to recover from this situation are surgical interventions, drug therapies, implant fixations, application of tractions, and autograft or allograft at the injured sites. However, for the treatment of segmental bone defects, scaffold is an alternative approach for overcoming it. Bone scaffolds are materials that are designed or engineered to contribute towards cellular interaction for the formation of new bone tissues. Hydoxyapatite (HA), the major constituent of human bone, is gaining huge interest in bone scaffold development due to its excellent bioactive, non-toxic, and osteoconductive properties (Tripathi and Basu 2012). HA can be synthesized by chemical synthesis processes, such as wet chemical process, sol–gel method, hydrothermal reaction, solid state reaction, co-precipitation reaction, microwave irradiation process, etc. HA can also be derived from natural sources, such as fish scales, fish bones, chicken bones, caprine bones, bovine bones, etc. using

[1] Department of Mechanical Engineering, National Institute of Technology Silchar, Assam-788010, India.
[2] Department of Mechanical Engineering, Parul Institute of Engineering and Technology, Parul University, Vadodara, Gujarat-391760, India.
Emails: imon18enator@gmail.com; sumitdaslala@gmail.com; ashishdeoghare@gmail.com
* Corresponding author: payeldebmech13@gmail.com

thermal degradation process (Deb and Deoghare 2019). Studies have shown that the HA extracted from natural sources have better bone bonding and cell attachment ability compared to synthetic HA (Boutinguiza et al. 2012).

Fish is a major food diet for the people of north eastern region, and fish scales contain high amounts of calcium (Ca) and phosphorus (P). Fish scales are generally dumped as bio-waste that can be harnessed to extract HA. The extracted HA can be a major constituent for the development of bone scaffold. A number of fish scales have been addressed in literature that are effectively utilized for the synthesis of HA. Pon-on et al. (Pon-On et al. 2016) extracted HA from P. jullieni fish scale to develop scaffold, and compared it with the chemically synthesized HA scaffold. It was found that the osteoblastic cell adhered with the fish scale derived HA scaffold more than the chemically synthesized HA scaffold. Tilapia (Oreochromis mossambicus) fish scales derived HA was used by Mondal et al. (Mondal et al. 2016) to develop scaffold. When compared with the synthetic HA scaffold, the fish scale derived HA scaffold shows better cell proliferation. Deb et al. (Deb et al. 2019) developed scaffold from HA synthesized from the fish scale of *Puntius conchonius*, which is suitable as a bone forming material for biomedical applications.

The present article will explore the concept of bioactivity and biodegradability of the bone scaffold developed from the *Labeo Rohita* fish scale derived HA. These are the essential criteria for scaffold formation which must be accomplished to enhance the success rate to treat critical segmental bone defect issues. The essential ingredient in the development of the scaffold is the biocompatible material. This material is extracted from the scales of the fish in the form of HA. The HA is derived from thermal degradation process of the fish scales. HA is utilized to develop a novel composite bone scaffold using Poly(methyl methacrylate) (PMMA) as the matrix. The developed scaffold is characterized using X-ray Diffraction (XRD) and Fourier Transform Infrared Spectrometry (FTIR) to confirm the presence of the HA and other constituent elements in the composite bone scaffold. Scanning Electron Microscopy (SEM) is used to investigate the morphology of the developed scaffold. The compressive property of the scaffold is also determined by executing uni-axial compression test. *In vitro* bioactivity and degradability studies are also conducted to assess the bioactive, bone forming ability, and degradation nature of the composite bone scaffold. The study reveals that HA synthesized from *Labeo Rohita* fish scales can be suitable bone substitute material in scaffold development.

Experimental procedure

Extraction of HA

Labeo Rohita fish scales are collected from the nearby fish market and are rinsed thoroughly with tap water for 3–4 times for the complete removal of debris and foreign materials attached to it. The fish scales are then pre-treated using 1(N) HCl solution for 24 hours. The pre-treatment process removes collagen, connective tissue, proteins, and organic matters from the fish scales, retaining the HA intact. Thereafter, the fish scales are thoroughly washed with distilled water to remove the HCl solution adhered to its surface, followed by drying in a hot air oven for 48 hours.

HA is then extracted from the fish scales by thermal degradation in a muffle furnace at a temperature of 900°C for 3 hours. The heating rate is maintained at 10°C/min.

Development of composite bone scaffold

Composite bone scaffold is developed by incorporating *Labeo Rohita* fish scale derived HA in Poly(methyl methacrylate) (PMMA) matrix. The weight percentage of HA and PMMA is taken as 70 wt% and 30 wt%, respectively. The weight percentage is chosen based on the fact that scaffold developed with HA:PMMA as 70:30 shows the best mechanical and physicochemical properties, as reported in literature (Deb et al. 2019). The polymer is dissolved in chloroform, and the HA is dissolved in ethyl alcohol separately. The polymer solution is stirred in a magnetic stirrer for 2 hr, whereas the HA solution is stirred for 8 hours. The individual solutions are then mixed, and the composite slurry is stirred at 1000 rpm in room temperature for 6 hours. NaCl (75% of PMMA) is added in the slurry as a porogen and stirred for 10 minutes. The slurry is then poured into the silicon rubber mold prepared using ASTM standards and allowed to dry in room temperature for 24 hours. The scaffolds are then soaked into distilled water of pH 7 for 48 hr for leaching out the NaCl particles, and every 12 hr, the water is changed. Then the scaffold are taken out from the water and oven dried for 50 hours. Figure 15.1 shows the pictorial representation of the entire process followed for scaffold development.

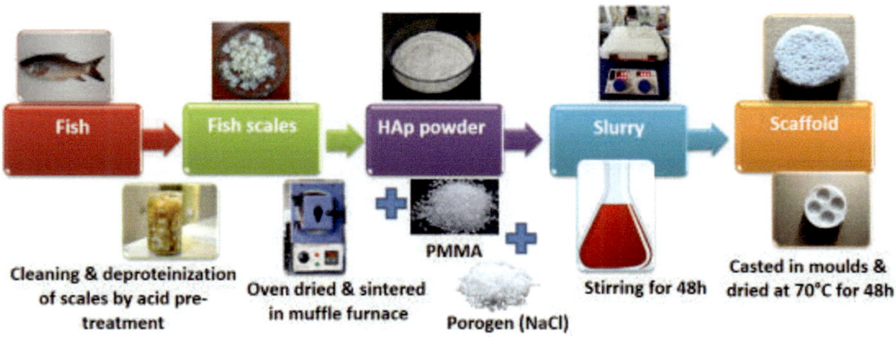

Figure 15.1. Complete process for bone scaffold development from HA synthesized from *Labeo rohita* fish scale.

Characterization techniques

One of the major aspects of material science is characterization of the developed material. Characterization of synthesized material exemplifies certain properties of the composition and structure of a material that are significant for a particular preparation. It is a scientific technique to investigate the basic fundamentals of engineering materials. Characterization may include—physicochemical, morphological, mechanical, and bioactivity study to understand the behavior of the developed material. The following characterization techniques have been executed to ascertain the properties of the developed scaffold.

X-ray diffraction (XRD)

It is a systematic investigation implemented to identify the phases in a crystal structure and to obtain dimension of crystals. The principle behind the XRD is the interference that takes place between the monochromatic X-rays and the crystalline materials, as illustrated in Fig. 15.2. The origin of the x-rays initiate from the cathode that are directed towards the samples, for which the diffraction needs to be carried out. The powdered samples that are held in a sample holder interact with the x-rays, which generate diffracted rays that are detected, counted, and processed. The spectrum arising out of the diffracted rays display the results in the form of peak intensities against the angular range values (2θ). Phase detection is executed by comparing the peaks obtained from XRD spectrum through experimentation and the standard complied by the International Centre for Diffraction Data (ICDD) using card no. 09-0432, 09-0169, and 13-0835 for hexagonal HA structure, rhombohedral β-TCP, and PMMA, respectively. ICDD is a non-profitable scientific association that is fully dedicated to collect, publish, and distribute the diffraction pattern for the determination of phases in a material. The presence of different crystalline phases in the scaffold and information about their crystallographic properties is obtained by performing XRD analysis. The analysis is conducted in an automatic computerized powder X-ray diffractometer (Model—X'Pert Pro and Make—PANalytical). The angular range (2θ) is measured within the range of 20° to 60° with a step size of 0.02°. The working voltage and current are maintained at 40 kV and 20 mA, with a counting time of 2 s per step. In the current study, identification of different phases present in the scaffolds is accomplished using XRD analysis. The average crystallite sizes of the scaffold are calculated from Scherer's equation, as depicted in equation 1 (Barua et al. 2019).

$$D = \frac{k\lambda}{\beta \text{COS}\theta} \tag{1}$$

where, D = crystallite size (nm),

k = broadening constant that varies with crystal habit, generally 0.9,
λ = wavelength of CuKα radiation, generally 0.154 nm,
β = full width at half of the maximum (radian),
θ = corresponds to peak position (°).

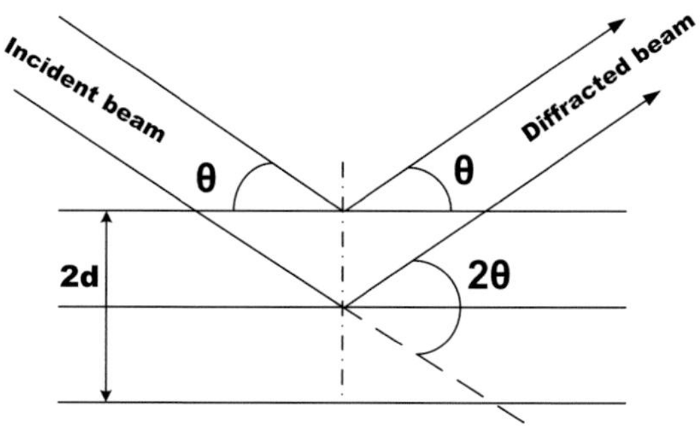

Figure 15.2. Working principle of XRD.

Fourier transform infrared spectrometry (FTIR)

FTIR is a method to acquire infrared spectrum of any material. FTIR spectrometer consists of globular and mercury vapor lamp as sources, an interferometer chamber comprising of KBr and mylar beam splitters, mirrors, and detector, as shown in Fig. 15.3. The spectrum is generated by vibrational motion of the different bonds present in the molecules. The nature of vibration can be either stretching or bending, which depends on the nature of molecule under investigation. The reason for the occurrence of stretching vibration is attributed to the change in inter-atomic distance along the axis of the bond, whereas bending vibration is attributed to the change in angle between two bonds in a molecule. The spectrometer accumulates data and generates curve with different peak intensities. The peaks correspond to different functional groups present in a material. In the current work, the FTIR analysis is conducted on FT-IR spectrometer (Make-Bruker and Model-3000 Hyperion Microscope) using KBr pellet technique. The range of scan varies from 400–4000 cm^{-1} with a resolution of 1.0 cm^{-1}. In the present study, FTIR analysis is executed to recognize the different functional groups present in the fabricated scaffold.

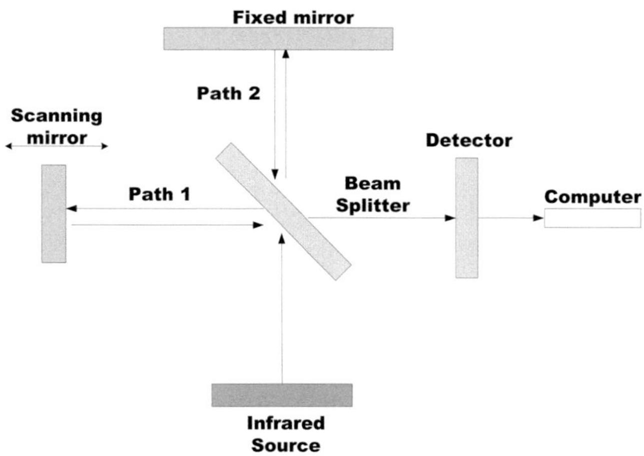

Figure 15.3. Basic principle of FTIR.

Scanning electron microscopy (SEM)

Scanning electron microscope is an electron microscope that generates micrographs of the material surface using a focused beam of electrons. The basic principle of SEM analysis, as shown in Fig. 15.4, is the formation of signals due to the interaction of electrons with the surface of the material. This produces signals that reveal the nature and topography of the material surface. SEM is used to evaluate grain size, particle size, and material homogeneity. A high resolution Scanning Electron Microscope (Quanta 200F model made by FEI) is used to evaluate surface morphology of the synthesized composite using either low vacuum (LV) or environment scanning electron microscope (ESEM) modes. The instrument can provide magnification

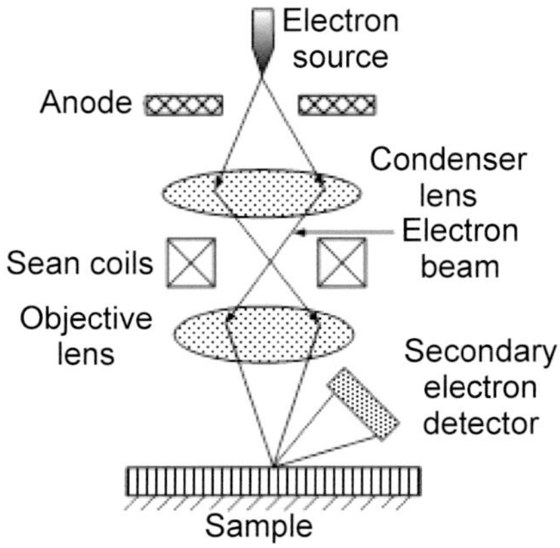

Figure 15.4. Basic principle of SEM.

range of minimum of 12X to maximum of 100000X. The biomaterials are frozen in liquid nitrogen and cryofractured and gold coated using Cressington-108 auto (Model No.: 7006-8) sputtering machine at 230 V, 50 Hz using a high speed, direct drive 2-stage vacuum pump. The gold coated samples are investigated at 20 kV by varying the magnification range from 150–30000X. SEM is executed to analyze the porous morphology of the developed scaffold. Pore size of the developed scaffold is accomplished from the SEM images using ImageJ software.

Compression test

Compression test determines the material behavior under compression loading. To conduct the test, cylindrical test specimen is placed between two plates that provide uniform distribution of load across both the surfaces of the specimen. The specimen is at first placed in the bottom face, and the upper face is slowly brought down and slightly adhered to its surface. The sample is compressed under loading condition, and the entire compression profile is obtained. The compressive strength of the developed scaffold is studied by performing uniaxial compressive test. Compression test is performed on servo-controlled micro-UTM Instron 5969. Specimens are prepared following ASTM D695 norms. The cross head movement was set at 1 mm/min with 5 kN load cell. The test is repeated on three identical samples to obtain the average result. The test facilitates the understanding of the behavior of materials under compressive loading and determines the corresponding stress, strain, and deformation. The test results will be obtained as ultimate compressive strength, compressive modulus, and yield strength of the material. Moreover, elastic limit, proportional limit, and yield point can also be evaluated from the test. In the current study, the test facilitates the ultimate compressive strength obtained for the developed scaffold.

Results and discussions

Physicochemical characterizations

XRD analysis

XRD is a characterization technique employed on a material that identifies the phase of a crystalline material and can provide detailed information on unit cell dimensions. The principle includes occurrence of constructive interference of monochromatic X-ray, thereby identifying the phases in the material. The analysis further determines the crystallite size and degree of crystallinity of a material. The spectrum of the developed scaffold, as shown in Fig. 15.5, reveals that most of the characteristic peaks correspond to HA since HA content is 70% in the developed scaffold. Peaks corresponding to PMMA are also obvious since the scaffold is developed using PMMA matrix. However, minor peaks of β-TCP are also observed in the spectrum, which appear due to the conversion of HA to β-TCP at high temperatures. Table 15.1 shows the 2θ values and their corresponding XRD planes obtained for the developed scaffold. The three highest intensity peaks are considered for the calculation of the average crystallite size. In the present study, the average crystallite size of the developed scaffold is found to be 109 ± 7 nm. The small crystallite size is favorable for bone tissue engineering applications (Bardhan et al. 2011).

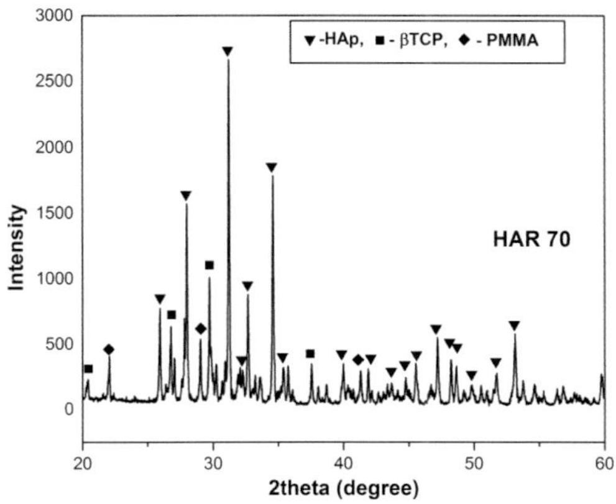

Figure 15.5. XRD spectrum of the developed scaffold.

FTIR analysis

It is a technique that enables the absorption or emission of infrared spectrum from any material. This absorption band corresponds to the bonds between molecules present in the material. FTIR spectrum of the developed scaffold, as represented in Fig. 15.6, shows characteristic peaks at 460, 560, 1026, 960 cm^{-1} corresponding to v2, v4, v3, and v1 stretching vibration, respectively of phosphate (PO_4^{3-}) group (Destainville et al. 2003). Vibration of carbonate (CO_3^{2}) group is found at 1650 cm^{-1} (Raynaud et al. 2002). The peaks at 1650 and 3420 cm^{-1} indicate the presence of hydroxyl (OH^-) ion (Han et al. 2006, Raynaud et al. 2002). The peak at 1730 cm^{-1} corresponds to the –COOH bond, whereas the peak at 2952 cm^{-1} corresponds to C–H

Table 15.1. XRD planes and corresponding 2 theta values of the developed scaffold.

	HAp		B-TCP		PMMA
Sample	**2 theta**	**Planes**	**2 theta**	**planes**	**2 theta**
HAR 70	16.89, 25.87, 28.11, 31.77, 32.19, 32.90, 34.05, 35.48, 40.45, 42.31, 43.81, 44.37, 45.30, 46.75, 48.11, 48.62, 49.46, 52.10, 53.14	101, 002, 102, 211, 112, 300, 202, 301, 221, 302, 113, 400, 203, 222, 311, 320, 213, 402, 004	20.29, 26.51. 29.66, 37.81	202, 122, 300 315	22.49, 29.45, 41.41
HAR 70 (Post SBF 28 days)	16.89, 25.87, 28.11, 31.77, 32.19, 32.90, 34.05, 35.48, 40.45, 42.31, 43.81, 44.37, 45.30, 46.75, 48.11, 48.62, 49.46, 52.10, 53.14	101, 002, 102, 211, 112, 300, 202, 301, 221, 302, 113, 400, 203, 222, 311, 320, 213, 402, 004	20.29, 26.51. 29.66, 37.81	202, 122, 300 315	22.49, 29.45, 41.41

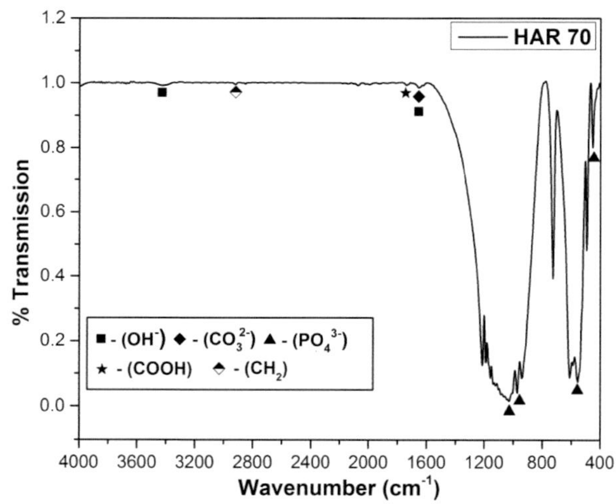

Figure 15.6. FTIR spectrum of the developed scaffold.

bond stretching vibrations of $-CH_2-$ group (Duan et al. 2008). All the peaks found in the IR spectra correspond to the presence of HA and PMMA in the scaffold.

SEM analysis

Surface topography of the developed scaffold is investigated from SEM micrographs, as shown in Fig. 15.7. SEM micrographs reveal a highly porous surface, with interconnected pores having maximum pore size of 128 ± 7.31 μm. It is reported in the literature that a minimum pore size of ~ 100 μm is required for efficient cell adhesion (Karageorgiou and Kaplan 2005). The morphology further reveals agglomerated HA particles having ultra-fine structure. Energy Dispersive X-Ray analysis (EDX) is a chemical microanalysis method generally applied in combination with SEM. The technique detects x-rays emitted from the material during bombardment by a beam of electron to characterize the constituent elements of the investigated volume. EDX analysis further confirms the presence of calcium (Ca), phosphorus (P), and

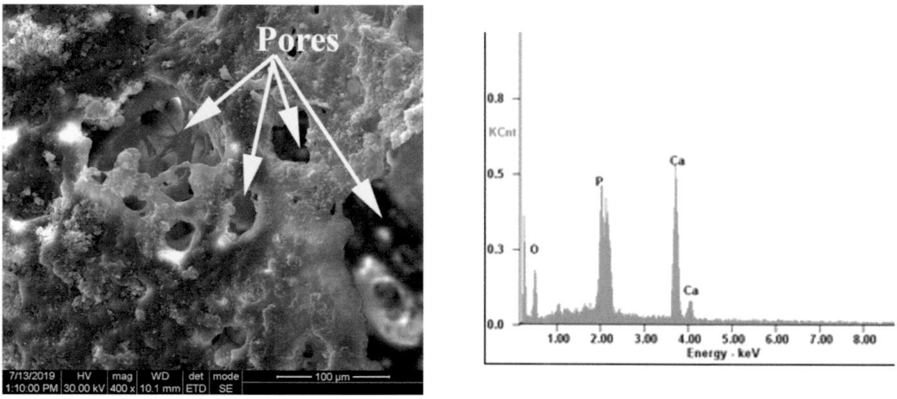

Figure 15.7. SEM-EDX micrograph of the developed scaffold.

oxygen (O) in the scaffold, which are in the form of carbonate and phosphate. The ratio of Ca/P is observed to be 1.69, which is similar to the stoichiometric ratio of apatite of human bone that is found to be 1.67.

Mechanical characterizations

Compression test analysis

In a compression test, a material is subjected to compressive forces. The material is either crushed or flattened depending upon its nature. The test is performed to identify the compressive strength of the fabricated scaffold. The specimen is prepared using ASTM D695 standards having 18 mm diameter and 21 mm length. The strength is reported up to the maximum value of stress strain curve where there is an initiation of first crack. The developed scaffold shows a compressive strength of 4.33 MPa. It is been reported by Yoshikawa and Myoui that the minimum compressive strength of bone scaffold required for biomedical application lies in the range of 1–10 MPa (Yoshikawa and Myoui 2005). Figure 15.8 shows the stress-strain curve

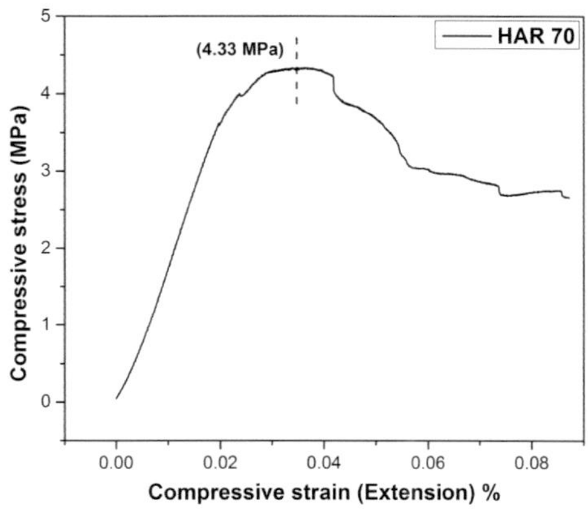

Figure 15.8. Stress-strain curve of the developed scaffold.

of the developed scaffold. Thus, the developed scaffold can be suitably applied in biomedical applications.

Bioactivity and degradability evaluations

In-vitro bioactivity analysis

The test is performed by immersing the HA/PMMA scaffold in Simulated Body Fluid (SBF) solution. To perform the test, HA/PMMA scaffolds of 13 mm diameter and 3 mm thickness are prepared using silicon rubber mold. The pellets are dipped in SBF solution for a period of 28 days. After the stipulated period, the pellets are taken out from the SBF solution and dried in a hot air oven. The post SBF treated scaffold is characterized using XRD and SEM analysis, as shown in Fig. 15.9. XRD analysis shows that there is a decrease in the intensity of the peaks for post-SBF treated scaffold as compared to the pre-SBF ones. Moreover, there are also merging of peaks due to the increase in the width of the peaks. It is observed from Fig. 15.9 that there are small peaks in the range of 25–30° in pre-SBF scaffold. However, post-SBF treated scaffold, after 28 days of immersion, shows the disappearance of the peaks and merging of the smaller peaks with the adjacent ones, as observed from the given figure. The result is accredited to the formation of hydroxy carbonate apatite (HCA) layer on the surface of the scaffold. HCA is amorphous in nature. Thus, there is a reduction in the crystalline nature of the post-SBF treated scaffold (Rolando et al. 2017). The observed peaks and corresponding 2 theta positions are listed in Table 15.1.

SEM-EDX micrograph of the post-SBF treated scaffold shows the formation of a structure similar to a spherical shiny region on the surface of the scaffold by blocking the pores. The structures correspond to the formation of hydroxy carbonate apatite layer on the surface of post-SBF treated scaffold. Figure 15.10 shows the SEM micrograph of post-SBF treated scaffold. EDX analysis of the post-SBF treated

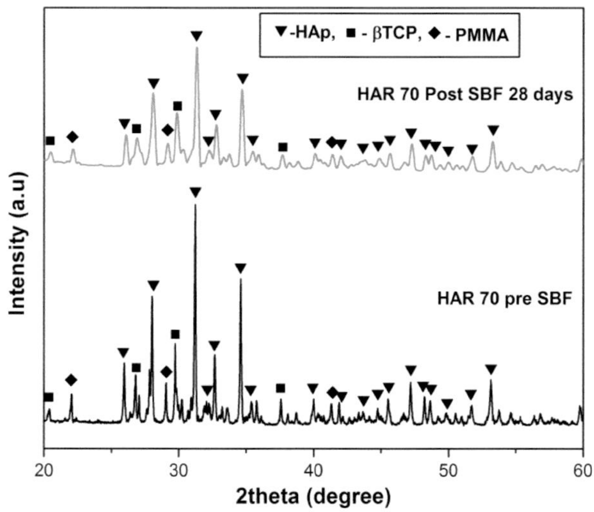

Figure 15.9. XRD spectra of pre and post SBF treated scaffold.

Figure 15.10. SEM-EDX spectra of the post-SBF treated scaffold.

scaffold shows the presence of Ca and P as major constituents, with a decrease in the Ca/P ratio corresponding to the pre-SBF ones. However, the ratio is found to be 1.65, which is fairly accurate to the Ca/P ratio of HCA. Moreover, traces of other minerals, such as Na and Mg are also observed in the developed scaffold. It is because the HA used in the present study for scaffold development is extracted from *Labeo Rohita* fish scale, which is a natural source for HA. It is found from the literature that the HA extracted from natural resources has an inherent content of trace minerals in them (Mu and Mu 2007). Thus, the presence of certain minerals is accomplished in the developed scaffold. Carbon is also found in post-SBF treated scaffold. The reason behind this may be the polymer or the carbonate substituted HCA (Feng et al. 2014). Hence, the developed scaffold shows highly bioactive properties suitable for biomedical applications.

In vitro biodegradability test

Degradation behaviors of the scaffolds are studied by performing *in-vitro* degradation test. The test is performed by immersing the scaffold in SBF solution at 37 ± 0.5°C, having a pH 7.4 for 7, 14, 21, and 28 days. To perform the test, the dry scaffold pellets are at first weighed, and thereafter dipped in SBF solution separately for the aforementioned time intervals. Scaffolds are taken from the SBF solution after 1, 7, 14, 21, 28 days, and washed with distilled water for a couple of times and dried. After drying, the scaffolds are weighed, and the weight loss is determined using the following equation:

$$\text{Weight loss\%} = \frac{\text{Ws} - \text{Wi}}{\text{Ws}} \times 100 \tag{2}$$

where W_s and W_i are the initial weight and weight of the dry scaffold at time I, respectively.

Figure 15.11 shows the variation of weight loss of the developed scaffold with time. It is noticed that with an increase in soaking time, there is an increase in degradation of the developed scaffold. The maximum degradation obtained is 11 ±

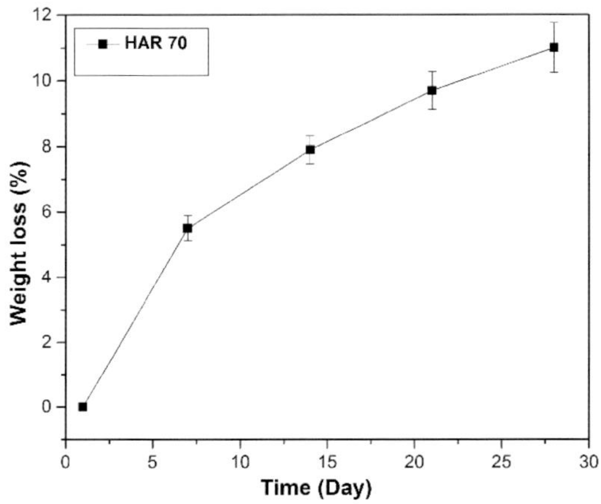

Figure 15.11. Degradation of the post-SBF treated scaffold.

0.76% for 28 days of immersion. This confirms the degradable nature of HA/PMMA scaffold, making it suitable for biomedical applications.

Conclusions

In the current study, *Labeo Rohita* fish scale derived HA/PMMA scaffold is developed using solvent casting particulate leaching technique. HA is extracted by thermal degradation of the fish scales, and is reinforced in the PMMA polymer matrix for the development of composite bone scaffold. The developed scaffold is characterized for physicochemical, mechanical, as well as bioactivity and biodegradability evaluations. The following conclusions are drawn for the successful development of scaffold:

- HA/PMMA scaffold is successfully developed, maintaining the weight percentage of HA and PMMA as 70:30.
- The developed scaffold shows porous morphology, with a maximum pore size of 128 ± 7.31 μm, and the Ca/P ratio is observed to be 1.69. The obtained pore size along with the Ca/P ratio is suitable for a bone scaffold to be used in segmental bone defect applications.
- Maximum compressive strength of 4.33 MPa is obtained for the fabricated scaffold, which lies in the range suitable for biomedical applications.
- The post-SBF treated scaffold shows the formation of bone-like apatite on its surface, confirming the bioactive nature of the developed scaffold.
- The scaffold shows 11 ± 0.76% degradation after immersing into SBF solution for 28 days, which confirms the degradable nature of the developed scaffold.

Acknowledgments

Authors acknowledge TEQIP III, CIF, and Indovation Lab of NIT Silchar for financial support, and IIT Bombay, IIT Madras, and IIT Guwahati for support in acquiring the data.

References

Bardhan, R., S. Mahata and B. Mondal. 2011. Processing of natural resourced hydroxyapatite from eggshell waste by wet precipitation method. Adv Appl Ceram 110(2): 80–86.

Barua, E., A.B. Deoghare, S. Chatterjee and P. Sapkal. 2019. Effect of ZnO reinforcement on the compressive properties, in vitro bioactivity, biodegradability and cytocompatibility of bone scaffold developed from bovine bone-derived HAp and PMMA. Ceram Int 45(16): 20331–45.

Bhat, S. and A. Kumar. 2012. Cell proliferation on three-dimensional chitosan–agarose–gelatin cryogel scaffolds for tissue engineering applications. J Biosci Bioengl14(6): 663–70.

Boutinguiza, M., J. Pou, R. Comesana, F. Lusquinos, A. De Carlos and B. Leon. 2012. Biological hydroxyapatite obtained from fish bones. Mater Sci Eng C 32(3): 478–86.

Deb, P., E. Barua, A.B. Deoghare and S.D. Lala. 2019. Development of bone scaffold using puntius conchonius fish scale derived hydroxyapatite: physico-mechanical and bioactivity evaluations. Ceram Int 45: 10004–12.

Deb, P. and A.B. Deoghare. 2019. Effect of pretreatment processes on physicochemical properties of hydroxyapatite synthesized from Puntius conchonius fish scales. Bull Mater Sci 42: 3.

Destainville, A., E. Champion, D. Bernache Assollant and E. Laborde. 2003. Synthesis, characterization and thermal behavior of apatitic tricalcium phosphate. Mater Chem Phys 80: 269–77.

Duan, G., C. Zhang, A.L. Xujie Yang, L. Lu and X. Wang. 2008. Preparation and characterization of mesoporous zirconia made by using a poly (methyl methacrylate) template. Nanoscale Res Lett 118–22.

Feng, P., P. Wei, C. Shuai and S. Peng. 2014. Characterization of mechanical and biological properties of 3-D scaffolds reinforced with zinc oxide for bone tissue engineering. PLOS One 9: (1).

Han, J.K., H.Y. Song, F. Saito and B.T. Lee. 2006. Synthesis of high purity nano-sized hydroxyapatite powder by microwave-hydrothermal method. Mater Chem Phys 99: 235–39.

Karageorgiou, V. and K. David. 2005. Porosity of 3D biomaterial scaffolds and osteogenesis. Biomaterials 26: 5474–91.

Mondal, S., U. Pal and A. Dey. 2016. Natural origin hydroxyapatite scaffold as potential bone tissue engineering substitute. Ceram Int 42(16): 18338–46.

Mu, F.A. and L. Mu. 2007. Preferred growth orientation of biomimetic apatite crystals preferred growth orientation of biomimetic apatite crystals. J Cryst Growth 304: 464–71.

Pon-On, W., P. Suntornsaratoon, N. Charoenphandhu, J. Thongbunchoo, N. Krishnamra and I. Ming Tang. 2016. Hydroxyapatite from fish scale for potential use as bone scaffold or regenerative material. Mater Sci Eng C 62: 183–89.

Raynaud, S., E. Champion, D. Bernache-Assollant and P. Thomas. 2002. Calcium phosphate apatites with variable Ca/P atomic ratio I. Synthesis, characterisation and thermal stability of powders. Biomaterials 23(4): 1065–72.

Rolando, T.C., T. Camille and P. Lech. 2017. Evaluation of the *in-vitro* behavior of nanostructured hydroxyapatite and zinc doped hydroxyapatite coatings obtained using solution precursor plasma spraying. J Biomed mater Res part B Appl Biomater 106: 2101–8.

Tripathi, G. and B. Basu. 2012. A porous hydroxyapatite scaffold for bone tissue engineering: physico-mechanical and biological evaluations. Ceram Int 38(1): 341–49.

Yoshikawa, H. and A. Myoui. 2005. Bone tissue engineering with porous hydroxyapatite ceramics. J Artif Organs 8(3): 131–36.

16

Physiological and Metabolic Aspects of Pesticides Bioremediation by Microorganisms

Murali Krishna Paidi,[1] *Praveen Satapute,*[2,*] *Shakeel Ahmed Adhoni,*[2] *Lakkanagouda Patil*[2] and *Milan V. Kamble*[2]

Introduction

Pesticides are synthetic foreign (man-made) chemicals, and have no obvious counterpart in the natural world (Tewari et al. 2012, Satapute and Kaliwal 2016a). A pesticide is intended to destroy any pest (Satapute et al. 2019a). The pest can be insects, weeds, molluscs, birds, mammals, fish, nematodes, bacteria, and fungi that can compete with the human for food (Kaur et al. 2019). Pesticides have been contributing to the spectacular increase in the agricultural crop yield, as well as controlling disease vectors in various fields. Developing countries use 26% of the total pesticides produced in the world (Dollacker 1991, Satapute and Kaliwal 2016b). After World War II, the pesticide usage rate has been increased in the agricultural field, leading to protect the crops, and crop losses may reach 45% of total food production worldwide (Radhika and Kannahi 2014). Currently, over 500 compounds have been recognised and used as potential pesticides (Satish et al. 2017). Toxicologically, pesticides are categorized into herbicides, insecticides, bactericides and fungicides, nematicides, rodenticides, acaricides, and molluscicides (Eldridge 2008). Various pesticide types, their modes of action, and major targets are listed in Table 16.1. Chemically, pesticides are diverse, including carbamates, organochlorines, organophosphates, aromatic rings, nitrogen-containing compounds,

[1] Department of Biotechnology and Phycology, CSIR-Central Salt and Marine Chemicals Research Institute (CSIR-CSMCRI), Bhavnagar, Gujarat, India.

[2] Department of Microbiology and Biotechnology, Karnatak University, Dharwad, Karnataka, India.

* Corresponding author: psatapute6@gmail.com

Table 16.1. Different pesticides classification and their mode of action on primary target pests.

Name of the pesticide	Chemical family and classification	Function	Primary target	Reference
2,4-D	Chlorinated phenoxy herbicide	Kill the target weeds by mimicking the plant growth hormone auxin	Broadleaf weeds (Bramble, Chickweed, milkweed, etc.)	(Song 2013)
Aldicard	Organochlorine Insecticide	Cholinesterase inhibitor	Aphids, Spider Mites, and Lygus	(Kok et al. 2002)
Cabetamide	Carbamate herbicide	Inhibition of mitosis or microtubule organization	Annual grass and suppresses broadleaf weeds	(Crovetto et al. 2009)
Cabofuran	Carbamate insecticide		Meloidogyne incognita (Root-knot nematode)	(Jada et al. 2011)
Chloropyrifos Or Dursban	Organophosphate pesticide	Inhibits the acetyl Cholinesterase	Light brown apple moth, western flower thrip, booldworm, cockroaches, termites, and fleas.	(Solomon et al. 2014, Lemus and Abdelghani 2000)
Chlorpropham	Carbamate herbicide	Inhibition of mitosis or microtubule organisation	Annual grass and suppresses broadleaf weeds	(Crovetto et al. 2009)
Cypermethrin	Pyrethroid insecticide	Inhibition of sodium channel	Lepidopterous pests including butterflies and moths	(Tan et al. 2005)
DDT	Organochlorine Insecticide	Act on voltage gated sodium channel proteins in insect nerve cell membrane	Housefly, cockroaches, and mosquitos	(Williamson et al. 1993, Silver et al. 2014)
Endosulfan	Organochlorine pesticide	Act as chloride channel antagonist, inhibits calcium and magnesium ATPase	Whiteflies, Aphids, leafhoppers, Colorado potato beetles and cabbage worms	(Satar et al. 2009)
Fipronil	Organochlorine Insecticide or Scabicides and pediculicides	Inhibit the chloride channels of insect $GABA_A$	Ants, beetles, cockroaches, ticks, termites, thrips, and rootworms	(Simon-Delso et al. 2015)
Lindane	Organochlorine Insecticide or Scabicides and pediculicides	Inhibit the chloride channels of insect $GABA_A$	Lice, mites (scabies)	(Abdullah and Kaki 2017)

Table 16.1 Contd. ...

...Table 16.1 Contd.

Name of the pesticide	Chemical family and classification	Function	Primary target	Reference
Profenofos	Organophosphate Insecticide	Inhibits the acetyl Cholinesterase	Tobacco budworm, cotton bollworm, armyworm, cotton aphid, shiteflies, spiders mites, plant bugs	(Kushwaha et al. 2016)
Propham	Carbamate herbicide	Inhibition of mitosis or microtubule organisation	Annual grass and suppresses broadleaf weeds	(Crovetto et al. 2009)
Solan				
Swep	Carbanilate pesticide			

pyrethroids, etc. Among these the carbamates, organophosphates derivatives, are the most commonly used pesticides, and they have replaced the organochlorines, such as DDT, aldrin, lindane, etc. (Palleschi et al. 1992, Cremisini et al. 1995).

Pesticides have a significant impact on the farmers' economy, and increase in agriculture productivity has resulted due to the use of synthetic pesticides. On the other hand, they have been accumulated, as their residues are found in soil, water, food, animals, and humans also (Sharma et al. 2016). The use of synthetic pesticides has increased 50-fold every year (Miller 2002). Many pesticides and their metabolites have medium to long term stability in an environment (Gouma et al. 2014). The abundant use of pesticides increases the transport, and accumulation of their residues at a considerable distance from the point of its original application (Miles and Pfeuffer 1997). Both insecticides and pesticides kill the target pests, but they also adversely kill non-target organisms, such as earthworm, natural predators, and pollinators (Edwards 1987). For example, a significantly increased amount of endosulfan is found in groundwater, as well as in sediment soils (Siddique et al. 2003, Menezes et al. 2017). Nevertheless, the persistence of pesticides depends on the nature of the chemical, pH, moisture, organic matter, clay, and nutrient contents, as well as on contaminant concentration and microbial metabolism (Yadav and Loper 2000, Karigar and Rao 2011). Organochlorine pesticides are the most commonly used pesticides in agriculture, as they persist in the environment for a long time. Hence, this leads to contamination of groundwater, food products, air, and soil (Kaur et al. 2019).

The impact of pesticides can be analyzed from two different points of view, such as environmental and public health. In the first view, when pesticides are introduced to food chains, there is decline of the population of phytoplankton and zooplankton. In the second view, they are carcinogenic, and lead to lower fertility and viability of invertebrates, fish, amphibians, insects, and mammals (Ortiz-Hernández et al. 2013). For examples, organochloride pesticides are cumulative in the organisms and pose chronic health effects, such as cancer and neurological and teratogenic effects (Vaccari et al. 2006). Many organochloride pesticides and their derivative metabolites are recalcitrant and resistant to degradation (Dua et al. 2002, Diaz 2004). Organophosphorus pesticides are more widely used worldwide, and are replacing the organochlorine pesticides, including DDT, aldrin, lindane, etc. (Palleschi et al. 1992). These pesticides block the prolonged inhibition of the cholinesterase enzyme activity (Yair et al. 2008), that affect the nervous system of insects, as well as humans (Colosio et al. 2009, Jokanovic and Prostran 2009). Similarly, carbamate pesticides are important due to their broad activity spectrum and wide range compounds, as well as having less toxicity towards humans (Wolfe et al. 1978).

The excessive use of organochlorine, organophosphorus, and carbamate pesticides in agriculture has originated serious problems in the environment (Singh and Walker 2006). In general, these synthetic organic compounds and their derivatives persist in the environment for many years (Ragnarsdottir 2000). They might be toxic and carcinogenic, and relatively inhibit the enzyme acetylcholinesterase and hydrolysis reaction of acetylcholine (Ach). This results in accumulation of ACH, which may lead to sweating, lacrimation, hypersalivation, and convulsion of extremities (Suzuki and Watanabe 2005). This chapter considers some microbial enzymes and their role in the degradation of organochlorine, organophosphorus, and carbamate pesticides bioremediation processes.

Principles and aim of bioremediation

The concept bioremediation is a process used to treat contaminated or polluted media, including water, soil, and other environments (Gouma et al. 2014). It is an option to destroy or render harmless various contaminants using normal biological activity (Vidali 2001). Bioremediation is gradually being accepted as the ultimate standard eco-friendly approach to detoxifying the contaminants as compared to other chemical and physical methods (Ekperusi and Aigbodion 2015, Ayangbenro and Babalola 2017). Moreover, bioremediation is one of the cheapest (Blaylock et al. 1997) and the most useful method for pesticide control by using microorganisms in targeting pesticides (Megharaj et al. 2011). Microbes are the key players in bioremediation that are live everywhere. Microorganisms, such as bacteria, fungi, and microalgae are ideally suited to the destruction of contaminants because they have various enzymes that attack the pollutants, convert them into harmless products, or utilize toxic pollutants to obtain energy as well as biomass production (Tang et al. 2007).

The main aim of bioremediation is encouraging the microbes to work by supplying optimal nutrients and concoctions for their metabolism to degrade or detoxify hazardous substances (Abatenh et al. 2017). The goal of bioremediation is to reduce pollutant levels to undetectable, nontoxic, or completely mineralized organic-pollutants to carbon dioxide (Pointing 2001). Cleaning up soil contaminated with a high concentration of atrazine and 4-chloroaniline by using combined biostimulation and bioaugmentation soil biological remediation techniques have proven effective (Silva et al. 2004, Tongarun et al. 2008). In these techniques, there is addition of nutrients and aeration to the contaminated site and encouragement of the presence of indigenous (biostimulation) (Kanissery and Sims 2011) or non-indigenous (bioaugmentation) microbes capable of degrading the toxicant (Pimmata et al. 2013). Without the activity of microorganisms, the earth would be buried in waste (Kensa 1970, Vidali 2001, Crawford and Crawford 2005).

In situ bioremediation of pesticide contamination

The treatment of pesticides contaminated soils or water in the location where they were found. In this process, the contaminated soil bioremediation is by using aerobic methods, such as bioventing and injection of hydrogen peroxide (Fekete-Kertész et al. 2013), permanganate (Xie and Barcelona 2003, Sahl et al. 2007) or Fenton reagent (Kulik et al. 2006, Cassidy and Hampton 2009), persulfate and ozone injections (Jung et al. 2005). Among them, the bioventing is a simple, inexpensive, and the most efficient procedure for pesticide bioremediation in targeted sites. In this method, the wells are injected into the contaminated soil, and air can be sucked or blown through wells. In addition, nutrients such as nitrogen and phosphorus may also be pumped through the injection wells to increase the growth rate of the microorganism in the contaminated soil (Singh 2008, Satish et al. 2017).

Ex situ bioremediation of pesticide contamination

Soil or water contaminated by pesticides can be treated once it has been excavated or pumped out of the location at which it was found (EPA 1998, Sutton et al. 2011, Satish et al. 2017). This can occur in two ways—Slurry phase bioremediation and Solid-phase bioremediation. In Slurry-phase bioremediation process, the contaminated soil is mixed with the water, oxygen, and other required nutrients in a large tank known as a bioreactor, in which microorganisms have an ideal environment to break down the contaminants (Gouma et al. 2014, Lal et al. 2010). However, in solid-phase bioremediation process, the contaminated soil is placed on above-ground treatment center by landfarming (Singh 2008, Lal et al. 2010), soil biopile, and composting process.

Enzyme and metabolic aspects of pesticide bioremediation

Nowadays, there is a significant increase of pesticide use in agriculture and commercial sites, including public gardens and houses to control the various type of pests that has led to contamination. There is public concern about the impact of pesticides on human health and in the environment. Under toxic environment, reactive oxygen species (ROS) production has been increased in several microbes (Satapute et al. 2019b). However, microbes have developed intracellular peroxidases (cytochrome peroxidase, ascorbate peroxidase, and catalase peroxidases), and extracellular peroxidases, including lignin peroxidase and manganese peroxidases enzymatic system. This may neutralize or utilize the ROS molecules in the oxidation of various phenolic compound degradation (Karigar and Rao 2011).

Bioremediation is generally the safest and least disturbing method to clean up contaminants in the environment using microorganisms (Sutherland et al. 2004). In general, energy chemical reactions are called oxidation-reduction reactions, in which organic compounds oxidation and reductions are carried by enzymes in microbes that lead to energy production for cell growth. Different enzymes and their reactive potential substrates or pesticides are described in Table 16.2. The degradation of 1,1,1-trichloro2,2-bis (p-chlorophenyl)ethane (DDT) in sediment is affected by moisture, pH, temperature, oxygen levels, bacterial dispersion, nutrient availability, DDT concentration, and its solubility. The degradation of DDT occurs naturally in the environment due to various organisms presented in Table 16.3. Mainly DDT can be detoxified in two main routes—oxidative degradation (in the presence of oxygen) and reductive degradation under oxygen-deficient conditions (Fialips et al. 2010). Under aerobic conditions, DDT is generally converted to 1,1-dichloro-2,2-bis (p-chlorophenyl)ethylene (DDE) by dehydrochlorination by *Acinetobacter radioresistance* and *Bacillus cereus* DQ207729 (Aislabie et al. 1997, Lawrence et al. 2005, Silver et al. 2014). Aerobic degradation occurs more quickly in warm and moist conditions. Although, a few studies have shown that DDE can be further degraded under anaerobic and aerobic conditions (Hay and Focht 1998).

Organophosphates (OPs) and their derivatives accumulation increased due to huge applications. Enzymes can be used to hydrolyze OP triesters into less or non-toxic compounds. These enzymes are possible for bioremediation because of their ability to

Table 16.2. Pesticides degrading enzymes and their reactive potential substrates/pesticide molecules.

Enzyme	Substrate or pesticide	Reaction	Reference
Laccase	Aminophenols, polyphenols	Oxidation, decarboxylation, and demethylation of substrate	(Karigar and Rao 2011)
Dehalogenases	DDT	DDT to DDE	(Fialips et al. 2010)
Nitroreductases	Carbamites (Nitroaromatic and nitroheterocyclic compounds)	Reduction of the nitro groups to hydroxylamino and/ or amino derivatives	(Roldán et al. 2008)
Nitrilases			
Cytochrome $_{450}$	Oraganophasphates	Catalyze monooxygenase reactions	(Tewari et al. 2012)
Manganese peroxidise	Phenols and halogenated phenols, polycyclic aromatic hydrocarbons (PAH)	Oxidoreductases of organic and inorganic compounds	(Bansal and Kanwar 2013)
Lingnin peroxidise	Lignin	Oxidative degradation of lignin to organic compounds	(Wang et al. 2018)
AtzA	Atrazine	Hydrolysis dechlorination of atrazine to non-toxic hydroxyl atrazine	(De Souza et al. 1996, Wackett et al. 2002)
AtzB	Atrazine	Dehydrochlorination of the hydroxy atrazine to produce N-isopropyl cyanuric amide	(De Souza et al. 1996, Wackett et al. 2002)
AtzC	Cyanuric acid and Isopropylamine	Cyanuric acid and isopropylamine to CO_2 and NH_3	(De Souza et al. 1996, Wackett et al. 2002)
α-ketoglutarate-dependent dioxygenase	2,4-D	2,4-D to 2,4-DCP	(Fukumori and Hausiner 1993, Streber et al. 1987)
Phenolhydrolase	2,4-DCP	2,4-DCP to dichlorocatechol	(Kumar et al. 2016)
1,2-dichlorocatechol dioxygenase	Dichlorocatechol	Dichlorocatechol to β-ketoadipate	(Van der Meer et al. 1992)
Organophosphotriesterases	Organophosphates (OP)	Hydrolysis Ops	(Sethunathan and Yoshida 1973)
Organophosphote hydrolases (OPH)	OP	Hydrolysis Ops	(Munnecke 1976)
Organophosphote degrading enzyme (OpdA)	OP	Hydrolysis Ops	(Ely et al. 2008, 2010)

Table 16.3. Microbial biotransformation of common pesticides into their metabolites compounds during microbial bioremediation procedure.

Name of the pesticide	Degrading microbes	Metabolites	References
Aldicarb	*Acinetobacter radioresistens*	Alsicarb sulfoxide and aldicarb sulfone	(Lawrence et al. 2005)
DDT	*Bacillus cereus* DQ207729, *Staphylococcus sciuri* AB233332 and *Stenotrophomonas maltophilia* AB294553	1,1-dichloro-2, 2-bis (p-chlorophenyl) ethane (DDD) and 1,1-dichloro-2, 2-bis (p-chlorophenyl) ethylene (DDE)	(Silver et al. 2014)
Endosulfan	*Achromobacter xylosoxidans* C8B, *Bacillus* sp., *Cladosporium oxysporum*, *Botryosphaeria laricina* JAS6, *Aspergillus tamarii* JAS6	Endosulfan diol, endosulfan sulphate, endosulfan ether, endosulfan hydroxyether, endosulfan lactone and endosulfan dialdehyde	(Silambarasan and Abraham 2013, Mir et al. 2017)
Lindane	*Bjerkandera adusta, Pleuro tusostratus, Trametes versicolor, Hypoxylon fragiforme and Chondrostereum purpureum* *Sphingobium japonicum*	1,2,4- trichlorobenzene (1,2,4-TCB), 2,5-dichlorophenol (2,5-DCP) and 2,5-dichlorohydroquinone (2,5-DCHQ)	(Nagata et al. 2007, Rigas et al. 2009, Ulčnik et al. 2012)
Fipronil	*Bacillus thuringiensis, Stenotrophomonas acidaminiphila*	Fipronil-sulfide, Fipronil-sulfone, fipronil-amide and 5-amino-1-(2,6-dichloro-4-trifluoromethylphenyl)-4-trifluoromethylsulfinyl pyrazole-3- carboxylic acid	(Mandal et al. 2013, Uniyal et al. 2016)
Carbetamide	*Pseudomonas striata Chester,* *Achromobacter* sp., and *Aspergillus ustus* *Fusarium oxysporum*	Aniline and isoprophyl alcohol	(Kaufman and Blake 1973)
Cypermethrin	*Pseudomonas aeruginosa,* *Streptomyces* sp., *Serratia marcescens* *Bacillus* sp. (SG2)	3-(2, 2-dichloro ethenyl)—2,2-dimethyl-cyclopropanecarboxylate, a-hydroxy-3-phenoxybenzeneacetonitrile, 3-phenoxybenzaldehyde and 4-propylbenzaldehyde	(Zhang et al. 2011, Lin et al. 2011, Chen et al. 2011, Cycon et al. 2014, Pankaj et al. 2016)
Cabofuran	*Sphingomonas* sp. strain SB5 *Pseudomonas* sp. 50432 *Novosphingobium* sp. strain FND-3	Cabofuran-7-phenol, 3-(2-hydroxy-2-methylpropyl)benzene-1,2-diol and methylamine	(Kim et al. 2004, Chaudhry et al. 2002, Yan et al. 2007)

Table 16.3 Contd. ...

...Table 16.3 Contd.

Name of the pesticide	Degrading microbes	Metabolites	References
Chloropyrifos Or Dursban	*Pseudomonas stutzeri* S7B4 and *Flavobacterium balustinum* S8B6. treptomyces sp. HP-11 *Agrobacterium radiobacter* P230	3, 5, 6-trichloro-2-pyridinol (TCP) and Diethyl Phosphorothioate (DETP)	(Lores et al. 1978, Supreeth et al. 2016)
Profenofos	*Pseudomonas aeruginosa, Burkholderia gladioli Bacillus subtilis* and *Stenotrophomonas* sp. G1	4-Bromo-2-chlorophenol and O-ethylS-propyl-O-phosphorothioic acid (EPPA)	(Jabeen et al. 2015, Kushwaha et al. 2016)
Carbetamide, Propham,	*Pseudomonas striata Chester, Achromobacter* sp., and *Aspergillus ustus*	Aniline and isoprophyl alcohol	(Kaufman and Blake 1973)
Solan	*Fusarium oxysporum*	3-chloro-p-toluidine	
Swep		3,4-dichloroaniline	

decontaminate OP-containing waters and soils (Raushel 2002, Ely et al. 2008). There are phosphotriesterases enzymes that have been isolated from *Flavobacterium* sp. A.T.C.C. 27551 (Sethunathan and Yoshida 1973), organophosphate hydrolase (OPH) from *Pseudomonas diminuta* (Munnecke 1976), and organophosphate degradation enzyme (OpdA) coding gene is identified in *Agrobacterium radiobacter* (Horne et al. 2002).

Carbamates are N-substituted esters of carbamic acid, and have been increased along with organophosphorus use in different kinds of crops all over the world. Carbamates also act like organophosphorus compounds, because they inhibit the acetylcholinesterase. The carbamates and their derivatives are suspected carcinogens and mutagens since they break down to aniline based derivatives (Nunes and Barcelo 1999, Wang and Lemley 2003). The methyl carbamates derivatives are considered non-genotoxic because of the inability of the methyl group to undergo metabolic degradation to an epoxide. However, the ethyl carbamates are considered as multispecies carcinogens, causing malignancies in different tissues (Bemis et al. 2015). The carboxylesterases are isolated from various bacteria, such as *Pseudomonas striata Chester, Achromobacter* sp., *Aspergillus ustus* and fungi, *Fusarium oxysporum* (Kaufman and Blake 1973). The first step in the metabolic degradation of pesticide carbamates is their hydrolysis catalyzed by carboxylester hydrolases family (Smith and Bucher 2012), including ester hydrolases, carboxylesters, thioesters, phosphoric and sulfuric esters.

The metabolic fate of any pesticides is dependent on both abiotic environmental conditions (temperature, moisture, soil pH, etc.), and biotic factors (microbial community or plant species as well as on pesticide chemical properties (Van Eerd et al. 2003). However, microbial transformation of organic pesticides normally occurs because they use organic pesticides as carbon sources, which is one of the basic building blocks of the cell, and they also provide electrons. Microbes gain energy by catalyzing chemical reactions that involve breaking the chemical bonds and transferring the electrons. Oxygenation is the most frequent first step in the biotransformation of pesticides. Many of these reactions are mediated by oxidative enzymes, e.g., cytochrome P450's (Satapute and Kaliwal 2016a). Oxidative enzymes are the most important enzymes in Phase I pesticide metabolism (Barrett 2000). Cytochrome P450s are hemethiolate proteins that have been characterized in microbes, such as bacteria, filamentous fungi, as well as in higher eukaryotic organisms (i.e., animals and plants). Cytochrome P450s often catalyze monooxygenase reactions, usually resulting in hydroxylation. Agrochemicals can influence cytochrome P450 systems by acting as effectors, thereby modifying pesticide metabolism, or by modulating the overall metabolism of an organism. These effects can increase or decrease physiological activities, which may affect the growth and development of an organism (Tewari et al. 2012).

Conclusion

Detoxification of organic pesticide contamination in soil, water, and other environments by using microbes is eco-friendly, safe, and economically cost-effective than other physical and chemical methods. In bioremediation methods, microbes, such as bacteria or fungi are the biological agents which detoxify the complex molecules into simple molecules. These molecules may participate in cellular catabolic reactions in the biosynthesis of biomolecules, such as proteins, lipids, nucleic acids. However, the oxidation-reductive enzymes, cytochrome P450s, phosphotriesterases, phosphohydrolases, peroxidases, and organophosphate degradation enzyme (OpdA) are the biocatalysts in various microbes. All these enzymatic systems enhance the toxicity tolerance of microbes and increase the biomass production, as well as actively transfer the chemical energy into the food web in a feedback manner.

References

Abatenh, Endeshaw, Birhanu Gizaw, Zerihun Tsegaya and Misganaw Wassie. 2017. Application of microorganisms in bioremediation—review. J Environ Microbiol 1(1): 2–9.

Abdullah, Nabeela Al and Reham Kaki. 2017. Lindane shampoo for head lice treatment among female secondary school students in Jeddah, Saudi Arabia: an interventional study. J Anc Disease & Preventive Remedies 05(04): 4-7. Doi: 10.4172/2329-8731.1000173.

Aislabie, J.M., N.K. Richards and H.L. Boul. 1997. Microbial degradation of DDT and its residues—a review. New Zeal J Agr Res 40(2): 269–282.

Ayangbenro, A.S. and O.O. Babalola. 2017. A new strategy for heavy metal polluted environments: a review of microbial biosorbents. Int J Env Res Pub He 14(1): 94.

Bansal, N. and S.S. Kanwar. 2013. Peroxidase(s) in environment protection. The Scientific World Journal.

Barrett, M. 2000. pp. 25–37. *In*: Cobbs, A.H. and R.C. Kirkwood (eds.). Herbicides and Their Mechanisms of Action. Sheffield, Great Britain: Sheffield Academic.

Bemis, J.C., C. Labash, S.L. Avlasevich, K. Carlson, A. Berg, D.K. Torous, Matthew Barragato, James T. MacGregor and S.D. Dertinger. 2015. Rat Pig-a mutation assay responds to the genotoxic carcinogen ethyl carbamate but not the non-genotoxic carcinogen methylcarbamate. Mutagenesis 30(3): 343–347.

Blaylock, M.J., D.E. Salt, S. Dushenkov, O. Zakharova, C. Gussman, Y. Kapulnik, Ensley, B.D. and I. Raskin. 1997. Enhanced accumulation of Pb in Indian mustard by soil-applied chelating agents. Environ Sci Technol 31(3): 860–865.

Cassidy, D. and D. Hampton. 2009. Hydrogen sulfide production from persulfate oxidation—implications for remediation. In Proceedings of the Tenth International *In Situ* and On-Site Bioremediation Symposium. Battelle, Baltimore, pp. K-15.

Chaudhry, G.R., A. Mateen, B. Kaskar, M. Sardessai, M. Bloda, A.R. Bhatti and S.K. Walia. 2002. Induction of carbofuran oxidation to 4-hydroxycarbofuran by Pseudomonas sp. 50432. FEMS Microbial Lett 214(2): 171–176.

Chen, S.H., M.Y. Hu, J.J. Liu, G.H. Zhong, L. Yang, M. Rizwan-ul-Haq and H.T. Han. 2011. Biodegradation of betacypermethrin and 3- phenoxybenzoic acid by a novel *Ochrobactrum lupini* DG-S-01. J Hazard Mater 187: 433–440.

Colosio, C., M. Tiramani, G. Brambilla, A. Colombi and A. Moretto. 2009. Neurobehavioural effects of pesticides with special focus on organophosphorus compounds: which is the real size of the problem? Neurotoxicology 30(6): 1155–1161.

Crawford, R.L. and D.L. Crawford. (eds.). 2005. Bioremediation: Principles and Applications (Vol. 6). Cambridge University Press.

Cremisini, C., S. Di Sario, J. Mela, R. Pilloton and G. Pallesch. 1995. Evaluation of the use of free and immobilized acetylcholinesterase for paraoxon detection with an amperometric choline oxidase based biosensor. Anal Chim Acta 311: 273–280.

Crovetto, G., A. Navalon, O. Ballesteros, J.L. Vílchez, F. Garcia-Herruzo and J.M. Rodríguez-Maroto. 2009. Mobility and fate of carbetamide in an agricultural soil. J Environ Sci Heal B 44(8): 764-71.

Cycoń, M., A. Zmijowska and Z. Piotrowska-Seget. 2014. Enhancement of delta methrin degradation by soil bioaugmentation with two different strains of Serratia marcescens. Int J Environ Sci Te 11: 1305–1316.

De Souza, M.L., M.J. Sadowsky and L.P. Wackett. 1996. Atrazine chlorohydrolase from Pseudomonas sp. strain ADP: gene sequence, enzyme purification, and protein characterization. J Bacteriol 178(16): 4894–4900.

Diaz, E. 2004. Bacterial degradation of aromatic pollutants: a paradigm of metabolic versatility.

Dollacker, A. 1991. Pesticides in the third World. Pflanzenschutz-Nachrichten Bayer (Germany, FR).

Dua, M., A. Singh, N. Sethunathan and A. Johri. 2002. Biotechnology and bioremediation: successes and limitations. Appl Microbiol Biot 59(2-3): 143–152.

Edwards, C.A. 1987. The environmental impact of insecticides. pp. 309–329. *In*: Delucchi, V. (ed.). Integrated Pest Management, Protection Integàee Quo vadis? An International Perspective. Parasitis 86, Geneva, Switzerland.

Ekperusi, O.A. and F.I. Aigbodion. 2015. Bioremediation of petroleum hydrocarbons from crude oil-contaminated soil with the earthworm: Hyperiodrilus africanus. 3 Biotech 5(6): 957–965.

Eldridge, B.F. 2008. Pesticide application and safety training for applicators of public health pesticides. California Department of Public Health, Vector-Borne Disease Section, 1616 Capitol Avenue, MS7307, P.O. Box 997377, Sacramento, CA.

Ely, F., J.L. Foo, C.J. Jackson, L.R. Gahan, D.L. Ollis and G. Schenk. 2008. Enzymatic bioremediation: organophosphate degradation by binuclear metallo-hydrolases. Curr Top Biochem Research (9): 63–78.

Ely, F., K.S. Hadler, L.R. Gahan, L.W. Guddat, D.L. Ollis and G. Schenk. 2010. The organophosphate-degrading enzyme from Agrobacterium radiobacter displays mechanistic flexibility for catalysis. Biochem J 432(3): 565–573.

EPA. 1998. Field applications of in situ remediation technologies: chemical oxidation. *In*: Solid Waste and Emergency Responses. EPA, Washington, DC, pp 37.

Fekete-Kertész, I., M. Molnár, Á. Atkári, K. Gruiz and É. Fenyvesi. 2013. Hydrogen peroxide oxidation for *in situ* remediation of trichloroethylene—from the laboratory to the field. Period Polytech Chem 57(1–2): 41–51.

Fialips, C.I., N.G. Cooper, D.M. Jones, M.L. White and N.D. Gray. 2010. Reductive degradation of P, p'-DDT by Fe (II) in nontronite NAu-2. Clay Clay Miner 58(6): 821–836.

Fukumori, F. and R.P. Hausinger. 1993b. Alcaligenes eutrophus JMP134 "2,4-dichlorophenoxyacetate monooxygenase" is an a-ketoglutarate- dependent dioxygenase. J Bacteriol 176: 2083–6.

Gouma, S., S. Fragoeiro, A.C. Bastos and N. Magan. 2014. Microbial Biodegradation and Bioremediation Bacterial and Fungal Bioremediation Strategies. Elsevier Inc. http://dx.doi.org/10.1016/B978-0-12-800021-2.00013-3.

Gunasekara, A.S., T. Truong, K.S. Goh, F. Spurlock and R.S. Tjeerdema. 2007. Environmental fate and toxicology of fipronil. J Pest Sci 32: 189–199. doi:10.1584/jpestics.R07-02.

Hay, A.G. and D.D. Focht. 1998. Cometabolism of 1, 1-dichloro-2, 2-bis (4-chlorophenyl) ethylene by Pseudomonas acidovorans M3GY grown on biphenyl. Appl Environ Microbiol 64(6): 2141–2146.

Horne, I., T.D. Sutherland, R.L. Harcourt, R.J. Russell and J.G. Oakeshott. 2002. Identification of an opd (organophosphate degradation) gene in an Agrobacterium isolate. Appl Environ Microbiol 68(7): 3371–3376.

Jabeen, H., S. Iqbal, S. Anwar and R.E. Parales. 2015. Optimization of profe-nofos degradation by a novel bacterial consortium PBAC using response surface methodology. Int Biodeg Bioremediat 100: 89–97.

Jada, M.Y., D.T. Gungula and I. Jacob. 2011. Efficacy of carbofuran in controlling root-knot nematode (meloidogyne javanica whitehead, 1949) on cultivars of bambara groundnut (Vigna Subterranea (L.) Verdc.) in Yola, Nigeria. Int J Agronomy 2011: 1–5.

Jokanovic, M. and M. Prostran. 2009. Pyridinium oximes as cholinesterase reactivators. Structure-activity relationship and efficacy in the treatment of poisoning with organophosphorus compounds. Curr Med Chem 16(17): 2177–2188.

Jung, H., Y. Ahn, H. Choi and I.S. Kim. 2005. Effects of *in-situ* ozonation on indigenous microorganisms in diesel contaminated soil: Survival and regrowth. Chemosphere 61(7): 923–932.

Kanissery, R.G. and G.K. Sims. 2011. Biostimulation for the enhanced degradation of herbicides in soil. Appl Environ Soil Sci 2011.

Karigar, Chandrakant S. and Shwetha S. Rao. 2011. Role of microbial enzymes in the bioremediation of pollutants: a review. Enzyme Res 2011.

Kaufman, D.D. and J. Blake. 1973. Microbial degradation of several acetamide, acylanilide, carbamate, toluidine and urea pesticides. Soil Biol Biochem 5(3): 297–308.

Kaur, Rajveer, Gurjot Kaur Mavi, Shweta Raghav and Injeela Khan. 2019. Pesticides classification and its impact on environment. Int J Curr Microbiol Appl Sci 1889–97.

Kensa, V.M. 1970. Bioremediation—an overview. I Control Pollution 27(2): 161–168.

Kim, I.S., J.Y. Ryu, H.G. Hur, M.B. Gu, S.D. Kim and J.H. Shim. 2004. Sphingomonas sp. strain SB5 degrades carbofuran to a new metabolite by hydrolysis at the furanyl ring. J Agr Food Chem 52(8): 2309–2314.

Kok, Fatma N., Faruk Bozoglu and Vasif Hasirci. 2002. Construction of an acetylcholinesterase-choline oxidase biosensor for aldicarb determination. Biosens Bioelectron 17(6–7): 531–39.

Kulik, N., A. Goi, M. Trapido and T. Tuhkanen. 2006. Degradation of polycyclic aromatic hydrocarbons by combined chemical pre-oxidation and bioremediation in creosote contaminated soil. J Environ Manage 78(4): 382–391.

Kumar, Ajit, Nicole Trefault and Ademola Olufolahan Olaniran. 2016. Microbial degradation of 2,4-dichlorophenoxyacetic acid: insight into the enzymes and catabolic genes involved, their regulation and biotechnological implications. Crit Rev Microbio 42(2): 194–208.

Kushwaha, M., S. Verma and S. Chatterjee. 2016. Profenofos, an acetylcholinesterase-inhibiting organophosphorus pesticide: a short review of its usage, toxicity, and biodegradation. J Environ Quality 45(5): 1478–1489.

Lal, R., G. Pandey, P. Sharma, K. Kumari, S. Malhotra, R. Pandey, V. Raina, H.P.E. Kohler, C. Holliger, C. Jackson and J.G. Oakeshott. 2010. Biochemistry of microbial degradation of hexachlorocyclohexane and prospects for bioremediation. Microbiol Mol Biol R 74(1): 58–80.

Lawrence, K.S., Y. Feng, G.W. Lawrence, C.H. Burmester and S.H. Norwood. 2005. Accelerated degradation of aldicarb and its metabolites in cotton field soils. J Nematology 37(2): 190–97.

Lemus, R. and A. Abdelghani. 2000. Chlorpyrifos: an unwelcome pesticide in our homes. Rev Environ Health 15(4): 421–433.

Lin, Q.S., S.H. Chen, M.Y. Hu, M. Rizwan-ul-Haq, L. Yang and H. Li. 2011. Biodegradation of cypermethrin by a newly isolated actinomycetes HU-S-01 from wastewater sludge. Int J Environ Sci Techol 8: 45–56.

Lores, E.M., G.W. Sovocool, R.L. Harless, N.K. Wilson and R.F. Moseman. 1978. A new metabolite of chlorpyrifos: Isolation and identification. J Agr Food Chem 26(1): 118–122.

Mandal, Kousik, Balwinder Singh, Monu Jariyal and V.K. Gupta. 2013. Microbial degradation of fipronil by Bacillus thuringiensis. Ecotox Environ Safe 93: 87–92. http://dx.doi.org/10.1016/j.ecoenv.2013.04.001.

Megharaj, M., B. Ramakrishnan, K. Venkateswarlu, N. Sethunathan and R. Naidu. 2011. Bioremediation approaches for organic pollutants: a critical perspective. Environ Int 37(8): 1362–1375.

Menezes, Ritesh G., T.F. Qadir, A. Moin, H. Fatima, S.A. Hussain, M. Madadin, S.B. Pasha, F.A. Al Rubaish and S. Senthilkumaran. 2017. Endosulfan poisoning: an overview. J Forensic Leg Med 51: 27–33.

Miles, C.J and R.J. Pfeuffer. 1997. Pesticides in canals of South Florida. Arch Environ Con Tox 32(4): 337–345.

Miller, G.T. 2002. Living in the environment. 12th Edition. Praeger Publishers, London.

Mir, Zahoor A. et al. 2017. Degradation and conversion of endosulfan by newly isolated Pseudomonas Mendocina ZAM1 strain. 3 Biotech 7(3): 1–12.

Munnecke, D.M. 1976. Enzymatic hydrolysis of organophosphate insecticides, a possible pesticide disposal method. Appl Environ Microbiol 32(1): 7–13.

Nagata, Y., R. Endo, M. Itro, Y. Ohtsubo and M. Tsuda. 2007. Aerobic degradation of lindane (c-hexachlorocyclohexane) in bacteria and its biochemical and molecular basis. Appl Microbiol Biot 76: 741–752.

Nunes, G.S. and D. Barceló. 1999. Analysis of carbamate insecticides in foodstuffs using chromatography and immunoassay techniques. Trac-Trend Anal Chem 18(2): 99–107.

Ortiz-Hernández, M.L., E. Sánchez-Salinas, E. Dantán-González and M.L. Castrejón-Godínez. 2013. Pesticide biodegradation: mechanisms, genetics and strategies to enhance the process. Biodegradation-life of Science 251–287.

Palleschi, G., M. Bernabei, C. Cremisini and M. Mascini. 1992. Determination of organophosphorous insecticides with a choline electrochemical biosensor. Sensor Actuator B 7: 513–517.

Pankaj, S.A., S. Gangola, P. Khati, G. Kumar and A. Srivastava. 2016. Novel pathway of cypermethrin biodegradation in a Bacillus sp. strain SG2 isolated from cypermethrin-contaminated agriculture field. 3 Biotech 6(1): 1–11.

Pimmata, P., A. Reungsang and P. Plangklang. 2013. Comparative bioremediation of carbofuran contaminated soil by natural attenuation, bioaugmentation and biostimulation. Int Biodeter Biodegr 85: 196–204.

Pointing, S. 2001. Feasibility of bioremediation by white-rot fungi. Appl Microbiol Biot 57(1-2): 20–33.

Radhika, M. and M. Kannahi. 2014. Bioremediation of pesticide (Cypermethrin) using bacterial species in contaminated soil. Int J Curr Microbiol App Sci 3(7): 427–35.

Ragnarsdottir, K.V. 2000. Environmental fate and toxicology of organophosphate pesticides. J Geol Soc 157(4): 859–876.

Raushel, F.M. 2002. Bacterial detoxification of organophosphate nerve agents. Curr Opin Microbiol 5(3): 288–295.

Rigas, F., K. Papadopoulou, A. Philippoussis, M. Papadopoulou and J. Chatzipavlidis. 2009. Bioremediation of lindane contaminated soil by Pleurotus ostreatus in non sterile conditions using multilevel factorial design. Water, Air, and Soil Pollution 197: 121–129.

Roldán, M.D., E. Pérez-Reinado, F. Castillo and C. Moreno-Vivian. 2008. Reduction of polynitroaromatic compounds: the bacterial nitroreductases. FEMS Microbiol Rev 32(3): 474–500.

Sahl, J.W., J. Munakata-Marr, M.L. Crimi and R.L. Siegrist. 2007. Coupling permanganate oxidation with microbial dechlorination of tetrachloroethene. Water Environ Res 79(1): 5–12.

Satapute, P. and B. Kaliwal. 2016a. Biodegradation of the fungicide propiconazole by Pseudomonas aeruginosa PS-4 strain isolated from a paddy soil. Ann Microbiol 66(4): 1355–1365.

Satapute, P. and B. Kaliwal. 2016b. Biodegradation of propiconazole by newly isolated Burkholderia sp. strain BBK_9. 3 Biotech 6(1): 110. doi:10.1007/s13205-016-0429-3.

Satapute, P., M.V. Kamble, S.S. Adhikari and S. Jogaiah. 2019a. Influence of triazole pesticides on tillage soil microbial populations and metabolic changes. Sci Total Environ 651: 2334–2344.

Satapute, P., M.K. Paidi, M. Kurjogi and S. Jogaiah. 2019b. Physiological adaptation and spectral annotation of Arsenic and Cadmium heavy metal-resistant and susceptible strain Pseudomonas taiwanensis. Environ Pollut 251: 555–563.

Satar, S., A. Sebe, N.R. Alpay, U. Gumusay and O. Guneysel. 2009. Unintentional endosulfan poisoning. Bratisl Med J 110(5): 301–303.

Satish, G. Parte, D. Mohekar Ashokrao and S. Kharat Arun. 2017. Microbial degradation of pesticide: a review. Afr J Microbiol Res 11(24): 992–1012.

Sethunathan, N. and T. Yoshida. 1973. A Flavobacterium sp. that degrades diazinon and parathion. Can J Microbiol 19(7): 873–875.

Sharma, Anita, Priyanka Khati, Saurabh Gangola and Govind Kumar. 2016. Microbial degradation of pesticides for environmental cleanup. In Bioremediation of Industrial Pollutants.

Siddique, Tariq, Benedict C. Okeke, Muhammad Arshad and William T. Frankenberger. 2003. Enrichment and isolation of endosulfan-degrading microorganisms. J Environ Quality 32(1): 47–54.

Silambarasan, Sivagnanam and Jayanthi Abraham. 2013. Mycoremediation of endosulfan and its metabolites in aqueous medium and soil by Botryosphaeria laricina JAS6 and Aspergillus tamarii JAS9. PLoS ONE 8(10): 1–10.

Silva, E., A.M. Fialho, I. Sá-Correia, R.G. Burns and L.J. Shaw. 2004. Combined bioaugmentation and biostimulation to cleanup soil contaminated with high concentrations of atrazine. Environ Sci Tech 38(2): 632–637.

Silver, Kristopher S., Yuzhe Du, Yoshiko Nomura, Eugenio E. Oliveira, Vincent L. Salgado, Boris S. Zhorov and Ke Dong. 2014. Voltage-gated sodium channels as insecticide targets. Adv In Insect Phys 46: 389–433.

Simon-Delso, N., V. Amaral-Rogers, L.P. Belzunces, J.M. Bonmatin, M. Chagnon, C. Downs, L. Furlan, D.W. Gibbons, C. Giorio, V. Girolami and D. Goulson. 2015. Systemic insecticides (neonicotinoids and fipronil): trends, uses, mode of action and metabolites. Environ Sci Pollut Res. 22(1): 5–34.

Singh, B.K. and A. Walker. 2006. Microbial degradation of organophosphorus compounds. FEMS Microbiol Rev 30: 428–471.

Singh, Dileep, K. 2008. Biodegradation and bioremediation of pesticide in soil: concept, method and recent developments. Indian J Microbiol 48(1): 35–40.

Smith, M.J. and G. Bucher. 2012. Tools to study the degradation and loss of the N-phenyl carbamate chlorpropham—A comprehensive review. Environ Int 49: 38–50.

Solomon, K.R., W.M. Williams, D. Mackay, J. Purdy, J.M. Giddings and J.P. Giesy. 2014. Properties and uses of chlorpyrifos in the United States. pp. 13–34. In Ecological Risk Assessment for Chlorpyrifos in Terrestrial and Aquatic Systems in the United States. Springer, Cham.

Song, Yaling. 2013. Insight into the mode of action of 2, 4-dichlorophenoxyacetic acid (2, 4-d) as an herbicide. J Integr Plant Biol 56(2): 106–13.

Streber, W.R., K.M. Timmis and M.H. Zenk. 1987. Analysis, cloning, and high-level expression of 2,4-dichlorophenoxyacetate monooxygenase gene tfdA of Alcaligenes eutrophus JMP134. J Bacteriol 169: 950–2955.

Supreeth, M., M.A. Chandrashekar, N. Sachin and N.S. Raju. 2016. Effect of chlorpyrifos on soil microbial diversity and its biotransformation by Streptomyces sp. HP-11. 3 Biotech 6(2): 147.

Sutherland, Tara D., I. Horne, K.M. Weir, C.W. Coppin, M.R. Williams, M. Selleck, R.J. Russell and J.G. Oakeshott. 2004. Enzymatic bioremediation: from enzyme discovery to applications. Clin Exp Pharmacol P 31(11): 817–21.

Sutton, Nora B., J. Tim C. Grotenhuis, Alette A.M. Langenhoff and Huub H.M. Rijnaarts. 2011. Efforts to improve coupled *in situ* chemical oxidation with bioremediation: a review of optimization strategies. J Soils Sediments 11(1): 129–40.

Suzuki, O. and K. Watanabe. 2005. Drugs and Poisons in Humans-A Handbook of Practical Analysis. pp. 559–570. *In*: Carbamate Pesticides. DOI: 10.1007/3-540-27579-7-62.

Tan, J., Z. Liu, R. Wang, Z.Y. Huang, A.C. Chen, M. Gurevitz and K. Dong. 2005. Identification of amino acid residues in the insect sodium channel critical for pyrethroid binding. Mol Pharmacol 67(2): 513–522.

Tang, C.Y., Q.S. Criddle, C.S. Fu and J.O. Leckie. 2007. Effect of flux (trans membrane pressure) and membranes properties on fouling and rejection of reverse osmosis and nano filtration membranes treating perfluorooctane sulfonate containing waste water. J Enviro Sci Tech 41: 2008–14.

Tewari, Lakshmi, Jitendra Saini and Arti. 2012. Bioremediation of pesticides by microorganisms: general aspects and recent advances. Bioremed Pollut 25–48.

Tongarun, R., E. Luepromchai and A.S. Vangnai. 2008. Natural attenuation, biostimulation, and bioaugmentation in 4-chloroaniline-contaminated soil. Curr Microbiol 56(2): 182–188.

Ulčnik, A., I. Kralj Cigić, L. Zupančič-Kralj, Č. Tavzes and F. Pohleven. 2012. Bioremediation of lindane by wood-decaying fungi. Drvna industrija: Znanstveni časopis za pitanja drvne tehnologije 63(4): 271–276.

Uniyal, Shivani, Rashmi Paliwal, R.K. Sharma and J.P.N. Rai. 2016. Degradation of fipronil by Stenotrophomonas acidaminiphila isolated from rhizospheric soil of Zea mays. 3 Biotech 6(1): 1–10.

Vaccari, D.A., P.F. Strom and J.E. Alleman. 2006. Environmental biology for engineers and scientists (Vol. 7, p. 242). New York: Wiley-Interscience.

Van der Meer, J.R. 1992. Molecular mechanisms of adaptation of soil bacteria to chlorinated benzenes. PhD diss., Landbouwuniversiteit te Wageningen.

Van Eerd, L.L., R.E. Hoagland, R.M. Zablotowicz and J.C. Hall. 2003. Pesticide metabolism in plants and microorganisms. Weed Sci 51(4): 472–495.

Vidali, M. 2001. Bioremediation. an overview. Pure Appl Chem 73(7): 1163–1172.

Wackett, L., M. Sadowsky, B. Martinez and N. Shapir. 2002. Biodegradation of atrazine and related s-triazine compounds: from enzymes to field studies. Appl Microbiol Biot 58(1): 39–45.

Wang, Q. and A.T. Lemley. 2003. Competitive degradation and detoxification of carbamate insecticides by membrane anodic Fenton treatment. J Agr Food Chem 51(18): 5382–5390.

Wang, X., B. Yao and X. Su. 2018. Linking enzymatic oxidative degradation of lignin to organics detoxification. Int J Mol Sci 19(11): 3373.

Williamson, Martin S., Ian Denholm, Caroline A. Bell and Alan L. Devonshire. 1993. Knockdown resistance (Kdr) to DDT and pyrethroid insecticides maps to a sodium channel gene locus in the housefly (Musca Domestica). Mol Gen Genet 240(1): 17–22.

Wolfe, N.L., R.G. Zepp and D.F. Paris. 1978. Use of structure-reactivity relationships to estimate hydrolytic persistence of carbamate pesticides. Water Res 12(8): 561–563.

Xie, G.B. and M.J. Barcelona. 2003. Sequential chemical oxidation and aerobic biodegradation of equivalent carbon number-based hydrocarbon fractions in jet fuel. Environ Sci Tech 37: 4751–4760.
Yadav, J.S. and J.C. Loper. 2000. Cytochrome P450 oxidoreductase gene and its differentially terminated cDNAs from the white rot fungus Phanerochaete chrysosporium. Curr Genet 37(1): 65–73.
Yair, S., B. Ofer, E. Arik, S. Shai, R. Yossi, D. Tzvika and K. Amir. 2008. Organophosphate degrading microorganisms and enzymes as biocatalysts in environmental and personal decontamination applications. Crit Rev Biotechnol 28(4): 265–275.
Yan, Q.X., Q. Hong, P. Han, X.J. Dong, Y.J. Shen and S.P. Li. 2007. Isolation and characterization of a carbofuran-degrading strain Novosphingobium sp. FND-3. FEMS Microbiol Lett 271(2): 207–213.
Zhang, C., S.H. Wang and Y.C. Yan. 2011. Isomerization and biodegradation of beta-cypermethrin by *Pseudomonas aeruginosa* CH7 with biosurfactant production. Bioresource Technol 102: 7139–7146.

17

Whole Effluent Toxicity Assessment of Sewage Discharge Water

Jun Jin[1,*] and *Takashi Kusui*[2]

Introduction

A wide variety of chemicals are used in industry and daily life, and the environmental impacts of trace chemicals in sewage discharge water are a source of concern. In Japan, current effluent standards aim to protect human health and the environment. As for the protection of the aquatic organisms, the only effluent standard for zinc has been established so far. However, one standard item is not enough to protect aquatic life from the stresses posed by many chemical substances present in water environment. Therefore, it is necessary to develop a more effective and comprehensive management system for chemicals in effluents. The whole effluent toxicity (WET) approach, which was established in the USA and Canada using biological responses to evaluate the comprehensive effect of chemicals in effluent, overcomes the shortcomings of traditional effluent regulation based on individual chemical compounds (Tatarazako 2006). Although the introduction of this approach has been discussed since 2009 in Japan, and a draft of test guidelines was released in March 2013, there are still few reports involved in WET results for sewage discharge water in Japan (Kusui and Blaise 1999, Yamamoto et al. 2010, Yamamoto et al. 2013, Takeda et al. 2016).

In contrast, the Japanese Pollutant Release and Transfer Register (PRTR) system promotes voluntary improvement of chemical management by business operators,

[1] Yokohama National University, 79-1 Tokiwadai, Hodogaya Ward, Yokohama, Kanagawa 240-8501, Japan.

[2] Toyama Prefectural University, 5180 Kurokawa, Imizu, Toyama 939-0398, Japan.
Email: kusui@pu-toyama.ac.jp

* Corresponding author: jinjun5134@foxmail.com

who notify the government of the amounts of specified chemical substances released into the environment (Ministry of the Environment, Government of Japan 2019). Previous research has shown that PRTR data are useful for effluent toxicity estimation in industrial effluents for screening purposes (Fukutomi et al. 2013). However, direct relationships between toxicity of the final effluent were not demonstrated.

The main purpose of this section was to assess the environmental effects of sewage discharge water, which include sewage effluent and combined sewage overflows (CSOs) using chemical analysis and biological tests, as these two approaches complement each other. The ecotoxicity of each effluent was also predicted based on PRTR data and a toxicity database. Finally, the results of both approaches were compared. The ecotoxicity of some samples was also prepared for toxicity identification evaluation (TIE) test to identify the toxic cause substances.

Sewage effluent toxicity assessment

Materials and methods

Analysis of water parameters

From November 2013 to December 2014, seven final effluent samples were collected from seven sewage treatment plants located in Toyama, Japan (Table 17.1). The sewage treatment methods consisted of conventional activated sludge processes at five plants and oxidation ditches at the remaining two plants. The samples (3.8 L) were immediately transported to the laboratory in amber glass bottles in a lightproof ice box, then filtered through a 60 μm nylon strainer, and stored in a dark refrigerator until analysis and bioassay. Biological tests were conducted within 36 hours of sampling. Table 17.2 shows the methods and measuring equipment of water parameters, such as temperature, pH, total organic carbon (TOC), and so on.

Table 17.1. Information about the sewage treatment plants.

Plant	Process	Sewage collection	Daily average flow (m^3/d)	Industrial wastewater (%)	Disinfectant
A	AS	Separate	53,000	3	Sodium hypochlorite
B	AS	Separate	71,000	27	Sodium hypochlorite
C	AS	Combined	100,000	17	Sodium hypochlorite
D	AS	Combined	50,000	62	Sodium hypochlorite
E	AS	Separate	5,000	4	Sodium hypochlorite
F	POD	Separate	350	0	Iso cyanuric ester
G	OD	Separate	13,000	12	Sodium hypochlorite

AS: activated sludge system, POD: prefabricated oxidation ditch, OD: oxidation ditch.

Whole effluent toxicity tests

In this section, the three short-term chronic biological tests proposed in the Japanese WET test guideline (NIES and MOE, Japan 2013) were performed. Final effluent samples were serially diluted (0, 5, 10, 20, 40, and 80%) to a range of concentrations.

Table 17.2. The methods and measuring equipment of water parameters.

Water parameters	Methods and measuring equipment	Water parameters	Methods and measuring equipment
Water temperature, pH, Electrical conductivity (EC)	pH-DO meter (pH/COND meter D-54, Horiba, Kyoto, Japan)	Total organic carbon (TOC)	The multi N/C2100S meter (Analytik Jena AG, Jena, Germany)
Dissolved oxygen (DO)	DO meter (ProODO, YSI Inc., OH, USA)	The total concentration of metals	Inductively coupled plasma mass spectrometry (ICP-MS) (Kusui et al. 2014)
Total hardness	The EDTA titrametric method in according to the Standard (Japan Water Works Association 2011)	Linear alkylbenzene sulfonate (LAS)	High performance liquid chromatography (HPLC) using fluorescence detection in accordance with the Survey Manual (Ministry of the Environment Government of Japan 2000)
Total and free residual chlorine	The diethyl-p-phenylenediamine method (DR2800, Hach Co., Loveland, USA)		
Total phosphorus (TP)	Molybdenum blue absorption method after decomposition by potassium peroxodisulfate (Japan Standards Association 2010)	Total nitrogen	Via UV spectrophotometry after decomposition by alkaline potassium persulfate (Japan Standards Association 2010)

The green microalga *Raphidocelis subcapitata* (NIES-35) was used for algal growth inhibition tests as per Organization for Economic Cooperation and Development (OECD) testing guideline TG201 with downsized minor modifications (OECD 2006). Precultured algae were inoculated at 5×10^3 cell/mL in an Erlenmeyer flask (200 mL capacity), and exposed to a range of effluent concentrations prepared with an OECD medium (60 mL). Six replicates were used for the control and three replicates were prepared for each effluent concentration. The algal cell density was determined after 0, 24, 48, and 72 hours with an electronic particle counter (detection range 3–12 μm; CDA-500, Sysmex, Kobe, Japan) in a 72 hour growth test. The median effective concentration (EC_{50}) and no observed effect concentration (NOEC) values were determined for the growth rate using analytical software (Ecotox-Statics ver. 2.6, Japanese Society of Environmental Toxicology, Tsukuba, Japan).

Crustacean reproduction tests, which consisted of three-brood renewal toxicity tests, were conducted with *Ceriodaphnia dubia* according to the Environment Canada method (Environment Canada 2007). First, neonate daphnia (< 24 hour old) were transferred to glass vessels containing 15 mL dilute effluents. Each dilute effluent required appropriate daily volumes of food, which consisted of yeast, Cerophyll, and trout chow (Recenttec K. K., Tokyo, Japan), and algae (*Chlorella*, Recenttec K. K.). Each test solution was renewed three times per week, and 10 replicates were performed per effluent concentration. EC_{50} and NOEC values were determined for fecundity using analytical software (Ecotox-Statics ver. 2.6).

Short-term toxicity tests were performed on *Danio rerio* fish embryo and sac-fry stages according to OECD TG212 (OECD 1998). Briefly, 10 fertilized eggs

(< 4 hours) were placed into a glass vessel containing 50 mL test solution. Each dilute effluent concentration comprised of four replicates, and was observed daily during the test period (9 days). The NOEC value was calculated using the survival and hatching rates with analytical software (Excel Tokei ver. 6.0, Esumi Inc., Japan).

The results of the toxicity tests were compared using chronic toxicity units (TUc), which were calculated as follows:

$$TUc = 100/NOEC\ (\%) \qquad (1)$$

TUc values of > 1 for a given effluent indicate that aquatic organisms might be adversely affected by the effluent; larger TUc values indicate higher toxicity. If the maximum concentration (80%) in this section has no toxic effect, the NOEC is 80% or more, and as a result, the TUc value is 1.25 or less, which is considered to have virtually no toxic effect.

Hazard quotient calculated from PRTR data

As the first step of calculating the hazard quotient using PRTR (Ministry of the Environment of Japan 2019), the average annual emissions (kg/yr) from each sewage treatment plant was divided by the number of annual days of operation (365 d/yr), and then divided by the average effluent discharge (m^3/d) to estimate the effluent concentration (*Ci*). The HQ*i* for chemical *i* was then calculated using the following formula:

$$HQi = Ci/NOECi \qquad (2)$$

The x% effect concentration (ECx) does not depend on the choice of test concentrations and number of replications, which might be more accurate than NOEC to evaluate toxic effect (Isnard et al. 2001). However, the NOECs have been widely used for statistical analysis of chronic ecotoxicity data, and are more accessible in terms of literature search from various risk assessment reports and toxicity databases (Chemicals Evaluation and Research Institute, Japan 2006, Ministry of the Environment, Government of Japan 2016, USEPA 2017). Therefore, we used NOECs in this section. In adopting toxicity values, values from tests using the same methods and types of test organisms as those used in the WET tests were prioritized. If more than one NOEC value was available, the lowest value (which would produce the most sensitive result) was used. Data was available for the *R. subcapitata* algal growth inhibition test. However, data that exactly matched the other two tests was not readily accessible. Thus, the toxicity value was selected using the following procedure—the lowest chronic value from the same test (same end point, same exposure period) and same family of freshwater species was prioritized. If no chronic data was available, an acute chronic ratio (ACR) of 10 (Zeeman 1995, Forbes and Callow 2002) was applied to estimate the chronic value from the acute value for the same or a similar organism. In cases featuring more than one PRTR chemical in a single effluent, the toxic effects were assumed to be additive (Marking 1985, Sarakinos et al. 2000), and the sum of the HQ*i* values was calculated. Toxicity due to PRTR substances might be expected when the sum of these ratios exceeds 1.0.

Results and discussion

Effects on aquatic organisms

Figure 17.1 shows the TUc values of the seven effluents. Effluents with TUc values of less than 1.25 were considered non-toxic. Toxicity to crustacean and algae was highest in effluent B. In particular, the TUc of algae measured 20; in other words, it would be necessary to dilute this effluent at least twenty times to avoid significant toxic effects. The second-highest toxicity to algae was observed in effluents F and G, which had TUc values of 10, followed by effluent D with a TUc value of 2.5. Of the seven effluents, effluent B, with a TUc of 2.5, was the most toxic to crustaceans. Only effluents A and F were subjected to the fish test, because conditions were not appropriate for fish spawning at the time of collection of other effluents. In this result, algae were the most sensitive to sewage effluent. However, perhaps due to the differences in water quality, daphnia were the most sensitive in previous studies on sewage effluent (Yamamoto et al. 2010) and industrial effluent (Kusui et al. 2014).

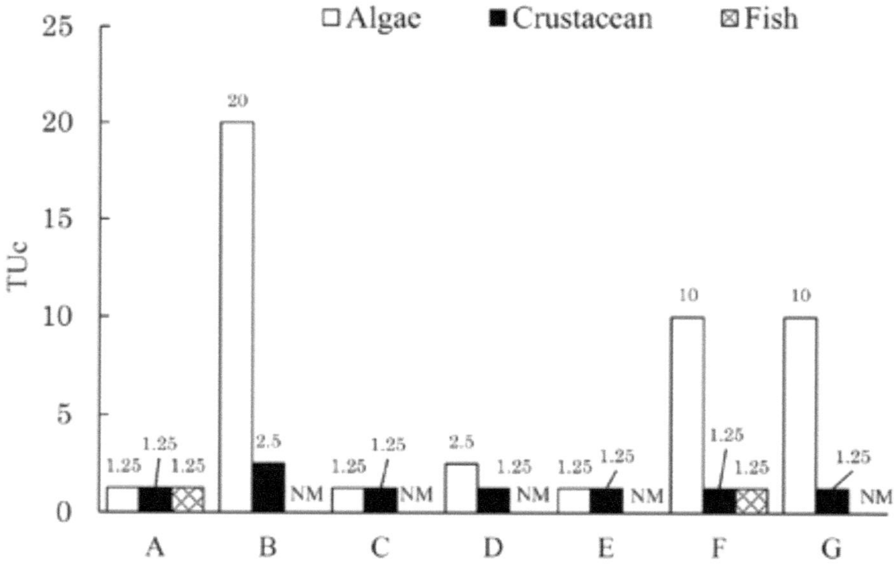

Figure 17.1. Chronic toxicity units (TUc) values for sewage effluent samples. NM: not measured.

Water parameters

The contributions of some parameters to observed toxicity are discussed below. The highest LAS concentration observed in the effluent samples was ~ 100 μg/L. The NOECs of C_{11}–C_{13} LAS for *R. subcapitata* (72 hr-NOEC), *C. dubia* (8 d-NOEC), and *D. rerio* (9 d-Hatching NOEC) were 1.8–11 mg/L, 0.07–1.9 mg/L and 0.28–4.5 mg/L, respectively (Tamura et al. 2017). It is revealed that the longer the chain length, the higher the toxicity. The concentrations of alkyl chains numbers C_{10}–C_{14} were measured for LAS concentrations in this study. Any of these concentrations is 0.05 mg/L or less. It was compared with the actually measured literature toxicity values of LAS concentration. The maximum toxicity value of LAS in this study were 12% of NOEC. Although LAS did not likely contribute to effluent toxicity in this section, the highest ratio of the predicted environmental concentration to the predicted no effect concentration (PEC/PNEC) was as high as 300 in the freshwater area in 2000 (Ministry of the Environment, Government of Japan, 2018). Therefore, this compound must be monitored due to its high risk to aquatic organisms.

The highest concentration of total residual chlorine was 0.39 mg/L in effluents B and G, followed by 0.21 mg/L in effluent D. The highest concentration of free chlorine was 0.04 mg/L in effluents B, D, and G. For combined residual chlorine (NH_2Cl, monochloramine), a 96 hour lowest observed effect concentration (LOEC) of 0.01 mg/L has been reported for *R. subcapitata* (Suzuki et al. 1996). However, the 96 hour NOEC value was not determined in this report. Herein, monochloramine was prepared using a previously reported method (Suzuki et al. 1996) and subjected to a *R. subcapitata* 72 hour algae growth inhibition test. The resultant NOEC was 0.01 mg/L, which is the same as the reported 96 hour LOEC. Monochloramine may be the main ingredient inhibiting algal growth in effluents B, D, and G.

Heavy metals, such as Ni, Cu, and Zn, are known to be toxic to aquatic organisms at relatively low concentrations. As shown in Table 17.3, Ni, which has a NOEC of algae value of 40 μg/L, featured concentrations of less than 10 μg/L in all plants. Therefore, Ni did not likely contribute to algae toxicity. The Cu concentrations were generally small, and only the Cu concentration of plant B and plant G were higher than the NOEC value of 10 μg/L, suggesting that Cu might contribute to toxicity in these effluents. Several tens of μg/L Zn were detected in all plants. The NOEC value of this metal for algae is 53 μg/L, and that for crustaceans is 25 μg/L, suggesting that Zn contributions to toxicity should be considered. However, because toxic effects were not observed even in effluents of plants C and E, where the concentration was higher than the reported NOEC, it was inferred that Zn did not contribute to toxicity. This reduction in metal toxicity might be explained by the coexisting organic matter and cations in the effluents, which have been reported to counteract the influence of metals (Di Toro et al. 2001, Santore et al. 2001).

In this study, there is a possibility that the toxicity to algae was likely derived from a combination of chlorine and other compounds, such as ammonia and triclosan, which have been reported to cause chronic sewage effluent toxicity in previous

Table 17.3. Metal analysis results (unit: μg/L) from the sewage treatment plants (A–G).

	Cr	Ni	Cu	Zn	Cd	Pb
A	0.3	1.8	6.5	62	0.1	0.7
B	2.1	4.5	12	84	0.3	2.8
C	0.2	9.5	3.7	26	< 0.01	0.2
D	0.2	0.8	3.2	33	< 0.01	0.1
E	0.7	2.4	8.7	63	0.1	41.8
F	< 0.01	< 0.01	7.0	27	< 0.01	< 0.01
G	< 0.01	< 0.01	21	52	< 0.01	< 0.01

studies (Orvos et al. 2002, Takeda et al. 2017), but not measured in this study. This might explain why algae was the most sensitive among three test organisms in this study. However, daphnia was reported to be the most sensitive in another study on sewage effluents, although causative substances were not identified (Yamamoto et al. 2010). As for the previous study on industrial effluents (Kusui et al. 2014), Ni of which concentration was higher than those of sewage effluents, was suspected to the cause of toxicity on *C. dubia* that is more sensitive to this metal than algae.

Further analysis and toxicity identification evaluation (USEPA 1992, 1993) are necessary to estimate the contributions of these compounds to toxicity. Moreover, emerging pollutants, such as pharmaceuticals and personal care products, in final effluents may pose risks to aquatic organisms (Takeda et al. 2015, Watanabe et al. 2016). Further monitoring is therefore necessary to evaluate environmental concentration trends for these chemicals.

Hazard quotient based on PRTR data

In addition to the 28 substances reported at plant F, per treatment plant reported an average of three PRTR substances discharged from each sewage treatment plant to aquatic environments in 2013 and 2014 (min = 0, max = 9). The six most abundant PRTR substances released from multiple sewage treatment plants were boron compounds, zinc compounds (water-soluble), arsenic and inorganic arsenic compounds, copper salts (water-soluble except for complex salts), dioxins, and manganese and its compounds.

The HQ results are shown in Fig. 17.2. Although slight variances are observed due to differences in estimation methods, HQ was predicted to exceed 10 at three treatment plants. In particular, relatively high HQ values were observed for all organisms at plant F, which also featured the largest number of PRTR substances. Of the three types of test organisms, the HQ values were the largest in crustaceans, followed by algae. This finding differs from the results of the biological tests, in which algae were the most affected and featured TUc values of 10 or more (Fig. 17.1).

HQ values based on PRTR data likely underestimate or overestimate toxicity, and that metal toxicity tends to be lower than that predicted by the measured concentration. Thus, results from both methods are inconsistent. Unlike industrial factories, where chemical substances in wastewater are measured appropriately, it is necessary to examine the accuracy of effluent PRTR data at sewage treatment plants that accept discharge from various sources.

Toxicity assessment of combined sewer overflows

Materials and methods

Analysis of water parameters and toxicity tests

The Takaoka treatment area (601.17 ha) of the combined sewer system is located in the central area of Takaoka City, which is in the northwestern part of Toyama Prefecture, Japan (Takaoka city 2010). The outfall C has the largest collection area (94.40 ha) among 21 CSO outfalls along the Senbo River, as shown in Fig. 17.3.

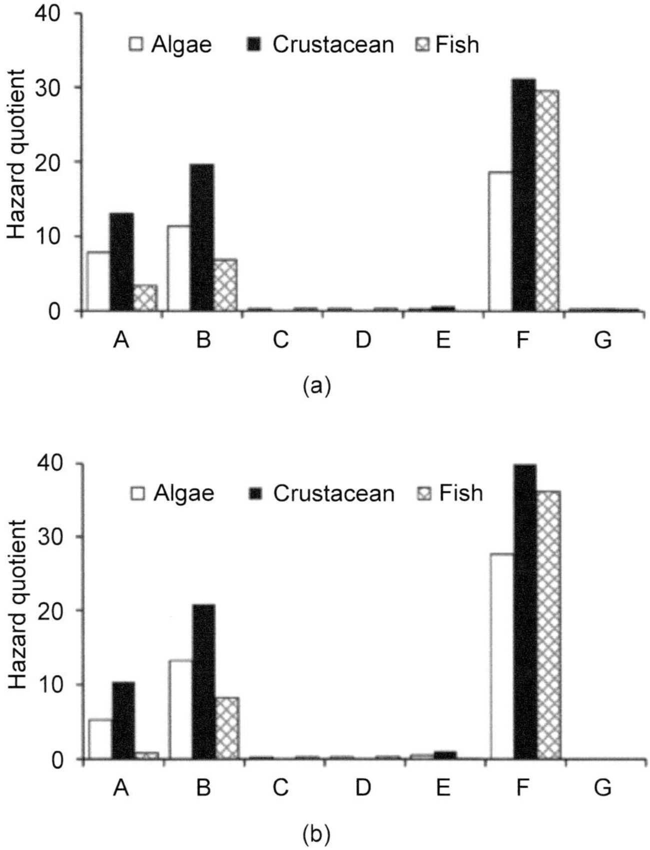

Figure 17.2. Hazard quotient for sewage treatment plants calculated using (a) average of Pollutant Release and Transfer Register (PRTR) data from 2013 and 2014, and (b) PRTR data from the sampling year.

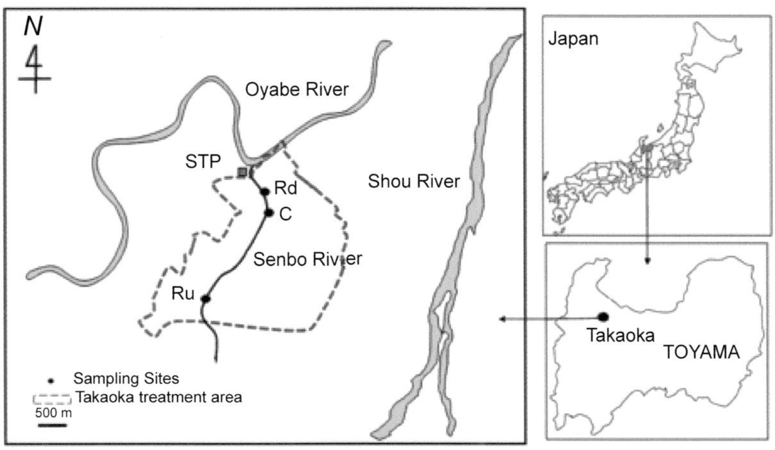

Figure 17.3. Sampling points and catchment location. Rd, Ru: river water sampling sites; C: CSOs; sampling site; STP: sewage treatment plant; dotted line: Takaoka treatment area.

From October to December 2016, seven samples were collected in three rainy events from outfall C. First flushes C1 and C2-1 were collected during the first and second sampling events, respectively. CSOs at different time points C2-1 (first flush), C2-2 (after 30 minutes), C2-3 (final flush, after approximately 1 hour) were collected during the second sampling event. During the third sampling event, overflow (C3) and river water samples upstream (Ru) and downstream (Rd) of the overflow outlet were collected at the end of overflow. The dry periods before these three sampling events lasted for approximately 60, 60, and 24 hours, respectively. The amount of rainfall in the three sampling events is shown in Fig. 17.4. To compare them with river water and raw sewage in dry weather, four river water samples and two raw

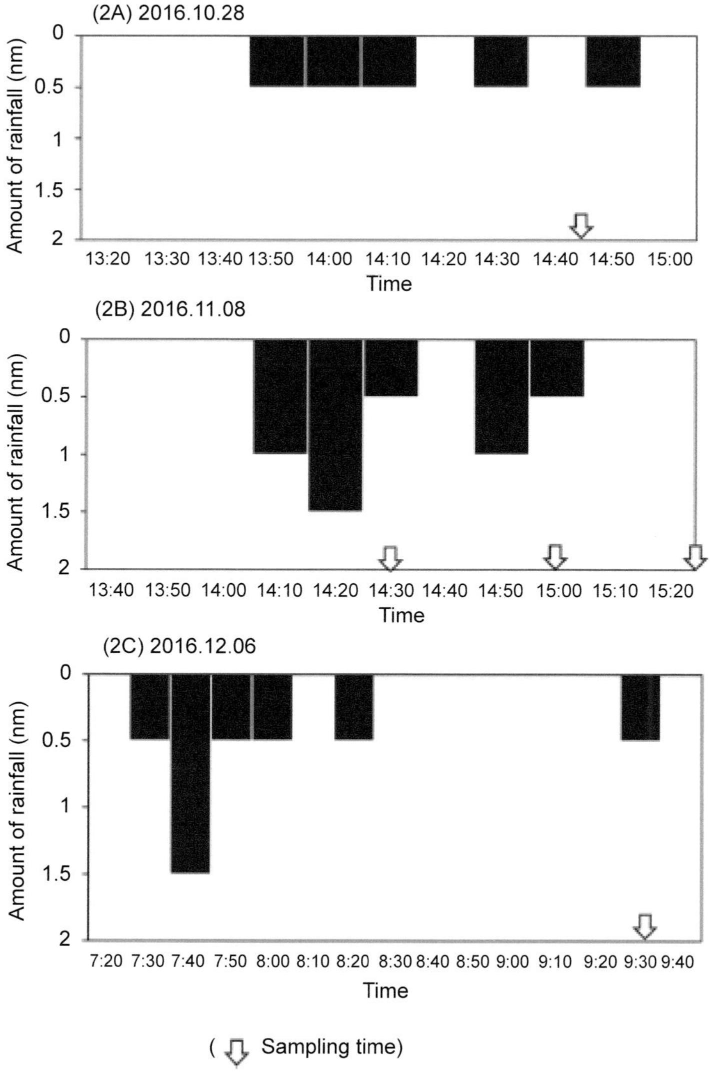

Figure 17.4. Amount of rainfall in three rainy events.

sewage water samples were also collected from Rd, Ru, and STP (Fig. 17.3). The water samples were transported to the laboratory and filtered through a 60-μm nylon strainer. The biological tests were conducted within 36 hours after sampling, and filtered samples were stored at 4°C.

At each sampling event, the water parameters of each sample, such as pH, EC, DO, etc. were measured by the same methods shown in Table 17.2 at phase 1.1.1.

Short-term chronic bioassays with *Danio rerio*, *Ceriodaphnia dubia*, and *Raphidocelis subcapitata* were conducted by the same method showing on 1.1.2 according to the Japanese WET test guideline (NIES and MOE, Japan 2013). If the toxicity was particularly strong, the sample was diluted to less than 5%, with a dilution factor of 2.

Characterization of toxic substances

Samples C3, Rd, Ru were subjected to phase I TIE, which included the physicochemical treatment of samples before biological tests (USEPA 1991, 1992), because they showed toxicity (see results below). The characteristics of toxic substances might be determined by comparing the results of biological experiments before and after sample treatment. The manipulations applied to the samples are listed in Table 17.4.

A daphnia reproduction test was used for TIE. Two concentrations, indicating the level of toxicity were used, and eight replicates were used in each concentration test group. The characteristics of the toxic substances were deduced by comparing the results of treated and untreated samples at the same concentration. Excel Tokei ver. 7.0 (Esumi Co. Ltd.) was used to analyze the statistical test results for the significant difference in one-way ANOVA.

Table 17.4. Effluent manipulations used in toxicity identification evaluation (TIE) (USEPA 1991, 1992).

Manipulation	Toxic substances examined
EDTA addition (3 mg/L)	Heavy metals
Sodium thiosulfate addition (10 mg/L)	Oxidants (i.e., chlorine)
pH adjustment (6.5, 8.5)	Acids or Bases
Post C18-solid phase extraction (SPE) column test	Non or low polar organic substances
Post anion exchange column test	Anions
Post cation exchange column test	Cations

Results and discussion

Effect of CSOs on aquatic organisms

Figure 17.5 shows the TUc and coliform bacteria levels in the samples collected at the first and second sampling events. The first flush samples, C1 and C2-1, had toxic effects on algae and daphnia (TUc = 2.5 and 10, respectively), but no toxic effect was detected in fish. In samples C2-1, C2-2, and C2-3, collected at the second sampling event, toxicity to daphnia generally decreased with overflow time. Sample C2-3 was not toxic for these freshwater organisms. The results obtained for coliform bacteria indicated that the concentrations of initial overflows C1 and C2-1 were relatively

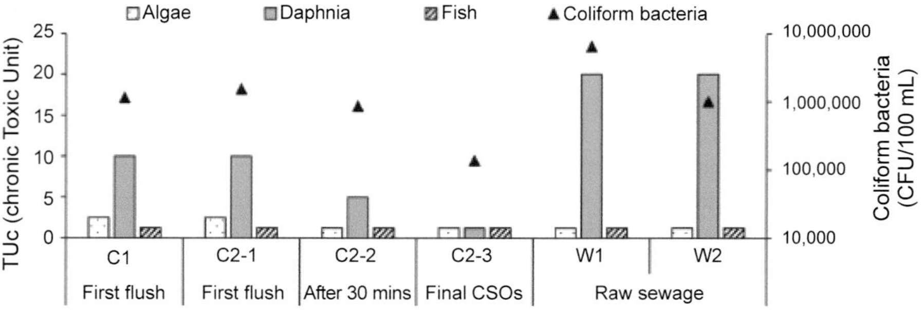

Figure 17.5. Chronic toxic unit (TUc) in raw sewage and samples taken at the first and second sampling events.

large: 1.15×10^6 colony forming units (CFU)/100 mL and 1.50×10^6 CFU/100 mL, respectively. The concentration of C2-1 as the initial overflow water was higher than the concentrations of C2-2 and C2-3. Similar to TUc, the number of coliform bacteria in C2-1, C2-2, and C2-3 samples decreased with the time of overflow. According to Fig. 17.5, which shows data on dry weather sewage in this district, the mean of coliform bacteria and TUc detected in daphnia tests were 3.72×10^6 CFU/100 mL and 20, respectively. The ratios of coliform counts for C2-1, C2-2, and C2-3 based on that of raw sewage, were 0.41, 0.23, and 0.04, respectively. These ratios suggest a tendency for the raw sewage in CSOs to be gradually diluted by urban runoff. Interestingly, TUc ratios for C2-1, C2-2, and C2-3 based on that of raw sewage were almost identical (0.5, 0.25, and 0.06, respectively) suggesting that toxic substances for daphnia were derived from household wastewater. The results of toxicity tests and coliform bacteria levels of samples from the third sampling event (Fig. 17.6) indicated that samples C3, Rd, and Ru were only toxic for daphnia. Their TUc were 20, 20, and 40, respectively. However, the number of coliform bacteria was lower than that observed in the first and second sampling events. The concentration of coliform bacteria in sample C3, which contained sewage, was the largest among

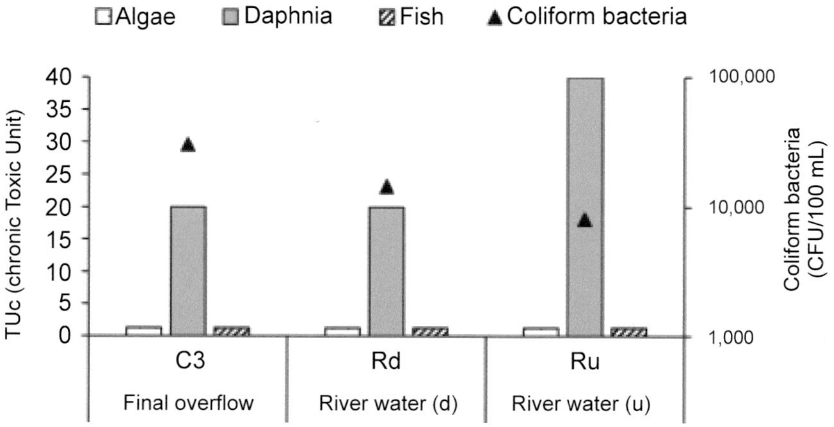

Figure 17.6. Chronic toxic unit (TUc) in samples taken at the third sampling event.

the third sampling event samples. The concentration range of coliform bacteria in dry weather river water was 1900–7300 CFU/100 mL (n = 4). Considering the large number of coliform bacteria mentioned before, and the high concentration of TN and TP in raw sewage (Table 17.5), the raw sewage in CSOs might be gradually diluted in the river. Therefore, the relatively strong toxicity observed in these samples did not derive from household wastewater, but from other sources, such as urban runoff or the effluents from industrial plants located upstream of this river. In summary, daphnia were the most sensitive organisms to CSOs and river water, which is in agreement with a previous study reporting that the *C. dubia* reproduction test was the most sensitive to overflow water among the bioassays used (Gooré et al. 2015).

Water quality

Water parameters and heavy metals from the seven samples, dry weather river water samples, and dry weather wastewater samples are shown in Tables 17.5 and 17.6, respectively. Water temperature, TOC, and coliform bacteria of samples C2-1, C2-2, and C2-3 gradually decreased with overflow time. The content of wastewater in the overflow was gradually diluted by rainwater. The concentration of TOC and coliform bacteria in overflow water were several times higher than in river samples. The hardness of the overflow water was about half that of the river. The average concentration of hardness in the river water in this study was high, at approximately 90 mg/L. This may be caused by river water receiving the effluent from the industrial plants located upstream of this river. Among the eight metals analyzed (Table 17.6), copper (Cu) and zinc (Zn) concentrations were higher in CSOs than in river water, and the highest concentrations were that of aluminum (Al), followed by Zn and Cu. This tendency was identical to that shown for CSOs samples from Longueuil, Canada (Gooré et al. 2015). However, the concentrations of Cu and Zn found in the present study were considerably lower than those reported for Cu (208–267 μg/L) and Zn (2,040–3,343 μg/L) in the CSOs of the Marais combined sewer system in Paris, France (Gromaire et al. 2002, Kafi et al. 2008). These high values are derived from the corrosion of zinc- and lead-bearing roofing, gutter materials, and automotive brake residues.

As shown in Table 17.6, the concentrations of Cu and Zn in river water samples in our study were lower than those observed in wastewater, suggesting that relatively high concentrations of Cu and Zn in CSOs might derive from wastewater. However, some of the metals, such as nickel, might come from upstream's urban runoff or the effluents from industrial plants located upstream of this river, which occur at a higher concentration in river water. Heavy metals in DWF (dry-weather flow) and WWF (wet-weather flow) were important contaminants in storm drain effluents, which come from permitted or illicit discharges, surface runoff from industrial facilities, atmospheric deposition, releases due to dissolution in water pipes, or releases from machine parts, especially automobiles (McPherson et al. 2005). While the survey area, i.e., Takaoka City, developed as a commercial and industrial city, the craft production of castings by traditional Takaoka copperware thrived and the Al industry developed. Therefore, the high concentrations of some metals in samples taken at the third sampling events might be due to the high rate of urban runoff

Table 17.5. Water quality of combined sewer overflows (CSOs) and river water samples.

Type	Sample	pH	DO	EC	Temperature	TOC	Hardness	Coliform bacteria	SS	TN	TP
			(mg/L)	(mS/m)	(°C)	(mg/L)	(mg/L)	(CFU/100 mL)	(mg/L)	(mg/L)	(mg/L)
CSOs	C1	7.38	4.84	10.8	17.4	13.5	47.0	1.15×10^6	84.2	34.3	3.43
	C2-1	7.15	5.18	30.5	17.1	23.0	38.4	1.54×10^6	574.0	29.3	3.15
	C2-2	7.30	7.86	12.9	16.3	6.2	38.7	8.60×10^5	61.0	18.6	1.96
	C2-3	7.23	8.22	25.2	14.3	6.1	39.2	1.37×10^5	30.0	7.96	1.25
	C3	8.04	7.93	99.3	13.0	14.4	40.0	3.03×10^4	27.8	26.9	1.73
RW	Rd	7.97	8.82	49.8	12.0	3.0	87.5	1.45×10^4	8.6	4.97	0.26
	Ru	7.76	9.22	35.8	12.5	2.0	88.2	8.01×10^3	5.9	4.52	0.23
DWRW	Rmin	7.12	9.06	24.0	13.9	1.3	80.0	1.90×10^3	2.3	1.98	0.14
	Rmax	8.01	9.98	32.9	22.2	5.5	95.0	7.30×10^3	4.0	4.79	0.99
DWW	W1	8.28	2.98	41.9	16.9	51.0	78.0	6.43×10^6	88.2	55.2	7.45
	W2	7.96	4.29	19.3	21.4	24.5	42.0	1.01×10^6	24.2	29.0	2.52

CSOs, combined sewer overflows; RW, river water; DWRW, dry weather river water; DWW, dry weather wastewater; R_{min}, the minimum concentration of dry weather river water (n = 4); Rmax, the maximum concentration of dry weather river water (n = 4); DO, dissolved oxygen; EC, electric conductivity; CFU, colony forming units; TOC, total organic carbon; SS, suspended solids; TN, total nitrogen; TP, total phosphorus.

Table 17.6. Heavy-metal levels (μg/L) detected in combined sewer overflows (CSOs) and river water samples.

Sample	Aluminum	Chromium	Nickel	Copper	Zinc	Arsenic	Cadmium	Lead
C1	164	1.2	5.3	11.0	74.5	2.3	0.2	0.9
C2-1	864	0.9	4.5	9.8	81.3	1.3	0.1	1.2
C2-2	191	0.5	1.4	9.8	55.2	1.4	0.1	0.8
C2-3	994	1.5	5.7	9.8	84.8	1.6	0.1	1.2
C3	654	31.7	22.4	18.5	83.7	2.1	0.2	22.7
Rd	297	32.2	18.9	3.9	19.7	1.3	0.1	21.6
Ru	191	31.3	19.7	2.9	11.7	1.1	0.1	21.2
Rmin	133	0.7	7.5	3.5	18.8	1.3	0.1	0.4
Rmax	165	9.2	33.9	4.4	38.7	1.4	0.1	2.0
W1	455	1.5	3.2	23.6	93.6	3.4	0.2	3.8
W2	328	0.8	2.2	13.2	66.0	2.5	0.2	1.9

waters carrying atmospheric particles originating from nearby industrial facilities. Atmospheric deposition is an important source of metal residues in urban areas, and may contribute up to 57–100% of the metal fluxes in urban runoff (57% for Zn; 100% for nickel (Ni) and lead (Pb)) in Los Angeles, USA (Sabin et al. 2005). In the combined sewage system of Morioka City, Japan, at least 86% of the manganese (Mn), Zn, and iron (Fe) is derived from rainfall runoff (Ishikawa 2016, Ishikawa et al. 2016). These heavy metals can accumulate in bottom sediments, bioaccumulate in animal tissues, and affect ecological systems in receiving waters (McPherson et al. 2005). To reduce the metal-derived effects of CSOs on aquatic organisms, it is necessary to consider relative pollutant contributions from urban runoff, and evaluate the relative contributions of each source.

Characterization of major toxicants

To examine the characteristics of toxic substances on daphnia from C3, Rd, and Ru samples, TIE tests were conducted. In C3 (Fig. 17.7), the cationic and anionic resin treatment significantly reduced toxicity in terms of the average number of neonates and mortality rate compared to CSOs (10% and 40% concentrations), suggesting that some cations and some anions were responsible for toxicity in C3. Toxicity was not observed after solid extraction by C_{18} resin (10%), but the 40% filtrates held the same toxicity as untreated samples. In addition, the 40% C_{18}-eluate (probably organic substances) isolated from the resin had significant toxicity compared with the control. These contradictory results suggest that the amount of toxic organic substances in the sample was probably that beyond the extraction capacity of the resin, or that part of causative hydrophilic organic matters cannot be removed by C_{18} resin. In summary, the mixture of cationic, anionic, and organic substances might be responsible for the toxicity of C3. The toxic effects of sample Rd at a concentration of 40% were slightly alleviated by anionic, cationic, and C_{18}-column treatments (Fig. 17.8). The EDTA addition significantly reduced toxicity at 40% concentrations,

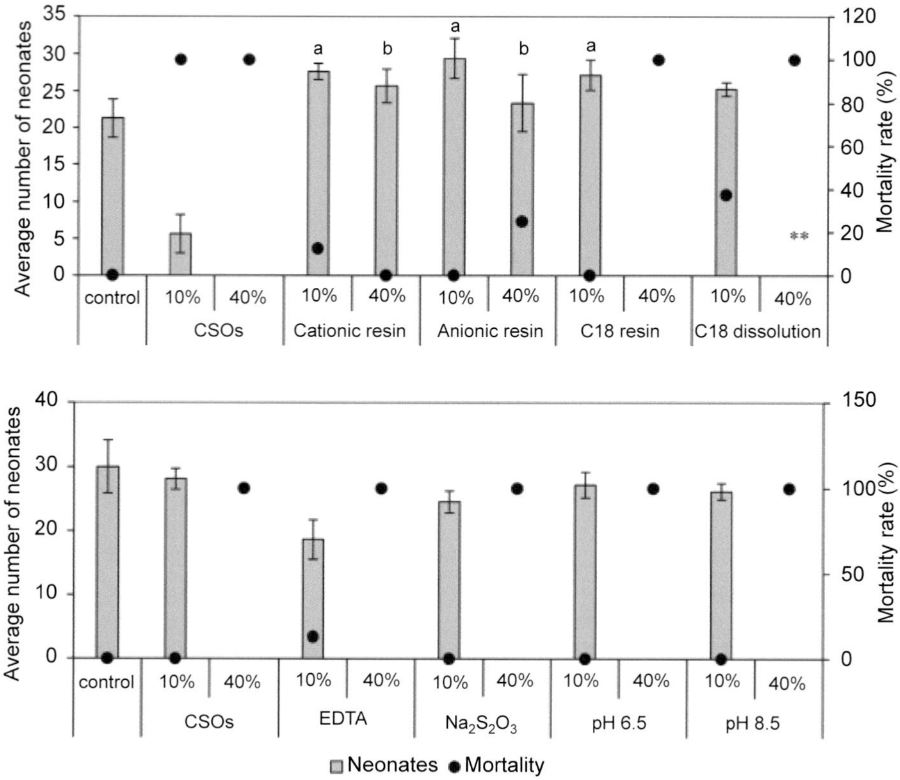

Figure 17.7. Results of the toxicity identification evaluation (TIE) test using *Ceriodaphnia dubia* in sample C3. CSOs correspond to untreated samples; Cationic resin, Anionic resin, and C_{18} resin correspond to samples subjected to solid-phase extraction; C_{18} dissolution corresponds to the methanol eluate from C_{18} resin; EDTA: samples treated with ethylenediaminetetraacetic acid; $Na_2S_2O_3$: samples treated with sodium thiosulfate solution; pH 6.5, pH 8.5: samples treated by pH adjustment. The error bars represent the standard error. In the TIE test of C3, the concentrations of 10% and 40%, in which toxicity was observed, were used. a: significantly different from untreated 10% sample ($p < 0.01$); b: significantly different from untreated 40% sample ($p < 0.01$); **: significantly different from control ($p < 0.01$).

suggesting that Rd toxicity might be partly due to metals. In Ru (Fig. 17.9), only the 20% concentration of cationic resin treatment and the addition of EDTA reduced sample toxicity. This result suggests that cationic compounds might be the toxic substances in this sample.

Metals' contribution to toxicity

To further examine the effect of heavy metals on samples' toxicity to daphnia, Hazard Quotients of heavy metals (HQm) (equation 3) were related to daphnia's TUc.

$$HQm = \Sigma Ci/NOECi \quad (3)$$

Ci: measured concentration of metal i
NOECi: no effect concentration of metal i

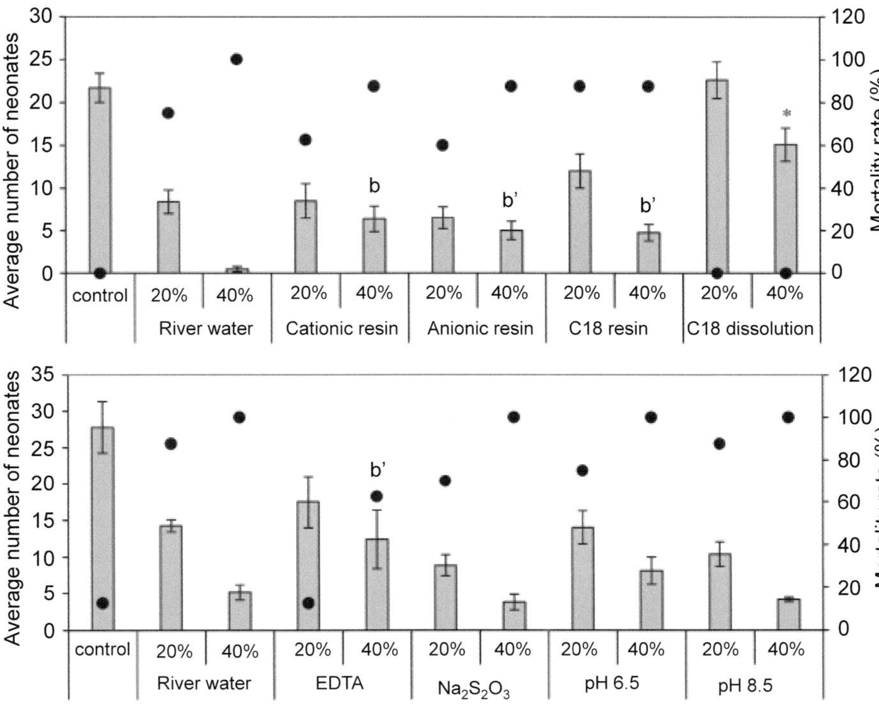

Figure 17.8. Results of the toxicity identification evaluation (TIE) test using *Ceriodaphnia dubia* in sample Rd. River water corresponds to untreated samples; Cationic resin, Anionic resin, and C_{18} resin correspond to samples subjected to solid-phase extraction; C_{18} dissolution corresponds to the methanol eluate from C_{18} resin; EDTA: samples treated with ethylenediaminetetraacetic acid; $Na_2S_2O_3$: samples treated with sodium thiosulfate solution; pH 6.5, pH 8.5: samples treated by pH adjustment. The error bars represent the standard error. In the TIE test of Rd, the concentrations of 20% and 40%, in which toxicity was observed, were used. b: significantly different from untreated 40% sample ($p < 0.01$); b': significantly different from untreated 40% sample ($p < 0.05$); *: significantly different from control ($p < 0.05$).

The NOEC of metals for *C. dubia* used in this formula were 5.66 μg/L for chromium (Cr) (DeGraeve et al. 1992), 1 μg/L for Ni (Itatsu et al. 2015), 25 μg/L for Zn (Itatsu et al. 2015), 12 μg/L for Cu (Itatsu et al. 2015), 1.0 μg/L for Cd (Suedel et al. 1997) and 36 μg/L for Pb (Spehar and Fiandt 1986). For the toxicity tests of 45 industries, the toxic action of priority contaminants was generally additive (Sarakinos et al. 2000). A review conducted in 1980 of 76 experiments on toxicity of general pollutants in wastewater in Europe on fish, reported that the general toxicity is additive effect (Marking 1985). Therefore, in this study, we considered that the metal interactions had a Concentration Addition (CA) for equation 2. The comparison between TUc and HQm (Fig. 17.10) showed that the toxicity of C3, Rd, and Ru was higher than in other samples. It is likely that heavy metals play an important role in toxicity. In addition, Cr, Cd, and Ni were probably the most toxic substances for daphnia. As Cu, Zn, Cr, Ni, Pb, etc. are the main-ubiquitous heavy metals in the urban environment, their highest concentrations were measured at the sites receiving CSOs water (Nakanishi 2008a, b, Komínková et al. 2016). The

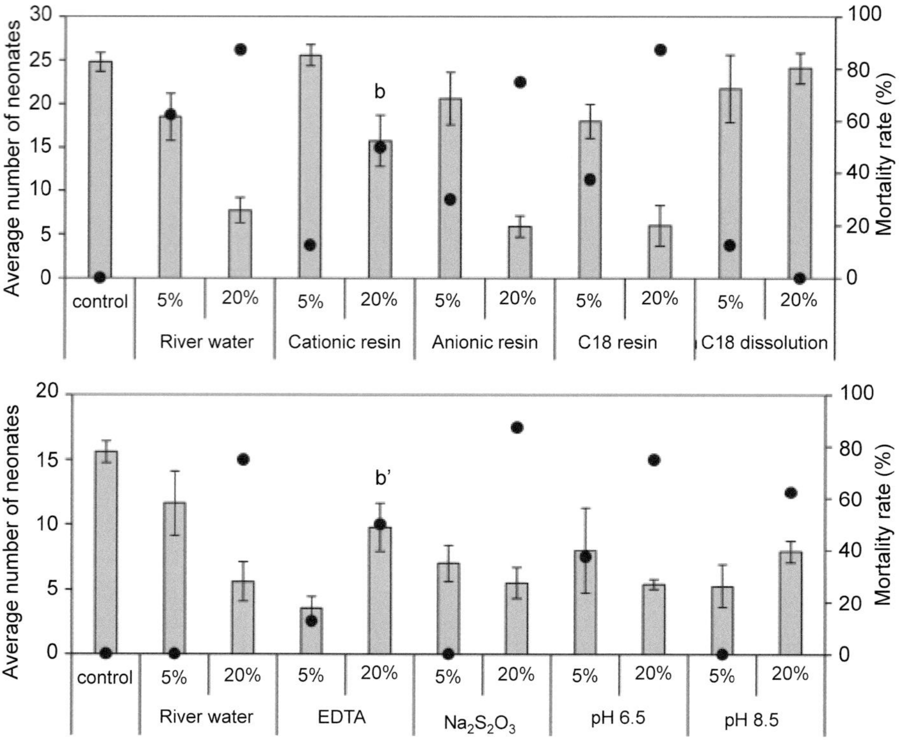

Figure 17.9. Results of the toxicity identification evaluation (TIE) test using *Ceriodaphnia dubia* in sample Ru. River water corresponds to untreated samples; Cationic resin, Anionic resin, and C_{18} resin correspond to samples subjected to solid-phase extraction; C_{18} dissolution corresponds to the methanol eluate from C_{18} resin; EDTA: samples treated with ethylenediaminetetraacetic acid; $Na_2S_2O_3$: samples treated with sodium thiosulfate solution; pH 6.5, pH 8.5: samples treated by pH adjustment. The error bars represent the standard error. In the TIE test of Ru, the concentrations of 5% and 20%, in which toxicity was observed, were used. b: significantly different from untreated 20% sample ($p < 0.01$). b': significantly different from untreated 20% sample ($p < 0.05$).

CSOs in different study areas combined different kinds of heavy metal pollutants from several sources, including domestic and industrial wastewater, as well as from parking and other impervious areas (Komínková et al. 2016).

The toxicity caused by the samples taken at the third sampling event on aquatic organisms seems to be mainly due to heavy metals. Metals are strongly complexed with organic ligands or absorbed to particles that may not be readily bioavailable. Consequently, the toxicity of metals will be alleviated due to the decrease of free metals, which are responsible for toxicity. The bioavailability of metals was influenced by the pH, hardness (Ca, Mg), and dissolved organic matter (Buck et al. 2007). Dissolved humic substances play an important role as organic matter bound to heavy metals in river water. Humic substances accounted for 50% of dissolved organic matter in river water (Martin-Mousset et al. 1997, Imai 2004). As a simulated sample of overflowing water, Kojima (Kojima 2010) mixed road dust collected from side groove (after mixing with rainwater) and sewage. The proportion of free ionic

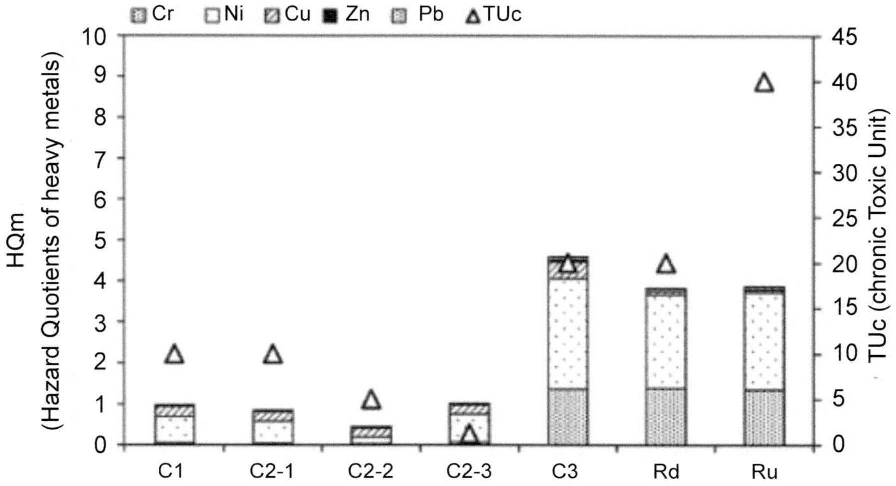

Figure 17.10. Comparison between hazard quotients (HQm) of heavy metals and daphnia chronic toxic unit (TUc) in all samples.

plus unstable complexed metals in these samples, which was measured by chelating disk cartridge, was 57–76% for Cu, 31–53% for Zn, 28–51% for Pb, and 19–25% for Ni. The similar trend is reported in effluent samples of wastewater treatment plants (WWTPs) (Chaminda et al. 2013). More than 55% of the dissolved metals were strongly complexed by ligands in the effluents. As these results suggest, the toxic effect of metals in this study might be greatly reduced. As shown in Fig. 17.10, the toxic effects of Ni may decrease if the proportion of free ions of Ni in the overflowing water was reduced to around 20 percent. Furthermore, the TU value of samples was generally higher than HQm value. Especially in case of C3, Rd and Ru samples, the average values of TU were about three times of those of HQm. Therefore, the major cause of toxicity might be other ingredients than heavy metals. The toxicity of CSO samples C1, C2-1, C2-3 could be interpreted in the same way. To evaluate the impact, it is necessary to accurately grasp the behavior of pollutant loads of trace harmful chemicals, besides heavy metals, such as polycyclic aromatic hydrocarbons, specific to CSOs.

Conclusions

In this section, WET testing and PRTR data have been applied on the sewage discharge water. The toxicity predicted using PRTR data was likely to be overestimated for daphnia and fish. There are two related factors that may help explain the overestimation. As for metals, there is a possibility that bioavailability might be reduced by coexisting matters. The other factor might be the estimation method of PRTR. It is necessary to examine the accuracy of PRTR data when evaluating sewage effluent toxicity based on this data.

Although phase I of TIE was used, it was not sufficient for determining toxicants or reducing toxicity. In the future, it will be necessary to utilize phase II and phase III of TIE to identify toxicants of concern and confirm suggested toxicants, respectively. Combining chemical analysis and TIE manipulations with toxicity tests is necessary to identify the characteristics of the substances causing sample toxicity. It is also necessary to consider relative pollutant contributions to CSOs from urban DWF and WWF. To reduce the influence of unique components in the overflow water on the environment of the discharge area, it is also worth developing new and effective projects to improve sewer water quality. We will continue using toxicity and TIE methods to assess the environmental effects of CSOs and to characterize the unique components of overflow water itself, exploring new ways to improve the water quality of combine overflow water and reduce its toxicity in the water receiving area.

Acknowledgments

First of all, I would like to thank Tomohiro Takano, Yasuyuki Itatsu, and Mamiko Fukutomi, who helped to analyze some of the parameters. I would also like to take this opportunity to thank the staff of sewage treatment plants that sampled and provided the related information. Moreover, this section was supported by the River Foundation Grant Aid Project 5311-Researcher/Research Organization Department (Survey/Research Grant, Grant Number: 285311012). The authors thank Ms Lifang Jin for helpful comments on the manuscript. Thank you all for your firm support.

References

Buck, K.N., J.R.M. Ross, A. Russell Flegal and K.W. Bruland. 2007. A review of total dissolved copper and its chemical speciation in San Francisco Bay, California. Environ Res 105(1): 5–19.

Chaminda, G.G.T., F. Nakajima, H. Furumai, I. Kasuga and F. Kurisu. 2013. Metal (Zn, Cu, Cd and Ni) complexation by dissolved organic matter (DOM) in wastewater treatment plant effluent. J Water Environ Technol 11.3: 153–161.

Chemicals Evaluation and Research Institute, Japan. 2006. Hazard Assessment Report, 2006. http://www.cerij.or.jp/evaluation_document/hazard_assessment_report_03.html Accessed on 18 October 2019 (in Japanese).

DeGraeve, G.M., J.D. Cooney, B.H. Marsh, T.L. Pollock and N.G. Reichenbach. 1992. Variability in the performance of the 7-d *Ceriodaphnia dubia* survival and reproduction test: An intra- and interlaboratory study. Environ Toxicol Chem 11(6): 851–866.

Di Toro, D.M., H.E. Allen, H.L. Bergman, J.S. Meyer, P.R. Paquin and R.C. Santore. 2001. Biotic ligand model of the acute toxicity of metals. 1. Technical basis. Environ Toxicol Chem 20(10): 2383–2396.

Environment Canada. 2007. Biological test method: test of reproduction and survival using the cladoceran *Ceriodaphnia dubia*, EPS 1/RM/21, 2nd edition. Environment Canada, Ottawa, Canada.

Forbes, V.E. and P. Callow. 2002. Extrapolation in ecological risk assessment: balancing pragmatism and precaution in chemical controls legislation: extrapolation is a practical necessity in ecological risk assessment, but there is much room for improvement in the extrapolation process. BioScience 52(3): 249–257.

Fukutomi, M., J. Jin, T. Kusui and Y. Itatsu. 2013. Consideration of the relevance of WET test and PRTR data of industrial wastewater. Proceedings of Environmental Science Association 2013 Annual Meeting, Tokyo, Japan, p. 94 (in Japanese).

Gooré, Bi E., F. Monette, J. Gasperi and Y. Perrodin. 2015. Assessment of the ecotoxicological risk of combined sewer overflows for an aquatic system using a coupled "substance and bioassay" approach. Environ Sci Pollut Res Int 22(6): 4460–4474.

Gromaire, M.C., G. Chebbo and A. Constant. 2002. Impact of zinc roofing on urban runoff pollutant loads: the case of Paris. Water Sci Technol 45(7): 113–122.

Imai, A. 2004. Characteristics and roles of humic substances in aquatic environments. J Jpn Soc Water Envron 27(2): 76–81 (in Japanese).

Ishikawa, N. 2016. Evaluation of impact on water area by direct discharge of confluent sewage reflecting rainfall situation. J Japan Sewage Works Assoc 53(644): 102 (in Japanese).

Ishikawa, N., D. Yoshida, A. Ito and T. Kaeda. 2016. Inflow load of pollutants to a confluent sewer system due to rainfall runoff. J Japan Sewage Works Assoc 53(645): 114 (in Japanese).

Isnard, P., P. Flammarion, G. Roman, M. Babut, P. Bastien, S. Bintein, L. Esserméant, J.F. Férard, S. Gallotti-Schmitt, E. Saouter, M. Saroli, H. Thiébaud, R. Tomassone and E. Vindimian. 2001. Statistical analysis of regulatory ecotoxicity tests. Chemosphere 45(4-5): 659–669.

Itatsu, Y., T. Takano, J. Jin, M. Fukutomi and T. Kusui. 2015. Ecotoxicological assessment of industrial effluents: Toxicity characterization and impact on receiving water. J Environ Chem 25(1): 19–26 (in Japanese).

Japan Standards Association. 2010. JIS K 0102:2010 Testing methods for industrial wastewater. Japanese Standards Association, Tokyo, Japan (in Japanese).

Japan Water Works Association. 2011. Water examination method, 2011 edition. Japan Water Works Association, Tokyo, Japan (in Japanese).

Kafi, M., J. Gasperi, R. Moilleron, M.C. Gromaire and G. Chebbo. 2008. Spatial variability of the characteristics of combined wet weather pollutant loads in Paris. Water Res 42(3): 539–549.

Kojima, K. 2010. Presence characteristics of heavy metals of sediment in the duct from combined sewer system and overflowing water in rainy weather. PhD Thesis, Graduate School of Engineering, The University of Tokyo (in Japanese).

Komínková, D., J. Nábělková and T. Vitvar. 2016. Effects of combined sewer overflows and storm water drains on metal bioavailability in small urban streams (Prague metropolitan area, Czech Republic). J Soils Sediments 16(5): 1569–1583.

Kusui, T. and C. Blaise. 1999. Ecotoxicological assessment of Japanese industrial effluents using a battery of small-scale toxicity tests. pp. 161–181. *In*: Rao, S.S. (ed.). Impact Assessment of Hazardous Aquatic Contaminants, Lewis Publisher, Chelsea, USA.

Kusui, T., Y. Takata, Y. Itatsu and J. Zha. 2014. Whole effluent toxicity assessment of industrial effluents in Toyama, Japan with a battery of short-term chronic bioassays. J Water Environ Technol 12(1): 55–63.

Marking, L.L. 1985. Toxicity of chemical mixtures. pp. 164–176. *In*: Rand, G.M. and S.R. Petrocelli (eds.). Fundamentals of Aquatic Toxicology. Hemisphere Publishing Corporation, Washington, DC, USA.

Martin-Mousset, B., J.P. Croue, E. Lefebvre and B. Legube. 1997. Distribution and characterization of dissolved organic carbon of surface waters. Water Res 31: 541–553.

McPherson, T.N., S.J. Burian, M.K. Stenstrom, H.J. Turin, M.J. Brown and I.H. Suffet. 2005. Trace metal pollutant load in urban runoff from a Southern California watershed. J Environ Eng 131(7): 1073–1080.

Ministry of the Environment, Government of Japan. 2000. Manual on required survey items. Ministry of the Environment, Government of Japan, Tokyo, Japan (in Japanese).

Ministry of the Environment, Government of Japan. 2016. Results of initial environmental risk assessment of chemicals, 2016. http://www.env.go.jp/chemi/risk/index.html Accessed on 18 October 2019 (in Japanese).

Ministry of the Environment, Government of Japan. 2018. A report about n-Alkylbenzenesulfonic acid and its salts on the Japanese Ministry of the Environment. http: www.env.go.jp/chemi/report/h19-03/pe/03-01.pdf Accessed in 18 January 2018 (in Japanese).

Ministry of the Environment. Government of Japan. 2019. PRTR Information Plaza Japan. http://www.env.go.jp/en/chemi/prtr/prtr.html Accessed on 24 October 2019 (in Japanese).

Nakanishi, J. and K. Ono. 2008a. Detailed risk assessment report series hexavalent chromium, Maruzen, Tokyo, Japan (in Japanese).

Nakanishi, J. and K. Tsunemi. 2008b. Series of detailed risk assessment 19: Nickel. The New Energy and Industrial Technology Development Organization (NEDO), the Research Center for Chemical Risk Management (CRM), Maruzen, Tokyo, Japan (in Japanese).

NIES (National Institute for Environmental Studies) and MOE (Ministry of the Environment), Japan: Examination method for effluent using biological response (draft), 2013. http://www.env.go.jp/press/101845.html Accessed on 1 October 2019 (in Japanese).

OECD. 1998. Test No. 212: Fish, short-term toxicity test on embryo and sac-fry stages. guideline for testing of chemicals. OECD, Paris, France.

OECD. 2006. Test No. 201: Freshwater alga and cyanobacteria, growth inhibition test. Guidelines for testing of chemicals. OECD, Paris, France.

Orvos, D.R., D.J. Versteeg, J. Inauen, M. Capdevielle, A. Rothenstein and V. Cunningham. 2002. Aquatic toxicity of triclosan. Environ Toxicol Chem 21(7): 1338–1349.

Sabin, L.D., J.H. Lim, K.D. Stolzenbach and K.C. Schiff. 2005. Contribution of trace metals from atmospheric deposition to stormwater runoff in a small impervious urban catchment. Water Res 39(16): 3929–3937.

Santore, R.C., D.M. Di Toro, P.R. Paquin, H.E. Allen and J.S. Meyer. 2001. Biotic ligand model of the acute toxicity of metals. 2. Application to acute copper toxicity in freshwater fish and Daphnia. Environ Toxicol Chem 20(10): 2397–2402.

Sarakinos, H.C., N. Bermingham, P.A. White and J.B. Rasmussen. 2000. Correspondence between whole effluent toxicity and the presence of priority substances in complex industrial effluents. Environ Toxicol Chem 19(1): 63–71.

Spehar, R.L. and J.T. Fiandt. 1986. Acute and chronic effects of water quality criteria-based metal mixtures on three aquatic species. Environ Toxicol Chem 5(10): 917–931.

Suedel, B.C., J.H. Rodgers and E. Deaver. 1997. Experimental factors that may affect toxicity of cadmium to freshwater organisms. Arch Environ Contam Toxicol 33(2): 188–193.

Suzuki, Y., R. Morishita and T. Maruyama. 1996. Toxicity evaluation of monochloramine and chlorinated sewage effluents by a phytoplankton growth test. J Jpn Soc Water Environ 19(11): 861–870 (in Japanese).

Takaoka City. 2010. Takaoka City combined sewer emergency improvement plan, Takaoka, Toyama, Japan (in Japanese).

Takeda, F., H. Mano, Y. Suzuki and S. Okamoto. 2015. Initial environmental risk assessment of Japanese PRTR substances in treated wastewater. J Water Environ Technol 13(4): 301–312.

Takeda, F., K. Komori, M. Minamiyama and S. Okamoto. 2016. Toxicity of wastewater with regard to ammonia evaluated by algal growth inhibition test: a case study using wastewater treatment pilot plant. Jpn J Water Treat Biol 52(4): 93–104.

Takeda, F., M. Minamiyama and S. Okamoto. 2017. Seasonal variation in ability of wastewater treatment for reduction in biological effects evaluated based on algal growth. J Water Environ Technol 15(3): 96–105 (in Japanese).

Tamura, I., Y. Yasuda, K. Kagota, S. Yoneda, N. Nakada, V. Kumar, Y. Kameda, K. Kimura, N. Tatarazako and H. Yamamoto. 2017. Contribution of pharmaceuticals and personal care products (PPCPs) to whole toxicity of water samples collected in effluent-dominated urban streams. Ecotoxicol Environ Saf 144: 338–350 (in Japanese).

Tatarazako, N. 2006. Bioassays for environmental water: a perspective in whole effluent toxicity. J Jpn Soc Water Environ 29(8): 426–432 (in Japanese).

USEPA. 1991. Methods for aquatic toxicity identification evaluations: phase I toxicity characterization procedures. EPA-600–6-91–003, United States Environmental Protection Agency, Duluth, USA.

USEPA. 1992. Toxicity identification evaluations: characterization of chronically toxic effluents, phase I, EPA-600–6-91-005F. EPA, Duluth, USA.

USEPA. 1993. Methods for aquatic toxicity identification evaluations: phase II toxicity identification procedures for samples exhibiting acute and chronic toxicity, EPA-600-R-92–080. EPA, Durham, USA.

USEPA. 2017. ECOTOX Knowledgebase, 2017. https://cfpub.epa.gov/ecotox/ Accessed on 28 October 2019.

Watanabe, H., I. Tamura, R. Abe, H. Takanobu, A. Nakamura, T. Suzuki, A. Hirose, T. Nishimura and N. Tatarazako. 2016. Chronic toxicity of an environmentally relevant mixture of pharmaceuticals to three aquatic organisms (alga, daphnid, and fish). Environ Toxicol Chem 35(4): 996–1006.

Yamamoto, H., K. Abe, K. Ikeda, Y. Yasuda, I. Tamura, Y. Nakamura and N. Tatarazako. 2010. Whole effluent toxicity test for the effluent of the selected sewage treatment plants in Tokushima, Japan. Environ Eng Res 47: 727–734 (in Japanese).

Yamamoto, H., Y. Yano, J. Morita, S. Nishie, Y. Yasuda, I. Tamua and N. Tatarazako. 2013. Toxicity identification evaluation for the effluent of sewage treatment plants focused on residual chlorine. J Jpn Soc Civil Eng Ser G Environ Res 69(7): III_375–III_384 (in Japanese).

Zeeman, M.G. 1995. Ecotoxicity testing and estimation methods developed under section 5 of the Toxic Substances Control Act (TSCA). pp. 703–716. *In*: Rand, G.M. (ed.). Fundamentals of Aquatic Toxicology, 2nd edition. Taylor & Francis, Philadelphia, USA.

18

Whole Effluent Toxicity Test for Ambient Water Monitoring

Takashi Kusui[1,*] and *Jun Jin*[2]

Introduction

Currently, many chemical substances are used in daily life and industry, and eventually discharged into the water environment. In order to prevent such trace chemicals from affecting health and the environment, various environmental standards and emission standards have been introduced. However, compared to the tens of thousands of chemical substances that are circulated on a daily basis, the number of standards is insufficient, from several tens to few hundreds. Furthermore, there are concerns about the effects of unregulated substances and the combined effect of trace chemicals. In order to address this, a test method using a biological response has attracted attention.

This test method, which is applied in wastewater management, is called whole effluent toxicity (WET) test, and has been introduced in North America, Canada, European countries, and South Korea, among others, and is used as an effluent standard (USEPA 1991a, Power and Boumphrey 2004, Tatarazako 2006). The advantages of this approach include aggregation of measured toxicity, detection of unknown toxins, and measurement of bioavailable toxins (Dorn and van Compernolle 1995). When toxicity exceeding the standard is detected, a method for characterizing, identifying, and confirming the causative substances by combining physicochemical fractionation, biological tests, and chemical analysis has been proposed as toxicity identification evaluation (TIE) procedure (USEPA 1991b, 1993a, b). Such a WET test has been reported to be effective for predicting the ecological impact of the river at the discharge destination (USEPA 1991a).

[1] Toyama Prefectural University, 5180 Kurokawa, Imizu, Toyama 939-0398, Japan.

[2] Yokohama National University, 79-1 Tokiwadai, Hodogaya Ward, Yokohama, Kanagawa 240-8501, Japan.
Email: jinjun5134@foxmail.com

* Corresponding author: kusui@pu-toyama.ac.jp

The Japanese Environmental Ministry established an advisory body in 2009 to examine the feasibility of introducing the WET approach into the present regulatory framework. In March 2013, draft guidelines for WET testing were released, which proposed three freshwater short-term chronic toxicity tests (algae, crustacean, and fish) (Subcommittee on bioassay technology for wastewater (ambient water) management 2013). As of March 2019, the introduction of the WET method as a means of self-management has been proposed (Committee on evaluation and management methods of water environment using aquatic organisms 2019). Using this proposed test protocol, industrial effluent (Kusui et al. 2014, 2018, Itatsu et al. 2015, Watanabe et al. 2015, Yamamoto et al. 2015), sewage effluent, and combined sewer overflows (CSOs) (Yamamoto et al. 2010, Yasuda et al. 2013, Takeda et al. 2016, Jin et al. 2018, Jin and Kusui 2019) have been evaluated. Additionally, there have been reports of cases where causative substances were studied using TIE (Yasuda et al. 2013, Yamamoto et al. 2015, Jin et al. 2018, Kusui et al. 2018).

In addition to the abovementioned application, the WET test has been applied for ambient water monitoring. Since 1986, the Central Valley Regional Water Quality Control Board, California, USA began testing ambient waters using WET procedures (de Vlaming et al. 2000). Using *Ceriodaphnia dubia* tests, toxicity caused by insecticides was detected in rivers and drainage from dormant orchards, rice fields, and urban areas. Toxicity to *Selenastrum* has been linked to copper and zinc from mines, and to the herbicide diuron in waters receiving agricultural or urban runoff. Ammonia-caused toxicity, originating from dairies and wastewater treatment plants, to fathead minnows has been identified. In Japan, however, there have been few reports regarding the application to ambient water monitoring (Yasuda et al. 2011, Morita et al. 2013). Further research is necessary in order to clarify the advantages and disadvantages of the Japanese WET tests.

In this chapter, we will present a case study of the application of Japanese WET tests to ambient water and identification of toxicity factors and their sources using TIE method.

Materials and methods

Survey area and sampling

To apply the WET test to ambient water monitoring, 10 samples from 9 rivers in Toyama, Japan were collected from May 2017 to November 2017. In the second survey in 2018, the samples were taken along Senbo river to find the distribution of toxicity and identify the source of toxicity.

Samples were collected using a plastic bucket. The samples were immediately transported to the laboratory in amber glass bottles in a lightproof ice box, then filtered through a 60 μm nylon strainer, and stored in a dark refrigerator (4°C) until analysis and bioassay; biological tests were conducted within 36 hours of sampling.

Measurement of water parameters

Among the water parameters, pH and DO were measured on-site by pH meter (pH/COND METER D-54, HORIBA) and DO meter (ProODO, YSI), respectively. In the laboratory, electrical conductivity (EC) were measured by pH/EC meter (pH/COND METER D-54, HORIBA). Metals in the samples were determined as follows—50 mL of filtered sample was mixed with 5 mL nitric acid (EL grade, Kantokagaku, Japan) in a metal-free PP tube (DigiTUBEs, SCP Science). The mixture was digested on a hot-plate (DigPREP, SCP Science) by wet digestion method. The digested sample was made up to 50 mL by adding ultrapure water (Milli-Q), and then analyzed by ICP-MS (Agilent 7700e, Agilent Technologies).

WET test

WET test was conducted according to the Japanese WET test guideline (Subcommittee on bioassay technology for wastewater (ambient water) management 2013). Table 18.1 summarize the test conditions of WET test.

Table 18.1. Test conditions of short-term chronic bioassays.

Test	**Algal growth inhibition test**	**Cladoceran reproduction test**	**Short-term toxicity test on fish embryo and sac-fry stages**
Test species	*Raphidocelis subcapitata* *	*Ceriodaphnia dubia*	*Danio rerio*
Organisms	exponential phase of growth	neonates (< 24 h old)	fertilized eggs (< 4 h)
Duration	3 days	max. 8 days	9 days
Test vessel	200-mL Erlenmeyer flask containing 60 mL of test solution	glass containing 15 mL of test solution	glass containing 50 mL of test solution
Replicates per treatment	6 replicates for control; 3 replicates for each test concentration	10 replicates	4 replicates
Light	continuous	16 h light: 8 h dark	16 h light: 8 h dark
Temperature	23 + 2°C	25 ± 2°C	26 ± 1°C
Test medium renewal	no	every two days	every two days
Endpoints	growth rate (NOEC)	reproduction (NOEC)	survival and hatching (NOEC)

*: formally known as *Pseudokirchneriella subcapitata.*

In each bioassay, organisms were exposed to samples (river water and effluent) in a series of dilutions (0, 5, 10, 20, 40, and 80%).

The algal growth inhibition test was conducted with the green microalgae *Raphidocelis subcapitata* (formally known as *Pseudokirchenriella subcapitata*) (NIES-35), in accordance with Organisation for Economic Co-operation and Development (OECD) test guideline TG201 (OECD 2006). Algal suspensions (60 mL) inoculated at 0.5 × 10^4 cells/mL in an Erlenmeyer flask (200 mL capacity) were exposed to a range of sample concentrations prepared with OECD medium. Samples were prepared in triplicate for each test concentration. The samples were incubated under continuous illumination from fluorescent lamps (ca. 60 μmol/m^2/s) at a temperature of 23 ± 2°C in an orbital shaking culture. Algal cell density was determined with a particle counter (detection range 3–12 μm, CDA-500, Sysmex, Japan) every 24 hours in the 72 hour growth test. The EC50 (50% of effective concentration) and NOEC (no observed effect concentration) for growth rate were determined with statistical software (Ecotox-Statics ver. 2.6, The Japanese Society of Environmental Toxicology, Japan).

Cladoceran reproduction tests were assessed via a three-brood renewal toxicity test using *Ceriodaphnia dubia* standardized by Environment Canada (Environment Canada 2007). At the beginning of the test, one neonate daphnia (< 24 hours old) was transferred to a glass containing 15 mL of a diluted samples. Each treatment consisted of 10 replicates of a particular test concentration or the control. During the test, the samples were incubated under illumination (light 16 hours/dark 8 hours) at a temperature of 25 ± 2°C. Appropriate volumes of food (YCT and algae) were added daily, and each test solution was renewed three times per week. The death of the first-generation daphnia and the number of live neonates produced by the first-generation daphnia were observed for 8 days. EC50 and the NOEC for fecundity were determined with statistical software (Ecotox-Statics ver. 2.6).

Short-term toxicity tests on fish sac-fry stages using *Danio rerio* were conducted according to OECD TG212 (OECD 1998). Briefly, 10 fertilized eggs (< 4 hours old) were placed in a glass containing 50 mL of test solution. Each treatment comprised four replicates of a particular test concentration or the control. During the test, embryos were incubated under illumination (light 16 hours/dark 8 hours) at a temperature of 26 ± 1°C. Hatching and survival rates were observed daily during the test period (10 days). Based on survival and hatching rates, the NOEC was calculated with statistical software (Excel Tokei ver. 6.0, Esumi Inc., Japan).

To compare the results of toxicity tests, chronic toxicity units (TUc) were calculated with the following formula:

TUc = 100/NOEC (%)

In case where the NOEC equals 5%, the value of TUc is 20. The smaller the NOEC is, i.e., the stronger toxicity is, the bigger the TUc.

TIE

In our study, we applied Phase I TIE (USEPA 1991b) to characterize the causative chemicals responsible for toxicity. Table 18.2 shows the series of manipulations

Table 18.2. Effluent manipulations for TIE.

Manipulation	Characterization of causative substances
EDTA addition test (3 mg/L))	heavy metals
Sodium thiosulfate test (10 mg/L)	oxidants (i.e., chlorine)
pH adjustment test (pH 6.5, 8.5)	acids or bases
Post C_{18} SPE column test	non or low polar organic substances
Post anion exchange column test	Anions
Post cation exchange column test	Cations

used in our study. By comparing the toxicity of samples treated chemically and/or physically with that of the original sample, the character of causative factors was estimated.

Results and discussions

Application of WET tests to major rivers in Toyama

Figure 18.1 shows the location of sampling points in the first survey. In the 70s, annual average Biochemical Oxygen Demand (BOD) exceeding 50 mg/L were documented at some rivers in Toyama Prefecture due to severe pollution from industrial effluents. However, since then, wastewater treatment and sewage facilities have been improved, and water quality has also improved. Annual average BOD of nine rivers in this study has improved to approximately < 0.5 to 2.8 mg/L (FY2017).

The toxicity of samples is shown in Fig. 18.2. Of the ten samples, seven samples did not show any effect on the three organisms. Three samples with inhibitory effect (F, Se-u, Se-d) showed a relatively high chronic toxicity on cladoceran (TUc $\geq$ 5). Among three samples, only sample F showed slight toxicity on algae (TUc = 2.5).

According to the results of water parameters, EC of sample F (1,006 mS/m) was significantly higher than those of other samples (6.8~28.4 mS/m). The sampling point of sample F was near the mouth of the canal to Toyama Bay, suggesting this sample contains seawater. High concentration of cations (Na^+, Mg^+, and Ca^+) and anions (SO_4^{2-}, Cl^-) in this sample support this assumption.

In order to verify this hypothesis, we conducted an experiment on crustacean reproduction using natural seawater and saline. Figure 18.3 shows the relationship

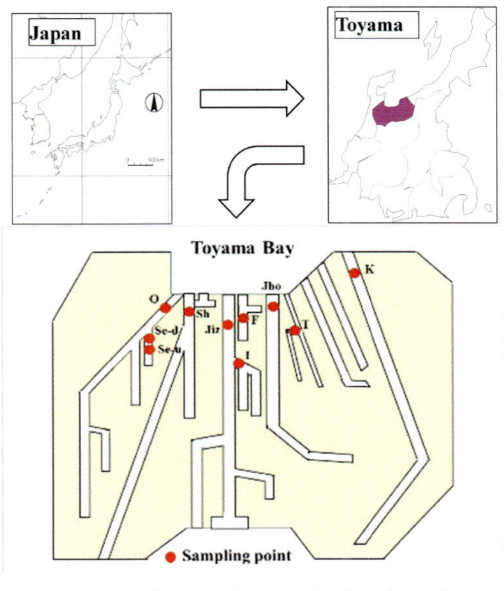

Figure 18.1. Location of study area. Sh: Sho River, O: Oyabe River, Jiz: Jinzu River, Jho: Jhoganzi River, K: Kurobe River, Se-u: Senbo River (upstream), Se-d: Senbo River (downstream), I: Itachi River, F: Fugan Canal, T: Totitsu River.

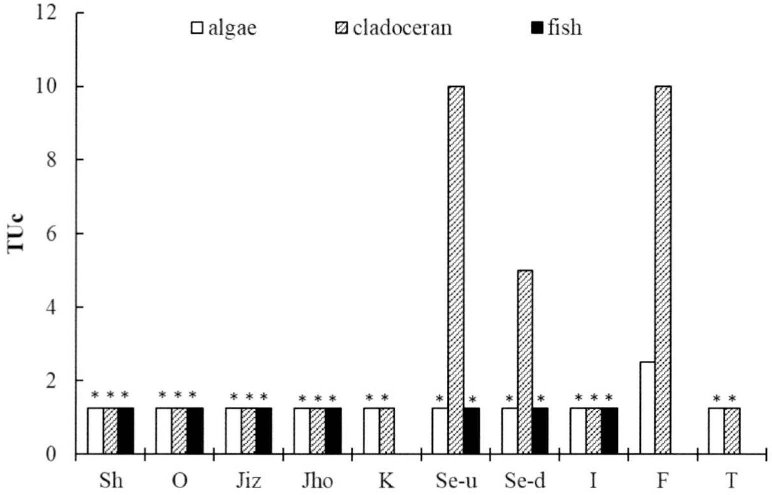

Figure 18.2. The toxicity of river samples. *: TUc ≤ 1.25 (virtually no toxicity), Sh: Sho River, O: Oyabe River, Jiz: Jinzu River, Jho: Jhoganzi River, K: Kurobe River, Se-u: Senbo River (upstream), Se-d: Senbo River (downstream), I: Itachi River, F: Fugan Canal, T: Totitsu River.

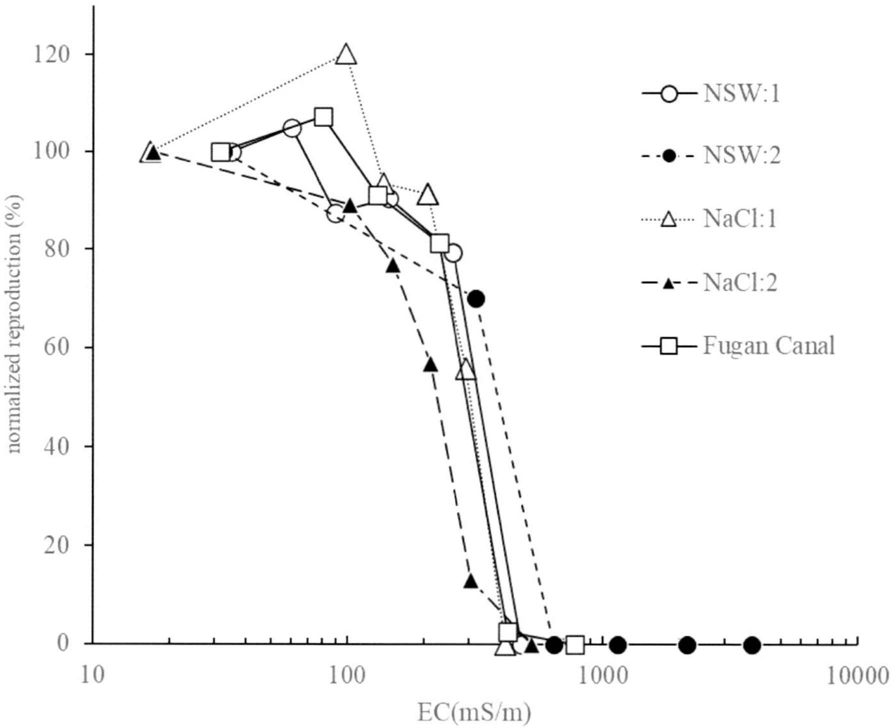

Figure 18.3. Effect of salinity on daphnia reproduction. NSW: natural seawater, NaCl: NaCl solution.

between EC and the effect on the reproduction in sample F, natural seawater (NW), and NaCl solution. Reproduction was normalized by the number of neonates in the control. With the increase of EC, the inhibitory effect on reproduction was observed in all samples. Interestingly enough, the relationship was almost similar. Therefore, it can be concluded that the toxicity of sample F is derived from seawater. Major inorganic cations (Na^+, Mg^+, and Ca^+) and anions (SO_4^{2-}, Cl^-), that are not toxic at low concentrations, have been reported to cause mortality and reproductive inhibition of *C. dubia* at high concentrations (Mount et al. 1997, 2016, Goodfellow et al. 2000). The results of our study are consistent with the results of these studies.

Distribution of toxicity along Senbo River

Among three samples with toxicity, two samples (Se-d and Se-u) were collected from Senbo River (Fig. 18.4). The Senbo River is 22 km long and flows through Takaoka City, Toyama Prefecture. It originates from Shogawa River and then flows to the west side of Takaoka City and joins Oyabe River. In the downstream of the Senbo River, a combined sewer system has been established, and there are factories and industrial parks along the river. In the 60s, this river was called the "death river", as it was heavily polluted with industrial and domestic wastewater, and no fish lived there. By 2017, the annual average BOD improved to 1.3 m/L.

In order to identify the cause of toxicity in the Senbo River, the distribution of toxicity and toxic factors were studied in the second survey. The sampling was conducted three times in March, April, and November 2018, and samples were collected within one day in each case. For the evaluation of toxicity, a crustacean chronic toxicity test which only showed inhibitory effect in the first survey was used. However, in order to be able to test a large number of samples simultaneously,

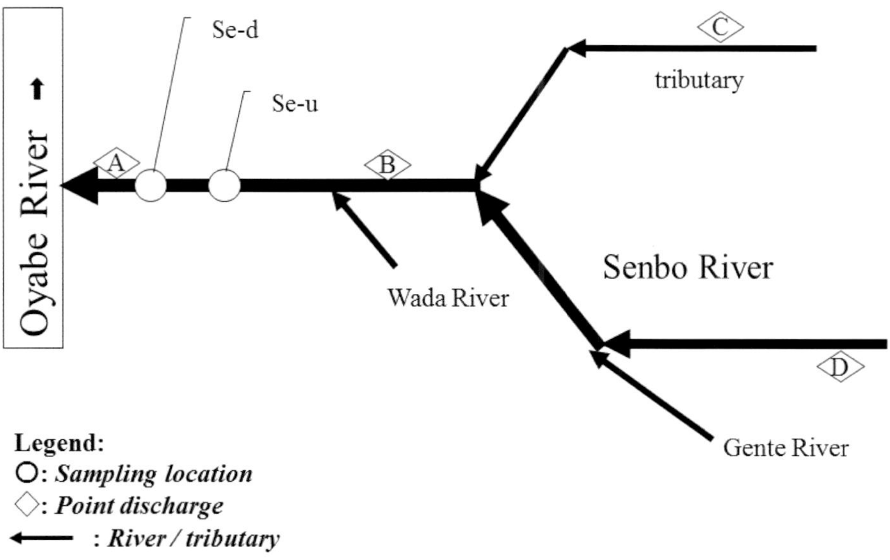

Figure 18.4. Schematic diagram of Senbo river system.

only 80% concentration solution of each sample was subjected to the test. As all the sampling was conducted in fine or cloudy weather, we assume that there was no impact from combined sewer overflow.

Figure 18.5 shows the results of the first sampling in March. Judging from the number of neonates per adult and the survival rate, there was almost no reproduction between the downstream D2 and D8, and the survival rate was 0 percent. However, because there was no effect on the sample D5 collected in the Wada River, which flows downstream, there is a high possibility that a toxic substance had flowed from the upstream from D8. Investigating the upstream part (U1 to U5), only sample U5 had a strong effect on reproduction and survival, suggesting the inflow of toxic substances from the point of discharge D or other points of discharge upstream. However, because samples upstream, except U5, were not affected, we assumed that the effect observed in U5 was diluted, as it flows and does not affect the downstream. However, no toxicity was observed in the samples collected at the tributaries (T4, T5, and T6). Based on these results, we assumed that the toxic inflow observed in the downstream part may have flowed in from the upstream part U1 or from the tributary part T4. In order to examine this argument in more detail, a second sampling was conducted. This revealed that the effluents from the point of discharge C in the tributaries flowed into the drainage channel, and were then discharged from the

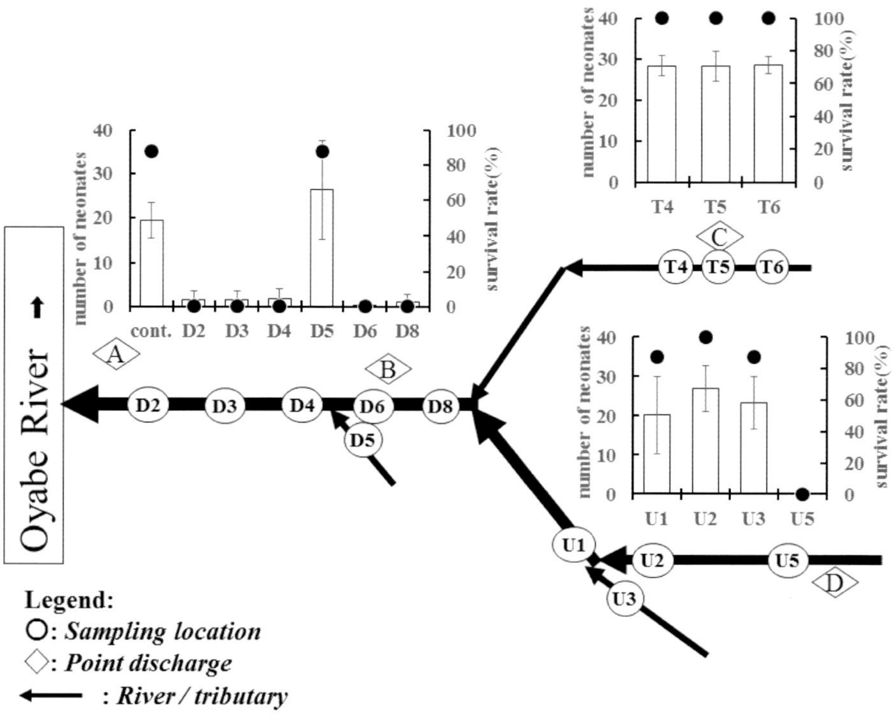

Figure 18.5. Effect on daphnia reproduction and survival: first sampling. cont.: control, white bar represents the mean number of three broods of neoneates per adult, standard deviation of neonates, and the black circle represents the survival rate.

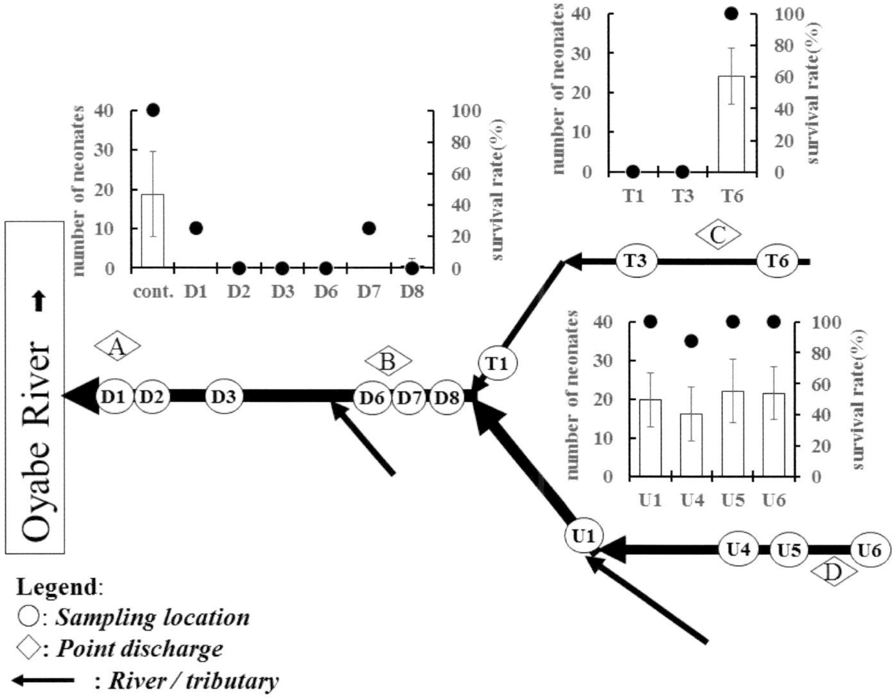

Figure 18.6. Effect on daphnia reproduction and survival: second sampling. cont.: control, white bar represents the mean number of three broods of neoneates per adult, standard deviation of neonates, and the black circle represents the survival rate.

downstream rather than the vicinity of the plant. Therefore, samples T4 and T5 did not contain the effluents from C.

The results of the second sampling in April are shown in Fig. 18.6. As in the first sampling, there was no reproduction of *C. dubia* in downstream samples D1 to D8, and the survival rate was 0 to 20%, showing strong toxicity. However, no effect was observed in upstream sample U1–U6. The downstream of U1 is a rural area and pesticides were not likely to be sprayed due to the season (April). In this sampling, strong chronic toxicity was observed at sample T1 and T3 in the tributaries, and no effect was observed in upstream sample T6. As samples T1 and T3 were collected downstream from the inflow point of the effluent C, the contribution of effluent C on toxicity was suggested.

Figure 18.7 shows the results of the third sampling conducted in November. In this sampling, no toxicity was observed in the downstream area (D1–D8). In addition, no toxicity was observed in the upstream part (U1 to U6). However, in the tributaries, toxicity was observed downstream from T6 (T3, T2, and T1). The tendency of the toxicity to decrease with the flow downstream from T3 suggests the impact of wastewater discharged from C plant.

Based on the abovementioned three samplings, it can be suggested that multiple sources are involved in defining toxicity in the target area of the Senbo River system.

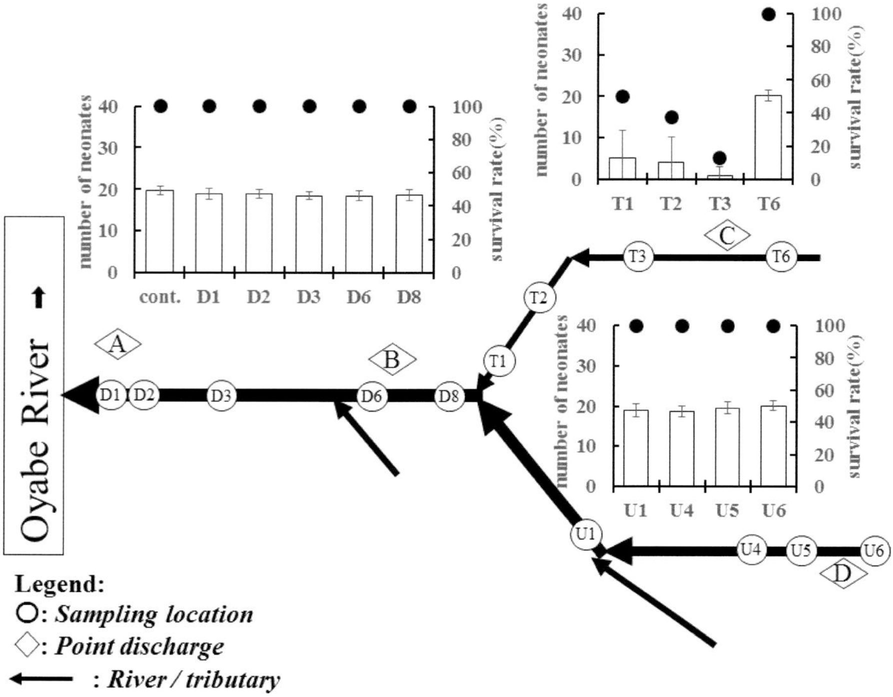

Figure 18.7. Effect on daphnia reproduction and survival: third sampling. cont.: control, white bar represents the mean number of three broods of neoneates per adult, standard deviation of neonates, and the black circle represents the survival rate.

Moreover, the involvement of the effluents was strong. At the same time, multiple samplings have shown that toxicity may change depending on flow rates and operating conditions at the factory.

Factors influencing toxicity

In order to search for factors influencing toxicity, TIE was performed using sample T3 from the second sampling, in which toxicity was observed (Fig. 18.8). In the first TIE, treatment with cation exchange resin did not mitigate the effects on reproduction; however, increased survival rates were observed. Additionally, we observed no significant change in toxicity by anion exchange resin and C18 solid phase extraction. Furthermore, the eluent obtained after solid phase extraction showed no toxicity. In the second TIE, significant mitigation of toxicity was observed only with the addition of EDTA. From the above results, it can be suggested that the contribution of organic substances was small, with the main contribution coming from heavy metals.

Table 18.3 shows the results of elemental analysis of sample T3 measured by ICP-MS and NOEC of reproduction of *C. dubia* (Hickey 1989, DeGraeve et al. 1992, De Schamphelaere et al. 2005, Naddy et al. 2007, Puttaswamy and Liber 2012). Among 7 elements, the concentrations of 6 elements (B, V, Cr, Ni, Cu, Ag) have a ratio of 1 or more to NOEC, and may contribute to toxicity. Specifically, V, Cr,

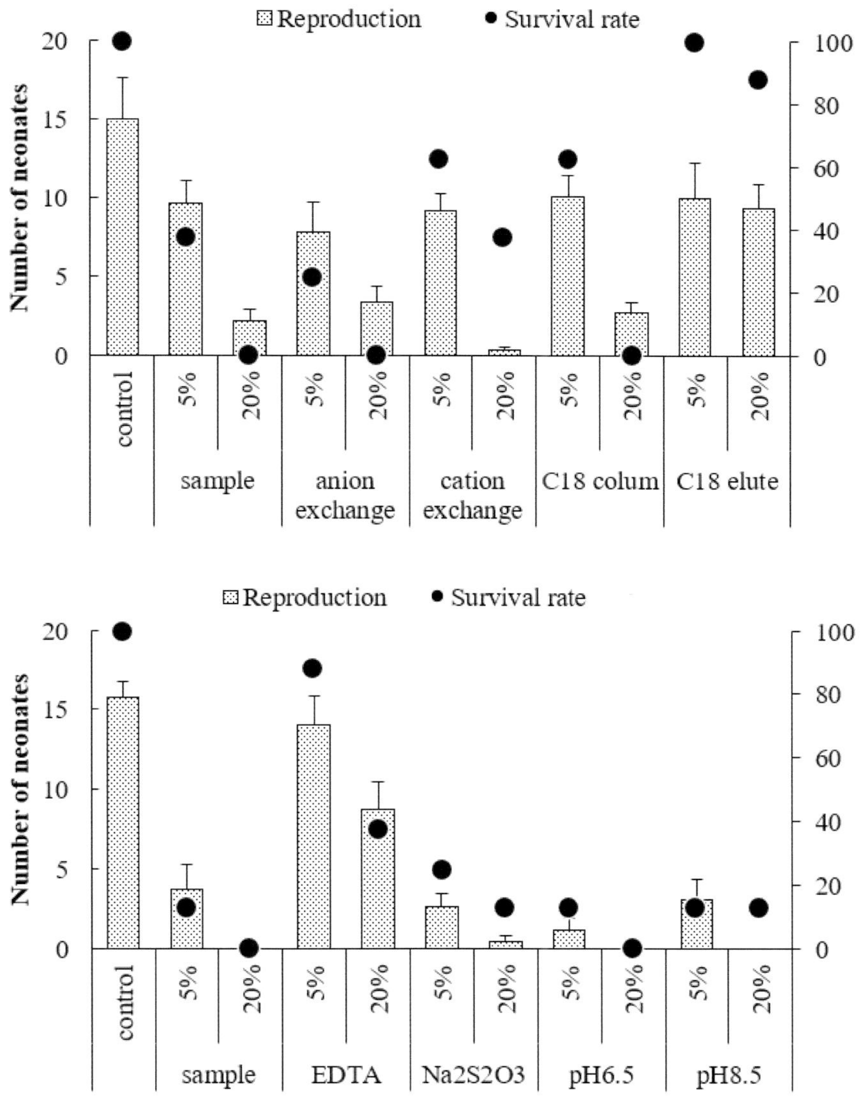

Figure 18.8. Results of TIE on sample T3 (second sampling). The dashed bar represents the mean number of three broods of neoneates per adult, standard deviation of neonates, and the black circle represents the survival rate (%).

Ni, and Cu have bigger ratios of 91.8, 79.3, 14.8, and 11.5, respectively, and the contribution to toxicity of these is considered high. V and Cr, which have a particularly high ratio, may exist in multiple valences, but NOECs in Table 18.3 are those of highly toxic form of V^{5+} and Cr^{6+}, respectively. Figure 18.9 shows the results of a *C. dubia* reproduction test using V^{5+} (V_2O_5) conducted in our laboratory. NOEC was 0.25 mg/L, almost the same as the reported value (Puttaswamy and Liber 2012).

Table 18.3. Measured concentration of elements in sample T3 (second sampling) and their NOECs.

	B	V	Cr	Ni	Cu	Zn	Ag
conc. (mg/L)	60.5	45.9	0.476	0.0148	0.173	0.0103	0.0095
NOEC (mg/L)	10[a]	0.5[b]	0.006[c]	0.001[d]	0.015[d]	0.053[e]	0.004[f]
conc./NOEC	6.05	91.8	79.3	14.8	11.5	0.19	2.37

References: a: Hickey 1989, b: Puttaswamy and Liber 2012, c: DeGraeve et al. 1992, d: in-house value, e: De Schamphelaere et al. 2005, f: Naddy et al. 2007.

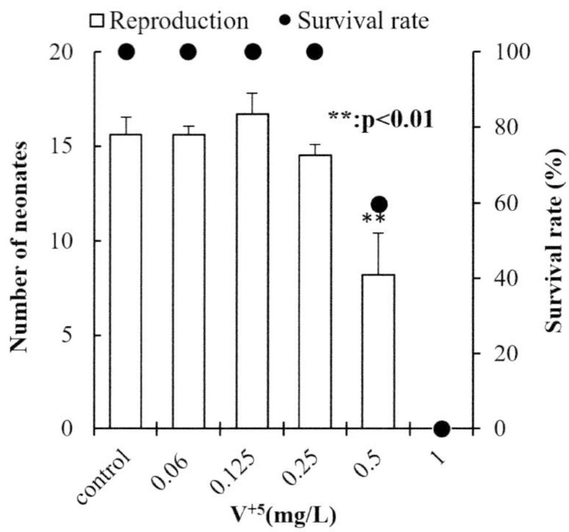

Figure 18.9. Chronic toxicity of V^{5+} (V_2O_5) on daphnia. The white bar represents the mean number of three broods of neoneates per adult, standard deviation of neonates, and the black circle represents the survival rate (%).

In order to examine the relationship between the observed toxicity in the second sampling and the six elements, the toxic factor (=measured concentration/NOEC) was calculated. The effect on *C. dubia* and the toxic factor of each sample is shown in Fig. 18.10. Immediately after the discharge of effluent C, the sum of toxic factors of the river water increases up to 200 or more, and tends to decrease as the water flows downstream. However, even at the most downstream point D1 in the study area, it decreases only to approximately 39. At all locations downstream of T3, which are affected by the effluent C, some samples show the survival rate of about 20 percent. However, there is little possibility of reproduction; therefore, it may be affected by elements such as V, Cr, and Ni. The total toxic factor was approximately 10 to 20 at other unaffected sites (T6, U1 to U6). Although these values may cause toxic effects, no effects were observed in this study. This is presumed to be due to the differences in the valence of the elements (V, Cr), differences in the specification of these elements, and mitigation effects due to coexisting components (Di Toro et al. 2001, Santore et al. 2001).

Looking at the factors influencing the toxicity of the toxic samples, V, Cr, and Ni are 23 to 91, 6 to 79, and 1.5 to 15, respectively. Therefore, it could be assumed that effluent C originally contained these elements.

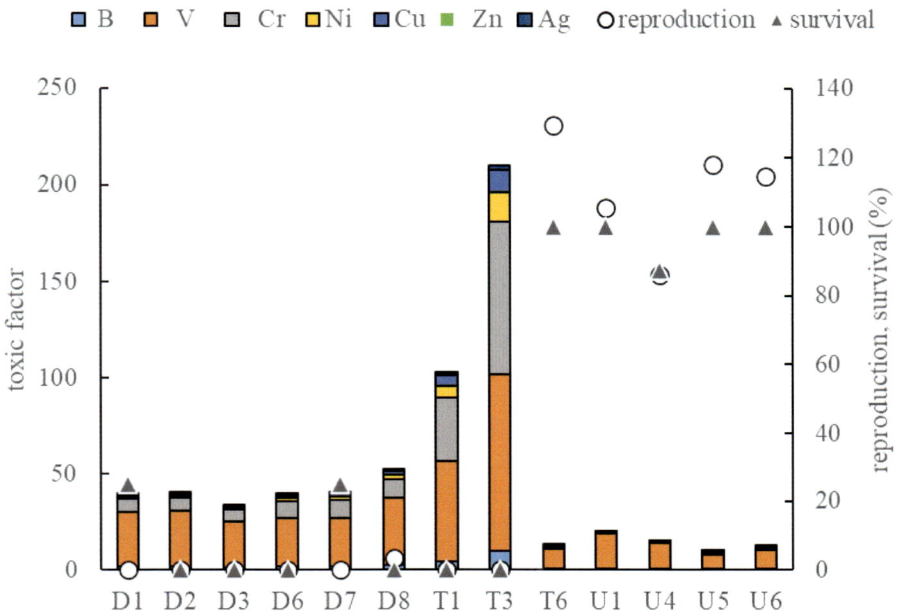

Figure 18.10. Toxic factor of elements and the effect on daphnia. The bar represents the toxic factor of each element; the white circle represents the normalized reproduction (%); the black triangle represents the survival rate (%).

Sources of toxicity

Prior to the field survey, the effects of chemical substances contained in the effluent from the point of discharge C located along the tributary were examined using the data obtained from the Japanese Pollutant Release and Transfer Register (PRTR) system (Ministry of the Environment, Government of Japan 2019). PRTR system promotes voluntary improvement of chemical management by business operators, who notify the government of the amounts of specified chemical substances released into the environment. Previous research has shown that PRTR data is useful for effluent toxicity estimation in industrial effluents for screening purposes (Fukutomi et al. 2013). According to PRTR data, the PRTR substances released into the water by this plant did not include the substances suspected to be toxic in the previous section. Two substances, sodium dodecyl sulfate and sodium poly (oxyethylene)= dodecyl ether sulfate ester, were reported. These substances are used to help the foaming of pulp liquid to adhere and remove ink in the deinking pulp (Deinked Pulp, DIP) manufacturing process using waste newspaper. As for the toxicity of these two substances on *C. dubia*, 6 day-NOEC (reproduction) 6.48 to 10.8 mg/L for sodium dodecyl sulfate (Cowgill et al. 1990), 2 day-EC50 (immobilization) 3.12 mg/L for (oxyethylene)=dodecyl ether sulfate (Warne and Schifko 1999) were reported. However, considering the amount of each substance discharged as well as the effluent flow, these substances were unlikely to contribute to the toxicity of the effluent.

In order to verify the hypothesis in the previous section, the effluent C was collected. Figure 18.11 shows the results of the chronic toxicity test with effluent C. NOEC was 2.5% with the survival rate of 0% at 10% or higher, suggesting high chronic and acute toxicity.

Table 18.4 shows the measured concentration in this effluent and NOEC of toxic element candidates. The higher ratio of measured concentration to NOEC of Ni and V suggest the contribution of these elements on toxicity. Compared to the previous T3, Ni concentration is higher, but the V concentration is lower, which reflects the fluctuations of the operational conditions in the factory. However, they are consistent in that both elements contribute to toxicity. Although the details are not shown, the TIE results for effluent C, like those of sample T3 mentioned earlier, showed that cation exchange resin treatment and EDTA addition alleviated toxicity, and the contribution of heavy metals was estimated. This result supports the hypothesis that toxicity is derived from V and Ni. However, despite the inclusion of "vanadium compounds", "nickel", and "nickel compounds" in the PRTR-designated substances for which reporting is mandatory, the question remains why effluent C contained both elements which were not reported in PRTR data of factory C.

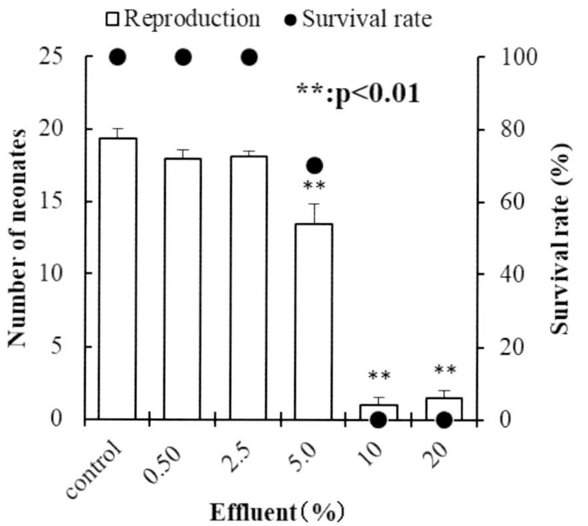

Figure 18.11. Effect of effluent C on daphnia reproduction and survival. The white bar represents the mean number of three broods of neoneates per adult, standard deviation of neonates; the black circle represents the survival rate (%).

Table 18.4. Measured concentrations of elements in effluent C and their NOECs.

	B	V	Cr	Ni	Cu	Zn	Ag
conc.(mg/L)	0.108	1.677	< 0.000	0.736	0.004	0.018	< 0.000
NOEC(mg/L)	10[a]	0.5[b]	0.006[c]	0.001[d]	0.015[d]	0.053[e]	0.004[f]
conc./NOEC	0.011	3.35	-	736	0.243	0.347	-

References: a: Hickey 1989, b: Puttaswamy and Liber 2012, c: DeGraeve et al. 1992, d: in-house value, e: De Schamphelaere et al. 2005, f: Naddy et al. 2007.

At the time of sampling, factory C uses petroleum coke as fuel for boiler turbines for power generation. Petroleum coke is a carbon-rich solid that is obtained as a final residue in petroleum refining. Vanadium and nickel are also present in organic substances, such as crude oil, oil shale, tar sand, asphalt, and bitumen, and exist as organometallic complexes. These are believed to be derived from local biogenic metals (Nakayama 2014). In crude oil distillation and refining processes, the majority of vanadium has been reported to remain in the residue (National Institute of Health Sciences 1988), and petroleum coke, which is the final residue, is expected to contain high concentrations of nickel and vanadium. According to the interview at the time of the survey, the staff of factory C said, "The water that contacted with the residue in the power plant furnace turns green. It probably contains vanadium". According to the flow chart of the wastewater treatment system of factory C, this "back water" from the power plant together with wastewater from other processes is mixed and treated by coagulating sedimentation, then discharged as a final effluent. However, it is assumed that this treatment method does not have enough capacity to remove vanadium and nickel from wastewater. Therefore, it seems that factory operators are not aware that the final effluent contains high concentration of V and Ni that are harmful to aquatic organisms, and that the effects remain downstream of Senbo River. As shown in this case study, WET test can be used to notify operators of unrecognized toxic factors.

In Japan, there are currently no environmental standards for the preservation of aquatic organisms for nickel and vanadium. However, standards for aquatic life conservation using species sensitivity distribution (SSD) have been proposed. The European Union has proposed 5.4 μg/l as the environmental standard for nickel in the Rhine River, obtained by dividing HC5 obtained from SSD by assessment factor (2) (Nickel institute 2010). In addition, for each of the European ecoregions (biogeographical regions), a value of 7.1–43.6 μg/l is proposed in consideration of the pH, hardness, and DOC (dissolved organic matter) of the water area. Schiffer and colleagues also sought acute and chronic standards for vanadium in the oil sands region of North Alberta, Canada (Schiffer and Liber 2017). As a result, they proposed 0.64 mg/l and 0.05 mg/l as acute values (HC5) and chronic values (HC5), respectively. If these criteria are applied to Japan in the future, further improvements of wastewater quality from industrial dischargers will be required for aquatic life conservation.

Conclusions

In this chapter, we applied the WET test to ambient water monitoring, and introduced the case study where toxicity factors and their sources were explored by using a combination of TIE methods and chemical analysis. The WET test was originally developed for the management of industrial effluents, but it is also effective for the monitoring of ambient water, and makes it possible to search for sources of contamination of point and non-point sources (e.g., land use). This study mostly shows the results of WET tests using freshwater species. However, in the case of tidal rivers, it is necessary to use WET tests using estuarine or marine organisms (USEPA 1988). As the number and amounts of chemical substances used increase, water quality management of ambient water may be insufficient when applying conventional environmental standards alone, and a more comprehensive approach using biological response tests will be needed.

Acknowledgements

The authors thank Mr Yasuhiro Uchida and Mr Yuma Watanabe for performing biological tests and water analysis. We would like to express our gratitude to the company that assisted us with effluent sampling. We would like to thank Editage (www.editage.com) for English language editing.

References

Committee on evaluation and management methods of water environment using aquatic organisms. 2019. Wastewater evaluation method using biological response tests (provisional name) and guidance for its use (interim report). http://www.env.go.jp/water/4.shiryo1%20.pdf. Accessed on 1 Nov 2019 (in Japanese).

Cowgill, U.M., D. Milazzo and B.D. Landenberger. 1990. The reproducibility of the three brood Ceriodaphnia test using the reference toxicant sodium lauryl sulfate. Arch Environ Contam Toxicol 19: 513–517.

DeGraeve, G.M., J.D. Cooney, B.H. Marsh, T.L. Pollock and N.G. Reichenbach. 1992. Variability in the performance of the 7-d *Ceriodaphnia dubia* survival and reproduction test: An intra- and interlaboratory study. Environ Toxicol Chem 11: 851–866.

De Schamphelaere, K.A.C., S. Lofts and C.R. Janssen. 2005. Bioavailability models for predicting acute and chronic toxicity of zinc to algae, daphnids, and fish in natural surface waters. Environ Toxicol Chem 24: 1190–1197.

de Vlaming, V., V. Connor, C. DiGiorgio, H.C. Bailey, L.A. Deanovic and D.E. Hinton. 2000. Application of whole effluent toxicity test procedures to ambient water quality assessment. Environ Toxicol Chem 19: 42–62.

Di Toro, D.M., H.E. Allen, H.L. Bergman, J.S. Meyer, P.R. Paquin and R.C. Santore. 2001. Biotic ligand model of the acute toxicity of metals. 1. Technical basis. Environ Toxicol Chem 20: 2383–2396.

Dorn, P.B. and R. van Compernolle. 1995. Effluents. *In*: Rand, G.M. (ed.). Fundamentals of Aquatic Toxicology. 2nd edn. Taylor & Francis, Washington.

Environment Canada. 2007. Biological test method: test of reproduction and survival using the cladoceran *Ceriodaphnia dubia*, EPS 1/RM/21, 2nd edition. Environment Canada, Ottawa, Canada.

Fukutomi, M., J. Jin, T. Kusui and Y. Itatsu. 2013. Consideration of the relevance of WET test and PRTR data of industrial wastewater. Proc Environ Sci Assoc 2013 Annual Meeting, 94 (in Japanese).

Goodfellow, W.L., L.W. Ausley, D.T. Burton, D.L. Denton, P.B. Dorn, D.R. Grothe, M.A. Heber, T.J. Norberg-King and J.H. Rodgers. 2000. Major ion toxicity in effluents: a review with permitting recommendations. Environ Toxicol Chem 19: 175–182.

Hickey, C.W. 1989. Sensitivity of four New Zealand cladoceran species and *Daphnia magna* to aquatic toxicants. NZ J Mar Freshwater Res 23(1): 131–137.

Itatsu, Y., T. Takano, J. Jin, M. Fukutomi and T. Kusui. 2015. Ecotoxicological assessment of industrial effluents: toxicity characterization and impact on receiving water. J Environ Chem 25(1): 19–26 (in Japanese).

Jin, J., X. Zhang and T. Kusui. 2018. Preliminary toxicity assessment of combined sewer overflows in Toyama, Japan. J Wat Environ Technol 16: 185–198.

Jin, J. and T. Kusui. 2019. Sewage effluent toxicity assessment: comparison between whole effluent toxicity (WET) and the Japanese pollutant release and transfer register (PRTR) data. J Wat Environ Technol 12: 55–63.

Kusui, T., Y. Takata, Y. Itatsu and J. Zha. 2014. Whole effluent toxicity assessment of industrial effluents in Toyama Prefecture with a battery of short-term chronic bioassays. J Wat Environ Technol 12: 55–63.

Kusui, T., Y. Itatsu and J. Jin. 2018. Whole effluent toxicity assessment of industrial effluents. pp. 331–347. *In*: Bidoia, E.D. and R.N. Montagnolli (eds.). Toxicity and Biodegradation Testing, Methods in Pharmacology and Toxicology. Springer Science+Business Media.

Ministry of the Environment, Government of Japan: PRTR Information Plaza Japan. http://www.env.go.jp/en/chemi/prtr/prtr.html Accessed on 1 Nov 2019 (in Japanese).

Morita, J., Y. Yasuda, K. Tamura, N. Tatarazako and H. Yamamoto. 2013. Characterization of toxicants for three aquatic organisms in ambient samples. J Jpn Soc Civil Eng Ser G (Environ Res) 69(7): IV_401–IV_410 (in Japanese).

Mount, D.R., D.D. Gulley, J. Hockett, T.D. Garrison and J.M. Evans. 1997. Statistical models to predict the toxicity of major ions to *Ceriodaphnia dubia*, *Daphnia magna* and *Pimephales promelas* (fathead minnows). Environ Toxicol Chem 16: 2009–2019.

Mount, D.R., R.J. Erickson, T.L. Highland, J.R. Hockett, D.J. Hoff, C.T. Jenson, T.J. Norberg-King, K.N. Peterson, Z.M. Polaske and S. Wisniewski. 2016. The acute toxicity of major ion salts to *Ceriodaphnia dubia*: I. influence of background water chemistry. Environ Toxicol Chem 35: 3039–3057.

Naddy, R.B., J.W. Gorsuch, A.B. Rehner, G.R. McNerney, R.A. Bell and J.R. Kramer. 2007. Chronic toxicity of silver nitrate to *Ceriodaphnia dubia* and *Daphnia magna*, and potential mitigating factors. Aquat Toxicol 84(1): 1–10.

Nakayama, K. 2014. Vanadium resources and their metallogenesis. J Soc Resource Geol 64(1): 31–53 (in Japanese).

National Institute of Health Sciences. 1988. Environmental health criteria 81-Vanadium. http://www.nihs.go.jp/hse/ehc/sum1/ehc081.html Accessed on 1 Nov 2019 (in Japanese).

Nickel institute. 2010. Fact sheets on the European Union environmental risk assessment of nickel. https://nickelinstitute.org/media/3714/eu-ni-ra-fact-sheet-1-2015-july.pdf Accessed on 1 Nov 2019.

OECD. 1998. Test No. 212: Fish, short-term toxicity test on embryo and sac-fry stages. Guideline for testing of chemicals. OECD, Paris, France.

OECD. 2006. Test No. 201: Freshwater alga and cyanobacteria, growth inhibition test. Guidelines for testing of chemicals. OECD, Paris, France.

Power, E.A. and R.S. Boumphrey. 2004. International trends in bioassay use for effluent management. Ecotoxicology 13(5): 377–398.

Puttaswamy, N. and K. Liber. 2012. Influence of inorganic anions on metals release from oil sands coke and on toxicity of nickel and vanadium to *Ceriodaphnia dubia.* Chemosphere 86(5): 521–529.

Santore, R.C., D.M. Di Toro, P.R. Paquin, H.E. Allen and J.S. Meyer. 2001. Biotic ligand model of the acute toxicity of metals. 2. Application to acute copper toxicity in freshwater fish and daphnia. Environ Toxicol Chem 20: 2397–2402.

Schiffer, S. and K. Liber. 2017. Estimation of vanadium water quality benchmarks for the protection of aquatic life with relevance to the Athabasca oil sands region using species sensitivity distributions. Environ Toxicol Chem 36: 3034–3044.

Subcommittee on bioassay technology for wastewater (ambient water) management. 2013. Examination method for effluent using biological response (draft). http://www.env.go.jp/water/files/sankou5.pdf. Accessed on 1 Nov 2019 (in Japanese).

Takeda, F., K. Komori, M. Minamiyama and S. Okamoto. 2016. Toxicity of wastewater with regard to ammonia evaluated by algal growth inhibition test: a case study using wastewater treatment pilot plant. Jpn J Water Treat Biol 52(4): 93–104.

Tatarazako, N. 2006. Bioassays for environmental water; a perspective in whole effluent toxicity. J Jpn Soc Water Environ 29(8): 426–432 (in Japanese).

USEPA. 1988. Short-term methods for estimating the chronic toxicity of effluents and receiving waters to marine and estuarine organisms. Cincinnati, Ohio. EPA-600-4-87-028.

USEPA. 1991a. Technical support document for water quality-based toxics control. Washington, D.C. EPA-505-2-90-001.

USEPA. 1991b. Methods for aquatic toxicity identification evaluations: phase I toxicity characterization procedures. EPA-600-6-91-003.

USEPA. 1993a. Methods for aquatic toxicity identification evaluations: phase II toxicity identification procedures for samples exhibiting acute and chronic toxicity. EPA-600-R-92-080.

USEPA. 1993b. Methods for aquatic toxicity identification evaluations: phase III toxicity confirmation procedures for samples exhibiting acute and chronic toxicity. EPA-600-R-92-081.

Warne, M.S.J. and A.D. Schifko. 1999. Toxicity of laundry detergent components to a freshwater cladoceran and their contribution to detergent toxicity. Ecotoxicol Environ Saf 44(2): 196–206.

Watanabe, H., T. Hayashi, I. Tamura, A. Nakamura, R. Abe, H. Takanobu, S. Ogino, M. Koshio and N. Tatarazako. 2015. Validation of a draft protocol of bioassays for effluent testing and a toxicity survey of industrial effluent. J Environ Chem 25(1): 43–53 (in Japanese).

Yamamoto, H., K. Abe, K. Ikeda, Y. Yasuda, I. Tamura, Y. Nakamura and N. Tatarazako. 2010. Whole effluent toxicity test for the effluent of the selected sewage treatment plants in Tokushima. Jpn Environ Eng Res 47: 727–734 (in Japanese).

Yamamoto, H., K. Ikebata, Y. Yasuda, I. Tamura and N. Tatarazako. 2015. Case study of toxicity identification evaluation (TIE) applied to the selected factory effluents in Tokushima, Japan. J Environ Chem 25(1): 11–17 (in Japanese).

Yasuda, Y., S. Yoneda, I. Tamura, K. Kagota, N. Nakada, S. Hanamoto, Y. Kameda, K. Kimura, N. Tatarazako and H. Yamamoto. 2011. Short-term chronic toxicity tests applied to river water contaminated by treated and untreated domestic sewage. J Jpn Soc Civil Eng, Ser G (Environ Res) 67(2): III_249–III_256 (in Japanese).

Yasuda, Y., Y. Yano, J. Morita, S. Nishie, Y. Yasuda, I. Tamura and N. Tatarazako. 2013. Toxicity identification evaluation for the effluent of sewage treatment plants focused on residual chlorine. J Jpn Soc Civil Eng, Ser G (Environ Res) 69(7): III_375–III_384 (in Japanese).

Index